Chlorophyll *a* Fluorescence in Aquatic Sciences

Methods and Applications

Developments in Applied Phycology 4

Series Editor:
Michael A. Borowitzka
School of Biological Sciences & Biotechnology
Murdoch University, Murdoch, Western Australia

For other titles published in this series, go to
www.springer.com/series/7591

David J. Suggett · Ondrej Prášil
Michael A. Borowitzka
Editors

Chlorophyll *a* Fluorescence in Aquatic Sciences

Methods and Applications

Springer

Editors
David J. Suggett
Department of Biological Sciences
University of Essex
CO4 3SQ Colchester
United Kingdom

Ondrej Prášil
Institute of Microbiology
Opatovciky mlyn
379 81 Trebon
Czech Republic

Michael A. Borowitzka
Algae R&D Center
School of Biological Sciences
and Biotechnology
Murdoch University
Murdoch 6150, WA
Australia

ISBN 978-90-481-9267-0 e-ISBN 978-90-481-9268-7
DOI 10.1007/978-90-481-9268-7
Springer Dordrecht Heidelberg London New York

Library of Congress Control Number: 2010932001

Printed on acid-free paper

Springer is part of Springer Science+Business Media (www.springer.com)

Preface

It is unquestionable that Chlorophyll *a* fluorescence is quite literally a global phenomenon. Fluorescence merely describes an optical phenomenon where light absorbed at one wavelength is re-emitted at another (longer) wavelength; it exists passively in nature and occurs wherever light exists to be absorbed by Chlorophyll *a* molecules. These molecules are a common property of all photoautotrophic organisms on land and in water; thus Chlorophyll *a* fluorescence is essentially ubiquitous in nature (Fig. 1). It is incredible that such a natural phenomenon has been exploited by such a wide variety of researchers and across the biological and environmental sciences, and perhaps is testament to the importance we place on understanding photoautotrophic activity. We have long known that Chlorophyll *a* fluorescence of photosynthetic organisms varies as a result of changes in the amount (biomass), as well as function (quantum yield), of Chlorophyll *a* present. At operational temperatures that exist in most natural environments, Chlorophyll *a* fluorescence is largely derived from the Chlorophyll *a* associated with photosystem II (PSII), i.e. the oxygen evolving complex; as such, changes in the quantum yield of fluorescence directly relate to changes in photosynthetic (O_2 evolving) capabilities. Thus, by actively inducing changes in Chlorophyll *a* fluorescence using an actinic light source, we can perturb the physiological *status quo* of (PSII) photoautotrophy itself. Packaging of technology to enable induction and measurement of such Chlorophyll *a* fluorescence perturbations has entirely made possible examination of processes associated with plant and algal ecology, physiology and productivity, and at scales from the single cell to the entire planet (van Kooten and Snel 1990). Therefore, it is hard to imagine a future that does not continue to exploit the properties of Chlorophyll *a* fluorescence, not only for research but also in how we continue to sustainably exploit our ever-changing environment.

The history of using fluorescence to investigate biomass, photosynthetic physiology and primary productivity has been covered in several comprehensive publications, most recently by Papageorgiou and Govindjee (2005) (and chapters therein); however, it is of course important to note the place of aquatic studies in this history, at least for the context of the following chapters. Whilst many major developments in using *variable* Chlorophyll *a* fluorescence have arguably come from studies on terrestrial (vascular) plants, free-living microalgae (chlorophytes in particular in particular) and cyanobacteria have proved to be important laboratory organisms in examining principal photobiological mechanisms. Examining such aquatic organisms under controlled laboratory conditions is a perhaps an obvious step; aside from the relative ease of probing photosynthetic machinery of single celled compared to multi-cellular organisms, microalgae and cyanobacteria dominate photosynthetic activity of much of the Earth's aquatic realm. However, in contrast to working on terrestrial plants, extending such laboratory-based observations to the 'real world' has proven to be the greatest challenge for aquatic scientists and one that has been largely led by technology and engineering. In overcoming the technical challenges, exciting and important discoveries, such as the confirmation of iron limitation of ocean productivity (Behrenfeld et al. 1996) and the discovery of aerobic anoxygenic bacteria (Kolber et al. 2001), have followed.

The earliest application of Chlorophyll *a* fluorescence to aquatic system research (*in situ*) is well recognized as from Carl Lorenzen (1966) who first pumped seawater through a shipboard fluorometer. Such a convenient, rapid approach was quickly adopted by both

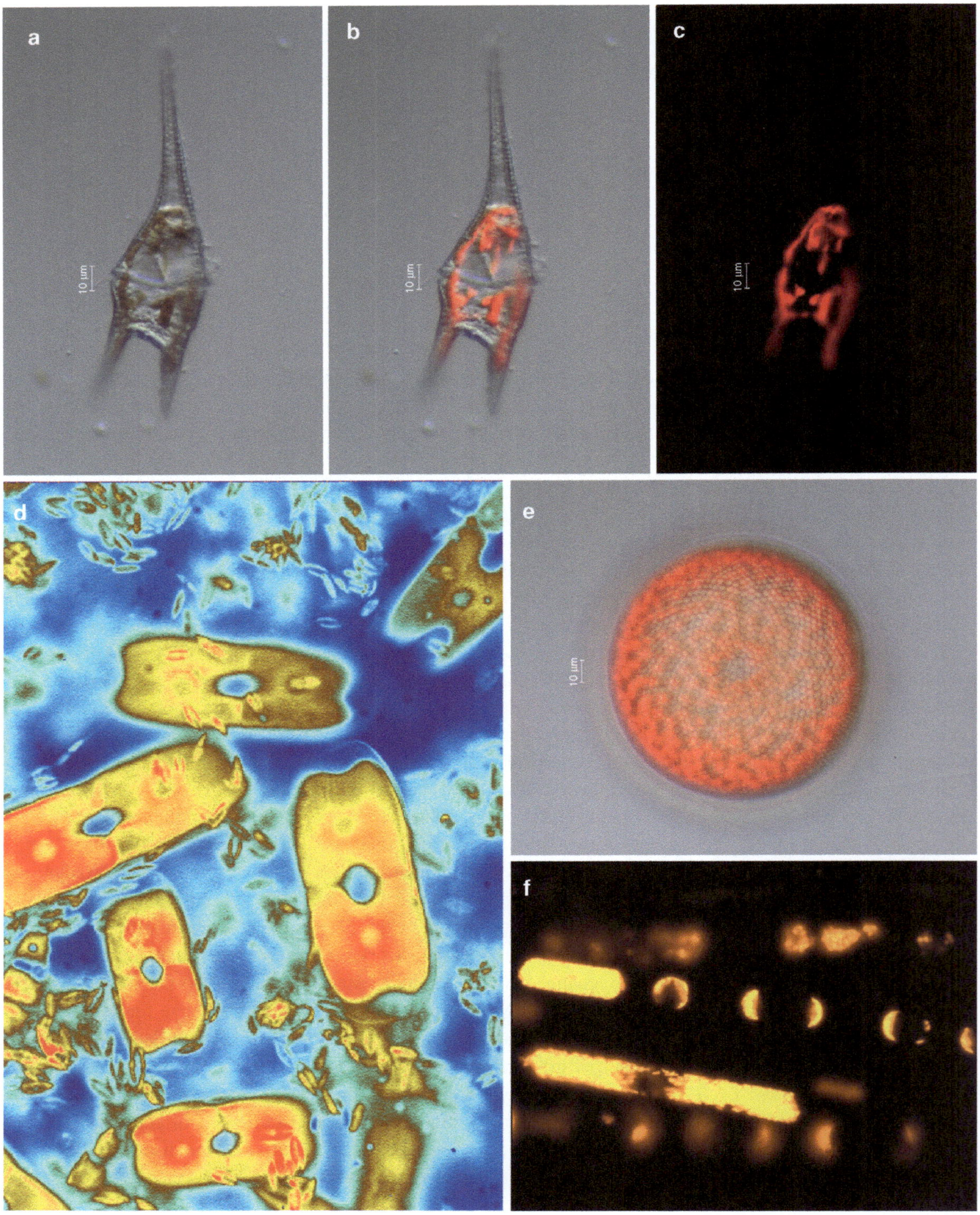

Fig. 1 Fluorescence in action: (**a–c**) Chloroplast fluorescence in the dinoflagellate *Ceratium* sp. (Photo: L. Novoveska); (**d**) False colour high resolution fluorescence image of cells of the diatom *Nitzschia dubia*. Fluorescence emanating from the chloroplasts becomes restricted to the area of the pyrenoid as light intensity increases (Photo: R. Perkins); (**e**) Chloroplast fluorescence in the centric diatom *Coscinodiscus* sp. (Photo: L. Novoveska); (**f**) Delayed fluorescence in the colonial diatom *Rhizosolenia* (Photo: M. Berden-Zrimek)

oceanographic and limnological communities; this, not surprisingly quickly led to a wealth of highly novel studies linking physical and biological processes, in particular, the distribution of phytoplankton with ocean turbulence (Platt 1972) and the discovery of the deep chlorophyll maximum of stratified waters (Cullen and Eppley 1981). The major challenge for aquatic scientists to evolve to *in situ* studies was 'simply' to package complex and innovative technology into a system that could withstand the constraints of working in water, especially in marine environments where salts and pressure rapidly build. It wasn't until the 1970s that technology caught up with concept and the first profilable *in situ* fluorometers were truly developed (see Falkowski and Kolber 1995). Ever since, such fluorometers have become smaller and better integrated to sensor arrays, and essentially a routine yet fundamental tool for aquatic scientists. However, despite their rapid adoption by the aquatic community, these fluorometers were still generally restricted to assaying a single chlorophyll fluorescence yield, which was set according to the excitation intensity of the instrument in question, and thus could only ever provide some approximate measure of Chlorophyll *a* biomass *in situ*. An important step to aquatic research was thus in producing fluorometers that induced a variable Chlorophyll *a* excitation (and hence fluorescence emission) protocol (Fig. 2).

Numerous laboratory studies by the 1970s and early 1980s had already demonstrated important concepts linking variable Chlorophyll *a* fluorescence to photosynthetic physiology in aquatic algae (e.g. Mauzerall 1972; Ley and Mauzerall 1982; but note an ISI Web of Science search yields >125 publications in the 1970s alone!), however, modification of these techniques to *in situ* aquatic studies to thus add a physiological component (the variable fluorescence 'transient') to measures of fluorescence yield (biomass) was not straightforward. Here, the development of actinic light sources that could deliver the intensity and/or frequency of excitation required to induce variable fluorescence remained an even greater technological challenge to the pre-existing in water operational constraints. Solving this problem essentially had to occur twice since variable fluorescence techniques have already evolved into two parallel but distinct paths (Chapter 3 by Huot and Babin, this volume): Pulse Amplitude Modulation (PAM; Schreiber et al. 1986), where fluorescence is induced by a weak modulated light source evaluated independently from a relatively long yet moderate intensity light pulse; and Pump and Probe (PP; Mauzerall 1972; Falkowski et al. 1986; but see also Kolber and Falkowski 1993), where variable fluorescence is measured by a weak 'probe' actinic flash before and after a saturating 'pump' flash. PP later evolved into Fast Repetition Rate (FRR; Kolber et al. 1998), where a complex fluorescence transient could be induced by initially delivering a series of subsaturating high intensity flashlets followed by a series of more widely spaced 'probing' flashlets that examined the subsequent fluorescence decay. All subsequent variable fluorometers have essentially followed one (or a combination) of these paths. Importantly, this new generation of fluorometers not only opened new possibilities for examining photoautotrophic physiology but also a potential revolution in how aquatic scientist would determine primary productivity (Kolber and Falkowski 1993, Kromkamp and Forster 2003; Suggett et al., Chapter 6, this volume).

Evolution of both PAM and FRR (PP) was originally driven from the pioneering laboratory work using microalgae; as such, the first *in situ* variable fluorometers in the 1980s and 1990s were essentially restricted to working on natural phytoplankton suspensions in lakes and oceans. Technical improvements in overall signal resolution since then has enabled researchers to investigate ever more oligotrophic waters of oceans and nutrient impoverished lakes. However, subtle technological changes in sensitivity and the optical configurations within a few years of PAM and FRR fluorometer introduction enabled the photophysiology of benthic autotrophs (corals, microphytobenthic mats, seagrasses and macroalgae) to be examined (see Chapter 9 by Enríquez and Borowitzka, Chapter 10 by Warner et al., and Chapter 11 by Shelly et al., this volume). More recent additional but relatively small optical alterations to the PAM and FRR 'model' to examine far red fluorescence (>800 nm) has introduced more new research opportunities, e.g. bacteriochlorophyll *a* (Kolber et al. 2001) and Photosystem I (PSI) variable fluorescence (Dual PAM, e.g. see Sukenik et al. 2009). Modification of the spectral quality of fluorescence excitation and emission detection has also added the potential for variable fluorometers to taxonomically discriminate bulk fluorescence properties (Schreiber 1998; Beutler et al. 2002; Chapter 7 by MacIntyre et al., this volume). All of these advances have unquestionably facilitated the explosion of interest in the use

Fig. 2 Fluorometers in action: (**a**) Diving PAM used to measure fluorescence signal in corals, Wakatobi Marine National Park, Indonesia (Photo: D. Smith); (**b**) Fluorometer comparisons at GAP Workshop, Eilat, Israel, 2008 (Photo: D. Suggett); (**c**) *In-situ* measurement of fluorescence quenching using the fluorometer PAM 101-103 (H.Walz, Germany) in the *Haematococcus* culture grown in solar photobioreactor at the Centre of Biological Technologies, University of South Bohemia in Nové Hrady, Czech Republic (Photo: J. Masojidek); (**d**) FRRF being deployed in winter, Bedford Basin, Canada (Photo: D. Suggett); (**e**) Fasttrack II attached to a CTD frame in water column sampling in Eilat, Israel (Photo: D. Suggett)

of fluorometers for aquatic research in recent years; arguably, compared to 20 years ago, aquatic research investigations are incomplete without some form of fluorescence examination.

Variable fluorometers have clearly provided a platform for aquatic scientists wishing to conveniently assay photosynthetic physiology non-invasively and more accurately scale changes of photosynthesis to the environment. Current acceleration of environmental variability via climate change perhaps provides very real justification for further investing in tools such as fluorometers that have the capacity to link ecosystem processes with environmental regulation. Fluorescence-based technological development (including delayed fluorescence; Chapter 14 by Berden-Zrimec et al., this volume) combined with research has publically produced a tool that can potentially inform stakeholders of the photosynthetic 'viability' (or 'health') of their associated aquatic environment; certainly, a tool that is less labour intensive and costly in the long term than conventional (and destructive) assays that require water or organisms to be removed and analysed in the laboratory. Such applications to those wishing to monitor and subsequently manage ecosystem function was an obvious step in exchanging the knowledge beyond pure research but also necessary for commercial manufacturers to invest further in instrument production. Key examples to date come from the monitoring of lakes and coastal waters for (harmful) algal blooms (Cullen et al. 1997) and coral reefs for pollution and coral bleaching (Jones 1999). PAM Fluorometry has also been demonstrated in action for two BBC documentaries – by Prof Ove Hoegh-Guldberg examining coral bleaching for the BBC documentary *State of the Planet* and by Dr Rupert Perkins investigating stromatolites for *Oceans*) Furthermore, recent developments of algae as biofuels will inevitably require application of fluorometers to optimize and also continually monitor yields (Kromkamp et al. 2009; Sukenik et al. 2009) and thus further move fluorometers from a purely ecological to an industrial monitoring tool.

Despite the potential growth industry that obviously exists for chlorophyll fluorescence, it is clear that the previous growth of fluorometer technological development and the subsequent array of commercially available fluorometers have somewhat superseded our fundamental understanding of the fluorescence signals generated. It is perhaps quite ironic that technological developments have already enabled us to collect vast fluorescence data sets, however, we are only recently arming ourselves with the key knowledge required to interpret and consequently apply these data into informed opinion. Examining the growth of citations for (variable) fluorescence-based papers over the past decade is perhaps more testament to our confidence in interpreting the data as opposed to reduced constraints in collecting the data itself. Armed with a decades worth of what is arguably 'fluorescence exploration', It is really only now that we are beginning to gain maximum benefit of using fluorescence as a tool to address fundamental research questions in the aquatic sciences.

Why the need for AQUAFLUO? — Rapid growth of using active fluorescence across the aquatic science disciplines has inevitably led to divergence in approach and terminology (see Chapter 1 by Cosgrove and Borowitzka, this volume, for recommended terminology). Even though many of us have been attempting to answer similar questions, this divergence has resulted in a lack of consistency required to facilitate information exchange; consequently, the field was not evolving as quickly as originally envisaged. Arguably, the aquatic sciences still communicate fluorescence-based studies in numerous dialects that are often not easily inter-comparable or reconcilable.

Using fluorescence as a non-invasive means for assaying processes, such as (harmful) bloom detection and primary productivity, is still heralded as a key breakthrough for aquatic research and not surprisingly has attracted much funding and research time investment. However, efforts to capitalise on these larger process-scale problems have somewhat overshadowed our need to understand the fundamental nuances of fluorescence measurements using different instrumentation, protocols and for the array of aquatic primary producers that exist. On many occasions, the interpretation of data sets has been confounded by what is real in nature *versus* an artifact of instrument use. Conversations amongst the aquatic sciences community over recent years have increasingly identified the need for conformity in the application and operation of (active) fluorometers, not only to standardise and reconcile existing data sets but also to ensure that fluorometry remained 'accessible' to the ever-growing new user community. Such a step is indeed critical if fluorometry is ever to evolve from a purely academic tool to an everyday, practical and informative management tool. However, despite attempts to call for a common set of approaches (the best example to date for the aquatic sciences is Kromkamp and Forster 2003), a few researchers have adapted their own approach and/or terms to fit. Perhaps the main limitation in adapting (amongst this rapidly growing and evolving field) has been where on earth should we start?

The concept of AQUAFLUO (AQUAtic FLUO rescence) was introduced in 2005 following breakout discussions between several of us at a meeting on modeling algal growth in Villefranche, France. It was then apparent that the immense popularity of fluorometry for aquatic studies was not a transient phenomenon. Both researchers and industry were increasingly investing in development of new fluorometers; however, the fundamental operational and conceptual issues that were limiting how confidently trends in variable fluorescence could be scaled to biology had become a bottleneck in supporting (and ultimately rationalising the need for furthering) this industry growth. Therefore, an organizing committee was established from a cross section of the aquatic community and these concerns translated into the ethos for an international meeting to be held at Nové Hrady in the Czech Republic in 2007 (Prášil et al. 2008). Initial priority questions were identified, including:

1. Is it time to step back from trying to focus on using variable fluorescence as a substitute for conventional productivity (^{14}C, ^{13}C, O_2) primary productivity techniques? Should we redistribute current research efforts and instead focus on the assessment of algal physiology and the general heterogeneity of algal physiology in nature?
2. How important (for the end users) are nuances in current fluorescence induction methodologies (e.g. single or multiple turnover flashes; single band *vs.* multi-spectral)? Can we constructively use these differences in experimental techniques to get better insight into algal physiology?
3. Can we expect successful scaling from direct to remotely sensed (LIDARs and satellites) variable fluorescence?

It was immediately clear that addressing such priority questions could only be achieved by bringing together researchers from across the aquatic disciplines (microalgae, macroalgae, submerged vascular plants, corals and aerobic anoxygenic photoheterotrophs (AAPs); lakes, rivers, coasts and oceans) as well as fluorometer manufacturers and engineers. By bringing this group together, the needs of all interested stakeholders could for the first time identify the common and complimentary needs required to move the field further forward and identify new opportunities; as well, to understand the individual needs of the various aquatic disciplines and develop specific approaches that may help to bridge gaps in consistency in the fluorescence yields that were being measured. A series of talks and workshops at the AQUAFLUO 2007 meeting led by leaders in key aquatic disciplines successfully laid the foundations in exploring these questions. Targeted research activities conducted since the meeting are already beginning to demonstrate that the original ethos of AQUAFLUO and the outcomes of the 2007 meeting are being adopted. However, given the continually evolving nature of using chlorophyll fluorescence in both physiological concept and technological approach, AQUAFLUO is seen as the beginning of a long-term relationship across the aquatic community and an idea with underlying goals and priority questions that will inevitably need to be continually revisited.

The chapters of this book communicate key components of the talks and workshops conducted during the AQUAFLUO 2007. Primarily, they address what measurements can be made and how; the common and aquatic discipline specific pitfalls that may be encountered in both performing measurements and interpreting the fluorescence yields themselves. In essence the book provides a guide to making Chlorophyll *a* fluorescence measurements in the various aquatic sciences and is certainly aimed at experienced and new users alike. *This book is certainly not intended to be a comprehensive review on the subject of Chlorophyll a fluorescence*; several key aspects are not considered in depth, notably remote sensing of variable fluorescence and examination of AAPs.

Each chapter summarises the progress specific to that discipline, the journey that Chlorophyll *a* fluorescence has taken in both approach and application; consequently, the scientific information that can be obtained. Importantly, these chapters also mark the output of that meeting: The first targeted effort to amalgamate the concerted efforts of leaders of the various fields/aquatic disciplines to identify what the next major conceptual (physiological, ecological, biogeochemical) questions that fluorescence measurements can contribute? What are the technological challenges we need to overcome to realize these contributions? The following chapters highlight fundamental areas for research focusing (a) on a range of organisms from corals and macroalgae to microphytobenthos, and (b) scales, from photosynthetic physiology from the cellular level to mass culture. Importantly, these chapters aim to not only target experienced users but also present best practice, which represents optimisation through past (and often frustrating) research, to those new to the field. Of course, since AQUAFLUO 2007 and preparation of this book, our understanding of active fluorescence will have inevitably evolved even further.

Finally, we would like to thank the contributers and the many reviewers of the chapters for their valuable input.

David J. Suggett
Ondrej Prášil
Michael Borowitzka

References

Behrenfeld MJ, Bale AJ, Kolber ZS, Aiken J, Falkowski PG 1996 Confirmation of iron limitation of phytoplankton photosynthesis in the equatorial Pacific Ocean. Nature 383:508–511

Beutler M, Wiltshire KH, Meyer B, Moldaenke C, Lüring C, Meyerhöfer M, Hansen U-P, Dau H (2002) A fluorometric method for the differentiation of algal populations *in vivo* and *in situ*. Photosynth Res 72:39–53

Cullen JJ, Ciotti A, Davis RF, Lewis MR (1997) Optical detection and assessment of harmful algal blooms. Limnol Oceanogr 42:1223–1239

Cullen JJ, Eppley RW (1981) Chlorophyll maximum layers of the southern California Blight and possible mechanisms of their formation. Oceanol Acta 4:23–32

Falkowski PG, Wyman K, Ley AC, Mauzerall DC (1986) Relationship of steady state photosynthesis to fluorescence in eukaryotic algae. Biochim Biophys Acta 849:183–192

Falkowksi PG, Kolber ZS (1995) Variations in chlorophyll fluorescence yields in phytoplankton in the world oceans. Aust J Plant Physiol 22:341–355

Jones RJ, Kildea T, Hoegh-Guldberg O (1999) PAM Chlorophyll Fluorometry: a new *in situ* technique for stress assessment in scleractinian corals, used to examine the effects of cyanide from cyanide fishing. Mar Poll Bull 38:864–874

Kolber ZS, Falkowski PG (1993) Use of active fluorescence to estimate phytoplankton photosynthesis *in situ*. Limnol Oceanogr 38:1646–1665

Kolber ZS, Plumley FG, Lang AS, Beatty JT, Blakenship RE, VanDover CL, Vetriani C, Koblížek M, Rathgeber C, Falkowski PG (2001) Contribution of aerobic photoheterotrophic bacteria to the carbon cycle in the ocean. Science 292:2492–2495

Kolber ZS, Prášil O, Falkowski PG (1998) Measurements of variable chlorophyll fluorescence using fast repetition rate techniques: defining methodology and experimental protocols. Biochim Biophys Acta 1367:88–106

Kromkamp JC, Forster RM (2003) The use of variable fluorescence measurements in aquatic ecosystems: differences between multiple and single turnover measuring protocols and suggested terminology. Eur J Phycol 38:103–112

Kromkamp JC, Beardall J, Sukenik A, Kopeký J, Masojidek J, van Bergeijk S, Gabai S, Shaham E, Yamshon A (2009) Short-term variations in photosynthetic parameters of *Nannochloropsis* cultures grown in two types of outdoor mass cultivation systems. Aquat Microb Ecol 56:309–322

Ley AC, Mauzerall DC (1982) Absolute absorption cross sections for photosystem II and the minimum quantum requirement for photosynthesis in *Chlorella vulgaris*. Biochim Biophys Acta 680:95–106

Lorenzen CJ (1966) A method for the continuous measurements of *in vivo* chlorophyll concentration. Deep Sea Res 13:223–227

Mauzerall DC (1972) Light induced changes in *Chlorella*, and the primary photoprotection for the production of oxygen. Proc Nat Acad Sci USA 69:1358–1362

Papageorgiou GC, Govindjee (2005) Chlorophyll *a* fluorescence: a signature of photosynthesis. Springer, Dordrecht, 818 pp

Platt T (1972) Local phytoplankton abundance and turbulence. Deep Sea Res 19:183–187

Prášil O, Suggett DJ, Cullen JJ, Babin M, Govindjee (2008) Aquafluo 2007: chlorophyll fluorescence in aquatic sciences, an international conference held in Nové Hrady. Photosynth Res 95:111–115

Schreiber U (1998) Chlorophyll fluorescence: New instruments for special applications. In: Garab G (ed) Photosynthesis: mechanisms and effects, vol 5. Kluwer, Dordrecht. pp 4253–4258

Schreiber U, Bilger W, Schliwa U (1986) Continuous recording of photochemical and nonphotochemical quenching with a new type of modulation fluorometer. Photosynth Res 10:51–62

Sukenik A, Beardall J, Kromkamp JC, Kopecký J, Masojídek J, van Bergeijk S, Gabai S, Shaham E, Yamshon A (2009) Photosynthetic performance of outdoor *Nannochloropsis* mass cultures under a wide range of environmental conditions. Aquatic Microbial Ecol 56:297–308

Van Kooten O, Snel JFH (1990) The use of fluorescence nomenclature in plant stress physiology. Photosynth Res 25: 147–150

Contents

Contributors

M. Babin
Université Pierre et Marie Curie-Paris 6,
Laboratoire d'Océanographie de Villefranche, UMR 7093,
France
babin@obs-vlf.fr

J. Beardall
Monash University, School of Biological Sciences, Clayton, Victoria 3800,
Australia
john.beardall@monash.edu

M. Berden-Zrimec
Institute of Physical Biology, Toplarniska 19, SI-1000 Ljubljana, Slovenia
maja@ifb.si

M. A. Borowitzka
Algae R&D Center, School of Biological Sciences and Biotechnology,
Murdoch University, Murdoch 6150, WA, Australia
M.borowitzka@murdoch.edu.au

J. Cosgrove
Algae R&D Center, School of Biological Sciences and Biotechnology,
Murdoch University, Murdoch 6150, WA, Australia

L. Drinovec
Institute of Physical Biology, Toplarniska 19, SI-1000 Ljubljana, Slovenia

Z. Dubinsky
The Mina & Everard Goodman Faculty of Life Sciences, Bar-Ilan University,
Ramat-Gan 52900, Israel

S. Enríquez
Laboratorio de Fotobiología, Unidad Académica Puerto Morelos, Instituto de
Ciencias del Mar y Limnología, Universidad Nacional Autónoma de México,
Apdo. Postal 1152, Cancún, Quintana Roo, México
enriquez@cmarl.unam.mx

K. Felcmanová
Laboratory of Photosynthesis, Institute of Microbiology ASCR,
Opatovický Mlýn, 379 01, Třeboň, Czech Republic

R. M. Forster
Ecosystem Interactions, Centre for Environment, Fisheries and Aquaculture Science (CEFAS), Pakefield Road, Lowestoft, Suffolk, NR33 0HT, UK

R. J. Geider
Department of Biological Sciences, University of Essex, Colchester, Essex CO4 3SQ, UK

D. Holland
Monash University, School of Biological Sciences, Clayton, Victoria 3800, Australia

Y. Huot
Département de Géomatique Appliquée, Université de Sherbrooke, Sherbrooke, J1K 2R1 Québec, Canada
yannick.huot@usherbrooke.ca

T. Jakob
Department of Plant Physiology, University of Leipzig, D-04103, Leipzig, Johannisallee 23, Germany

B. Jesus
Centro de Oceanografia, Faculdade de Ciências da Universidade de Lisboa, Campo Grande, 1749-016 Lisboa, Portugal

O. Komárek
Laboratory of Photosynthesis, Institute of Microbiology ASCR, Opatovický Mlýn, 379 01, Třeboň, Czech Republic
komarek@alga.cz

E. Kotabová
Laboratory of Photosynthesis, Institute of Microbiology ASCR, Opatovický Mlýn, 379 01, Třeboň, Czech Republic

S. A. Kranz
Alfred Wegener Institute for Polar and Marine Research, Am Handelshafen 12, 27570, Bremerhaven

J. C. Kromkamp
NIOO-KNAW Centre for Estuarine and Marine Ecology, P.O. Box 140, 4400 AC Yerseke, The Netherlands

S. R. Laney
Biology Department, Woods Hole Oceanographic Institution, Woods Hole, MA 02543, USA
slaney@whoi.edu

J. Lavaud
CNRS UMR6250 'LIENSs', Institute for Coastal and Environmental Research (ILE), University of La Rochelle, France

E. Lawrenz
Marine Science Program, University of South Carolina, Columbia, SC 29208, USA

S. Lefebvre
Laboratoire de Biologie et Biotechnologies Marine, Université de Caen, Esplanade de la Paix, 14032 Caen cedex, France

M. P. Lesser
Department of Molecular, Cellular, and Biomedical Sciences, University of New Hampshire, Durham, NH 03824, USA

H. L. MacIntyre
Department of Oceanography, Dalhousie University, Halifax, NS B3H 4J1, Canada
hugh.macintyre@dal.ca

J. Masojídek
Institute of Microbiology, Academy of Science, Opatovickýmlýn CZ-37981 Třeboň, Czech Republic; Institute of Physical Biology, University of South Bohemia, Zámek 136 CZ-37333 Nové Hrady, Czech Republic
masojidek@alga.cz

C. M. Moore
National Oceanography Centre, Southampton, University of Southampton Waterfront Campus, European Way, Southampton SO14 3ZH, UK

J. L. Mouget
Laboratoire de Physiologie et de Biochimie végétales, MMS EA 2160, Université du Maine, Av. O. Messiaen 72085 Le Mans Cedex 9, France

R. J. Olson
Biology Department, Woods Hole Oceanographic Institution, Woods Hole, MA, 02543, USA

R. G. Perkins
School of Earth and Ocean Sciences, Cardiff University, Main Building, Park Place, Cardiff, CF10 3YE, UK
PerkinsR@cf.ac.uk

K. Petrou
Plant Function Biology & Climate Change Cluster, University of Technology, PO Box 123, Broadway, Sydney, NSW, Australia

Y. P. Grinblat
The Mina & Everard Goodman Faculty of Life Sciences, Bar-Ilan University, Ramat-Gan 52900, Israel
pinhasy2@zahav.net.il

O. Prášil
Laboratory of Photosynthesis, Institute of Microbiology ASCR, Opatovický Mlýn, 379 01, Třeboň, Czech Republic

P. J. Ralph
Plant Function Biology & Climate Change Cluster,
University of Technology, PO Box 123, Broadway, Sydney, NSW, Australia
Peter.Ralph@uts.edu.au

T. L. Richardson
Marine Science Program, University of South Carolina, Columbia, SC 29208, USA; Department of Biological Sciences, University of South Carolina, Columbia, SC 29208, USA

J. Serôdio
Departamento de Biologia and CESAM – Centro de Estudos do Ambiente e do Mar, Universidade de Aveiro, Campus de Santiago, 3810-193, Aveiro, Portugal

E. Šetlíková
Laboratory of Photosynthesis, Institute of Microbiology ASCR, Opatovický Mlýn, 379 01, Třeboň, Czech Republic

K. Shelly
Monash University, School of Biological Sciences, Clayton, Victoria 3800, Australia

H. M. Sosik
Biology Department, Woods Hole Oceanographic Institution, Woods Hole, MA, 02543, USA
hsosik@whoi.edu

D. J. Suggett
Department of Biological Sciences, University of Essex, Colchester, Essex CO4 3SQ, UK
dsuggett@essex.ac.uk

G. Torzillo
Istituto per lo Studio degli Ecosistemi, CNR, Via Madonna del Piano 10, 50019 Sesto Fiorentino (Firenze), Italy

M. Trtílek
Photon Systems Instruments Ltd, Koláčkova 39, Řečkovice, Brno 621 00, Czech Republic

E. V. Armbrust
School of Oceanography, University of Washington, Seattle, WA, 98105, USA

A. Vonshak
Ben Gurion University J. Blaustein Institutes for Desert Research, Sede Boqer 84990, Israel

M. E. Warner
College of Earth, Ocean, and Environment, University of Delaware, 700 Pilottown Rd. Lewes, DE 19958, USA
mwarner@udel.edu

C. Wilhelm
Department of Plant Physiology, University of Leipzig, D-04103, Leipzig, Johannisallee 23, Germany

A. Zrimec
Institute of Physical Biology, Toplarniska 19, SI-1000 Ljubljana, Slovenia

Chapter 1
Chlorophyll Fluorescence Terminology: An Introduction

Jeff Cosgrove and Michael A. Borowitzka

1 Introduction

The terminology used to describe the various components of variable chlorophyll fluorescence has evolved as our understanding of variable chlorophyll fluorescence has increased. For the newcomer to *in vivo* chlorophyll *a* (chl-*a)* fluorescence studies one of the most confusing aspects can be the large number of terms and notations used, many of which often refer to the same parameter. Several proposals to standardise fluorescence notation, most notably by van Kooten and Snel (1990) and Maxwell and Johnson (2000), have reduced the extent of this variation in more recent publications, but some variation still occurs (Baker and Oxborough 2004). Furthermore, in recent years the notation used necessarily has become more complex as new instruments have allowed researchers to apply several techniques within a single study. On such occasions it is essential that notation also distinguishes between techniques (e.g. single turnover vs. multiple turnover) or method (e.g. steady-state light curve vs. non-steady-state light curve).

This chapter introduces the basic nomenclature associated with the study of variable chl-*a* fluorescence and provides some background on the parameters measured and/or derived. Here we present guidelines for the application and interpretation of chl-*a* fluorescence terminology, with the aim of enhancing communication and translation between methods.

J. Cosgrove and M.A. Borowitzka (✉)
Algae R&D Center, School of Biological Sciences and Biotechnology, Murdoch University, Murdoch 6150, WA, Australia
e-mail: M.borowitzka@murdoch.edu.au

Since chlorophyll fluorescence is reliant on the prior absorption of photon energy, a section outlining the basics of light absorption and its estimation has been included. The accurate estimation of the amount of photosynthetically active radiation absorbed by LHCII and funnelled to RCII is essential for the calculation of fluorescence parameters such as electron transport rate through PSII.

2 Light and Absorption

Radiant energy, or irradiance, is expressed as energy incident per unit time and area; in earlier studies of photosynthetic irradiance was usually denoted with the symbol '*I*', however, as *I* is also used for radiant intensity (W sr^{-1}), the symbol *E* is recommended by the International Union of Pure and Applied Chemistry (IUPAC) to denote irradiance (W m^{-2}) (Braslavsky 2007). Incident radiance in the 400–700 nm waveband is generally considered the photosynthetically active component of total spectral irradiance ($E_{(\lambda)}$) and is termed Photosynthetically Active Radiation (PAR or E_{PAR}) or photosynthetic photon flux density (PPFD or E_{PPFD}). Algae can use irradiance at wavelengths as low as 350 nm for photosynthesis, however, difficulties of measuring the 350–400 nm waveband and its very small contribution (~5–7%) to total irradiance (either solar or from commonly used emission sources) means this is usually ignored (Geider and Osborne 1991; Sakshaug et al. 1997). Since a photon of any wavelength between 400 and 700 nm is equally competent at generating charge separation, E_{PAR} is generally given in units of mol quanta m^{-2} s^{-1} or mol photons m^{-2} s^{-1} rather than

D.J. Suggett et al. (eds.), *Chlorophyll* a *Fluorescence in Aquatic Sciences: Methods and Applications*, Developments in Applied Phycology 4, DOI 10.1007/978-90-481-9268-7_1, © Springer Science+Business Media B.V. 2011

Table 1 Summary of various terms and abbreviations used (see also Tables 2 and 3 for terms relating to the Kautsky curve and variable fluorescence)

Parameter	Other notations	Definition	Units
E	I	Irradiance	$W.m^{-2}$
$E_{(\lambda)}$	$I_{(\lambda)}$	Spectral Irradiance	$W.m^{-2}$
$\overset{o}{E}$		Scalar Irradiance	$W.m^{-2}$
PAR	PPFD	Photosynthetically active radiation – or	mol quanta m^{-2} s^{-1}
E_{PAR}	E_{PPFD}	-Photosynthetically active photon flux density[a]	mol photons m^{-2} s^{-1}
PSII	PS2	Photosystem II	
PSI	PS1	Photosystem I	
LHC I	LHC2	Light Harvesting Complex II	
RCII	RC2	Reaction Center II	
Φ_{PSII}		Quantum yield (efficiency) of electron transport through PSII	
Φ_{O2}		Quantum yield (efficiency) of oxygen evolution	mol O_2 evolved. absorbed quanta^{-1}
Φ_{CO2}		Quantum yield (efficiency) of carbon fixation	mol CO_2 evolved. absorbed quanta^{-1}
Φ_f		Quantum efficiency of fluorescence	
a^{chl}		Chlorophyll-*a* specific absorption coefficient	m^{-2} (mg chl-*a*)$^{-1}$
$\bar{a}^{chl}$		Average chlorophyll-*a* specific light absorption coefficient	m^{-2} (mg chl-*a*)$^{-1}$
PUR^{chl}		Absorbed, chlorophyll-specific, photosynthetically usable radiation	μmol quanta (mg chl-*a*)$^{-1}$ s^{-1}
Q_{phar}		Absorbed photosynthetically usable radiation	μmol quanta m^{-2} s^{-1}
σ_{PSII}		Functional absorption cross-section for PSII	$Å^2$ quantum^{-1}; or m^2 mol RCII^{-1}
τ		PSII turnover time	s
η_{PSII}		Density of functional PS II centres	mol PSII (mol chl *a*)$^{-1}$
$a_{PSII}{}^{chl}$		chl *a*–specific light absorption for PSII photochemistry	m^2 (mg chl *a*)$^{-1}$

[a] The Joint Global Ocean Flux Study (JGOFS) Task Team for Photosynthetic Measurements recommends avoidance of the use 'photon flux density' (Sakshaug et al. 1997)

Einsteins m^{-2} s^{-1} (For easily quotable unit sizes standard practice is to convert from mol to μmol; e.g. 70 μmol photons m^{-2} s^{-1}). For phytoplankton cells which collect radiant energy equally for all sides the photosynthetically active scalar irradiance, $\overset{o}{E}_{PAR}$, measured with a spherical (4π) sensor should be used.

In order to obtain accurate estimates of the quantum yields (efficiency) of electron transport through PSII (Φ_{PSII}), oxygen evolution (Φ_{O2}) or carbon fixation (Φ_{CO2}), one must know the amount of E_{PAR} that is absorbed by the study organism. Difficulties associated with accounting for the scattering component of attenuation in optical measurements have limited the determination of absorbed irradiance as distinct from incident irradiance (Geider and Osborne 1991).

Spectrophotometric techniques are used for the determination of optical absorption coefficients and a number of techniques have been developed to estimate absorption of light by phytoplankton cells or macrophyte tissue.

Conventional spectrophotometers measure transmittance (T) and/or absorbance (A, also often referred to as O.D. (optical density)):

$$A(\lambda) = \log_{10}\left(\frac{E_\lambda^0}{E_\lambda}\right) = -\log_{10} T(\lambda) \qquad (1)$$

where E_λ^0 is the incident (prior to absorption) spectral irradiance and E_λ is the transmitted spectral irradiance.

Working with phytoplankton techniques that have negligible losses due to scattering, such as the opal glass method (Shibata et al. 1954), use of a diffusing plate and minimal sample-detector distance (Bricaud et al. 1983), or an integrating sphere (Bricaud et al. 1983; Maske and Haardt 1987), allows absorptance (A) to be calculated as follows:

$$A = -\log_{10}(1-\alpha) \qquad (2)$$

This parameter, used extensively for characterising the absorption of a cell suspension, is the absorption coefficient (a) with units of m^{-1}. The absorption coefficient is expressed as an exponential function of the absorptance (α) and the path length (l) in metres through the suspension:

$$a = -(1/l)\log_e(1-\alpha) \tag{3}$$

Rearranging the above equations allows a to be calculated directly from A:

$$a = 2.303 \cdot A/l \tag{4}$$

When expressed per unit mass of chlorophyll-a, the chlorophyll-a specific absorption coefficient a^{chl} is:

$$a^{chl} = a/[\text{chl-}a] \tag{5}$$

where [chl-a] is the chlorophyll-a concentration ($mg.m^{-3}$). a^{chl} has been found to vary from 0.004–0.043 m^2 (mg chl-a)$^{-1}$ (Geider and Osborne 1991).

The E_{PAR} that is absorbed and photosynthetically useable can be estimated by a number of different methods: two of the more commonly applied methods will be described here. Calculation of the absorbed, chlorophyll-specific, photosynthetically-usable radiation (PUR^{chl}) with units of μmol quanta (mg chl-a)$^{-1}$ s^{-1} can be calculated from Eq. 6. However, the parameter Q_{phar} (Eq. 7), which provides an estimate of absorbed photosynthetically usable radiation (PUR) in units of μmol quanta m^{-2} s^{-1} is used preferentially for the calculation of absolute Electron Transport Rate (ETR) (Gilbert et al. 2000a, b; Toepel et al. 2004; Wilhelm et al. 2004; Jakob et al. 2005). Estimation of either PUR^{chl} or Q_{phar} requires knowledge of both emission and absorption spectra.

$$PUR^{chl} = \sum_{\lambda=400}^{\lambda=700} E(\lambda)a^{chl}(\lambda) \tag{6}$$

$$Q_{phar} = \int_{400nm}^{700nm} E(\lambda) - E(\lambda)\cdot e^{(-a^{chl}(\lambda)[\text{Chl-}a]\cdot d)}\,d\lambda \tag{7}$$

where E(λ) = photosynthetically available (incident) spectral radiation (μmol quanta m^{-2} nm^{-1} s^{-1}); $a^{chl}(\lambda)$ = Chl-a specific *in vivo* absorption coefficient of the cell suspension at wavelength λ in [m^2 mg^{-1}(chl-a)]; [chl-a] = Chl-a concentration in [mg (chl-a) m^{-3}]; d = optical path length (m).

Not all of the absorbed photosynthetically usable radiation is necessarily actively used in photosynthesis. Absorption by photosynthetic pigments is described by the photosynthetic cross-section (Suggett et al. 2003, 2004; Macintyre and Cullen 2005). The functional absorption cross-section for PSII (σ_{PSII}; Å^2 quanta^{-1}) is the product of the light-harvesting capability (optical absorption coefficient) of the photosynthetic pigments and the efficiency of excitation transfer to the reaction centre, and provides a measure of the effectiveness of incident light capture and conversion to electron transfer in PSII (Mauzerall and Greenbaum 1989; Kolber and Falkowski 1993; Wood and Oliver 1995). Decreasing growth irradiance and the onset of nutrient limitation lead to increases in σ_{PSII} (Kolber et al. 1988; Falkowski and Kolber 1995; Wood and Oliver 1995). On the other hand, non-photochemical dissipation of excitation energy, such as heat dissipation by the xanthophyll cycle, will decrease σ_{PSII} (Olaizola et al. 1994; Babin et al. 1996; Barranguet and Kromkamp 2000).

Algae may experience threefold changes in magnitude of σ_{PSII} without any changes in quantum efficiency of PSII as they adapt to different growth irradiances, although the quantum yield of O_2-evolution may be altered (Olaizola et al. 1994). All in all, measurements of σ_{PSII} in natural phytoplankton communities are highly variable, with data from Kolber and Falkowski (1993) suggesting a range from 250–1,000 Å^2 quanta^{-1} and an average of ~500 Å^2 quanta^{-1}. Relatively few measurements of σ_{PSII} have been obtained from individual taxa under controlled conditions (Suggett et al. 2004). However, a recent analysis of multiple ship cruise and controlled culture data-sets (Suggett et al. 2009) has highlighted the influence of phytoplankton community structure, including taxonomic composition (pigments) and cell size, on recorded σ_{PSII}.

As σ_{PSII} reflects light harvesting capacity of the PSII antenna and exciton transfer efficiency, the time interval between exciton arrivals at the PSII reaction centre can be calculated by:

$$\tau = (\sigma_{PSII} \times E_{PAR})^{-1} \tag{8}$$

where τ is the PSII turnover time (s). τ can increase markedly as cells adapt to lower growth irradiance and the ability of the dark carbon fixation pathways to match the pace of the light-driven Electron Transport Chain (ETC) (when in saturating irradiances) is reduced (Kolber and Falkowski 1993).

Importantly, for those studying chlorophyll fluorescence, σ_{PSII} multiplied by incident irradiance and the density of functional PS II centres, η_{PSII} (mol PSII (mol chl *a*)$^{-1}$), provides a measure of total photon flow available for charge separation at PSII (E_{PSII}) (Kromkamp and Forster 2003).

$$E_{PSII} = E_{PAR} \times \sigma_{PSII} \times \eta_{PSII} \qquad (9)$$

Therefore the parameters σ_{PSII} and η_{PSII} combine to provide the chl *a*–specific light absorption for PSII photochemistry (a^{chl}_{PSII}; m^2 (mg chl *a*)$^{-1}$) (Suggett et al. 2004):

$$a^{chl}_{PSII} = \sigma_{PSII} \times \eta_{PSII} \times 0.00674 \qquad (10)$$

where the constant 0.00674 is a conversion factor based on conversion of σ_{PSII} units to m^2 (mol PSII)$^{-1}$ and η_{PSII} to mol PSII (mg chl-*a*)$^{-1}$ (i.e. $\sigma_{PSII} \times 6023$ and $\eta_{PSII}/893490$).

Unfortunately η_{PSII} is difficult or impractical to measure and assumed values are commonly used (Suggett et al. 2004). Values of 1.6×10^{-3}–2×10^{-3} mol PSII (mol chl *a*)$^{-1}$ based on the measurements of total photosynthetic unit (PSU) size by Emerson and Arnold (1932) and Gaffron and Wohl (1936) are common. These authors measured between 2,000 and 2,500 chlorophyll molecules associated with each PSU and η_{PSII}^{-1} was considered to be one quarter of the PSU size due to the requirement for four electrons to pass through PSII for each O_2 molecule evolved (Ley and Mauzerall 1982). However, η_{PSII}^{-1} values ranging from 260 to 800 have been used (Kromkamp and Forster 2003 and references therein).

Adopting an assumed value for η_{PSII} may produce inaccurate results since η_{PSII} is known to change as a result of photoacclimation, photoinhibition and nutrient limitation (Suggett et al. 2004).

3 Fluorescence

Fluorescence is the re-emission of energy in the form of a photon (light) as an electron returns to ground state from a singlet excited state. In the case of chl-*a* fluorescence a chlorophyll molecule can become excited and achieve singlet state 1 (S_1) after absorbing a photon of less than 670 nm wavelength (Bolhàr-Nordenkampf and Öquist 1993). If the energy is not utilised in charge separation, heat dissipation, or resonance energy transfer, fluorescence will occur as the electron drops out of the excited state. As some energy is also given off as heat, the photon is red-shifted with an emission peak of ~685 nm. If the absorbed photon is of a shorter wavelength (e.g. blue light at about 420 nm) the extra energy excites the chlorophyll molecule to the singlet state 2 (S_2) and heat is emitted as it rapidly decays to the S_1 state. It is commonly considered that, at ambient temperatures, nearly all fluorescence (~90–95%) originates from PSII at 685 nm (Krause and Weis 1991; Papageorgiou et al. 2007) and represents 0.6% to ~10% of the absorbed light (Nicklisch and Köhler 2001). This is because the 680 nm absorption peak of chl-*a* molecule at the core of PSII is red-shifted only 5 nm from the absorption peak of the lowest singlet excited state of chl-*a* in most antenna systems. With such a small difference, energy can escape from RCII back into the pigment bed, hence PSII is known as a "shallow trap" and most fluorescence is emitted from the PSII antennae molecules (Krause and Weis 1988). The absorption maximum for PSI is sufficiently red-shifted (25 nm) relative to its antenna so that there is much less chance for energy to escape (Falkowski and Raven 1997).

Carl Lorenzen introduced the technique of *in vivo* chlorophyll fluorescence analysis to biological oceanography in 1966 (Lorenzen 1966). Since then a large number of different fluorometers have been designed to measure variable chl-*a* fluorescence under a wide range of conditions and for various applications. Each of these fluorometers is based on one of a few basic operational principals and can be classified as one of the following:

1. Pulse Amplitude Modulation fluorometer (PAM)
2. Fast Repetition Rate Fluorometer (FRR)
3. Fluorescence Induction and Relaxation System (FIRe)
4. Pump and Probe Fluorometer (PandP)
5. Induction Fluorometer/Continuous Excitation Fluorometer

The application and underlying theory of most of these methods for the *in situ* measurement of phytoplankton fluorescence has recently been reviewed (Babin 2008; Huot and Babin Chapter 3).

Light energy absorbed by a photosystem and its LHC can be used/dissipated through one of three competing pathways: (1) photochemistry (primary charge separation and photosynthetic electron transfer), (2) thermal dissipation (non-radiative decay) or, (3) fluorescence emission (Falkowski et al. 1986; Seaton and Walker 1990; Kolber and Falkowski 1993; Nicklisch and Köhler 2001). It is assumed that the sum of the quantum yields of each of these processes is unity. Thus, changes in fluorescence yield reflect changes in the complementary pathways. It is important to note that the above-mentioned fluorometers measure fluorescence yield (which may vary up to a factor of 5 or 6), not fluorescence intensity (which may by a factor of several thousand), as it is the former that carries information on photosynthesis (Dau 1994; Schreiber 2004).

The quantum efficiency of fluorescence (Φ_f) is simply the ratio of quanta fluoresced to total quanta absorbed:

$$\Phi_f = Q_f / Q_a \tag{11}$$

where Q_a is the quanta absorbed and Q_f the quanta fluoresced. If Φ_f is known, then this can be used to determine fluorescence emission on a per unit chlorophyll basis:

$$F = E \times \bar{a}^{chl} \times \Phi_f \tag{12}$$

where F is the chlorophyll-specific fluorescence, E = incident irradiance, $\bar{a}^{chl}$ = average chl-specific light absorption coefficient (m^{-2} (mg chl)$^{-1}$) (Estrada et al. 1996).

Higher Φ_f, and therefore higher F, is correlated with a closure of RCIIs since energy must then be dissipated to a greater extent by pathways other than primary charge separation (Ralph and Gademann 2005). As the redox state of the quinone Q_A determines whether RCII is open (Q_A) or closed (Q_A^-), it is also the main controlling factor determining chl-*a* fluorescence yield (Schreiber et al. 1998).

The Kautsky curve, also referred to as the fluorescence induction curve or the fluorescence transient, describes the characteristic changes in chl-*a* fluorescence yield upon illumination of a dark-adapted alga or leaf (Fig. 1). This pattern was first described by Kautsky and Hirsch (1931), but has been elaborated on since then (see Govindjee and Papageorgiou 1971; Govindjee 1995 for details) The fluorescence induction curve is a complex phenomenon that can be broken down into two primary phases: a *fast phase* (up to ~1 s)

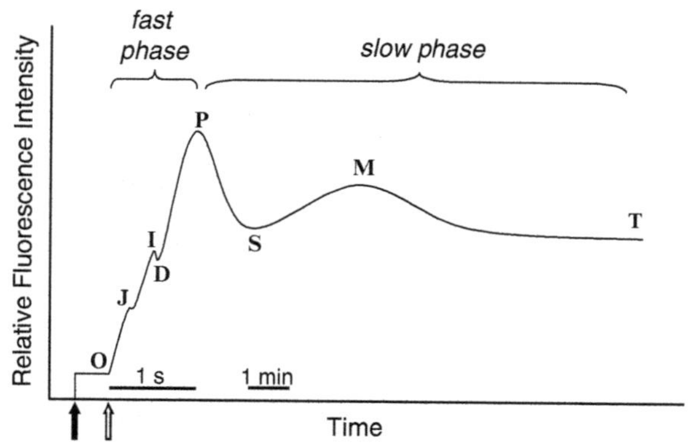

Fig. 1 Stylised representation of the chl-*a* fluorescence induction curve. Closed arrow represents activation of non-actinic measuring light. On application of strong actinic light (open arrow) fluorescence rises from the origin (O) to a peak (P) via two inflections (J and I). A dip (D) may occur after I. This O-P rise is known as the *fast phase* and reflects primary photochemistry and redox state of Q_A. After P fluorescence declines due to formation of a transthylakoid pH gradient and associated thermal quenching. The remainder of the transient (S-M-T) is called the *slow phase* and is the result of induction of Calvin cycle enzyme activity and its subsequent interaction with the electron transport chain (via NADPH) and photochemical and non-photochemical quenching

followed by a *slow phase* (up to several minutes) (Krause and Weis 1984; Govindjee 1995) (Fig. 1).

3.1 Fast Phase (O-J-I-P)

The fast phase of the fluorescence rise begins upon illumination of dark acclimated samples as fluorescence rises rapidly from the origin (O) to a peak (P) via an inflection (I) and dip (D) of variable magnitude (Fig. 1, Table 2). If the actinic light is strong another inflection (J) can be observed between O-I (Govindjee 1995; Strasser et al. 2004). The fast phase is often named after these cardinal points and referred to as the O-J-I-P curve, although other features may appear under certain experimental conditions (see *additional features* below), and reflects changes in the redox state of the RCII coinciding with the primary processes of photosynthesis (Govindjee and Papageorgiou 1971; Büchel and Wilhelm 1993). The fast phase has also been labelled the ‘polyphasic fluorescence rise’.

The origin (O) of the fluorescence induction curve represents a minimum value and is fluorescence emitted from excited chl-*a* molecules in the antennae complex

Table 2 Nomenclature used in characterising features of the fluorescence induction curve

Parameter	Synonym	Definition	Time (ms)
O		Origin, Constant fluorescence, minimum fluorescence, initial fluorescence	0 or 0.05
J	I_1		~2–5
I	I_2	Inflection or intermediate	~30–50
D		Dip	
P	M	Peak	~500
L			~0.1–0.2
K			~0.3–0.4
H		Hump	
G			
S		Semi-steady state , stationary level	
M		Maximum	
T		Terminal steady-state	
F_i		Fluorescence yield at *i*, where *i* equals any of the parameters above (e.g. F_P equals yield at P)	

before excitons have migrated to the reaction centre (Krause and Weis 1984). This point is also given the terms *dark fluorescence*, *constant fluorescence*, *initial fluorescence* or *fluorescence minimum*. Fluorescence yield at O is commonly given the notation F_o (subscript o representing "origin"), F_0 (subscript 0 representing minimum) or simply F_{min}.

The rapid O-J rise in fluorescence represents the photochemical phase (Govindjee 1995), with fluorescence yield increasing proportionally with reduction of Q_A. Consequently the slope and height of this phase is dependent on incident light intensity (Lazár 2006). The inflection J occurs after ~2–5 ms of illumination (Lazár 2006; Papageorgiou et al. 2007), reflects a momentary maximum of Q_A reduction and is equivalent to the less common notation of I_1, of Neubauer and Schreiber (1987) and Schreiber and Neubauer (1987).

The remaining rise in fluorescence yield from J to its peak (P) represents the thermal phase, as it is impacted by temperature within the physiological range (slower at lower temperatures) (Lazár 2006). This phase of the fast fluorescence rise is shaped by the two-step reduction of Q_B ($Q_B \rightarrow Q_B^- \rightarrow Q_B^{2-}$) and a heterogeneity in the reduction of the plastoquinone (PQ) pool (Lazár et al. 1999). The second inflection I, also known as I_2 (Neubauer and Schreiber 1987; Schreiber and Neubauer 1987), occurs some 30–50 ms after illumination (Lazár 2006; Papageorgiou et al. 2007) and is thought to reflect a temporary maximum of $Q_A^-Q_B^{2-}$ (Govindjee 2004) , however other interpretations are presented in recent literature (Heredia and Rivas 2003; Ilík et al. 2006; Lazár 2006). The fluorescence yield may dip (D) after the J and/or I inflection(s) as electrons begin to move from one quencher to the next (e.g. Q_A^- to Q_B), resulting in transient reoxidation of the primary quencher (Krause and Weis 1984; Schreiber et al. 1998). If multiple dips occur it is standard to label each dip with a numeric subscript (e.g. O-J(D_1)-I(D_2)-P) (Govindjee 1995). Fluorescence yield then continues to rise as a lack of reductants on the PS II acceptor side becomes limiting, reaching the peak (P) when the PQ pool becomes fully reduced and $Q_A^-Q_B^{2-}$ concentration reaches a second maximum. In this situation, under saturating excitation irradiance, P is equivalent to F_m (maximum fluorescence yield).

3.1.1 Additional Features

The O, J, I, D and P features of the fast fluorescence transient are those that are most commonly observed under normal conditions. However, additional features can be observed in certain experimental conditions. Labelling of each of these features has been conducted such that they run in reverse alphabetical order between O and P.

The earliest feature is the L-band. This covers an area of the initial rise from O about 0.1–0.2 ms after the onset of illumination. Curvature of this portion of the fluorescence transient is related to PS II connectivity, with a shift to a more hyperbolic transient indicating greater connectivity (Oukarroum et al. 2007).

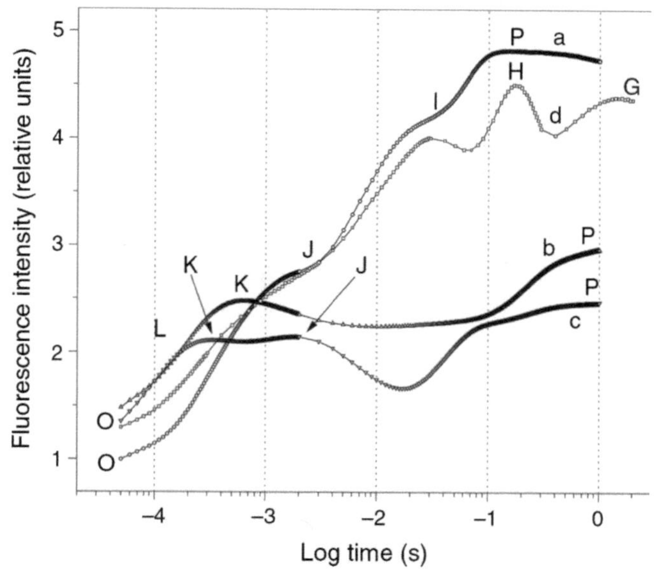

Fig. 2 Chl-*a* fluorescence rise measured with dark adapted pea leaves (curve a, not treatment; curve b, leaf incubated at 47°C in water for 5 min), potato leaf (curve c, leaf incubated at 44°C in water for 13 min), and the lichen *Umbilicaria hirsuta* (curve d, no treatment) by PEA fluorometer under high intensity of excitation light [3,400 μmol photons m^{-2} s^{-1} of red light]. Steps O, L, K, J, I, H, G and P are labelled (Figure modified from Lazár 2006)

Samples under stress, such as heat or drought stress or nitrogen limitation (Strasser et al. 2004; Oukarroum et al. 2007), can experience donor-side inhibition of PS II. When this occurs the K-band becomes apparent, typically ~0.2–0.4 ms after the onset of illumination. The K-band may even be associated with dissociation of the oxygen evolving complex (Oukarroum et al. 2007).

In some taxa P is effectively split into two separate peaks, H and G (curve d, Fig. 2). Ilík et al. (2006) proposed that this might be due to the combined actions of rapidly activated PS I reoxidation (ferredoxin-$NADP^+$ oxidoreductase activation) or an active Mehler-peroxidase reaction causing transient reoxidation of the PQ pool, Q_B^{2-} and Q_A^- (dip from H) before limitation of electron flow at cyt b_6/f causes a rise in fluorescence to the secondary peak, G.

3.2 *Slow Phase (S-M-T)*

Following the fast phase there is a polyphasic decline in chl-*a* fluorescence that ends at a terminal steady state level (T). This phase of the curve is known as the *slow phase* and may include a secondary peak (M) (refer to Walker (1981) for a discussion on the occurrence and dynamics of the M-peak) after a trough (S) (Fig. 1). This slow phase is primarily related to the balancing of a number of processes and several oscillations may be observed as overcompensation occurs in the regulation of reductive power and thylakoid energisation (Renger and Schreiber 1986). This results in successive S troughs and M peaks, each labelled with successive numeric subscripts (e.g. $S_1M_1S_2M_2$…T). In dark-adapted cyanobacteria, the S-M-T transient may dominate the fluorescence induction pattern, with fluorescence yield increasing above P as a result of State 2 → State 1 transition (Papageorgiou et al. 2007).

The decay in fluorescence from P → S is a complex phenomenon that has been described as the least understood part of the fluorescence transient (Govindjee and Papageorgiou 1971). Current understanding suggests that fluorescence yield decreases with the combined effects of enhanced PSI activity and ΔpH formation (Govindjee and Satoh 1986; Krause and Weis 1991). The faster passage of electrons through PSI, with the activation of ferredoxin-$NADP^+$ oxidoreductase results in a build up of qP as Q_A^- becomes partially reoxidised (Renger and Schreiber 1986). Formation of a transthylakoid pH gradient (ΔpH) results in protonation of acidic amino acids leading to conformational changes in core PSII antenna complexes that promote non-radiative dissipation of excess energy (Strasser et al. 2004; Holub et al. 2007; Papageorgiou et al. 2007). State transition quenching ($q(T_{1\rightarrow2})$) is also thought to contribute to the P-S decline . Such processes, and therefore the P → S feature of the fluorescence induction curve, depend upon an intact chloroplast envelope (Papageorgiou et al. 2007).

The S-M rise, when present, has been explained by LHC II dephosphorylation and a return to State 1 (Holub et al. 2007) and a lack of reductant ($NADP^+$) on the acceptor side of PSI (Govindjee and Satoh 1986; Renger and Schreiber 1986; Krause and Weis 1988). In phycobilisome-containing cyanobacteria the State 2 → 1 change, as the photosynthetic electron transport chain become reoxidised after reduction in the dark by components of the respiratory electron transport chain, can result in the S-M rise being the dominant feature of the fluorescence induction curve (see Figure 2 in Papageorgiou et al. 2007).

The decline from M to T is generally recognised as a reflection of activation of the Calvin-Benson Cycle enzymes. The resulting increase in CO_2 fixation rate and NADPH reoxidation has a flow-through effect back up the photosynthetic electron transport chain, yielding a higher qP (Renger and Schreiber 1986).

Due to the complex nature of the numerous interactions influencing the chl-*a* fluorescence induction curve there has been some question over the level of interpretation of the data that is possible (Holzwarth 1993; Trissl et al. 1993; Falkowski 1994; Trissl 1994), however, it is generally acknowledged that analysis of the induction curve remains a useful tool that will only improve as our knowledge of photosynthetic (and interacting) processes advances (Govindjee 1995).

3.3 The Saturation Pulse Method

The fast phase of the induction curve is commonly used to estimate photochemical quantum yield. The method applied is the *saturation pulse method* (Fig. 3) and involves analysis of the quenching components. Among the first to apply this method were Bradbury and Baker (1984) with refinement by Schreiber et al. (1986), and the technique has been further described (Schreiber et al. 1995b, c; Schreiber 2004). Minimum fluorescence yield (F_0) will occur when all RCIIs are open (Q_A in all RCIIs is oxidised) and quantum energy reaching the reaction centre has the maximal chance of being utilised photochemically and a negligible chance of being dissipated as heat or fluorescence. This state is generally considered to be achieved after adaptation to the dark and dissipation of any transthylakoid pH gradient

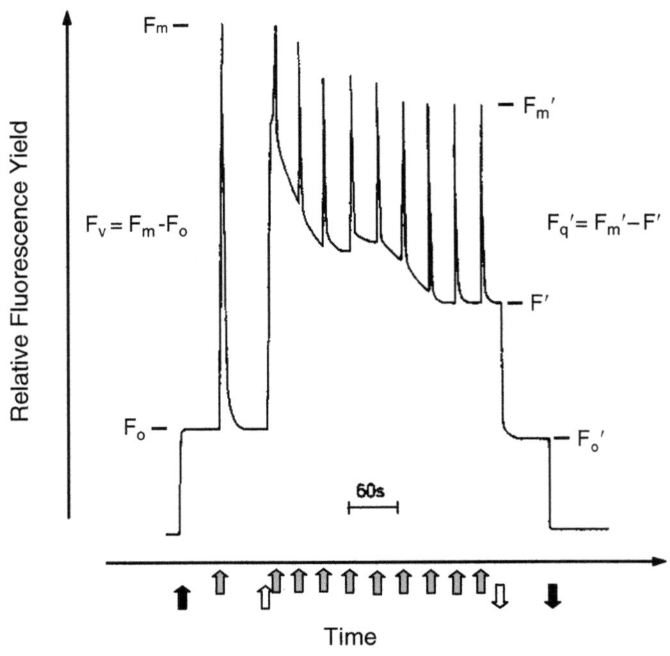

Fig. 3 Fluorescence induction kinetics including application of the saturation pulse method and associated nomenclature (refer to text). Dark arrows indicate measuring light on (*up*) and off (*down*); Grey arrows indicate application of a short pulse of saturating light; open arrows indication activation (*up*) and deactivation (*down*) of actinic light (Modified from Büchel and Wilhelm 1993)

(see *Quenching* below). In order to measure such a state the fluorometer's measuring light must be weak enough (<0.5 μmol quanta m^{-2} s^{-1}) so as not to induce reduction of Q_A and closure of any reaction centres. When a pulse of high intensity light sufficient to close all RCIIs (reduce all Q_A) is applied to a sample, a condition is induced where photochemistry is reduced to zero and fluorescence yield is maximal. If the sample was dark-adapted prior to application of the saturation pulse, non-photochemical quenching will be negligible and fluorescence yield will reach its true maximum (F_m). However, if the sample was not dark-adapted, non-photochemical quenching will act to quench the fluorescence yield and the achieved maximum value will be lower (F_m'; where the ′ denotes that the sample was not dark adapted). Hence a drop in F_m to F_m' can be used as a measure of non-photochemical quenching. These measures assume that no non-photochemical quenching is induced by the short saturation pulse (Schreiber et al. 1995a; Schreiber 2004). We can therefore use these principles to estimate photochemical quenching (see Fig. 3 and Table 3). It should be

Table 3 Fluorescence parameters, their definition and their synonyms (following Baker et al. (2001)

Parameter	Synonyms	Definition	Derivation
ETR	J_e	Electron Transport Rate (through PSII)	
ETR_m	ETR_{max}	Maximum ETR	
rETR		Relative Electron Transport Rate (through PSII)	$F_q'/F_m' \times E_{PAR}$ ($F_q'/F_m' \times E_{PAR} \times 0.5$)
$rETR_m$	$rETR_{max}$	Maximum rETR	
F_o	F_0, F_{min}, Φ_{Fo}, Φ_f^{min}	Fluorescence yield at O; Minimum fluorescence yield; dark fluorescence yield (dark adapted, all RCIIs open)	
F_m	F_{max}, Φ_{Fm}, Φ_f^{max}, (F_P)	Maximum fluorescence yield (dark adapted, all RCIIs closed with no NPQ), Fluorescence yield at point P (F_P) of the fluorescence induction curve is equivalent to F_m if irradiance is saturating.	
F_v	Φ_{Fv}	Maximum variable fluorescence yield (qN = 0)	$F_m - F_o$
F_v/F_m	Φ_{Po}, Φ_P^{max}, Φ_{PSII}^{max}, $\Delta\Phi_m$	Maximum photochemical efficiency (quantum yield) of open RCIIs. [The term $\Delta\Phi_m$ is the equivalent term for single turnover saturation pulse measurements.]	$(F_m - F_o) / F_m$
F'	F, F_t, $\Phi_{F'}$, F_v	Fluorescence yield in actinic light; fluorescence yield at time *t* (F_v is from Schreiber et al. (1986))	
F_s	Φ_{Fs}, F_T	Steady-state fluorescence yield in actinic light. F_T represents fluorescence yield at point T of the fluorescence induction curve.	
F_o'		Minimum fluorescence yield in light-acclimated state (Usually measured with the application of far-red light)	
F_m'		Maximum fluorescence yield in actinic light	
F_v'		Variable fluorescence yield in actinic light	$F_m' - F_o'$
$F_{m\,m}'$		The maximum value of F_m'	
F_q'	ΔF	Difference between fluorescence yields F_m' and F'	$F_m' - F'$
F_q'/F_m'	$\Delta F / F_m'$, Φ_{PSII}	Effective photochemical efficiency of RCIIs in actinic light	$(F_m' - F') / F_m'$
F_v'/F_m'		Maximum PSII photochemical efficiency or quantum yield in actinic light.	$(F_m' - F_o') / F_m'$
NPQ		(Stern-Volmer) Non-photochemical quenching	$(F_m - F_m') / F_m'$
q_E	qE, q(E)	Energy-dependent non-photochemical quenching	$(F_v - F_v') / F_v$
q_I	qI, q(I)	Photoinhibitory non-photochemical quenching	
q_N	qN, q(N)	Non-photochemical quenching	$(F_m - F_m') / (F_m - F_o)$
q_P	q_Q, F_q' / F_v', qP, q(P)	Photochemical quenching	$(F_m' - F') / (F_m' - F_o')$
q_T	qT, $q(T_{1\rightarrow 2})$	State-transition related non-photochemical quenching	

All parameters are dimensionless

noted here that there is the potential for induction of non-photochemical quenching pathways in the dark (possibly linked to chlororespiration) (Jakob et al. 2001) which would impact particularly the F_m value. The potential error caused by this can often be overcome by applying a short pulse of far-red light to ensure full reoxidation of the PQ pool (P. Ralph, personal communication) or by correcting F_0 and F_m based on measured non-photochemical quenching (Ting and Owens 1993). However, in some experimental situations it may be necessary to use the inhibitor 3-(3,4-dichlorophenyl)-1,1-dimethylurea (DCMU) to be confident of measuring true F_m.[1]

The measured F_m (or F_m') is dependent on the measuring technique. Some fluorometers, such as the FRR type, apply a rapid flash (also called a "flashlet") that is short enough (10–100 μs) to result in only a single reduction of Q_A to close the RCII and cause a rise in fluorescence through the photochemical phase of the induction curve, but not the thermal phase. Thus, F_m measured using this 'single turnover' technique is roughly equivalent to F_J of the fluorescence induction curve, although with the application of DCMU it can be significantly higher than F_J due to a reduction in quenching by Q_B and the PQ pool. Other fluorometers, such as the PAM type, apply an extended flash (usually between 0.4 and 1.0 s) of high intensity light that results in multiple turnovers of the reaction centre to saturate Q_A, Q_B and PQ to close all RCIIs (i.e. both the photochemical and thermal phases are induced) and F_m is equivalent to F_P of the fluorescence induction curve. The Fluorescence Induction and Relaxation (FIRe) system by Satlantic employs both single- and multiple-turnover protocols. A comprehensive comparison of these two saturation techniques and their respective advantages and disadvantages has been provided by Kromkamp and Forster (2003) and Babin (2008), however, it is fundamentally important to note that direct comparison of results between single turnover and multiple turnover techniques cannot be made as they yield different results. Even direct comparison of results within the same fluorometer type, PAM fluorometers for example, should be done with some caution since system geometry and the position of the detector relative to the cuvette/sample (Mouget and Tremblin 2002) or stirring in the Water-PAM (Cosgrove and Borowitzka 2006) can affect results. It is for these reasons, amongst others, that researchers should carefully detail their equipment and methodology used and nomenclature should, where confusion could occur, specify the method or technique used.

Examples of the used of well applied nomenclature aimed at distinguishing methodologies are included here. Rascher et al. (2000) compared effective quantum yield ($\Delta F/F_m'$) calculated from saturation pulse measurements taken under ambient conditions with those calculated from a light curve protocol. An extra subscript term (i.e. $\Delta F/F_m'{}_{amb}$ or $\Delta F/F_m'{}_{LC}$) was used to indicate which method was used. This approach was also employed by Kromkamp and Forster (2003) and Kromkamp et al. (2008) to distinguish parameters derived from single turnover (ST) flashes from those derived from multiple turnover (MT) flashes, although on this occasion the added descriptor was enclosed in brackets when other subscript notation was present (e.g. $F_{m(ST)}$ vs. $F_{m(MT)}$ or ETR_{ST} vs. ETR_{MT}). This nomenclature has, in principle, been followed by others such as (Röttgers 2007), who removed the brackets from the method descriptor (e.g. $F_{m,ST}$ *vs.* $F_{m,ST}$; $rETR_{m,ST}$ vs. $rETR_{m,MT}$). For clarity we recommend the use of brackets for the method descriptor.

Similarly, a number of authors (Kühl et al. 2001; Serôdio et al. 2006; Ulstrup et al. 2007; Cruz and Serôdio 2008) have calculated fluorescence parameters and photosynthetic electron transport from both Rapid Light Curves (RLCs) and steady-state light curves (LCs; also referred to as SSLCs by Ulstrup et al. (2007). Serôdio et al. (2006) and Cruz and Serôdio (2008) used the notation to clearly differentiate between the methods (e.g. $ETR_{m,RLC}$), while other workers have chosen to do so in the text.

3.4 Quantum Yield for PSII (Φ_{PSII})

The efficiency of a light-dependent process is referred to as the *quantum yield* or *quantum efficiency*. In basic terms the quantum yield (Φ) can be described as the ratio of product output to gathered quanta:

$$\Phi = \frac{mol\ product\ out}{mol\ quanta\ in} \tag{13}$$

Saturation pulse analysis, as described above, can be used to estimate the quantum yield, or efficiency, of

[1]DCMU must be added in total darkness and the sample should not be exposed to any light before measurements are made: Since DCMU functions by displacing Q_B, even low light can cause quick net formation of Q_A^- artificially raising the measured F_0 as Chl fluorescence is high when Q_A^- is present (Govindjee 2004; Huot and Babin Chapter 3).

PSII photochemistry, Φ_{PSII} (Genty et al. 1989; Schreiber et al. 1995c). In this circumstance the end product can be described as primary charge separation and the passage of an electron through PSII.

When in the dark, with Q_A in a fully oxidised state and no non-photochemical quenching, the *maximum* quantum yield of PSII (Φ_{PSII}^{max}) can be estimated by normalising the variable fluorescence (F_v) to the maximum fluorescence yield (F_m) (Table 3). In the light the photochemical efficiency of PSII and closure of reaction centres (reduced Q_A) decreases due to the induction of non-photochemical quenching (and possibly photoinhibition in high light) and the parameter F_q' ($F_q' = F_m' - F'$; see Oxborough et al. (2000)), where subscript "q" signifies quenched fluorescence, is normalised to the light adapted maximum fluorescence yield (F_m') (Table 3). Thus, by applying the saturation pulse methods the quantum yield of PSII can be rapidly measured on a virtually real-time basis with high sensitivity (Kolber et al. 1994).

Maximum theoretical values for F_v/F_m are ~0.65 (Kolber and Falkowski 1993) for single turnover saturation pulses and ~0.83 for multiple turnover pulses (Magnusson 1997)}. In practice, maximum achievable F_v/F_m is known to vary between taxa as a result of differences in pigment composition and cell structure (Koblížek et al. 2001; Suggett et al. 2009). For example, smaller taxa appear to have lower F_v/F_m values (as low as 0.3–0.4 for the smallest pico-eukaryotes) along with higher σ_{PSII} values (Suggett et al. 2009). Environmental factors that impact upon PSII, directly or indirectly, will also impact measures of F_v/F_m (Greene et al. 1992). Dominant factors in this regard include light, nutrient status and temperature (Wozniak et al. 2002), however Brand (1982) also found that many marine phytoplankton species exhibit endogenous diel patterns in fluorescence parameters and suggested this may be the result of changes in cellular metabolism to "predict" environmental condition. All these factors combine to confound interpretation of F_v/F_m and other fluorescence parameters (Kroon et al. 1993).

3.5 Quenching

Fluorescence, or radiative decay, is one of three competitive pathways for the de-activation of chl-excited states in the photosynthetic reaction centres and their antennae. Thus, the other two pathways, photochemistry and non-radiative decay (heat dissipation), act to *quench* the fluorescence signal. These processes are called photochemical- and non-photochemical quenching respectively. To quantify these quenching pathways various quenching coefficients have been defined. As mentioned previously, the saturation pulse method can be used to measure each of the quenching components.

Photochemical quenching, q_P, estimates the percentage of RCIIs that are open (Magnusson 1997) or the capacity for photochemistry to compete for trapped quantum energy (Ting and Owens 1993). When all reaction centres are open $q_P = 1$ and when all centres are closed $q_P = 0$ (Schreiber et al. 1986). The RCII is considered 'open' when Q_A is oxidised and capable of accepting an electron from RCII via pheophytin. It is important to note that energy transfer between RCIIs, or "*connectivity*", modifies the linear relationship between q_P and the fraction of RCIIs that are open (Suggett et al. 2003; Schreiber 2004). Thus, q_P more accurately represents the redox state of Q_A. The quinones Q_A and Q_B where labelled "Q" because they act to *quench* fluorescence, Q_B is a secondary quencher as it acts to re-oxidise Q_A, thereby returning it to its quenching state. Similarly, the redox status of PQ may influence q_P. For example, q_P can be increased by increasing ambient dissolved inorganic carbon (DIC) concentrations as this acts to favour the Rubisco carboxylase reaction and increase the rate of linear electron flow, resulting in partial reoxidation of the PQ pool (Carr and Björk 2003). Given that these primary components influencing q_P are highly conserved across taxa, the mechanism of q_P are likely to be similar (Ting and Owens 1993).

Based on the work of Bilger and Schreiber (1986) who found that F_o could be quenched to F_o', q_P was defined by Van Kooten and Snel (1990) with standardised nomenclature (see Table 3). F_o' may be hard to measure and on occasion F_o has been used for the calculation of q_P instead (e.g. Weis and Berry (1987) or Ralph and Gademann (2005)). This risks overestimation of q_P and, as a consequence, an alternative method for deriving F_o' has been described (Eq. 14) (Oxborough and Baker 1997).

$$F_o' = \frac{F_o}{F_v / F_m + F_o / F_m'} \tag{14}$$

The total non-photochemical quenching coefficient (q_N) is a measure of the fraction of maximum dark-adapted variable fluorescence ($F_m - F_o$) that is quenched in the light (Eq. 15). This coefficient has two disadvantages in that: (1) it involves the estimation of F_o', and (2) there is evidence that it may be influenced directly by the rate constant of photochemistry (Krause and Jahns 2004). Another measure of total non-photochemical quenching (NPQ) can be calculated using the Stern-Volmer equation (see Table 3). In this case NPQ represents the relative increase in the sum of the rate constants of the non-photochemical deactivation processes (fluorescence emission, heat dissipation and spillover of excitation energy from PSII to PSI) relative to the dark-adapted state (assuming no non-photochemical reduction of the PQ pool in the dark) (Krause and Jahns 2004). The parameter NPQ (Eq. 16) is considered more robust and is often used in preference to q_N (Ralph and Gademann 2005).

$$q_N = 1 - (F_m' - F_o')/(F_m - F_o) \quad (15)$$

$$NPQ = (F_m - F_m')/F_m' \quad (16)$$

The primary site for the development of non-photochemical quenching is thought to be the light harvesting antennae and is largely independent of Q_A redox state (Ting and Owens 1993; Oxborough and Baker 1997). Given the diversity in composition of light harvesting complexes one may expect the mechanisms and response of non-photochemical quenching to vary between taxa.

Non-photochemical quenching of chl fluorescence has three major components, each of these components can be distinguished by careful analysis of dark relaxation kinetics, as described by Horton and Hague (1988). The dominant component of non-photochemical quenching is *energy dependent* quenching (q_E). However, the relative contribution of each NPQ component is dependent upon the light history of the sample and conditions under which the measurements were taken (Ting and Owens 1992). Energy dependent quenching is also the quickest NPQ component to relax upon return to darkness and reports on q_E relaxation time vary from 30–60 s (Ralph and Gademann 2005) to 2–3 min (White and Critchley 1999) or a $t_{½}$ of <1 min (Masojidek et al. 1999).

Energy dependent quenching can account for up to 90% of the decay in F_v on exposure to light (Krause and Weis 1991). This down-regulation of PSII photochemistry may act as a photoprotective mechanism by reducing the potential for the formation of triplet state chlorophyll ($^3chl^*$) in the RCII and the formation of reactive oxygen species (Krause and Jahns 2004) and there is overwhelming evidence indicating that photoinhibition is diminished by the development of a large q_E (Krause and Weis 1991). An early adaptive response by plants and algae in unfavourable conditions is to increase q_E (Schreiber et al. 1995b) and plants adapted to high light have been shown to have more active q_E and less PSII closure than low light adapted plants (Ralph and Gademann 2005).

The next NPQ component has a relaxation half-time of ~5–10 min and is known as state transitional quenching, or q_T (Masojidek et al. 1999). As state transitions (state I → state II) involve the movement of LHCII*b* from PSII to PSI and as comparatively little fluorescence escapes from PSI, fluorescence is consequently quenched (Schreiber et al. 1995b). In this scenario F_v and F_o are quenched by the same proportion (Krause and Weis 1991). While q_T may be of particular importance in low light environments, as light increases acidification of the lumen appears to inhibit LHCII*b* phosphorylation and the role of q_T may become negligible (Krause and Jahns 2004). However, there remain many question as to the mechanisms and function of state transitions, especially in the Chromista, hence the role and activity of q_T in many taxa remains unclear.

The slowest NPQ component to relax, taking >10 min to hours, is photoinhibitory quenching (q_I) which is related to photoinhibitory damage of PSII. There is general agreement that the primary source of q_I is damage of the PSII core protein D1 as a result of donor- or acceptor-side photoinhibition (Hill et al. 2005). However, since q_I formation may precede inactivation of D1, other mechanisms such as the presence of persistent levels of de-epoxidised xanthophyll cycle pigments in the dark (Krause and Jahns 2004) must be considered. Photoinhibitory quenching relaxes only as PSII repair mechanisms take place, such as the synthesis and replacement of damaged D1 proteins (Masojidek et al. 1999). Due to the longer recovery time q_I can be indicated by a reduction in F_v/F_m after exposure to high light. As a rule, this reduction in F_v/F_m is the result of diminished F_v, however the contribution of the F_o component to this decrease may vary (Krause and Weis 1984, 1991).

While studying state transitions in cyanobacteria Mullineaux and Allen (1986) reported quenching of fluorescence in the dark and linked this to the influence of the respiratory electron transport chain on the photosynthetic electron transport chain and state transitions. A little over a decade ago Ting and Owens (1993) reported the unexpected observation that DCMU revealed quenching of fluorescence in dark adapted cultures of the diatom *Phaeodactylum tricornutum*. Since then other researchers have encountered the same phenomenon in *P. tricornutum* (Geel 1997; Jakob et al. 2001), *Thalassiosira weissflogii* (Dijkman and Kroon 2002), benthic diatom mats (McMinn et al. 2004; Serôdio et al. 2005) and the brown alga *Fucus serratus* (Mouget and Tremblin 2002). In these circumstances quenching of F_m (and to a much smaller extent F_o) relaxes in low light (generally <100 μmol quanta m^{-2} s^{-1}, although Serôdio et al. (2005) recorded minimal quenching at 178.5 μmol quanta m^{-2} s^{-1}) and recorded F_m' values are higher than F_m (Mouget and Tremblin 2002; Kromkamp and Forster 2003). This may also yield F_q'/F_m' measurements higher than F_v/F_m. When F_m' > F_m Serôdio et al. (2005) termed the maximum F_m'value $F_m'_m$. The quenching is known to be induced by the build up of a ΔpH across the thylakoid membrane and therefore in the form of q_E. The most commonly proposed mechanism for this dark induced quenching is chlororespiration (Ting and Owens 1993; Schreiber et al. 1995b, 1997; Kromkamp and Forster 2003; McMinn et al. 2004) as it has, until recently, been linked with proton transport(Beardall et al. 2003). This hypothesis has been questioned since dark-induced quenching can occur in anaerobic conditions, which inhibits chlororespiration (Ting and Owens 1993), and by proposals that chlororespiration does not transfer protons across the thylakoid (Peltier and Cournac 2002; Bennoun 2002; Beardall et al. 2003). Another proposed mechanism for F_m' > F_m has been light-induced reorganisation of the antennae (Mouget and Tremblin 2002) and in cyanobacteria it is well known that quenching in the dark is a function of transition to state 2 brought about by a reduction of the PQ pool by the respiratory ETC (Mullineaux and Allen 1986). It is clear, however, that the capacity for NPQ to develop in the dark requires further investigation across a range of taxa. This capacity may be of particular importance for phytoplankton in mixed turbid environments where light levels may rapidly oscillate from darkness to full sunlight and an already-active NPQ could prevent photodamage.

Calculation of NPQ can be problematic when F_m' is observed to increase above the measured F_m value. To date, researchers have calculated NPQ based on one of two premises: (1) where $F_m'_m$ > F_m NPQ equals zero, or (2) quenching was present in the dark-adapted state and $F_m'_m$ is a closer representation of the true unquenched state. Working on the second of these premises, Serôdio et al. (2005) estimated NPQ by replacing F_m with $F_m'_m$; thus:

$$NPQ = (F_m'_m - F_m')/F_m' \qquad (17)$$

4 Conclusion

Although the terminology of the various aspects of variable Chlorophyll *a* fluorescence is complex, reflecting the complexity of the fluorescence signal and the various ways of measuring it, the degree of standardisation in nomenclature in recent years has made it easier to read the papers and to avoid misinterpretations. An agreed common use of terms and symbols in scientific communication is essential, however it must also be recognised that terminology and symbolic representation evolves as our understanding and technology evolves.

References

Babin M (2008) Phytoplankton fluorescence: theory, current literature and *in situ* measurement. In: Babin M, Roessler CS, Cullen JJ (eds) Real-time coastal observing systems for ecosystem dynamics and harmful algal blooms: theory, instrumentation and modeling. UNESCO Publishing, Paris, pp 237–280

Babin M, Morel A, Claustre H, Bricaud A, Kolber ZS, Falkowski PG (1996) Nitrogen- and irradiance-dependent variations of the maximum quantum yield of carbon fixation in eutrophic, mesotrophic and oligotrophic marine systems. Deep-Sea Res 43:1241–1272

Baker NR, Oxborough K (2004) Chlorophyll fluorescence as a probe of photosynthetic productivity. In: Papageorgiou G, Govindjee (eds) Chorophyll *a* fluorescence: a signature of photosynthesis. Springer, Dordrecht, pp 65–82

Baker NR, Oxborough K, Lawson T, Morrison JIL (2001) High resolution imaging of photosynthetic activities of tissues, cells and chloroplasts in leaves. J Exp Bot 52:615–621

Barranguet C, Kromkamp J (2000) Estimating primary production rates from photosynthetic electron transport in estuarine microphytobenthos. Mar Ecol Prog Ser 204:39–52

Beardall J, Quigg A, Raven JA (2003) Oxygen consumption: photorespiration and chlororespiration. In: Larkum AWD, Douglas SE, Raven JA (eds) Photosynthesis in algae. Kluwer, Dordrecht, pp 157–181

Bennoun P (2002) The present model for chlororespiration. Photosynth Res 73:273–277

Bilger W, Schreiber U (1986) Energy-dependent quenching of dark-level chlorophyll fluorescence in intact leaves. Photosynth Res 10:303–308

Bolhàr-Nordenkampf HR, Öquist G (1993) Chlorophyll fluorescence as a tool in photosynthesis research. In: Hall DO, Scurlock JMO, Bolhàr-Nordenkampf HR, Leegood RC, Long SP (eds) Photosynthesis and production in a changing environment: a field and laboratory manual. Chapman & Hall, London, pp 193–206

Bradbury M, Baker NR (1984) A quantitative determination of photochemical and non-photochemical quenching during the slow phase of the chlorophyll fluorescence induction curve of bean leaves. Biochim Biophys Acta 765:275–281

Brand LE (1982) Persistent diel rhythms in the chlorophyll flourescence of marine phytoplankton species. Mar Biol 69:253–262

Braslavsky SE (2007) Glossary of terms used in photochemistry 3rd edition. Pure Appl Chem 79:293–465

Bricaud A, Morel A, Prieur L (1983) Optical efficiency factors of some phytoplankters. Limnol Oceanogr 28:816–832

Büchel C, Wilhelm C (1993) *In vivo* analysis of slow chlorophyll fluorescence induction kenetics in algae: progress, problems and perspectives. Photochem Photobiol 58:137–148

Carr H, Björk M (2003) A methodological comparison of photosynthetic oxygen evolution and estimated electron transport rate in tropical *Ulva* (Chlorophyceae) species under different light and inorganic carbon conditions. J Phycol 39:1125–1131

Cosgrove J, Borowitzka M (2006) Applying pulse amplitude modulation (PAM) fluorometry to microalgae suspensions: stirring potentially impacts fluorescence. Photosynth Res 88:343–350

Cruz S, Serôdio J (2008) Relationship of rapid light curves of variable fluorescence to photoacclimation and non-photochemical quenching in a benthic diatom. Aquat Bot 88:256–264

Dau H (1994) Short-term adaptation of plants to changing light intensities and its relation to phytosystem II photochemistry and fluorescence emission. J Photochem Photobiol B 26:3–27

Dijkman NA, Kroon B (2002) Indications for chlorespiration in relation to light regime in the marine diatom *Thalassiosira weissflogii*. J Photochem Photobiol B 66:179–187

Emerson R, Arnold W (1932) The photochemical reactions in photosynthesis. J Gen Physiol 16:191–205

Estrada M, Marrasé C, Salat J (1996) *In vivo* fluorescence/Chlorophyll *a* ratio as an ecological indicator in oceanography. In: Figueroa FL, Jiménez C, Pérez-Lloréns JL, Niell FX (eds) Underwater light and algal photobiology. International Centre for Coastal Resources Research, Barcelona, pp 317–325

Falkowski PG (1994) The role of phytoplankton photosynthesis in global biogeochemical cycles. Photosynth Res 39:235–258

Falkowski P, Kolber Z (1995) Variations in chlorophyll fluorescence yields in phytoplankton in the world oceans. Aust J Plant Physiol 22:241–355

Falkowski PG, Raven JA (1997) Aquatic photosynthesis. Blackwell Science, Massachusetts

Falkowski P, Wyman K, Ley A, Mauzerall D (1986) Relationship of steady state photosynthesis to fluorescence in eukaryotic algae. Biochim Biophys Acta 849:183–192

Gaffron H, Wohl K (1936) Zur Theorie der Assimilation I and II. Naturwissenschaften 24:81–90

Geel, C (1997) Photosystem II electron flow as a measure for phytoplankton gross primary production. Ph.D. thesis, Wageningen Agricultural University, Wageningen, p 110

Geider RJ, Osborne BA (1991) Algal photosynthesis: the measurement of algal gas exchange. In: Dring MJ, Melkonian M (eds) Current Phycology. Chapman & Hall, New York, pp 256

Genty B, Briantais J, Baker N (1989) The relationship between the quantum yield of photosynthetic electron transport and quenching of chlorophyll fluorescence. Biochim Biophys Acta 990:87–92

Gilbert M, Domin A, Becker A, Wilhelm C (2000a) Estimation of primary productivity by chlorophyll *a in vivo* fluorescence in freshwater phytoplankton. Photosynthetica 38:111–126

Gilbert M, Wilhelm C, Richter M (2000b) Bio-optical modelling of oxygen evolution using in vivo fluorescence: comparison of measured and calculated photosynthesis/irradiance (P/I) curves in four representative phytoplankton species. J Plant Physiol 157:307–314

Govindjee, Papageorgiou G (1971) Chlorophyll fluorescence and photosynthesis: fluorescence transients. In: Geise AC (ed) Photophysiology. Academic Press, New York, pp 1–45

Govindjee (1995) Sixty-three years since Kautsky: chlorophyll *a* fluorescence. Aust J Plant Physiol 22:131–160

Govindjee (2004) Chlorphyll *a* fluorescence: a bit of basics and history. In: Papageorgiou G, Govindjee (eds) Chlorophyll *a* fluorescence: a signature of photosynthesis. Springer, Dordrecht, pp 1–42

Govindjee, Satoh K (1986) Fluorescence properties of chlorophyll b- and chlorophyll c-containing algae. In: Govindjee, Amesz J, Fork DC (eds) Light emission by plants and bacteria. Academic Press, Orlando, USA, pp 497–537

Greene R, Geider R, Kolber Z, Falkowski P (1992) Iron-induced changes in light harvesting and photochemical energy conversion process in eukaryotic marine algae. Plant Physiol 100:565–575

Heredia P, Rivas JDL (2003) Fluorescence induction of Photosystem II membranes shows the steps till reduction and protonation of the quinone pool. J Plant Physiol 160:1499–1506

Hill R, Frankart C, Ralph PJ (2005) Impact of bleaching conditions on the components of non-photochemical quenching in the zooxanthellae of a coral. J Exp Mar Biol Ecol 322:83–92

Holub O, Seufferheld MJ, Gohlke C, Govindjee HGJ, Clegg RM (2007) Fluorescence lifetime imaging microscopy of *Chlamydomonas reinhardtii*: non-photochemical quenching mutants and the effect of photosynthetic inhibitors on the slow chlorophyll fluorescence transient. J Microsc 226:90–120

Holzwarth A (1993) Is it time to throw away your apparatus for chlorophyll fluorescence induction? Biophys J 64:1280–1281

Horton P, Hague A (1988) Studies on the inducion of chlorophyll fluorescence in isolated barley protoplasts. IV. Resolution of non-photochemical quenching. Biochim Biophys Acta 932:107–115

Ilík P, Schansker G, Kotabová E, Váczi P, Strasser RJ, Barták M (2006) A dip in the chlorophyll fluorescence induction at 0.2–2 s in *Trebouxia*-possessing lichens reflects a fast reoxidation of photosystem I. A comparison with higher plants. Biochim Biophys Acta - Bioenergetics 1757:12–20

Jakob T, Goss R, Wilhelm C (2001) Unusual pH-dependence of diadinoxanthin de-epoxidase activation causes chlororespiratory induced accumulation of diatoxanthin in the diatom *Phaeodactylum tricornutum*. J Plant Physiol 158:383–390

Jakob T, Schreiber U, Kirchesch V, Langner U, Wilhelm C (2005) Estimation of chlorophyll content and daily primary production of the major algal groups by means of multiwavelength-excitation PAM chlorophyll fluorometry: performance and methodological limits. Photosynth Res 83:343–361

Kautsky H, Hirsch A (1931) Neue Versuche zur Kohlensãureassimilation. Naturwissenschaften 19:964

Koblížek M, Kaftan D, Nedbal L (2001) On the relationship between the non-photochemical quenching of the chlorophyll fluorescence and the Photosystem II light harvesting efficiency. A repetitive flash fluorescence induction study. Photosynth Res 68:141–152

Kolber Z, Falkowski P (1993) Use of active fluorescence to estimate phytoplankton photosynthesis *in situ*. Limnol Oceanogr 38:1646–1665

Kolber, ZS, Barber RT, Coale KH, Fitzwater SE, Greene RM, Johnson KS, Lindley S, Falkowski PG (1994) Iron limitation of phytoplankton photosynthesis in the equatorial Pacific Ocean. Nature 371:145–149

Kolber Z, Zehr J, Falkowski P (1988) Effects of growth irradiance and nitrogen limitation on photosynthetic energy conversion in photosystem II. Plant Physiol 88:923–929

Krause GH, Jahns P (2004). Non-photochemical energy disipation determined by chlorophyll fluorescence quenching: characterization and function. In: Papageorgiou G, Govindjee (eds) Chlorophyll *a* fluorescence: a signature of photosynthesis. Springer, Dordrecht, pp 463–495

Krause GH, Weis E (1984) Chlorophyll fluorescence as a tool in plant physiology. II. Interpretation of fluorescence signals. Photosynth Res 5:139–157

Krause GH, Weis E (1988) The photosynthetic apparatus and chlorophyll fluorescence: an introduction. In: Lichtenthaler HK (ed) Applications of chlorophyll fluorescence. Kluwer, Dordrecht, pp 3–12

Krause G, Weis E (1991) Chlorophyll fluorescence and photosynthesis: the basics. Annu Rev Plant Physiol Plant Molec Biol 42:313–349

Kromkamp J, Forster R (2003) The use of variable fluorescence measurements in aquatic ecosystems: differences between multiple and single turnover measuring protocols and suggested terminology. Europ J Phycol 38:103–113

Kromkamp JC, Dijkman NA, Peene J, Simis SGH, Gons HJ (2008) Estimating phyoplankton primary production in Lake Ijsselmeer (The Netherlands) using variable fluorescence (PAN-FRRF) and C-uptake techniques. Eur J Phycol 43:327–344

Kroon B, Prézelin B, Schofield O (1993) Chromatic regulation of quantum yields for photosystem II charge separation, oxygen evolution, and carbon fixation in *Heterocapsa pygmaea* (Pyrrophyta). J Phycol 29:453–462

Kühl M, Glud RN, Borum J, Roberts R, Rysgaard S (2001) Photosynthetic performance of surface-associated algae below sea ice as measured with a pulse-amplitude-modulated (PAM) fluorometer and O_2 microsensors. Mar Ecol Prog Ser 223:1–14

Lazár D (2006) The polyphasic chlorophyll *a* fluorescence rise measured under high intensity of exciting light. Funct Plant Biol 33:9–30

Lazár D, Pospíšil P, Nauš J (1999) Decrease of fluorescence intensity after the K step in chlorophyll *a* fluorescence induction is suppressed by electron acceptors and donors to photosystem 2. Photosynthetica 37:255–265

Ley A, Mauzerall D (1982) Absolute absorption cross sections for photosystem II and the minimum quantum requirement for photosynthesis in *Chlorella vulgaris*. Biochim Biophys Acta 680:95–106

Lorenzen CJ (1966) A method for the continuous measurement of in vivo chlorophyll concentration. Deep Sea Res Oceanogr Abst 13:223–227

Macintyre HL, Cullen JJ (2005) Using cultures to investigate the physiological ecology of microalgae. In: Andersen R (ed) Algal culturing techniques. Elsevier/Academic Press, Boston, pp 287–326

Magnusson G (1997) Diurnal measurements of Fv/Fm used to improve productivity estimates in microalgae. Mar Biol 130:203–208

Maske H, Haardt H (1987) Quantitative in vivo absorption spectra of phytoplankton: Detrital absorption and comparison with fluorescence excitation spectra. Limnol Oceanogr 32:620–633

Masojidek J, Torzillo G, Koblížek M, Kopecky J, Bernardini P, Sacchi A, Komenda J (1999) Photoadaptation of two members of the Chlorophyta (*Scenedesmus* and *Chlorella*) in laboratory and outdoor cultures: changes in chlorophyll fluorescence quenching and the xanthophyll cycle. Planta 209:126–135

Mauzerall D, Greenbaum NL (1989) The absolute size of a photosynthetic unit. Biochim Biophys Acta 974:119–140

Maxwell K, Johnson GN (2000) Chlorophyll fluorescence – a practical guide. J Exp Bot 51:659–668

McMinn A, Runcie JW, Riddle M (2004) Effect of seasonal ice breakout on the photosynthesis of benthic diatom mats at Casey, Antarctica. J Phycol 40:62–69

Mouget JL, Tremblin G (2002) Suitability of the fluorescence monitoring ystem (FMS, Hansatech) for measurement of photosynthetic characteristics of algae. Aquat Bot 74:219–231

Mullineaux CW, Allen JF (1986) The state 2 transition in the cyanobacterium *Synechococcus* 6301 can be driven by respiratory electron flow into the plastoquinone pool. FEBS Lett 205:155–160

Neubauer C, Schreiber U (1987) The polyphasic rise of chlorophyll fluorescence upon onset of strong continuous illumination. I. Saturation characteristics and partial control by the photosystem II acceptor side. Z Naturforsch 42c: 1246–1254

Nicklisch A, Köhler J (2001) Estimation of primary production with Phyto-PAM-Fluorometry. Ann Rep Inst Freshw Ecol Inland Fish Berlin 13:47–60

Olaizola M, La Roche J, Kolber Z, Falkowski P (1994) Nonphotochemical quenching and the diadinoxanthin cycle in a marine diatom. Photosynth Res 41:357–370

Oukarroum A, Madidi SE, Schansker G, Strasser RJ (2007) Probing the responses of barley cultivars (*Hordeum vulgare* L.) by chlorophyll a fluorescence OLKJIP under drought stress and re-watering. Env Exp Bot 60:438–446

Oxborough K, Baker NR (1997) Resolving chlorophyll *a* fluorescence images of photosynthetic efficiency into photochemical and non-photochemical components – calculation of qP and Fv'/Fm' without measuring Fo'. Photosynth Res 54:135–142

Oxborough K, Hanlan ARM, Underwood GJC, Baker NR (2000) In vivo estimation of the photosystem II photochemical efficiency of individual microphytobenthic cells using high-resolution imaging of chlorophyll *a* fluorescence. Limol Oceanogr 45:1420–1425

Papageorgiou G, Tsimilli-Michael M, Stamatakis K (2007) The fast and slow kinetics of chlorophyll *a* fluorescence induction in plants, algae and cyanobacteria: a viewpoint. Photosynth Res 94:275–290

Peltier G, Cournac L (2002) Chlororespiration. Annu Rev Plant Biol 53:523–550

Ralph PJ, Gademann R (2005) Rapid light curves: a powerful tool for the assessment of photosynthetic activity. Aquat Bot 82:222–237

Rascher U, Liebig M, Lüttge U (2000) Evaluation of instant light-response curves of chlorophyll fluorescence parameters obtained with a portable chlorophyll fluorometer on site in the field. Plant Cell Env 23:1397–1405

Renger G, Schreiber U (1986). Practical applications of fluorometric methods to algae and higher plant research. In: Govindjee, Amesz J, Fork DC (eds) Light emission by plants and bacteria. Academic Press, Orlando, USA, pp 589–619

Röttgers R (2007) Comparison of different variable chlorophyll a fluorescence techniques to determine photosynthetic parameters of natural phytoplankton. Deep Sea Res I 54:437–451

Sakshaug E, Bricaud A, Dandonneau Y, Falkowski P, Kiefer D, Legendre L, Morel A, Parlsow J, Takahashi M (1997) Parameters of photosynthesis: definitions, theory and interpretation of results. J Plankt Res 19:1637–1670

Schreiber U (2004). Pulse-Amplitude-Modulation (PAM) fluorometry and saturation pulse method: an overview. In: Papageorgiou G, Govindjee (eds) Chorophyll *a* fluorescence: a signature of photosynthesis. Springer, Dordrecht, pp 279–319

Schreiber U, Neubauer C (1987). The polyphasic rise of chlorophyll fluorescence upon onset of strong continuous illumination. II. Partial control by the photosystem II donor side and possible ways of interpretation. Z Naturforsch 42c:1255–1264

Schreiber U, Bilger W, Schliwa U (1986) Continuous recording of photochemical and non-photochemical quenching with a new type of modulation fluorometer. Photosynth Res 10:51–62

Schreiber U, Endo T, Mi H, Asada K (1995a) Quenching analysis of chlorophyll fluorescence by the saturation pulse methods: particular aspects relating to the study of eukaryotic algae and cyanobacteria. Plant Cell Physiol 36:873–882

Schreiber U, Hormann H, Neubauer C, Klughammer C (1995b) Assessment of photosystem II photochemical quantum yield by chlorophyll fluorescence quenching analysis. Aust J Plant Physiol 22:209–220

Schreiber U, Gademan R, Ralph PJ, Larkum AWD (1997) Assessment of photosynthetic performance of *Prochloron* in *Lissoclonium patella in hospite* by chlorophyll fluorescence measurements. Plant Cell Physiol 38:945–951

Schreiber U, Bilger W, Hormann H, Neubauer C (1998) Chlorophyll fluorescence as a diagnostic tool: basics and some aspects of practical relevance. In: Raghavendra A (ed) Photosynthesis: a comprehensive treatise. Cambridge University Press, Cambridge, pp 320–335

Seaton GGR, Walker DA (1990) Chlorophyll fluorescence as a measure of photosynthetic carbon assimilation. Proc Roy Soc London, B 242:29–35

Serôdio JS, Cruz S, Vieira S, Brotas V (2005) Non-photochemical quenching of chlorophyll fluorescence and operation of the xanthophyll cycle in estuarine microphytobenthos. J Exp Mar Biol Ecol 326:157–169

Serôdio J, Vieira S, Cruz S, Coelho H (2006) Rapid light-response curves of chlorophyll fluorescence in microalgae: relationship to steady-state light curves and non-photochemical quenching in benthic diatom-dominated assemblages. Photosynth Res 90:29–43

Shibata K, Benson AA, Calvin M (1954) The absorption spectra of suspensions of living micro-organisms. Biochim Biophys Acta 15:461–470

Strasser RJ, Tsimilli-Michael M, Srivastava A (2004). Analysis of the Chlorophyll *a* fluorescence transient. In: Papageorgiou G, Govindjee (eds) Chorophyll *a* fluorescence: a signature of photosynthesis. Springer, Dordrecht, pp 321–362

Suggett DJ, Oxborough K, Baker NR, Macintyre HL, Kana TM, Geider RJ (2003) Fast repetition rate and pulse amplitude modulation chlorophyll *a* fluorescence measurements for assessment of photosynthetic electron transport in marine phytoplankton. Eur J Phycol 38:371–384

Suggett DJ, Macintyre HL, Geider RJ (2004) Evaluation of biophysical and optical determinations of light absorption by photosystem II in phytoplankton. Limnol Oceanogr Methods 2:316–332

Suggett DJ, Moore CM, Hickman AE, Geider RJ (2009) Interpretation of fast repetition rate (FFR) fluorescence: signatures of phytoplankton community structure versus physiological state. Mar Ecol Prog Ser 376:1–19

Ting CS, Owens TG (1992) Limitations of the pulse-modulated technique for measuring the fluorescence characteristics of algae. PlantPhysiol 100:367–373

Ting CS, Owens TG (1993) Photochemical and nonphotochemical fluorescence quenching processes in the diatom *Phaeodactylum tricornutum*. Plant Physiol 101:1323–1330

Toepel J, Gilbert M, Wilhelm C (2004) Can chlorophyll-*a* in-vivo fluorescence be used for quantification of carbon-based primary production in absolute terms? Arch Hydrobiol 160:515–526

Trissl H (1994) Response to Falkowski et al. Biophys J 66:925–926

Trissl H, Gao Y, Wulf K (1993) Theoretical fluorescence induction curves derived from coupled differential equations describing the primary photochemistry of photosystem II by an exciton-radical pair equilibrium. Biophys J 64:974–988

Ulstrup K, van Oppen M, Kühl M, Ralph P (2007) Inter-polyp genetic and physiological characterisation of *Symbiodinium* in an *Acropora valida* colony. Mar Biol 153:225–234

Van Kooten O, Snel J (1990) The use of chlorophyll fluorescence nomenclature in plant stress physiology. Photosynth Res 25:147–150

Walker DA (1981) Secondary fluorescence kinetics of spinach leaves in relation to the onset of photosynthetic carbon assimilation. Planta 153:273–278

Weis E, Berry JA (1987) Quantum efficiency of Photosystem II in relation to 'energy'-dependent quenching of chlorophyll flourescence. Biochim Biophys Acta 894:198–208

White AJ, Critchley C (1999) Rapid light curves: a new fluorescence method to assess the state of the photosynthetic apparatus. Photosynth Res 59:63–72

Wilhelm C, Becker A, Toepel J, Vieler A, Rautenberger R (2004) Photophysiology and primary production of phytoplankton in freshwater. Physiol Plantarum 120:347–357

Wood M, Oliver R (1995) Fluorescence transients in response to nutrient enrichment of nitrogen- and phosphorus-limited *Microcystis aeruginosa* cultures and matural phytoplankton populations: a measure of nutrient limitation. Aust J Plant Physiol 22:331–340

Wozniak B, Dera J, Ficek D, Ostrowska M, Majchrowski R (2002) Dependence of the photosynthesis quantum yield in oceans on environmental factors. Oceanologia 44: 439–459

Chapter 2
In Situ Measurement of Variable Fluorescence Transients

Samuel R. Laney

1 Introduction

Chlorophyll variable fluorescence provides considerable insight into the photosynthetic physiology of plants and algae, in particular the structure and function of Photosystem II (PSII). A longstanding method for measuring variable fluorescence relies on the addition of DCMU, an herbicide, which blocks electron flow through PSII and eliminates photochemistry as a quencher of fluorescence (Malkin and Kok 1966; Trebst 1980). When the photochemical pathway is blocked by DCMU a sample's fluorescence yield F is greater than when it is not blocked, and this variable fluorescence difference between F measured before and after the addition of DCMU is a valuable indicator of photochemistry in photosynthetic organisms. Unfortunately the DCMU method is not well suited for use in the field. It is possible to measure variable fluorescence using DCMU on discrete or continuous samples of natural phytoplankton assemblages (e.g., Cullen and Renger 1979; Roy and Legendre 1979; Vincent 1981) but it is difficult to do so *in situ* under the ambient light and nutrient conditions that phytoplankton experience in the natural environment.

An alternative is to force F from its minimum to its maximum photochemically using brief actinic flashes of light. Several variations of this basic approach have been developed to date including the pump and probe fluorometric method (Mauzerall 1972; Falkowski et al. 1986), the pulse amplitude modulation (PAM) method (Schreiber 1986), the fast repetition rate (FRR) method (Kolber and Falkowski 1992; Kolber et al. 1998), the pump during probe (PDP) method (Olson et al. 1996), and several others that are functionally similar (e.g., Koblížek et al. 2001; Fuchs et al. 2002; Gorbunov and Falkowski 2004; Johnson 2004; Chekalyuk and Hafez 2008). At least two of these methods – pump and probe fluorometry and FRR fluorometry – have been performed directly on phytoplankton assemblages in the natural aquatic environment using specialized submersible instruments (e.g., Kolber and Falkowski 1992; Antal et al. 2001; Fujiki et al. 2008). These *in situ* "variable fluorometers" are now widely used in oceanographic and lacustrine field research in large part because they are easy to operate and because they readily integrate into standard platforms for profiling or towing instrumentation. These fluorometers have greatly advanced our ability to examine phenomena such as the vertical structure of primary production in the ocean (e.g., Boyd et al. 1997; Melrose et al. 2006), the distribution of photophysiology over meso- and basin scales (e.g., Behrenfeld and Kolber 1999; Holeton et al. 2005), succession in natural assemblages (e.g., Strutton 1997; Suggett et al. 2001), and seasonal and inter-annual cycles in primary production (e.g., Corno et al. 2006; Kaiblinger and Dokulil 2006; Suggett et al. 2006; Fujiki et al. 2008). The ease of operating and deploying these instruments is somewhat deceiving, however, because proper stimulation, measurement, and interpretation of variable fluorescence kinetics is anything but simple.

A number of physiological, optical, instrumental, and computational factors can each introduce considerable error into those photophysiological properties of PSII that can be estimated from variable fluorescence

S.R. Laney (✉)
Biology Department, Woods Hole Oceanographic Institution, Woods Hole, MA 02543, USA
e-mail: slaney@whoi.edu

D.J. Suggett et al. (eds.), *Chlorophyll a Fluorescence in Aquatic Sciences: Methods and Applications*, Developments in Applied Phycology 4, DOI 10.1007/978-90-481-9268-7_2, © Springer Science+Business Media B.V. 2011

transients $F(t)$, sometime in subtle and counterintuitive ways. Some of these sources of error have been examined before (e.g., Cullen and Davis 2003; Laney 2003; Laney and Letelier 2008) but many have not. In the best case scenario such errors can be identified, minimized, or even made negligible by carefully examining variable fluorescence techniques and the resulting measured fluorescence transients. In the worst case scenario, however, these errors remain unexamined and unquantified and have an unknown effect on PSII photophysiological estimates. This introduces uncertainty into any metabolic or ecological inference drawn from these estimates and thus robust interpretation of variable fluorescence transients in a physiological, metabolic, or ecological context requires accurate *in situ* stimulation and measurement of $F(t)$, as well as the accurate application and fitting of an appropriate biophysical model. This ultimately requires that the dominant sources of error in $F(t)$ be identified and their effects estimated.

A primary goal of this discussion is to provide an overview of some of the primary factors specific to *in situ* measurement of $F(t)$ that introduce error into the estimates of phytoplankton PSII photophysiology derived from variable fluorescence transients. Many aspects of the natural aquatic environment can potentially alter measured $F(t)$ considerably, as can issues related to how *in situ* variable fluorometers are typically deployed in field studies (Table 1). Expert users have composed guidelines for using particular *in situ* instruments in the field (e.g., Suggett DJ, Moore CM, Oxborough K, and Geider RJ, 2005, personal communication) and developed software for identifying sources of artifact or bias in measured $F(t)$ transients (Laney 2002), yet at present a general review of *in situ* measurement of variable fluorescence remains lacking. This is partly due to an inherent difficulty with identifying or quantifying many of the artifacts that arise when measuring $F(t)$ *in situ*, as it is not always easy to determine which of several factors may introduce the largest error or even gauge the extent to which any single factor affects $F(t)$. A secondary goal of this overview is to identify avenues for minimizing or perhaps even eliminating altogether the influence of some of these environmental or operational factors. Although this overview focuses on phytoplankton assemblages primarily, many of the factors examined here may also be encountered when using variable fluorescence approaches on other aquatic photoautotrophs such as benthic algae, corals, and seagrasses. A third goal of this overview is to identify some of those similarities and encourage the examination of these factors in aquatic photosynthesis research beyond that concerning phytoplankton alone.

Table 1 Some environmental and operational factors that can introduce error into estimates of PSII photophysiology when using variable fluorescence techniques *in situ*. Also noted is whether or not each particular factor may be more or less relevant when measuring variable fluorescence in aquatic systems that are classified optically as Case II or Case I

Sources of error	Case II vs. Case I
Environmental	
Presence of other fluorophores/absorbers/scatterers (e.g., Chekalyuk and Hafez 2008)	Case II > Case I typically
Algal assemblage taxonomic composition (e.g., Suggett et al. 2004)	Assemblage specific
Heterogeneity of PSII properties within a sample (e.g., Suggett et al. 2004)	Assemblage specific
Vertical structure of phytoplankton biomass (e.g., Corno et al. 2006)	Case II ≈ Case I typically
Signal degradation by scattered ambient sunlight	Acts deeper in Case I
Operational	
Relative motion of water sample past instrument	Deployment specific
Shading of the sample by instrument	"
Orientation of instrument vis-à-vis light field	"

2 Phytoplankton Variable Fluorescence *In Situ*

2.1 Dynamical Protocols for Stimulating Variable Fluorescence

The *in vivo* fluorescence of chlorophyll molecules in living organisms is a very different phenomenon than that of isolated chlorophyll molecules in solution. In some sense there is no such thing as *in vivo* "chlorophyll" fluorescence because *in vivo F* is not determined by chlorophyll molecules alone but by a complex system of pigments and proteins whose yield is determined by its inherent structure and by its immediate photochemical state. All of the different optical methods listed earlier (e.g., pump-and-probe, FRR, PDP, etc.) use a brief actinic flash of light to stimulate this system and photochemically close all functional PSII reaction centers. Because this system is dynamical, this flash forces F to rise from some initial level F_0 to a maximal, saturated level F_m. For this discussion the yields F_0 and F_m will be used in the most general sense and subtleties in nomenclature as to whether or not they are measured in the dark- or light-regulated state, e.g., as discussed elsewhere by Kromkamp and Forster (2003) and in Chapter 1 (Cosgrove and Borowitzka) will be neglected for simplicity.

A brief review of some different variable fluorescence protocols may be instructive. The pump-and-probe fluorometric approach uses a single actinic "pump" flash short enough and intense enough to saturate all PSII reaction centers almost instantaneously. As a result F is forced very quickly from F_0 to F_m, on the scale of one to several microseconds. Measurements of F using weak probe flashes before and after this strong pump flash provide measurements of F_0 and F_m which are used to compute the variable fluorescence yield as F_v/F_m, where $F_v \equiv F_m - F_0$. With other protocols the light in this actinic flash is distributed over a longer time scale, on the order of tens of microseconds in the case of FRR or PDP fluorometry to up to hundreds of thousands of microseconds in the case of PAM fluorometry. The same amount of photons delivered over a longer period effectively results in a slower rate of excitation for this dynamical system, and consequently the apparent F rises gradually from F_0 to F_m with kinetics $F(t)$ that reflect dynamical process related to PSII light harvesting and electron transport. These kinetics apparent in this $F(t)$ transient are physiologically more informative than the near-instantaneous transition from F_0 to F_m induced by a single pump-and-probe measurement, which is partly why most *in situ* variable fluorometers used in field studies are those that close PSII gradually and not instantaneously (i.e., most are FRR fluorometers instead of pump-and-probe instruments).

Because this gradual closure of PSII is inherently a manipulation of a dynamic physiological system it should be evident that there is no single "correct" protocol for stimulating variable fluorescence in order to obtain a physiologically meaningful fluorescence transient. Rather, any number of different protocols can be designed to elicit dynamical responses that emphasize particular photophysiological aspects of interest. A very simple idealized protocol is shown in Fig. 1 in which a single light flash of uniform intensity is used to saturate PSII in a phytoplankton sample over a time scale of order 100 µs. The initial saturation portion of this protocol is similar in principle to the PDP approach used by Olson et al. (1996) and Koblížek et al. (2001). In this protocol the subsequent relaxation of F following this actinic flash, from F_m back down toward F_0, is monitored periodically over the many-ms time scale of a single photochemical turnover using weak light flashes that (ideally) only negligibly re-close PSII that are gradually reopening.

Fluorescence yield F is, strictly speaking, not the same as the observed fluorescence emission EM but for the purposes of this discussion EM and F will be considered equivalent for simplicity and to maintain consistent notation with prior treatments of fluorescence excitation and emission (Laney and Letelier 2008). A measured variable fluorescence transient $EM(t)$ only becomes useful in a physiological or ecological context if its kinetics can be robustly interpreted using a physiological model. A simple physiological model for the transient in Fig. 1 is one of a cumulative one-hit Poisson function that describes this ideal fluorescence transient $EM(t)$ in terms of its initial F_0, its final F_m, and a mean functional cross section of individual PSII (σ_{PSII}).

$$F(t)_{t<100\mu s} = F_o + (F_m - F_o)\cdot(1 - e^{-\sigma_{PSII}\cdot\int_0^t Edt}) \tag{1}$$

More complex physiological models can include a term to account for other factors that cause this saturation phase of $F(t)$ to deviate from a simple cumulative one-hit Poisson such as sharing of exciton energy among individual PSII (Joliot and Joliot 1964; Paillotin 1976; Laney 2003). Since the relaxation kinetics of $F(t)$ (right panel) reflect the action of a pool of electron acceptors (e.g., Q_A and others downstream) these are often described with one or several exponential decay constants which may or may not be weighted (e.g., Kolber et al. 1998), e.g.,

$$F(t)_{t>100\mu s} = F_o + (F_m - F_o)\cdot e^{-\frac{t}{\tau}} \tag{2}$$

2.2 The Practical Relevance of the Single-turnover Time Scale *In Situ*

The idealized fluorescence emission transient that would be stimulated by the protocol shown in Fig. 1 serves as a useful model for empirically interpreting the variable fluorescence kinetics measured in an

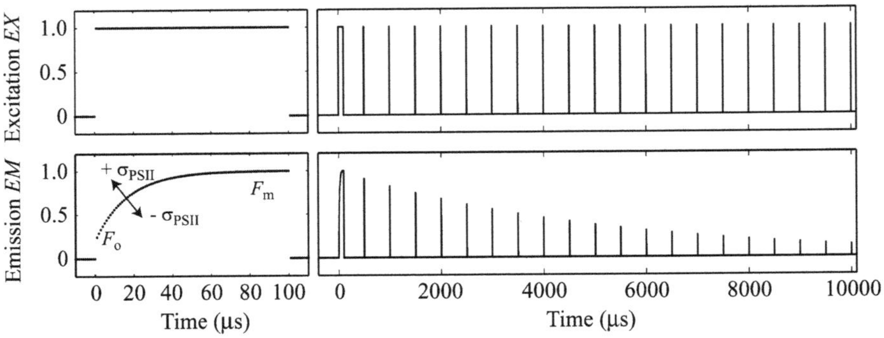

Fig. 1 An example of a simple, idealized variable fluorescence protocol that uses a single light flash of many tens of microsecond in duration (*EX*(*t*), *top left*) to gradually drive fluorescence emission *EM*(*t*) from an initial level F_o to a final saturated level F_m (*bottom left*). This saturation flash is the first in a longer train of shorter flashes that measure the relaxation in PSII photochemistry from this saturated state (*right panels*)

idealized phytoplankton assemblage. The kinetics of actual assemblages, however, are more complex than this idealization suggests. Under physiologically realistic irradiances a small number of photosystems are always open for photochemistry and thus absolute saturation is never achieved *sensu stricto*. This reflects the continual reopening of photochemically closed photosystems by downstream acceptors, even at the very beginning of the saturation portion of the excitation protocol. Thus the relaxation kinetics of a PSII population are always and unavoidably convolved with those of photochemical saturation during the actinic flashes.

If the energy delivery rate needed to saturate phytoplankton of a given species growing under particular conditions is already known, or if there is time to determine this iteratively using trial and error *in situ*, then this convolution effect can be minimized considerably and variable fluorescence can be stimulated in a way that results in *F*(*t*) very similar to those in Fig. 1. Yet, when performing variable fluorescence measurements *in situ*, especially when profiling vertically through assemblages of strongly dissimilar taxonomic compositions, it is not always possible to know ahead of time which particular protocol is appropriate, or even if there is a single protocol which will stimulate appropriate *F*(*t*) across strongly differing assemblages. An example of this effect is shown in Fig. 2, a fluorescence transient measured in the top few meters of the highly oligotrophic North Pacific subtropical gyre 100 nautical miles northeast of Oahu. In this region surface phytoplankton assemblage are dominated by cyanobacteria such as *Synechococcus* and deeper assemblages are more typically dominated by nano- or picoeukaryotes. The former are inherently difficult to saturate photochemically with the blue-green excitation wavelengths used by available *in situ* variable fluorometers (Suggett et al. 2001; Raateoja et al. 2004; Kaiblinger and Dokulil 2006) whereas the light harvesting apparatus of the latter will typically saturate in a shorter period of time given the same excitation wavelengths and delivery rates. If only a single excitation delivery rate is to be used when profiling an instrument vertically through these two different phytoplankton assemblages it cannot be so large that the deeper assemblages saturate too quickly, which would make their variable fluorescence kinetics difficult to interpret. Therefore a delivery rate is chosen in which the near-surface assemblages are saturated over a somewhat longer time scale, which may begin to overlap with the time scale of relaxation processes. This may result in an apparent decrease in *F* during the latter part of the actinic saturation flash (e.g., Fig. 2 left panel between 200 and 450 μs) and possibly only near-saturation of *F* instead of full saturation. As long as this phenomenon is recognized, physiological models for *F*(*t*) that include Q_A reoxidation or other processes that drive *EM* relaxation can be used to interpret these measured transients (e.g., Kolber et al. 1998, Eq. 9). An added complication in this particular example is that these near-surface cyanobacteria assemblages are presumably acclimated to the high-irradiance, upper optical depths and thus their light-harvesting apparatus may exhibit enhanced photochemical turnover rates (e.g., Kana and Glibert 1987; MacIntyre et al. 2002). These faster rates would exert even more influence on *EM* during the latter part of a longer actinic flash and contribute to an even greater

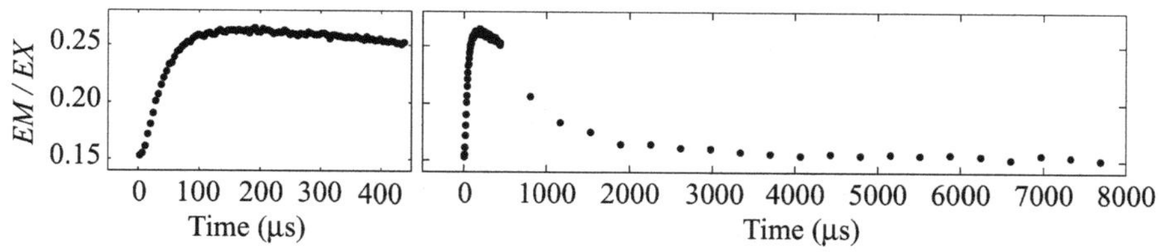

Fig. 2 An example variable fluorescence transient showing the effect of photochemical relaxation during longer time scales of photochemical saturation. Here, a longer-than-usual train of excitation flashlets is required to saturate (or near-saturate) a phytoplankton sample over ~450 μs, as the excitation source is not bright enough to do so more rapidly in these surface, oligotrophic phytoplankton samples. The time scale of this longer excitation train overlaps that of the relaxation processes in PSII which introduce an apparent decrease in fluorescence yield during the latter half of the saturation transient. Even if the actual emission did not attain the theoretical F_m, this parameter can still be retrieved by using a more appropriate dynamical model of PSII physiology than the simple one shown in Fig. 1, provided that the relaxation time scales are known accurately

apparent decrease during this period. The saturation kinetics of the fluorescence transient in Fig. 2 can be better described by an equation that incorporates both the saturation and relaxation process that are operating on the several-hundred microsecond time scale of this saturation protocol.

$$F(t) = F_o + (F_m - F_o)\cdot(1 - e^{-\sigma_{PSII}\cdot\int_0^t E dt})\cdot\sum_{i=1}^{n}\alpha_i e^{-\frac{t}{\tau_i}} \quad (3)$$

The need to use such a representation when measuring $F(t)$ *in situ* on assemblages with strongly different photo-physiologies underscores the importance of considering the measurement time scale when using variable fluorescence methods *in situ*.

The fact that $F(t)$ *in situ* is measured on unconstrained samples of phytoplankton involves a somewhat similar convolution and is also of importance. Laboratory measurements of variable fluorescence are typically made on small sample volumes that are fixed in space relative to the volume that is illuminated by the excitation flash, e.g., in a cuvette. When the sample is constrained in this manner it is generally possible to stimulate and record fluorescence transients over virtually any time scale that is desired, including those of "multiple turnovers" much longer than those of in Figs. 1 and 2. These protocols with longer time scales (e.g., Kolber et al. 1998; Kromkamp and Forster 2003; Gorbunov and Falkowski 2004), which also include those of PAM fluorometry (Schreiber et al. 1993, 1995), would be difficult to make robustly *in situ* because phytoplankton samples in those situations are typically unconstrained and thus can move through the illuminated sample volume during the period in which variable fluorescence transient is being stimulated and recorded.

This effect is not easy to assess empirically but can be estimated using numerical models. The results of a simple simulation is shown in Fig. 3, where $F(t)$ is stimulated on a spatially uniform phytoplankton sample by an instrument moving along this sample while stimulating with the protocol shown in Fig. 1. Any degree of motion of sample past this volume means that some of the cells that were initially in the sample volume at the beginning of the excitation protocol have exited the sample volume, being replaced by new cells that arrived into the volume at some point in time $t > 0$. Thus $F(t)$ appears to change over time simply due to relative motion of sample and instrument. This simulation examined three different cases of relative motion: one with no relative motion of the sample past the instrument, one with an instrument passing through a uniform volume of phytoplankton at a vertical profiling rate of 60 m min^{-1} (100 cm s^{-1}, or 1×10^{-5} cm μs^{-1}), and one with an instrument being towed through this volume at 10 knots (≈ 5 m s^{-1}, or 0.5×10^{-3} cm μs^{-1}). In each case the instrument is given a sample volume having a characteristic length of 1 cm. It is apparent that with these model parameters relative motion between instrument and sample has no discernable effect on the saturation kinetics (left column) when profiling vertically at typical rates, and only a negligible effect is seen in the apparent relaxation kinetics. Thus a variable fluorescence protocol like the one in Fig. 1 would likely remain robust when profiling a variable fluorometer vertically through the water column at these rates. However, in the case where the same instrument is towed behind a ship at a typical speed of 10 knots, this relative motion would introduce a considerable apparent effect on the relaxation scale kinetics (right column) that would

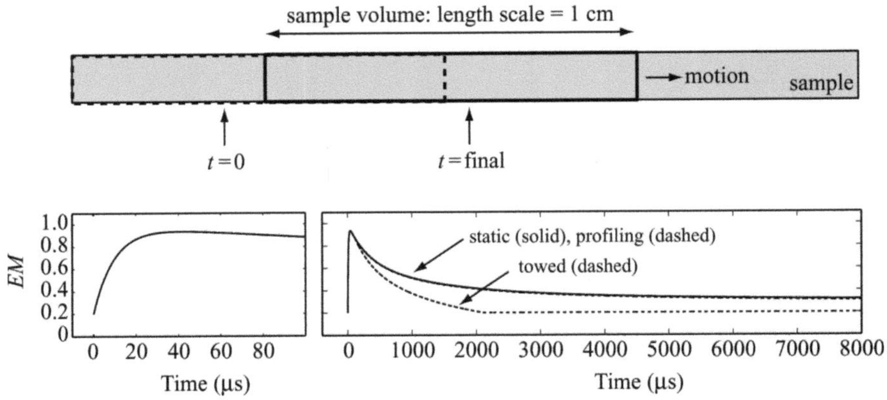

Fig. 3 Results of a simulation examining how relative motion between a phytoplankton sample and an instrument's sample volume affects a measured fluorescence transient *EM* on scales of photochemical saturation and relaxation. In this simulation an idealized sample window of length scale 1 cm (top diagram) stimulates and observes *F*(*t*) in a uniformly distributed sample, while the instrument moved lengthwise (here, to the right) through the sample volume during the measurement. The fluorescence transients computed by this model (bottom panels) reflect the expected behavior of *EM* in the absence of any relative motion (solid trace) and when the instrument moves relative to the sample (dashed traces) at 1e-5 cm μs^{-1} (typical of a vertically profiled instrument) and 5e-4 cm μs^{-1} (typical of a towed instrument). Towing an instrument through a given water mass will result in *EM* transients that suggest a much faster rate of photochemical relaxation, an artifact due to the relative motion of the instrument past the sample

consequently cause the estimated electron throughput rates of these phytoplankton to be overestimated.

A specific conclusion from this modeling exercise is that studies that deploy variable fluorescence instruments on towed vehicles (e.g., Aiken et al. 1998; Berman and Sherman 2001; Allen et al. 2002; Moore et al. 2003; Melrose 2005) should exercise caution when estimating photochemical relaxation time scales from these apparent fluorescence transients. For completeness it should be noted that the same effect of relative motion may apply to variable fluorescence measurements performed on high-volume flows such as those made by attaching variable fluorometers to continuous seawater supplies on ships, depending on the flow rate and the characteristic length scales of the sample volume. A similar effect might be noted when measuring variable fluorescence on coral or benthic samples, e.g., using a fiber-probe variable fluorometer, if the instrument's sensing head moves substantially relative to the sample during the time scale of the transient measurement. Beyond these specific conclusions, a more general one is that typical degrees of relative motion between sample and instrument *in situ* may make it difficult to use multiple turnover protocols on small volumes of unconstrained phytoplankton if there is any meaningful degree of relative motion between instrument and sample. Seawater handling solutions might be devised to circumvent this limitation and thus perform multiple turnover measurements robustly *in situ*, e.g., using a stop-flow approach or similar, but these avenues have yet to be well explored.

2.3 Issues Related to the Marine Light Field

One obvious benefit of measuring variable fluorescence *in situ* is that the photophysiological properties of phytoplankton are assessed in the actual light fields in which these microbes are growing. This capability is especially important for studies aimed at estimating primary production from PSII physiology, given that most of the relevant photophysiological properties are extremely sensitive to the ambient light intensity and that any change in the ambient light field will cause these properties to change rapidly from their *in situ* values. However, measurement of phytoplankton *F*(*t*) in the ambient marine light field can be complicated by several factors related to the light field itself and to the manner in which the instrument that is used to measure *F*(*t*) also changes the ambient light conditions of the phytoplankton under study.

A direct effect of the underwater light field is through the scattering or redirection, into the instrument's fluorescence detector, of some portion of the ambient light field that spectrally overlaps the chlorophyll fluorescence bandwidth. This effect would be particularly pronounced in samples taken at midday in the top optical depth and would be difficult to distinguish from other sources of apparent fluorescence that affect the $F(t)$ baseline, such as the background fluorescence from dissolved organic matter. This effect can be mitigated to some extent by physically blocking the ambient sunlight from entering the emission detector or by restricting the solid angle observed by the fluorescence detector, both of which act to reduce the relative contribution of solar scatterance. Another common strategy for avoiding this source of artifact is to reject very near-surface measurements of $F(t)$. Yet it is not easy to determine a threshold depth at which this effect becomes negligible, or what its magnitude would be when it is non-negligible. A strategy of rejecting very near-surface measurements may be unwise regardless if the physiological responses of interest occur predominantly in the well-lit, top optical depth (see Chapter 6 – Suggett et al.) as is the case when variable fluorescence measurements are used to inform remote sensing studies of ocean color.

If this ambient scatterance were effectively constant during a variable fluorescence measurement then its effect on $F(t)$ could be considered static over the measurement time scale and could potentially be corrected for using some independent estimate of ambient irradiance, such as from a PAR sensor. The corrective framework proposed by Laney and Letelier (2008) would be one way to incorporate these measurements of ambient irradiance, using a network diagram like the one shown in Fig. 4. It is also possible that the underwater light field cannot be considered invariant on these short time scales (≈ 10 ms) in some cases, especially over the small spatial scales of the samples of interest (≈ 1 cm), for example in the top optical depths where surface wave focusing may be considerable (Stramska and Dickey 1998; Zaneveld et al. 2001). In that situation the same or a similar network diagram may be used, which includes a description for how short-term fluctuations in ambient sunlight may introduce nonrandom fluctuations in $F(t)$ on the measurement time scale.

Engineering approaches such as restricting the view angle of the instrument, or numerical approaches such

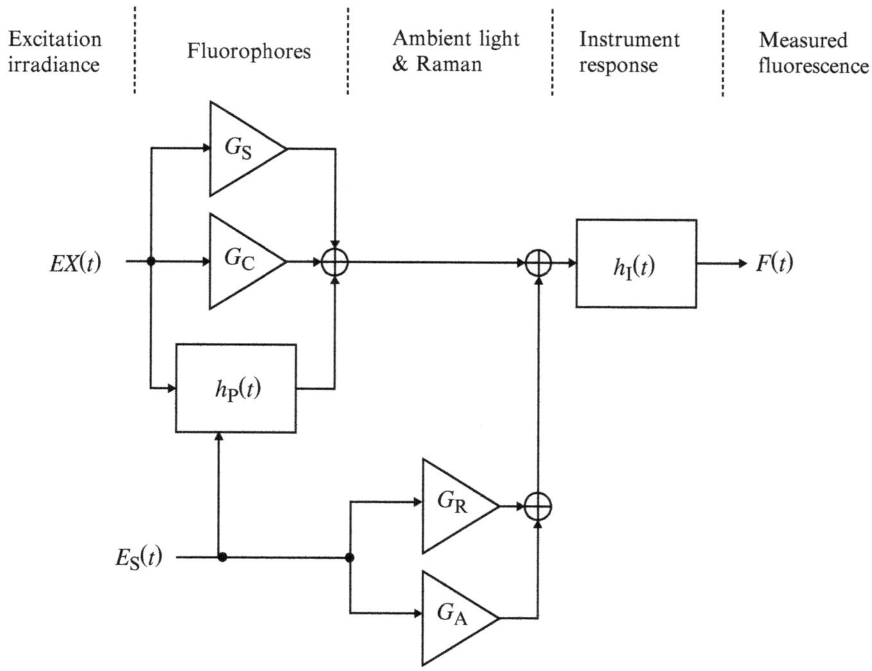

Fig. 4 A dynamical model showing how a time-varying incident solar irradiance $E_s(t)$ affects both the photosynthetic physiology of the sample $h_p(t)$ and contributes elastic and Raman scattering (G_A and G_R) to the measured fluorescence transient. Such a dynamical representation may be necessary to examine the effects of certain factors that may be time-variant on the time scales of a fluorescence transient measurement (Laney and Letelier 2008)

as dynamical corrective frameworks, can both potentially decrease an instrument's sensitivity to ambient solar scatterance and its effect on measured $F(t)$. Yet there still remains an unavoidable physiological impact of the instrument shading the phytoplankton sample under study, from some solid angle of the ambient light field. Because *in situ* variable fluorometers are large compared to the volume of water they sample, the phytoplankton being examined experience a transient in ambient irradiance at some point before the actual $F(t)$ measurement itself. Simply pointing the instrument "up" toward the surface will not eliminate this effect completely, as a non-negligible amount of the underwater light field has a upward component. This instrument shading effect may or may not affect the photophysiology of the sample meaningfully, but this depends on the degree of the shading, the directional structure of the underwater light field at a given depth, the geometry of the instrument and sample volume, and the time scales with which this shading acts on the cells passing through the instrument's (shaded) sample volume.

For a profiling instrument that is relatively free from ship shading a rough estimate of the time a phytoplankton experiences this ambient irradiance transient before the actual $F(t)$ measurement might be on the order of a second or less. The photophysiological effect on cells passing through this modulated ambient light field will presumably not be seen in photosynthetic responses that have time scales of a few seconds or so (Horton and Ruban 2005) but they may be apparent in processes such as those in the thermal phase (e.g., Samson et al. 1999) whose characteristic time scales are closer to this perturbation in ambient irradiance. This effect has not been examined in detail and so its actual impact on measured $F(t)$ in realistic light environments *in situ* is difficult to predict. An analysis approach similar to that used to examine relative motion between sample and instrument may potentially provide some estimate of the degree of this effect in actual $F(t)$ measurements.

2.4 Apparent Effects Resulting from Assemblage Composition

The analysis of variable fluorescence transients typically involves two assumptions, (a) that the phytoplankton sample of interest can be idealized as a population of individual PSII, and (b) that any physiological property of these PSII can be reasonably well represented by the value observed in a bulk fluorescence transient measurement on a volume containing a large number of cells. As was demonstrated earlier, gross differences in the taxonomic composition of a natural phytoplankton assemblage can introduce complications when using variable fluorescence methods *in situ*, in vertical profiles, if those different assemblages are examined using the same excitation protocol. Yet taxonomic factors can also be important in any single sample volume if there is a significant difference in PSII photophysiologies among cells in the sample volume, either between-species or within-species or both. With the simple $F(t)$ transient of Fig. 1 it can be shown that measurements of F_m and F_o from a single phytoplankton assemblage of differing photosynthetic physiologies will accurately reflect the average F_m and F_o of all cells in the sample but that bulk estimates of σ_{PSII} made on an assemblage do not reflect the average cross section of the individual cells (Fig. 5). With the latter, the total fluorescence transient of the bulk assemblage during the saturation phase is larger than what would be expected of an assemblage of the same number of cells with cross sections of the average of all cells in the bulk volume. The sum $F(t)$ of all cells in the sample will saturate faster than the transient that is computed from the arithmetic average of σ_{PSII} among all cells and as a result the functional cross section of a mixed assemblage will appear larger than the average cross section of all cells.

This predicted weighting of σ_{PSII} toward cells of larger functional cross section is supported empirically by laboratory binary mixing experiments performed with cultures (Suggett et al. 2004), but such studies are unfortunately uncommon and so the effect of mixed assemblages on PSII properties such as F_o, F_m, σ_{PSII} (and other physiological aspects such as PSII connectivity and relaxation time constants) remains largely unexamined. Laboratory and field studies can help elucidate this effect in natural assemblages and the degree to which it may affect the PSII properties of interest. Direct measurement of the taxonomic composition of micro- and nanophytoplankton assemblages in conjunction with $F(t)$, such as was done by Olson and coworkers using a flow cytometric and microscopy approach (Olson et al. 1996) may provide important insight into the degree which taxonomic variability affects measurements of $F(t)$ by examining these transients in individual cells.

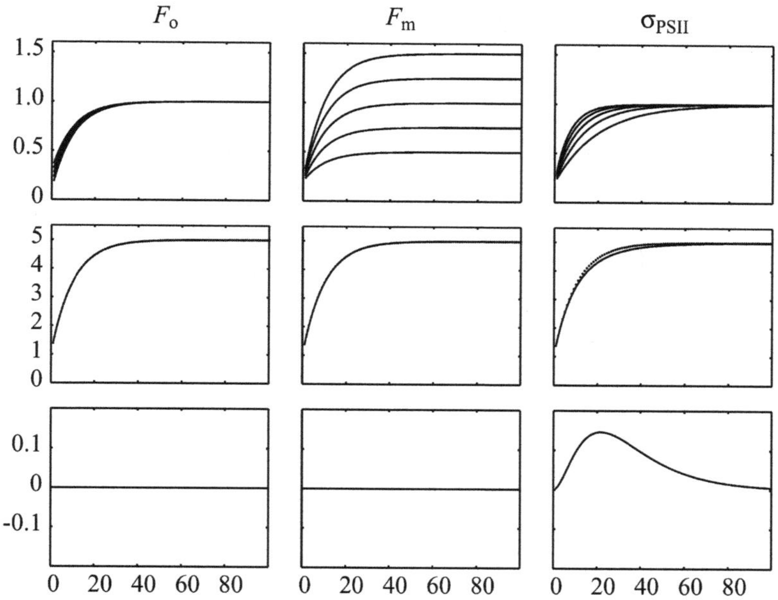

Fig. 5 Model results showing how the distribution of photophysiological parameters among individual cells in a phytoplankton assemblage affects the variable fluorescence transient of the bulk assemblage. For each of these three parameters (F_o, F_m, and σ_{PSII}), the top row shows the transient behavior of five individual cells. The second row shows the sum of these five transients (dashed lines) as well as their average, scaled by a factor of 5 (solid). For F_o and F_m there is no difference between the scaled average and the sum (bottom row), indicating that these properties are conservative. For σ_{PSII} the scaled average is less than the sum, particularly during the initial part of the saturation phase, indicating that it is not conservative in bulk

2.5 Effects Due to Optical Properties of Natural Waters

In situ measurement of $F(t)$ by definition involves stimulating and measuring variable fluorescence transients in natural, environmental water samples. Natural waters typically contain a number of optically active constituents that, depending on the optical design of the instrument itself, can either add photons to or subtract photons from the $F(t)$ transient that is being stimulated and measured. These photons may come from the fluorescence of colored dissolved organic matter (CDOM) including the degradation products of chlorophyll, or from particulate or molecular scattering of the excitation flash itself back into the emission detector. The amount of scatterance or non-algal fluorescence that reaches the instrument is itself decreased by the absorption properties of water and other optically active constituents, again such as dissolved organic matter. Water also has several Raman bands that couple excitation irradiance at ≈ 545 to 565 nm into the chlorophyll fluorescence wavelengths around 685 nm (Bartlett et al. 1998).

Since these same factors typically affect shipboard measurements of variable fluorescence in the same way as they do *in situ*, a number of workers have already examined these factors and their effect on F (Fuchs et al. 2002; Cullen and Davis 2003; Laney 2003; Laney and Letelier 2008). In most *in situ* situations these sources and detriments of variable fluorescence are unavoidable and so all *in situ* measurements of $F(t)$ will contain some degree of bias due to them. When working with discrete samples of natural assemblages in the laboratory or shipboard there is typically more opportunity to determine how these various factors affect the measured $F(t)$, but when working *in situ* there is generally little opportunity to make comparable assessments. Thus gauging the relative effect of these disparate optical properties of natural water samples is a fundamental challenge that is not easily addressed.

Methodological or instrumentation approaches may provide the most direct ways to identify the effect of these non-phytoplankton sources of apparent F and determine how best to correct for them. If the apparent variable fluorescence of filtered natural waters can be obtained, either through intermittently sampling filtered water using automated valves and pumps, or by performing duplicate vertical profiles with and without filters, appropriate corrective procedures can be identified for removing the influence of dissolve contributors to apparent *F* (Laney and Letelier 2008). A different approach, recently described by Chekalyuk and Havez (2008), merges a PDP-like fluorometric protocol with a laser source in order to measure directly the contribution of scatterance and CDOM fluorescence in a wide range of natural waters. Generally speaking the effects of CDOM and particulates will be stronger in estuarine and coastal waters compared to the open ocean but even in dense open ocean blooms CDOM absorption can be large enough to warrant attention (Nelson et al. 2007). This issue of how the non-variable fluorescence contributors to *F(t)* vary in time and space in natural waters remains largely unexamined and should be one of the main focuses of future *in situ* investigations.

3 Conclusions and Future Directions

The appropriate *in situ* use of extant or future variable fluorescence techniques is not simply a matter of performing standard laboratory protocols in the aquatic environment. Rather, proper *in situ* application of variable fluorescence methods requires careful attention to a number of operational and environmental factors that are not encountered in the laboratory, many of which have not been well examined and are not currently possible to assess directly. Simulations and models provide a means to estimate the effect of some of these factors, and characterization studies under controlled circumstances can also shed light on their magnitudes. The current challenge with using variable fluorescence methods *in situ* isn't so much a technical one of performing such measurements or a physiological one of appropriately interpreting measured *F(t)* kinetics, but rather one of determining how different sources of apparent *F* contribute to measured transients and how to eliminate their effect on the photophysiological properties of interest.

To some readers the challenges that these various factors introduce may bring to mind the comment by Holzwarth and colleagues who questioned somewhat rhetorically whether or not it was time to "throw away" their fluorescence induction instruments given the ambiguities associated with that technique (Holzwarth 1993). The concern was whether or not the physiological inferences that were being drawn from fluorescence induction measurements were strongly supported by the biophysical bases of the measurement technique being used. An analogous question with respect to modern *in situ* variable fluorescence methods is: with a given *in situ* measurement of *F(t)*, is this transient signal understood well enough so that photophysiological parameters of PSII can be derived from it with confidence? If the answer is no, then the question becomes: what are the limits on the physiological, metabolic, and ecological inferences that can be drawn from *in situ* measurements of phytoplankton variable fluorescence? The factors discussed in this overview represent only a few of the issues now known to affect variable fluorescence use *in situ*, primarily with respect solely to phytoplankton assemblages. Continuing to review the accuracy of these variable fluorescence approaches in phytoplankton, and identifying similar challenges when using these approaches with corals, seagrasses, and benthic algae, will remain an important part of using variable fluorescence techniques *in situ*.

References

Aiken, J, Cummings DG, Gibb SW, Rees NW, Woodd-Walker R, Woodward EMS, Woolfenden J, Hooker SB, Berthon J-F, Dempsey CD, Suggett DJ, Wood P, Donlon C, Gonzalez-Benitez N, Huskin I, Quevedo M, Barciela-Fernandez R, de Vargas C, McKee C (1998) AMT-5 Cruise Report. In: Hooker SB, Firestone ER (eds) NASA Tech Memo 1998-206892, Vol. 2. NASA Goddard Flight Center, Greenbelt, Maryland, 113 pp

Allen JF, Cornell V, Moore CM, Crisp N, Dunning J (2002) Operational oceanography using the new SeaSoar undulator. Sea Technol 43:35–40

Antal TK, Venediktov PS, Matorin DN, Ostrowska M, Woźniak B, Rubin AB (2001) Measurement of phytoplankton photosynthesis rate using a *pump-and-probe* fluorometer. Oceanologia 43:291–313

Bartlett JS, Voss KJ, Sathyendranath S, Vodacek A (1998) Raman scattering by pure water and seawater. Appl Optics 37:3324–3332

Behrenfeld MJ, Kolber ZS (1999) Widespread iron limitation of phytoplankton in the south Pacific Ocean. Science 283:840–843

Berman MS, Sherman K (2001) A towed body sampler for monitoring marine ecosystems. Sea Technol 42:48–52

Boyd PW, Aiken J, Kolber Z (1997) Comparison of radiocarbon and fluorescence based (pump and probe) measurements of phytoplankton photosynthetic characteristics in the Northeast Atlantic Ocean. Mar Ecol Prog Ser 149:215–226

Chekalyuk A, Hafez M (2008) Advanced laser fluorometry of natural aquatic environments. Limnol Oceanogr Meth 6:591–609

Corno G, Letelier RM, Abbott MR, Karl DM (2006) Assessing primary production variability in the North Pacific Subtropical

Gyre: a comparison of fast repetition rate fluorometry and 14C measurements. J Phycol 42:51–60

Cullen JJ, Davis RF (2003) The blank can make a big difference in oceanographic measurements. Limnol Oceanogr Bull 12:29–35

Cullen JJ, Renger EH (1979) Continuous measurement of the DCMU-induced fluorescence response of natural phytoplankton populations. Mar Biol 53:13–20

Falkowski PG, Wyman KD, Ley AC, Mauzerall DC (1986) Relationship of steady state photosynthesis to fluorescence in eukaryotic algae. Biochim Biophys Acta 849:183–192

Fuchs E, Zimmerman RC, Jaffe JS (2002) The effect of elevated levels of phaeophytin in natural waters on variable fluorescence measured from phytoplankton. J Plankton Res 24:1221–1229

Fujiki T, Hosaka T, Kimoto H, Ishimaru T, Saino T (2008) In situ observation of phytoplankton productivity by an underwater profiling buoy system: use of fast repetition rate fluorometry. Mar Ecol Prog Ser 353:81–88

Gorbunov MY, Falkowski PG (2004) Fluorescence Induction and Relaxation (FIRe) technique and instrumentation for monitoring photosynthetic processes and primary production in aquatic ecosystems. In: van der Est A, Bruce D (eds) 13th International Congress of Photosynthesis, vol 2. Allen Press, Montreal, pp 1029–1031

Holeton CL, Nedelec F, Sanders R, Brown L, Moore CM, Stevens DP, Heywood KJ, Statham PJ, Lucas CH (2005) Physiological state of phytoplankton communities in the Southwest Atlantic sector of the Southern Ocean, as measured by fast repetition rate fluorometry. Polar Biol 29:44–52

Holzwarth AR (1993) Is it time to throw away your apparatus for chlorophyll fluorescence induction? Biophys J 64:1280–1281

Horton P, Ruban A (2005) Molecular design of the photosystem II light-harvesting antenna: photosynthesis and photoprotection. J Exp Bot 56:365–373

Johnson Z (2004) Development and application of the background irradiance gradient - single turnover fluorometer (BIG- STf). Mar Ecol Prog Ser 283:73–80

Joliot P, Joliot A (1964) Études cinétique de la réaction photochimique libérant l oxygene au cours de la photosynthese. C R Acad Sci Paris 258:4622–4625

Kaiblinger C, Dokulil M (2006) Application of fast repetition rate fluorometry to phytoplankton photosynthetic parameters in freshwaters. Photosynth Res 88:19–30

Kana TM, Glibert PM (1987) Effect of irradiances up to 2000 μE m^{-2} s^{-1} on marine *Synechococcus* WH7803 – II. Photosynthetic responses and mechanisms. Deep Sea Res 34:497–516

Koblížek M, Kaftan D, Nedbal L (2001) On the relationship between the non-photochemical quenching of the chlorophyll fluorescence and the Photosystem II light harvesting efficiency. A repetitive flash fluorescence induction study. Photosynth Res 68:141–152

Kolber ZS, Falkowski PG (1992) Fast Repetition Rate fluorometer for making in situ measurements of primary productivity. In: Proceedings of IEEE Ocean 92 Conference. Newport, Rhode Island, pp 637–641

Kolber ZS, Prasil O, Falkowski PG (1998) Measurements of variable chlorophyll fluorescence using fast repetition rate techniques: defining methodology and experimental protocols. Biochim Biophys Acta 1367:88–106

Kromkamp JC, Forster RM (2003) The use of variable fluorescence measurements in aquatic ecosystems: differences between multiple and single turnover measuring protocols and suggested terminology. Eur J Phycol 38:103–122

Laney SR (2002) Documentation for v5 fluorescence yield analysis software. Available via ORSOO. http://picasso.oce.orst.edu/ORSOO/FRRF/code/v5.zip.

Laney SR (2003) Assessing the error in photosynthetic properties determined by fast repetition rate fluorometry. Limnol Oceanogr 48:2234–2242

Laney SR, Letelier RM (2008) Artifacts in measurements of chlorophyll fluorescence transients, with specific application to fast repetition rate fluorometry. Limnol Oceanogr: Meth 6:40–50

MacIntyre HL, Kana TM, Anning T, Geider RJ (2002) Photoacclimation of photosynthesis irradiance response curves and photosynthetic pigments in microalgae and cyanobacteria. J Phycol 38:17–38

Malkin S, Kok B (1966) Fluorescence induction studies in isolated chloroplasts, I: number of components in the reaction and quantum yields. Biochem Biophys Acta 126

Mauzerall D (1972) Light-induced fluorescence changes in *Chlorella* and the primary photoreactions for the production of oxygen. Proc Nat Acad Sci USA 69:1358–1362

Melrose DC (2005) Underway profiling of photosynthesis and dissolved oxygen in Narragansett Bay, Rhode Island. Ph.D. thesis, University of Rhode Island. p257. http://digitalcommons.uri.edu/dissertations/AAI3186913

Melrose DC, Oviatt CA, O'Reilly JE, Berman MS (2006) Comparisons of fast repetition rate fluorescence estimated primary production and ^{14}C uptake by phytoplankton. Mar Ecol Prog Ser 311:37–46

Moore CM, Suggett D, Holligan PM, Sharples J, Abraham ER, Lucas MI, Rippeth TP, Fisher NR, Simpson JH, Hydes DJ (2003) Physical controls on phytoplankton physiology and production at a shelf sea front: a fast repetition-rate fluorometer based field study. Mar Ecol Prog Ser 259:29–45

Nelson NB, Siegel DA, Carlson CA, Swan C, Smethie WM, Khatiwala S (2007) Hydrography of chromophoric dissolved organic matter in the North Atlantic. Deep Sea Res I 54:710–731

Olson RJ, Chekalyuk AM, Sosik HM (1996) Phytoplankton photosynthetic characteristics from fluorescence induction assays of individual cells. Limnol Oceanogr 41:1253–1263

Paillotin G (1976) Movement of excitations in the photosynthetic domains of Photosystem II. J Theor Biol 58:237–252

Raateoja M, Seppälä J, Ylöstalo P (2004) Fast repetition rate fluorometry is not applicable to studies of filamentous cyanobacteria from the Baltic Sea. Limnol Oceanogr 49:1006–1012

Roy S, Legendre L (1979) DCMU-enhanced fluorescence as an index of photosynthetic activity of phytoplankton. Mar Biol 55:93–101

Samson G, Prášil O, Yaakoubd B (1999) Photochemical and thermal phases of chlorophyll *a* fluorescence. Photosynthetica 37:163–182

Schreiber U (1986) Detection of rapid induction kinetics with a new type of high frequency modulated chlorophyll fluorometer. Photosynth Res 9:261–272

Schreiber U, Hormann H, Neubauer C, Klughammer C (1995) Assessment of photosystem II photochemical quantum yield by chlorophyll fluorescence quenching analysis. Aust J Plant Physiol 22:209–220

Schreiber U, Neubauer C, Schloiwa U (1993) PAM fluorometer based on medium-frequency pulsed Xe-flash measuring light: A highly sensitive new tool in basic and applied photosynthesis research. Photosynth Res 36:65–72

Stramska M, Dickey TD (1998) Short term variability of optical properties in the oligotrophic ocean in response to surface waves and clouds. Deep Sea Res I 45:1393–1410

Strutton P (1997) Phytoplankton patchiness: quantifying the biological contribution using Fast Repetition Rate Fluorometry. J Plankton Res 19:1265–1274

Suggett D, Kraay G, Holligan P, Davey M, Aiken J, Geider R (2001) Assessment of photosynthesis in a spring cyanobacterial bloom by use of a fast repetition rate fluorometer. Limnol Oceanogr 46:810

Suggett DJ, MacIntyre HL, Geider RJ (2004) Evaluation of biophysical and optical determinations of light absorption by photosystem II in phytoplankton. Limnol Oceanogr Meth 2:316–332

Suggett DJ, Moore CM, Maranon E, Omachi C, Varela RA, Aiken J, Holligan PM (2006) Photosynthetic electron turnover in the tropical and subtropical Atlantic Ocean. Deep Sea Res II 53:1573–1592

Trebst A (1980) Inhibitors in electron flow. Methods Enzymol 69:675–715

Vincent WF (1981) Photosynthetic capacity measured by DCMU-induced chlorophyll fluorescence in an oligotrophic lake. Freshw Biol 11:61–78

Zaneveld JRV, Boss E, Barnard A (2001) Influence of surface waves on measured and modeled irradiance profiles. Appl Optics 40:1442–1449

Chapter 3
Overview of Fluorescence Protocols: Theory, Basic Concepts, and Practice

Yannick Huot and Marcel Babin

At the request of the editors, this chapter is based on the book section written by Babin (2008), however, much of the manuscript has been revised and updated to address different readerships. While Babin (2008) is aimed at a more general audience interested in understanding the basis of the measurement and the current instruments available, this chapter is aimed at those who will use the fluorescence tool and are interested in understanding more of its underlying theory, as well as the assumptions associated with it. In short, while the first chapter was aimed more at a beginning user, this one is aimed more at an intermediate user. Nevertheless, some sections have seen little changes.

1 Introduction

When chlorophyll molecules absorb light, a fraction of the energy absorbed is reemitted as fluorescence. This property of the chlorophyll molecule – which has no ecological or physiological raison d'être – has, with little doubt, been the measure that has supported the most discoveries in biological limnology and oceanography over the past 40 years. Its measurement *in vivo* was introduced in the aquatic sciences in 1966 (Lorenzen 1966) to monitor changes in phytoplankton biomass. It gained rapid acceptance and it is now routinely measured at sea using various commercial benchtop and *in situ* fluorometers. In 1977 (Morel and Prieur 1977; Neville and Gower 1977), the first reports suggesting that fluorescence emission could be observed using spectroradiometers in the natural upwelling light field of the ocean were published. Twenty-two years later, the MODIS spectroradiometer on the Terra satellite was launched with two spectral bands dedicated to measuring this signal, allowing synoptic global observations of the fluorescence emitted from the ocean surface. It was followed a few years later by the MODIS Aqua and MERIS instruments. In the early 1980's the flow cytometer was introduced to oceanography (Yentsch et al. 1983) thus allowing the measurement of the fluorescence from individual cells and initiating new approaches to study allometric relationships in the ocean. Major oceanographic cruises now rarely leave port without at least one flow cytometer on board.

Over time, chl *a* fluorometry has thus become a major tool for rapidly mapping the spatial distribution of phytoplankton at various scales, and for monitoring temporal fluctuations in phytoplankton biomass. However, its impact goes beyond the estimates of biomass. In the late 1970s and during the 1980s, protocols previously developed in plant physiology for measuring the so-called "variable fluorescence" were introduced to aquatic sciences to study and monitor phytoplankton physiology (Cullen and Renger 1979; Vincent 1981). This fluorometric approach allows an observation of the photophysiological status of the cells. Based

Y. Huot (✉)
Département de Géomatique Appliquée, Université de Sherbrooke, Sherbrooke, J1K 2R1 Québec, Canada
e-mail: yannick.huot@usherbrooke.ca

M. Babin
Université Pierre et Marie Curie-Paris 6, Laboratoire d'Océanographie de Villefranche, UMR 7093, France
e-mail: babin@obs-vlf.fr

D.J. Suggett et al. (eds.), *Chlorophyll* a *Fluorescence in Aquatic Sciences: Methods and Applications*, Developments in Applied Phycology 4, DOI 10.1007/978-90-481-9268-7_3, © Springer Science+Business Media B.V. 2011

on theoretical development (Butler 1978; Falkowski and Kiefer 1985; Genty et al. 1989) made in photosynthesis research, variable fluorescence was later applied to estimate primary production in aquatic sciences (e.g. Kolber and Falkowski 1993). Since their introduction, following improvement in fluorescence theory, these techniques have been refined (Kolber et al. 1998) and today commercial variable-fluorescence sensors are available that are easily deployed in the field providing non-intrusive assessment of physiological status.

This chapter reviews the basic theory underlying the fluorescence emission and the current protocols used in aquatic sciences to exploit this property and obtain information about the photosynthetic processes and biomass. While the content is broadly applicable to plants and algae (including their symbiotic associations), our focus, especially when protocols are presented, will be on phytoplankton cells (other articles in this special issue discuss in details other algae). In many sections, we will, however, often borrow from the more abundant literature on higher plants, recognizing that phytoplankton often show stark differences in their fluorescence characteristics.

The creative use of chlorophyll fluorescence in aquatic sciences is bewildering. The excitation of the signal has been carried out with nearly every light sources available to mankind and it seems like every combination of instrument and protocols has been used. For example: Sun-induced chl *a* fluorescence provided estimates of primary production (Kiefer et al. 1989) and the flow cytometer was coupled to the variable fluorescence protocol (Gorbunov et al. 1999; Olson et al. 1999). It will thus be clearly impossible to review all these advances herein but we hope that our approach based on simple theoretical consideration will allow the reader to examine the techniques we overlooked from a new perspective. We invite the reader interested in furthering the concepts reviewed herein to consult the primary literature cited and see previous more focused reviews and monographs by Blankenship (2002); Clayton (1980); Krause and Weis (1991); Falkowski and Raven (2007); Hall and Rao (1999); Baker (2008) and Whitmarsh and Govindjee (1999); Owens (1991). Since our focus will be mostly on eukaryotic phytoplankton, we direct the reader interested in cyanobacteria to the nice review by Campbell et al. (1998).

2 Theoretical Background

2.1 *The Fluorescing Properties of Chlorophyll* a

Chlorophyll *a* absorbs light in the blue and red parts of the visible electromagnetic spectrum (Fig. 1a). When chlorophyll *a* is dissolved in acetone, the red absorption band, is centred at 662 nm. This band corresponds to an electronic transition from the ground to the first excited state of the chl *a* molecule (Fig. 2). In acetone, the blue absorption band is centred at 430 nm and corresponds to an electronic transition from the ground to the second excited state (Fig. 2). Each electronic state is characterized by a number of substates that correspond to different molecular vibrational energy levels thence leading to the wide absorption bands rather than distinct lines. A transition down from the second to the first excited state generally occurs through thermal decay (no fluorescence) within less than a picosecond. In solution, the transition from the first excited state to the ground state occurs whether by thermal decay or by the re-emission of a photon. The re-emission of a photon is called *fluorescence* or *phosphorescence*. Because fluorescence occurs much faster (ca. 15 ns) than phosphorescence (ca. 1 ms), it accounts for nearly all of photon re-emission. In acetone, fluorescence by chl *a* is spectrally characterized by a main peak centred at 668 nm and a secondary shoulder near 730 nm (Fig. 1a). The quantum yield of fluorescence in solution (ϕ_F^{sol}) is defined as the fraction of photons absorbed by chl *a* re-emitted as fluorescence. It is ca. 30% for chl *a* in acetone, which means that the energy of the remaining 70% of absorbed photons is dissipated as heat. It can be expressed mathematically by the ratio of the first order rate constant for deexcitation by fluorescence (k_F) to the sum of the rate constants of fluorescence and heat (k_H): $\phi_F^{sol} = k_F / (k_F + k_H)$. In algae, chl *a* is not in solution but bound to one of two protein complexes: photosystem I and photosystem II; as will be described later, this strongly modifies its fluorescence characteristics.

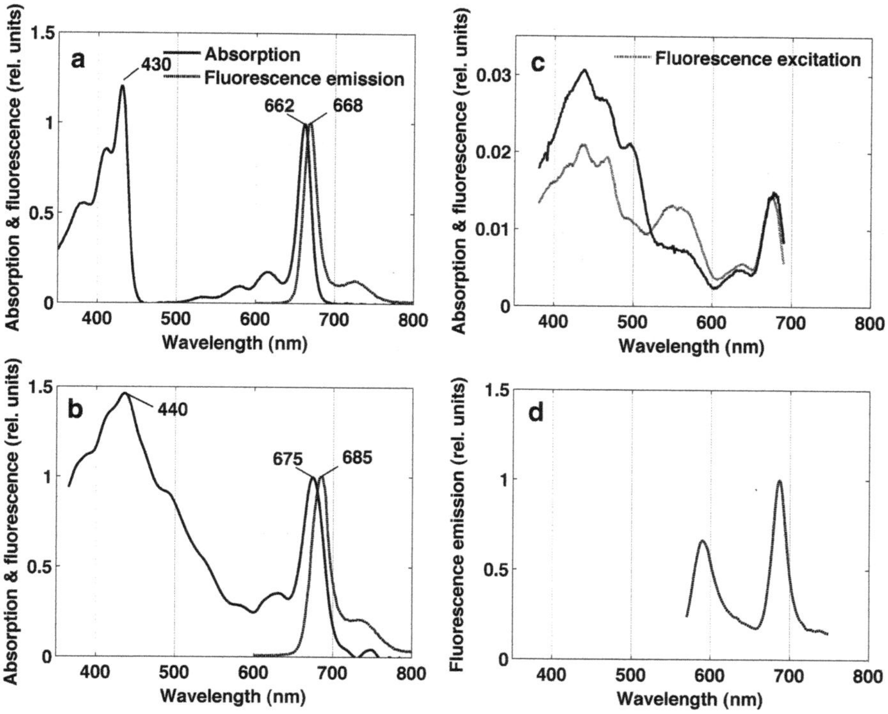

Fig. 1 Fluorescene and absorption spectra. (**a**) Spectra of light absorption and of fluorescence emission by pure chlorophyll *a* dissolved in acetone. The fluorescence spectrum is more or less a mirror image of the absorption spectrum in the red spectral range, shifted toward longer wavelengths. This phenomenon is called the Stokes shift and results from the fact that: (1) the vibrational energy levels are similarly positioned with respect to the minimum energy levels of the first excited state and ground states, and (2) fluorescence emission always occurs between the minimum vibrational level of the first excited state and any level of the ground state. The result is that the emission is thus at lower energy than the absorption since the latter always starts from the ground state (see Fig. 2 and related text). (**b**) Example of *in vivo* absorption and fluorescence emission spectra measured on the diatom *Thalassiosira pseudonana.* Note that, *in vivo*, absorption and fluorescence bands are systematically shifted toward longer wavelengths relative to the spectra for pigments in solution (see *Panel A*). (**c**) Absorption and fluorescence excitation spectra of *Cryptomonas sp*. Fluorescence as detected around 730 nm (data obtained during the study described by Sciandra et al. 2000). Both spectra are scaled to equal 1 at 675 nm. (**d**) Fluorescence emission spectrum of *Cryptomonas sp.* Fluorescence was excited around 545 nm (data obtained during the study described by Sciandra et al. 2000)

2.2 Source of Fluorescence in Seawater and Mathematical Description of Fluorescence Emission

For many applications, especially those where information about physiology is necessary, we are interested in measuring the fluorescence emission from photosystem II.[1] However, in seawater a suite of fluorescing compounds that can potentially influence the measurement are present.

When broadband irradiance is impinging upon a volume of seawater, a fraction goes through the volume without any interaction, a part of this light is elastically scattered out of that volume in every direction at the same wavelength as the incident photons and a part is absorbed within the volume (e.g. Mobley 1994). A fraction of this absorbed light is reemitted at a different wavelength by two main processes: Raman scattering and fluorescence. The total flux emitted per unit volume (F_{tot}, μmol photon m^{-3} s^{-1}) can be represented by:

$$F_{tot} = F_{PSII} + F_{PSI} + F_{phyco} + F_{phaeo} + F_{CDOM} + F_{Raman}. \quad (1)$$

Where the subscripts on the terms on the right hand side represent successively photosystem II (PSII), photosystem I (PSI), phycobilins (phyco), phaeopigments (phaeo), colored dissolved organic matter (CDOM), and Raman emission (Raman). By design, chlorophyll fluorometers observe only the red part of the spectrum

[1] Photosystem II is the location of the first photochemical reaction in the photosynthetic apparatus and location of water splitting and oxygen evolution (see Section 2.3.1).

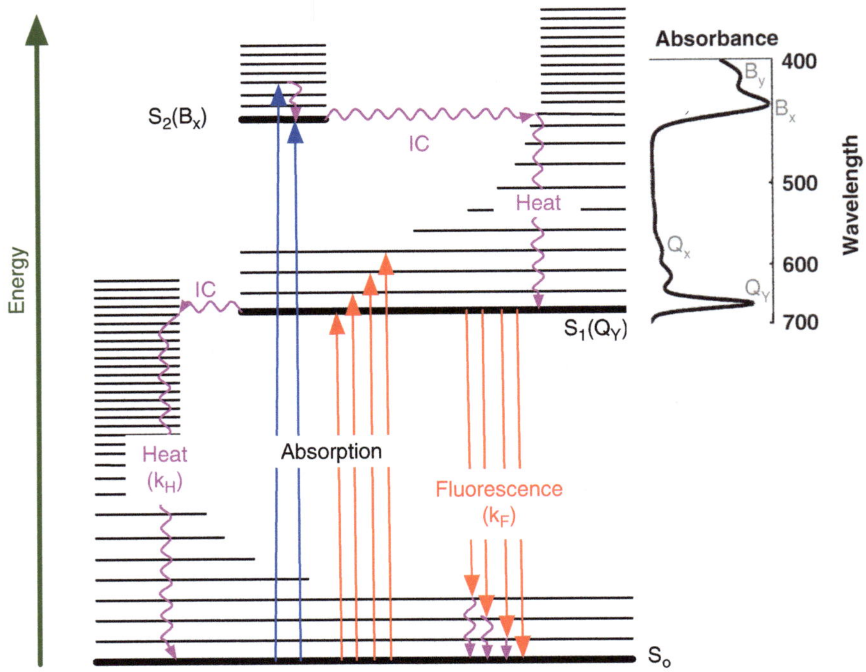

Fig. 2 Diagram showing the energy levels and different types of transitions between electronic excited states, when a photon is absorbed by chlorophyll *a*. Two electronic absorption bands are shown: B_x in the blue part of the spectrum (corresponds excited state S_2) and Q_Y in the red (corresponds excited state S_1). The energy levels corresponding to weaker B_Y and Q_X absorption bands are not shown. In acetone, the energy of an absorbed photon is whether released as heat or fluoresced. The absorption of blue photons corresponds to a transition of an electron from the lowest vibrational energy level of the ground state (So) to the second excited state (S_2). Following radiationeless decay to the lowest vibrationnal level of S_2, and internal conversion (IC, transition between two excited state with same spin) to the first excited state (S_1) and radiationless decay to the lowest vibrational level, the remaining excitation energy can whether be dissipated as heat (with rate constant k_H) after IC to ground state (So) or lead to the emission of a photon (fluorescence with rate constant k_F). The absorption of red photons corresponds to a transition of an electron from the lowest vibrational energy level of the ground state (So) to the S_1 level. Vibrational decay to the lowest vibrational level of S_1 follows, and the remaining excitation energy can whether lead to fluorescence or be dissipated as heat (after international conversion to S_0). On the right hand side is illustrated the absorption spectrum of chlorophyll *a* with absorption bands approximately matched to the electronic levels. Not shown are the energy levels for intersystem crossing (usually denoted T)

where CDOM emission is weak under most oceanic conditions. This condition is not, however, true of all coastal and inland waters which can contain significant amount of CDOM (Exton et al. 1983; Chekalyuk and Hafez 2008). Under most illumination used in fluorescence research (though it is not the case for Sun-induced fluorescence), significant Raman emission occurs only at a shorter waveband than chlorophyll fluorescence and can be separated spectrally. If the phaeopigment concentration in the water is low relative to chl *a*, its fluorescence is also negligible (but this is not always the case e.g. Yentsch and Menzel 1963; SooHoo and Kiefer 1982; Fuchs et al. 2002). When F_{phaeo}, F_{CDOM}, and F_{Raman} are negligible, fluorescence observed by a fluorometer originates mostly from algal cells ($F_{tot} = F_{PSII} + F_{PSI} + F_{phyco}$). If the concentration of phycobilin containing algae is low or if the emission is distinguished based on its spectral characteristics, the fluorescence from chl *a* can be observed almost exclusively. In the shorter waveband (near 685 nm) most of the in vivo emission originates from PSII. At longer wavebands (above ~700 nm), the contribution from PSI is greater (e.g. Franck et al. 2002). For the mathematical description of fluorescence we will assume that we are under ideal conditions and that we are observing fluorescence emission solely from PSII. We will return on the validity of these assumptions when describing specific methods.

The flux of *in vivo* chl *a* fluorescence from PSII emitted by an elementary volume (F_{PSII}; µmol photons m^{-3} s^{-1} nm^{-1}) at wavelength λ_{em} can be expressed by the following equations (an identical equation can be written for PSI) for monochromatic excitation at wavelength λ:

$$F_{PSII}(\lambda,\lambda_{em}) = Q^*_{abs}(\lambda_{em})\overset{o}{E}(\lambda)\,{}^{PS}a_{PSII}(\lambda)\eta_{PSII}(\lambda\rightarrow\lambda_{em})\Delta\lambda; \quad (2)$$

and by the following one for polychromatic light

$$F_{PSII}(\lambda_{em}) = Q^*_{abs}(\lambda_{em})\int_{\lambda_{min}}^{\lambda_{em}} \overset{o}{E}(\lambda)\,{}^{PS}a_{PSII}(\lambda)\eta_{PSII}(\lambda\rightarrow\lambda_{em})d\lambda; \quad (3)$$

where $\overset{o}{E}(\lambda)$ (μmol photons m^{-2} s^{-1} nm^{-1}) is the incident scalar irradiance at the excitation wavelength λ, ${}^{PS}a_{PSII}$ (m^{-1}) is the absorption of photosynthetic pigments (those that transfer their excitation energy to the reaction center) of photosystem II, Q^*_{abs} (dimensionless) is the intracellular reabsorption factor for fluorescence, and η_{PSII} [mol photons emitted (mol photons absorbed)$^{-1}$ nm^{-1}] is the spectral fluorescence quantum efficiency function describing the fraction of photons absorbed at waveband λ by photosynthetic pigment in PSII and reemitted at λ_{em} within the cell. The bounds on the integral correspond to the minimum wavelength where light is present, λ_{min}, and to the emission wavelength λ_{em}. In these equations, the product $\overset{o}{E}(\lambda)\,{}^{PS}a_{PSII}(\lambda)$ represents the rate of light absorption by PSII per unit volume of water at wavelength λ, $\eta_{PSII}(\lambda\rightarrow\lambda_{em})$ indicates the fraction that is converted to fluorescence and reemitted at λ_{em}, and $Q^*_{abs}(\lambda_{em})$ the fraction of fluorescence that is leaving the cell at wavelength λ_{em} (i.e. that is not re-absorbed within the cells, see Collins et al. 1985).

While Eqs. 2 and 3 provide a fully spectral description of the fluorescence emission; it is often useful and sufficient to simplify them. This implies some assumptions. First, if we assume that the quantum yield is independent of the excitation wavelength (that is PSII photosynthetic pigments transfer their absorbed energy with the same efficiency to the fluorescing chlorophyll molecule) we can represent $\eta_{PSII}(\lambda\rightarrow\lambda_{em})$ as the product of a constant and a function of the emission wavelength. The constant, which represent *the fraction of absorbed photons at all wavelength by photosynthetic pigments in PSII that are reemitted over the whole emission band*, is the quantum yield of fluorescence in PSII, ϕ_F [mol photons emitted (mol photons absorbed)$^{-1}$]. The function $\Phi_F(\lambda_{em})$(nm^{-1}) represents the spectral dependence of the emission (it must be scaled to have an integral of 1): $\eta_{PSII}(\lambda_{em}) = \phi_F\Phi_F(\lambda_{em})$. In aquatic sciences, the function $\Phi_F(\lambda_{em})$ is often represented (e.g. Gordon 1979; Ostrowska et al. 1997; Maritorena et al. 2000) by a Gaussian centred near 685 nm with a half-height width of around 25 nm. Generally, however, a secondary shoulder at a longer wavelength is present (see Fig. 1b). The maximum emission wavelength can also vary significantly in particular in dinoflagellates (Millie et al. 2002) because of the different light harvesting systems. Furthermore, because most of the absorbed energy by photosynthetic pigments is in the 400 to 700 nm domain (often referred to as PAR, for photosynthetically available radiation; see Fig. 1b), and since λ_{em} the upper bound on the integral (Eq. 3), is always near 700 nm, we can replace $\overset{o}{E}(\lambda)$ by $\overset{o}{E}(PAR)$ which is the sum of the irradiance over the PAR domain. However, to keep the information on the spectral matching between $\overset{o}{E}(\lambda)$ and ${}^{PS}a_{PSII}(\lambda)$ this representation requires defining an irradiance weighted absorption coefficient as

$${}^{PS}\overline{a}_{PSII} = \frac{\int_{400}^{700} {}^{PS}a_{PSII}(\lambda)\,\overset{o}{E}(\lambda)\,d\lambda}{\overset{o}{E}(PAR)} \quad (4)$$

The variations in the ${}^{PS}a_{PSII}(\lambda)$ spectrum are due to changes in pigment composition and packaging effects. As expressed by Eq. 4, the changes in ${}^{PS}\overline{a}_{PSII}$ are due to changes both in the magnitude of ${}^{PS}a_{PSII}(\lambda)$, and the spectral matching between the spectra of ${}^{PS}a_{PSII}(\lambda)$ and $\overset{o}{E}(\lambda)$.

With these assumptions and simplifications, we can now rewrite Eq. 3 as

$$F_{PSII}(\lambda_{em}) = \overset{o}{E}(PAR)\,{}^{PS}\overline{a}_{PSII}\,\phi_F\Phi_F(\lambda_{em})Q^*_{abs}(\lambda_{em}) \quad (5)$$

For application where the chlorophyll concentration [*chla*] (mg m^{-3}) is desired we can use the relationship ${}^{PS}\overline{a}_{PSII} = {}^{PS}\overline{a}^*_{PSII}[chla]$ to express Eq. 5 as:

$$F_{PSII}(\lambda_{em}) = \overset{o}{E}(PAR)\,{}^{PS}\overline{a}^*_{PSII}[chla]\,\phi_F\Phi_F(\lambda_{em})Q^*_{abs}(\lambda_{em}) \quad (6)$$

where ${}^{PS}\overline{a}^*_{PSII}$ is the chlorophyll specific absorption coefficient of PSII (note that [*chla*] is the total chlorophyll *a* associated with both PSI and PSII). We will use Eqs. 2, 5 and 6 in the following sections as a support for the description of the different protocols.

The parameter Q^*_{abs} can be further expressed as (Morel and Bricaud 1981):

$$Q^*_{abs}(\lambda_{em}) = \frac{a_\phi(\lambda_{em})}{a_{sol}(\lambda_{em})} \quad (7)$$

where a_ϕ is the phytoplankton total absorption coefficient, and a_{sol} is the absorption coefficient of unpackaged phytoplankton pigments. In practice a_{sol} is obtained by extraction of phytoplankton in a solvent and spectrally correcting the measured absorption to match in vivo absorption (Bidigare et al. 1990; Bricaud et al. 2004). Intracellular reabsorption of fluorescence results from the overlap between the red absorption band (maximum centred at ca. 675 nm) and the fluorescence emission band of chl *a* (see Fig. 1b and Collins et al. 1985). The Q^*_{abs} term is necessary when describing the fluorescence emission of phytoplankton as we cannot practically define a volume smaller than the cell. Therefore, we must describe the characteristics of the emission just outside the cell, instead of the fluorescence leaving the chl *a* molecule itself. In theory, Q^*_{abs} can vary between 0 and 1, but in reality limits are imposed by the maximal intracellular pigment concentration. For phytoplankton Q^*_{abs} is generally larger than 0.2 at 685 nm.

2.3 The Functional Organization of the Photosynthetic Apparatus

The light reactions of photosynthesis for eukaryotic algae (see Campbell et al. 1998; Ting et al. 2002 for cyanobacteria) occur within the thylakoid membrane located within the chloroplasts. We focus here first on the description of photosystem II and follow by a more general description of the rest of the photosynthetic chain.

2.3.1 Photosystem II

Photosystem II can be thought of as an enzyme which catalyzes the light-driven oxidation of the water molecule. Doing so, it allows electrons to enter the photosynthetic electron transfer chain and provides part of the free energy necessary for ultimately reducing inorganic carbon. Photosystem II also binds many of the first electron acceptors in this chain. It is composed of several proteins to which hundreds of pigment molecules are bound in an extremely precise configuration that can be regulated according to environmental conditions. Recent technical advances have allowed tremendous progress in our understanding of the organisation of photosystem II (see for e.g., Liu et al. 2001; Zouni et al. 2001; Barber 2006; Nelson and Yocum 2006). For practical applications relevant to aquatic scientists such detailed description is, however, usually unnecessary; a representation, such as the one given in Fig. 3a, provides sufficient details for describing most processes relevant to our use. Photosystem II spans the thickness of the thylakoid membrane; it is composed of (see, Hankamer et al. 1997; Barber 2006):

1. A peripheral set of pigment-proteins complexes (LHCII), which serves almost exclusively as a light-harvesting group, forming part of the so-called "light-harvesting antenna". In eukaryotes, these pigment-protein complexes can readily separate from the rest of PSII thus diminishing its absorption cross section during so-called "state-transitions" (see section "Quenching due to state transitions")
2. An inner antenna (CP24, CP26, CP29) whose function appears to be mostly the transfer (and its regulation) of absorbed energy from the peripheral antenna to the core proteins CP43 and CP47. Since the inner antenna contains light harvesting pigments to transfer the energy, these pigments also play a role in light harvesting.
3. A core complex of four proteins (D1, D2, CP47, CP43). The proteins D1 and D2 (an heterodimer) bind the first electron acceptors, the site of water splitting (the manganese cluster) and the charge separation site (chlorophyll cluster P680) whereas CP47 and CP43 appear to mostly act in energy transfer from the inner antenna towards the charge separation site.
4. A few proteins (pale grey on Fig. 3a) form the lumenally exposed side of the oxygen evolving complex and appears to have mostly a structural function aiding the efficient oxidation of water at the manganese cluster site.
5. Several low molecular weight proteins (dark grey rod-like structures on Fig. 3a) many with largely unknown functions. One of these (PsbS highlighted in green on Fig. 3a) appears to play a significant role in the dissipation of excess energy and the regulation of energy transfer from the antenna to the core (see Section 2.9.1).

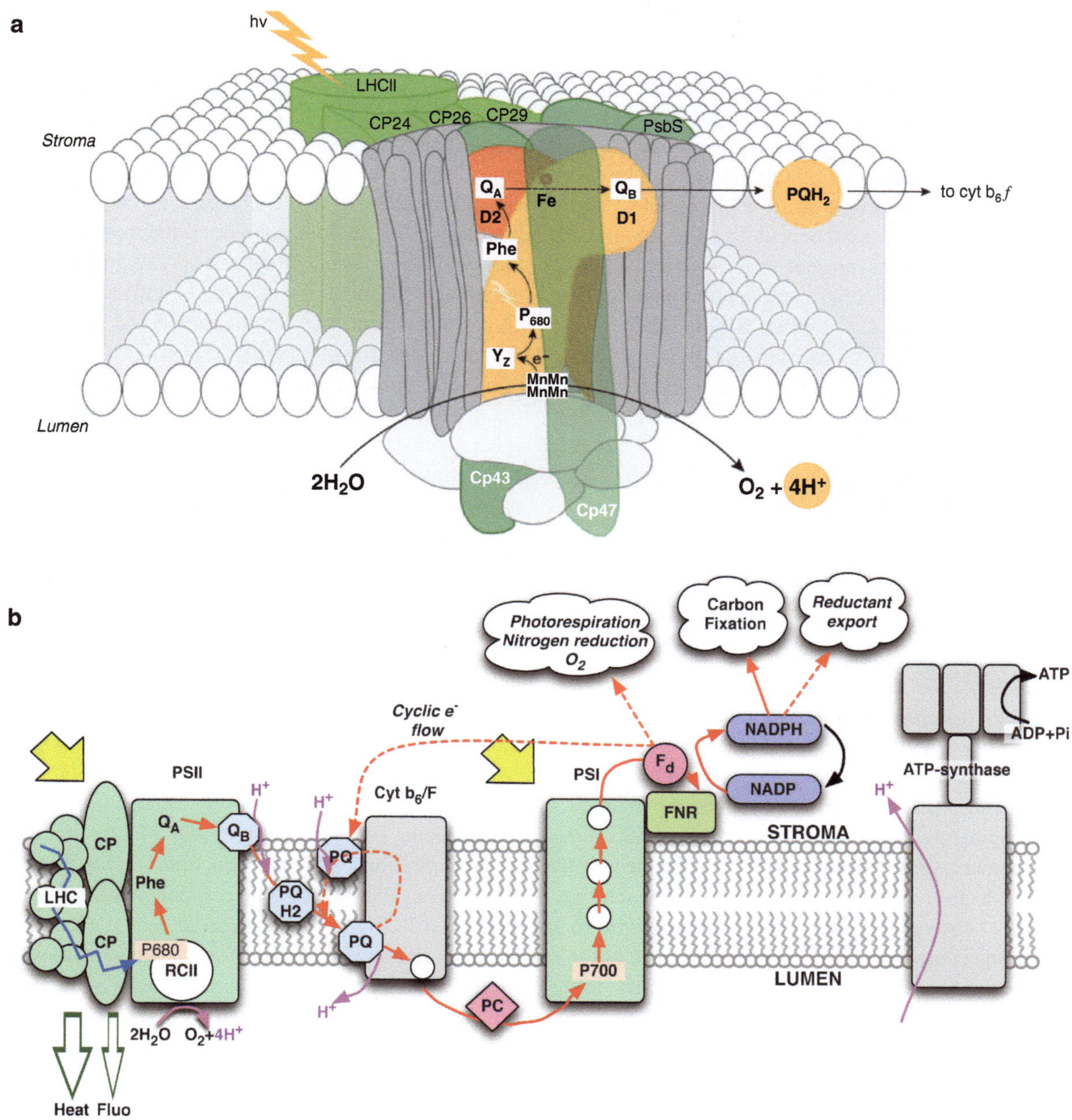

Fig. 3 Organization of photosystem II and electron transport chain. (**a**) Organization of photosystem II. See text in section 2.3.1. (**b**) Schematic representation of the light reactions of photosynthesis in plants. From left to right are represented four protein complexes: photosystem II (PSII), cytochromes B6/F (Cyt b_6/f), photosystem I (PSI), and ATP synthase. The continous red arrows indicate the linear electron flow while the dashed red arrows represent alternative paths. On photosystem II are represented schematically the light harvesting complex (LHC) and two CP proteins, the reaction center (RCII), the phaeophytin (Phe), the primary electron acceptor quinone Q_A and a bound secondary electron acceptor quinone Q_B. Between PSII and the Cyt b_6/F is represented the plastoquinone pool. The electron transporter between the Cyt b_6/F and the P700 chlorophyll *a* dimer in PSI is a plastocyanin (PC). The terminal electron acceptor downstream of PSI is a ferrodoxin (Fd) which with the ferrodoxin-NADP reductase (FNR) reduces NADP to NADPH. Alternative sinks for the electrons from ferrodoxin and NADPH are shown in the white clouds. Also represented are the sources of protons (H^+) to the lumen from the splitting of the water molecule and the transport by the plastoquinones, as well as the main sink of H^+ from the transport through the ATP-synthase protein complex

The LHCII presented here is consistent with those of higher plants and green algae. In other photosynthetic plankton this structure is complemented or replaced depending on taxa. As examples, we can mention: most cyanobacteria and rhodophytes, where the LHCII complexes are absent and are replaced by phycobilisomes structure; the prochlorophytes, where the phycobilisomes are absent and a chlorophyll binding protein "pcb" is present (e.g. Ting et al. 2002; Bryant and Frigaard 2006); and the dinophytes where soluble peridinin-Chl*a* complexes also attach to the peripheral antenna. An overview of light-harvesting systems in other microalgal taxa is provided in Prezelin and Boczar (1986). It is now widely accepted that PSII in vivo forms a dimer with both cores likely sharing excitation from both light-harvesting antennae.

Typically, a photon is absorbed within the peripheral antenna which leads to a pigment molecule being in an excited state. This excited state propagates to the chlorophyll complex P680 which becomes excited and transfers an electron to a pheaophytin (Phe) in the first step of photosynthetic electron transfer referred to as "charge separation". The second step is the passage of the electron from the phaeophytin to the first quinone (Q_A), this second step is called charge stabilization as the probability of recombination of the electron with P680 is then much reduced. The P680 is then reduced from an electron originating from the tyrosine (Yz). These first three steps occurs within ~200 ns. The next step, the transfer of an electron from Q_A to the second quinone Q_B, requires between ~150 and 600 μs. Because it is much longer than the previous steps, under high light, this step acts as the bottleneck in the transfer of electron to the rest of the photosynthetic chain. At roughly the same time a tyrosine is reduced with an electron from the Mn cluster which is in turn reduced by a water molecule. Four such charge separation events (minimum of four photons) are required to oxidize two water molecules leading to the release of an oxygen molecule (O_2) and four hydrogen ions (H^+) in the lumen (see Fig. 3a).

2.3.2 The Photosynthetic Chain

The transport of electron through the photosynthetic chain is represented in Fig. 3b. The Q_B, in fact a loosely bound plastoquinone, is fully reduced after accepting two electrons from Q_A (the so-called "two electrons gate") and thus after two photosystem II turnovers. The reduced Q_B binds two protons (H^+) from the stroma and is then released within the thylakoid membrane and becomes part of the plastoquinone pool. The plastoquinone then diffuses within the thylakoid membrane (being hydrophobic it does not leave the membrane made up of a lipid bilayer), reaches the cytochrome b6-f (cyt b_6/f) complex and attaches to a binding site on the lumenal side of the membrane. At this point the plastoquinone releases its protons into the lumen, leading to the creation of a pH and electrochemical gradient between the lumen and stroma. At the same time, the plastoquinone transfer its electrons to cyt b_6/f. A plastocyanin transfer the electrons from the cyt b_6/f to the PSI reaction centre composed of a special chlorophyll pair referred to as P700. For this transfer to occur, the P700 must be oxidized. This oxidation occurs through a second light reaction whereby a photon is absorbed in PSI and leads to the transfer of an electron to the first electron acceptor of PSI. The electron from the plastocyanin then reduces P700. The electron from P700 is then transferred through four molecules to finally reduce a ferrodoxin. From there, several path are possible: two electrons are required to reduce one NADP molecule (to form $NADPH_2$) through the ferrodoxin-NADP-oxidoreductase enzyme which can then be used in the dark reactions for the fixation of carbon or alternatively a series of non-carbon reducing path are possible[2] (e.g. Mehler reaction, photorespiration, nitrogen reduction, cyclic electron flow around PSI). Since both reaction centres II and I (RCII and RCI) must work in series (equal number of charge separation/photons) for continuous oxygen evolution and since four electrons (four photons in PSII) are required to oxidise two H_2O, a minimum of eight photons are required for the production of each O_2. For each carbon fixed, two molecules of $NADPH_2$ (four electrons) are required thus the minimum quantum requirement for the fixation of one carbon atom is also eight photons.

In addition to chl *a*, the pigment antenna of PSII contains other pigments referred to as accessory pigments: other chlorophylls, carotenoids and phycobilins. The composition is taxa specific (reviewed by

[2] Variation in the fluxes going to these alternative paths is the proposed mechanisms for the rapid fluorescence transients observed upon nutrient supply of nutrient stressed algae (see Chapter 11 – Shelly et al.).

Jeffrey and Vesk 1997). Among accessory pigments, the light-harvesting or photosynthetic pigments transfer nearly 100% of their excitation energy to chl *a* when they absorb a photon. Non-photosynthetic pigments are those pigments that absorb light but do not transfer energy to chl *a*. They include pheopigments, β-carotene,[3] and zeaxanthin[4] in prokaryotic phytoplankton (e.g. cyanobacteria and *Prochlorococcus* sp.). Among non-photosynthetic pigments, the xanthophylls, a family of carotenoids, mostly located in the inner antenna (Bassi et al. 1993), generally play a role of photoprotection against excessive irradiance in plants (Demmig-Adams and Adams 1992). They participate to a process known as energy dependent non-photochemical quenching (see section 2.9 for more details) which leads to an increase in the energy dissipation in the form of heat within the PSII antenna in the light.

2.4 Adaptation, Acclimation, Regulation of Phytoplankton

Phytoplankton must cope with a wide range of incident irradiance. Even without the fluctuation due to clouds, surface wave focusing, or vertical mixing, a phytoplankton near the water surface will experience a change from complete darkness to around 2000 μmol photons m^{-2} s^{-1} throughout the day. Three processes allow phytoplankton to cope with this light regime: adaptation, acclimation and regulation (sensu Falkowski and LaRoche 1991; Raven and Geider 2003). As a result of changes in their genetic make up, different species are best suited for growth in different light environment (e.g. near the surface or at depth) – this is photoadaptation. A single cell does not change its genetic make up and does not photoadapt. A species adapted for growth in a given environment may need to change its macromolecular composition (e.g. add or remove pigments) to improve growth efficiency or reduce potential damage (e.g. MacIntyre et al. 2002; Geider et al. 2009) if the light field changes, such as during a cloudy day: this is termed photoacclimation. In the case of photoaclimation, the genetic make up does not change. Finally a cell that is photoacclimated to a specific light regime may need to rapidly tune their photosynthetic efficiency due to rapid changes in the light field – this is termed photoregulation. As opposed to photoacclimation, photoregulation does not require de novo synthesis or break down of molecules.

Fluorescence allows us to probe different aspects of these three processes. However, to provide results that can be interpreted accurately, protocols require specific regulation states. Two states are particularly important. The dark regulated state – ideally, in this state the cell has downregulated all the means that allows it to cope with the light but has not changed its macromolecular composition from the acclimation state in which it was before sampling.[5,6] The light regulated state – in this state all the photoregulation that accompany the presence of light are active; in general processes are probed after the cells have reached a steady-state. Many authors have referred to these two states alternatively as dark and light acclimated or adapted we prefer to keep these terms for different processes.

2.5 Fates of Absorbed Photons Within PSII

There are three main fates for the energy of a photon absorbed within phytoplankton. It can either be used for charge separation, dissipated as heat, or lead to fluorescence. An example of the distribution of the energy to the different fates is presented in Fig. 4 for an undamaged, high light grown phytoplankton in both the light and dark regulated state (numbers are typical but can vary significantly). It shows that in the dark

[3] β-carotene is thought to be involved in the de-excitation of active forms of oxygen that occur under high light and produce damage to the cell.

[4] In prokaryotic phytoplankton, zeaxanthin is often present in larger concentration than chl *a*. However, it is located in the cell wall and, though it absorbs light, it does not appear to have a direct link to photosynthetic processes.

[5] The rapid switch between xanthophylls associated with the so-called xanthophylls cycle is not considered a break down or de novo synthesis and is included within the process of regulation.

[6] The presence of photodamage, which can arguably be considered a form of photoacclimation/photoprotection, should be conserved in this state. Removal of photodamage indeed requires break down and de novo synthesis of proteins.

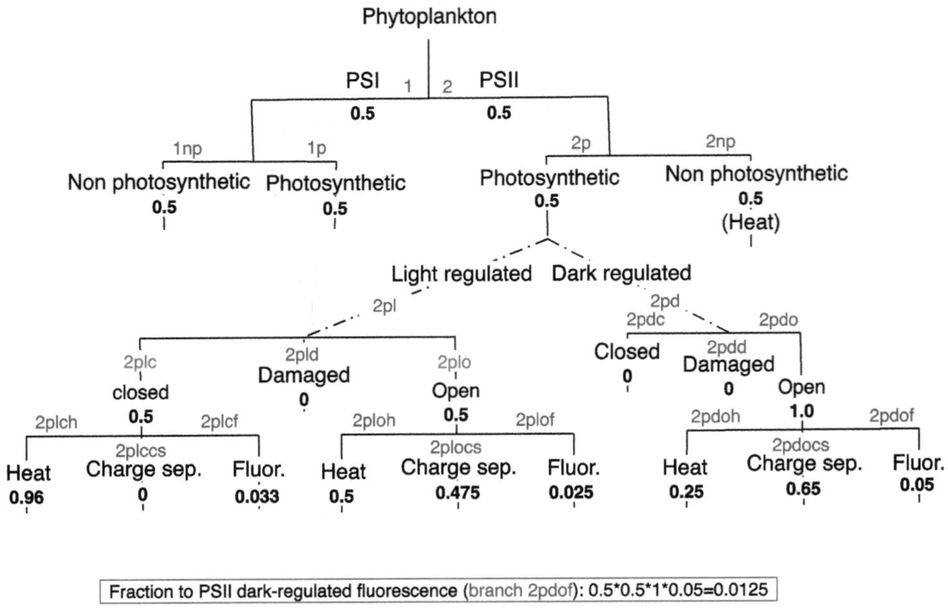

Fig. 4 Probability of the different fates for an absorbed photon for undamaged phytoplankton in the light or dark regulated state. Numbers in black indicate the probability that the energy of a photon will follow that path. Gray characters are identifiers for the different paths. Dashed lines are for two different light regulated states (i.e. these paths do not exist under the same light condition)

regulated state, roughly 25% of the light absorbed by phytoplankton is intercepted by photosynthetic pigments within PSII[7]; of this 25% roughly 5% is emitted as fluorescence. Assuming negligible emission from PSI, the fraction of the total absorbed photon reemitted, which we will denote, ϕ_F^{app}, is around 0.0125. Note that ϕ_F^{app} is ~50% lower than ϕ_F as it accounts for the absorption of all pigments, not only photosynthetic ones. The former is better termed the *apparent* quantum yield of chl *a* fluorescence as it includes the effects of chromophores that do not participate in the fluorescence emission (Miller et al. 2002).

Within PSII the potential fates can be further broken down as follows (see Fig. 4) for a light regulated state (in the dark regulated state, the path to closed reaction centres does not exist):

1. The photon is absorbed by a photoprotecting pigment molecule and its energy is dissipated as heat (branch 2np, Fig. 4).
2. The photon that is absorbed by chl *a* or whose energy is transferred to chl *a* from a light harvesting pigment is either: (1) dissipated as heat directly or transferred to photoprotecting pigment and dissipated as heat (branches 2ploh and 2plch); (2) reemitted as a red photon (fluorescence, branches 2plof and 2plcf); or (3) transferred to P_{680} in the RCII.
3. The energy transferred to P_{680} is whether (1) used to transfer an electron to (i.e. to reduce) the first electron acceptor (*Phe*) when Q_A is not in a reduced state (branch 2plocs, it is then said to be "open", e.g. Schatz et al. 1988); (2) back-transferred to antenna chl *a* when Q_A is already in a reduced state (it is then said to be "closed", branch 2plccs); or (3) dissipated as heat if RCII is damaged (branch 2pld).
4. The fate of an exciton back-transferred from RCII to the antenna is as in 2.

This description is for a single antenna attached to a single RCII (i.e. the so-called puddle model). In the dimeric organisation of PSII with shared antenna or with multiple RCII sharing antennas, an exciton can also be transferred (or "feel") another RC antenna before being fluoresced, dissipated as heat or involved in charge separation.

[7] In low-light acclimated/adapted cells, this fraction can increase to near 50% since the fraction going to non-photosynthetic pigments is reduced.

2.6 A Simple Model of In Vivo Processes In PSII At Room Temperature

2.6.1 Quantum Yield of Fluorescence

While the quantum yield of chl a fluorescence in solution could be described by two rate constants, in vivo other deexcitation paths exist, and thus require more rate constant to describe the quantum yield. Furthermore, the presence or absence of rate constants will depend on the state of the reaction centre. A simple representation of PSII reaction centres can be achieved with two states: opened and closed. Within the simplified framework of the puddle model[8] again, the quantum yield of fluorescence originates from different fractions of reaction centres in these two states and can be expressed as follows (but see Havaux et al. 1991 for more detailed models of fluorescence ; Lavergne and Trissl 1995).

$$\phi_F' = \mathbb{R}_O \frac{k_F}{k_F + k_P + k_D + k_{qe}} + \mathbb{R}_C \frac{k_F}{k_F + k_D + k_{qe}} = \mathbb{R}_O \phi'_{Fopen} + \mathbb{R}_C \phi'_{Fclosed} \quad (8)$$

where, k_F, k_P, k_D, k_{qe}, are respectively the rate constants (s^{-1}) of fluorescence; photochemistry (i.e. charge separation); basal thermal dissipation in the dark regulated state of an undamaged reaction centre; the dissipation of energy associated with the light regulated state (more specifically "energy dependent quenching", see section 2.9.1). This model is equivalent to the representation used by Rohácek[9] (2002). The $\mathbb{R}_O$ and $\mathbb{R}_C$ are respectively the fractions of reaction centres in the open and closed state ($\mathbb{R}_O + \mathbb{R}_C = 1$). The ϕ'_{Fopen} and $\phi'_{Fclosed}$ are respectively the quantum yield of fluorescence for open and closed reaction centres in the light regulated state. In this model, the k_{qe}, is not strictly a constant but a variable which is allowed to take values from 0 (in the dark regulated state) to some maximum (which depends on species and physiology) in the light regulated state.[10] The prime (′) symbol indicates that the yield is not in the dark regulated state (i.e. that $k_{qe} > 0$); when $k_{qe} = 0$ then the quantum yield is denoted ϕ_F.

In the field, a large fraction of reaction centres can be damaged. While this is generally not included in most models of fluorescence, we propose a simplified explicit representation here based on the puddle model[11] which allows us to examine the potential effect of damage on the different protocols used. To extend Eq. 8, we thus include a fraction of damaged reaction centres, $\mathbb{R}_{dam}$ (now $\mathbb{R}_O + \mathbb{R}_C + \mathbb{R}_{dam} = 1$). We define these damaged reaction centres as those that are incapable of charge separation following light induced (oxidative) damage.[12] We also include a representation of the quantum yield of these reaction centres that uses a new rate constant k_I (s^{-1}), which represents the dissipation of energy at damaged reaction centres:

$$\phi_F' = \mathbb{R}_O \frac{k_F}{k_F + k_P + k_D + k_{qe}} + \mathbb{R}_C \frac{k_F}{k_F + k_D + k_{qe}} + \mathbb{R}_{dam} \frac{k_F}{k_F + k_D + k_{qe} + k_I} = \mathbb{R}_O \phi'_{Fopen} + \mathbb{R}_C \phi'_{Fclosed} + \mathbb{R}_{dam} \phi'_{Fdam} \quad (9)$$

The terms on the right hand side of Eqs. 8 and 9 are similar but differ by the factor multiplying them (fractions $\mathbb{R}_O$, $\mathbb{R}_C$ and $\mathbb{R}_{dam}$) while the term representing the open and damaged reaction centres have an additional deexcitation path, represented by an additional constant in the denominator (k_D and k_I respectively), which diminishes the fluorescence coming from these reaction

[8] The model for the completely connected case (or lake model) is also expressed very simply as $\phi_F' = \frac{k_F}{k_F + \mathbb{R}_O k_P + k_D + k_{qe}}$. It thus leads to equivalent expressions for the fluorescence yield when the reaction centres are all open ($\mathbb{R}_O = 1$) or all closed ($\mathbb{R}_O = 0$)

[9] Rohacek uses a multiplicative factor to represent the increased heat dissipation under high light as dk_D where d is termed the "dissipation factor". In our representation (which we used to portray two different types of dissipation) we use $k_D + k_{qe}$ to describe the same processes. It follows that $d = 1 + k_{qe}/k_D$.

[10] The k_{qe} is equivalent to what Oxborough and Baker (2000) call the "non-radiative decay through Stern-Volmer quenching", instead of using a "variable rate constant", they express it as a rate constant k_{SV} multiplied by a concentration of quencher [SV]. The equivalence is $k_{qe} = k_{SV}[SV]$.

[11] The completely connected model is probably not appropriate in the case of damaged reaction centres as these are generally separated from the complex of connected PSII in the grana (when it exist) and transported to the stroma for resynthesis and reconstruction (Aro et al. 2005).

[12] We assume that these photosystems are still capable of increased heat dissipation due to energy-dependent quenching as it seem to be regulated by events that are relatively far from the reaction centre, furthermore, there is no obvious reason why they couldn't "feel" the transthylakoid pH gradient (see section 2.9.3 for more details).

centres relative to the closed reaction centre. Such that $\phi'_{Fclosed} > \phi'_{Fopen} \approx \phi'_{Fdam}$ (implying $k_P \approx k_I$; but keeping in mind that ϕ'_{Fdam} remains to be fully characterized).

2.7 Charge Separation at PSII

Because fluorescence is used to probe photosynthetic rates, it is useful to describe some basic relationships related to charge separation. First, we introduce the photosynthetic optical cross section ($\bar{\sigma}^{opt}_{PSII}$, m² PSII⁻¹) of a single photosystem II as:

$$\bar{\sigma}^{opt}_{PSII} = \bar{a}^{PS}_{PSII} / N_{PSII}. \tag{10}$$

where N_{PSII} is the total number of PSII (open, closed or damaged) per unit volume of water (PSII m^{-3}). The $\bar{\sigma}^{opt}_{PSII}$ represents the area upon which all the light impinging is absorbed by photosynthetic pigments and available to photosystem II (i.e. the light going through branch 2p in Fig. 4). The fraction of this cross section used for photosynthesis is called the effective absorption cross section for photosynthesis and is equal to

$$\bar{\sigma}'_{PSII} = \bar{\sigma}^{opt}_{PSII}\phi'_{pO} \tag{11}$$

where ϕ'_{pO} is the quantum yield for charge separation at an open PSII (electron photon^{-1}), that is the maximum quantum yield attainable under the light regulated state if all reaction centres were open, and is given by

$$\phi'_{pO} = \frac{k_P}{k_F + k_P + k_D + k_{qe}}. \tag{12}$$

An equivalent set of equations can be written for the dark regulated state (when $k_{qe} = 0$). The effective absorption cross section becomes

$$\bar{\sigma}_{PSII} = \bar{\sigma}^{opt}_{PSII}\phi^{max}_{pO}. \tag{13}$$

The quantum yield of charge separation is then maximal and expressed as

$$\phi^{max}_{pO} = \frac{k_P}{k_F + k_P + k_D}. \tag{14}$$

The actual or realized quantum yield of photosynthesis (including open and closed reaction centre), ϕ'_p, is obtained by multiplying the quantum yield of open reaction centres by the fraction of open reaction centres:

$$\phi'_p = \mathbb{R}_O\phi'_{pO} \tag{15}$$

It follows (using Eqs. 10–15) that the rate of charge separation (e^{-1} s^{-1} m^{-3}) in a given volume of water can be expressed using different equivalent representations:

$$\begin{aligned} P_e &= \phi'_p\ {}^{PS}\bar{a}_{PSII}\ \mathring{E}(PAR) \\ &= \mathbb{R}_O\phi'_{pO}\ {}^{PS}\bar{a}_{PSII}\ \mathring{E}(PAR) \\ &= \phi'_p\ \bar{\sigma}^{opt}_{PSII} N_{PSII}\ \mathring{E}(PAR) \\ &= \mathbb{R}_O\bar{\sigma}'_{PSII} N_{PSII}\ \mathring{E}(PAR) \\ &= \mathbb{R}_O\bar{\sigma}_{PSII}\left(\phi'_{pO}/\phi^{max}_{pO}\right) N_{PSII}\ \mathring{E}(PAR) \end{aligned} \tag{16}$$

The fraction of open RCIIs, $\mathbb{R}_O$, varies as a function of ambient light. In the absence of damage (and for non-connected reaction centres), if a flash of light is sufficiently intense to close reaction centers before any reopen (shorter than ~100 μs) the relationship between $\mathbb{R}_O$ and the radiant exposure of the flash (ε, photons m^{-2}) follows a cumulative one-hit Poisson function according to the so-called "target theory" (see Mauzerall and Greenbaum 1993):

$$\mathbb{R}_O = e^{-\bar{\sigma}'_{PSII}\varepsilon} \tag{17}$$

Still in the absence of damage (see Han 2002 for the case with damage), if light is delivered continuously, $\mathbb{R}_O$ can be expressed as (Zonneveld 1997; Han 2001):

$$\mathbb{R}_O = \frac{1}{1 + \tau_p\bar{\sigma}'_{PSII}\mathring{E}} \tag{18}$$

where τ_P (s) is defined as the turnover time for electron transport through the electron carrier chain.

2.8 Photochemical Quenching of Fluorescence

According to Eq. 8 (since $\phi'_{Fopen} < \phi'_{Fclosed}$) when the fraction of open reaction centers ($\mathbb{R}_O$) increases, ϕ'_F (or ϕ_F) decreases. This decrease involves the competition between the charge separation (k_p) and the fluorescence path (k_F). Because it is related to photochemistry (i.e. k_p) this decrease is generally referred to as "*photochemical quenching*" of fluorescence. This expression reflects the use of the maximal quantum yield of fluorescence as the reference level in fluorescence work. Any process that decreases fluorescence quantum yield from this maximal level is said to "quench" it. In algae, this maximal level is generally achieved when all reaction centres are closed ($\mathbb{R}_C = 1$) and when the phytoplankton is in a dark regulated state ($k_{qe} = 0$).[13]

2.9 Non-photochemical Quenching of Fluorescence

The decrease of ϕ'_F (or ϕ_F) that is not due to photochemistry is referred to as "*non-photochemical quenching*" of fluorescence. Non-photochemical quenching (NPQ) involves different causes and mechanisms that lead to an increase in thermal dissipation of absorbed energy within the PSII pigment antenna and/or RCII. A qualitative distinction between different types of non-photochemical quenching is often made from the relaxation kinetics of non-photochemical quenching when a sample is placed in the dark or dim light after exposure to bright light (see Fig. 13a, described later, for example). Three relaxation phases are generally observed (see Krause and Weis 1991).

2.9.1 Energy-dependent Non-photochemical Quenching

The first relaxation phase lasts from seconds to a few minutes and is dependent on the build-up of a proton gradient across the thylakoid membrane (or "ΔpH") under high irradiance. This build-up occurs when sources of H^+ in the lumen (PSII water splitting and plastoquinone transport, see Fig. 3) are greater than the sinks (mostly ATP-synthase; see Fig. 3b). This type of non-photochemical quenching is referred to as "energy-dependent" quenching, and denoted q_E. It is correlated with the changes in the epoxidation state of xanthophylls. In chlorophytes, three xanthophylls, zeaxanthin, antheraxanthin and violaxanthin associated with the PSII pigment antenna are involved in a cycle; they are enzymatically inter-converted back and forth by de-epoxidation and epoxidation following the sequence zeaxanthin ↔ antheraxanthin ↔ violaxanthin (Gilmore and Govindjee 1999). The build-up of a proton gradient across the thylakoid membrane (ΔpH, see Fig. 3b), stimulates the de-epoxidation pathway that leads to the formation of zeaxanthin. In chromophytes (chlorophyll-c containing phytoplankton), an analogue cycle with two xanthophylls, diadinoxanthin and diatoxanthin (where the former is the high-light end-member), has been shown to be involved in photoprotection (Olaizola and Yamamoto 1994). Much less is known about this cycle.

While the change in pigment stoichiometry is the most easily measured characteristic that covaries with the formation of NPQ, the exact cause of energy-dependent non-photochemical quenching is still actively investigated. A necessary step in the formation of q_E is the protonation of the PsbS molecule (see Niyogi 1999 and Fig. 3a) and resulting changes in the conformation of PSII antenna (Horton et al. 1996). This step does not involve the xanthophylls. In addition, two, non-exclusive, molecular mechanisms are presently suggested as main quenching mechanisms. In the first, energy dependent quenching in the LHCII complex does not directly require zeaxanthin (energy is transferred to a lutein, Ruban et al. 2007). In the second, occurring in the CP29 protein, the zeaxanthin plays a central role in a process that involves (charge separation) quenching at a chlorophyll-zeaxanthin heterodimer (Ahn et al. 2008).

The q_E can lead to a decrease of up to 90% in ϕ_F and, thereby, serve as a highly efficient and dynamic way to down-regulate transfer of excess energy to RCII under high irradiance. It can be particularly efficient in diatoms (Ruban et al. 2004). This type of quenching plays a role in photoprotection of the photosystem II and avoids the formation of potentially dangerous radicals near the reaction centres by dissipating the excitation energy rapidly.

Whichever the exact process, it seems that the xanthophylls cycle provides a proxy of the quenching; the response of the xanthophyll cycle to irradiance changes can be fast through epoxidation/de-epoxidation (minutes), and slow through variations in the pool size of xanthophylls-cycle pigments (hours to days) (see review by Lavaud et al. 2004). An increase in the pool size providing a greater capacity to the fast response (Horton et al. 1996). This quenching is represented in Eq. 8 by k_{qe}.

[13] Exception do arise, however, such as when the plastoquinone pool is reduced in the dark by chlororespiration, the maximal level of fluorescence is then observed in low light (see Kromkamp and Forster 2003 and references therein).

2.9.2 Quenching Due to State Transitions

The second phase of NPQ relaxation, denoted q_T, lasts between 5 and 20 min and is generally associated with so-called "state transitions". Historically, the term "state transition" has been used to designate the relative changes in the absorption cross sections of PSI and PSII observed in response to changes in the spectrum of ambient light light. The mechanism behind this phenomenon consists in the detachment and attachment of LHCII as well as CP29 under high- and low-irradiance conditions, respectively (Kargul and Barber 2008). Detachment and attachment would result from the enzyme-driven phosphorylation and dephosphorylation of the antenna proteins, respectively, regulated by the reduction level of the plastoquinone pool (reviewed by Aro and Ohad 2003). Under high irradiance, the greater reduction state of the plastoquinone pool would stimulate phosphorylation of antenna proteins. When state transitions occurs under high light, such changes in σ_{PSII}^{opt} are probably not associated with inverse complementary changes in PSI absorption cross section (e.g. Kargul and Barber 2008). No rate constant for this process are associated with Eqs. 8 and 9. Nevertheless, a reduction in F_{PSII} is observed due to the reduction in σ_{PSII}^{opt} or ${}^{PS}\overline{a}_{PSII}$ (see Eqs. 5 and 10). The decrease F_{PSII} has been reported to be on the order of 10–20% (Falkowski and Raven 2007) for microalgae. Its impact could be greater in certain algae; in response to changes in the spectral light quality, *Chlamydomonas* can move 80% of the PSII antenna to PSI (Finazzi et al. 2002).

2.9.3 Quenching Linked to Inhibition

Finally, the third phase of NPQ relaxation, denoted q_I, lasts several tens of minutes to a few hours and was named "inhibition" or "photoinhibition" quenching as it seems to be related to photoinhibition of photosynthesis (i.e. a reduced quantum yield of photosynthesis) and results at least in part from photodamage in the photosynthetic apparatus. However, other sources for this persistent quenching have also been proposed (see Krause and Jahns 2004). According to the current understanding about the nature of photoinhibition, the D1 protein in RCII would be the main target (see the reviews by Aro et al. 1993; Melis 1999). The rate of q_I relaxation would be a function of the rate of D1 reactivation (which depends on irradiance Chow 2001). In the marine environment, q_I may be responsible for decreases in ϕ_F by up to 40% (Falkowski et al. 1995). In our simplified model (Eq. 9), this quenching is related to the presence of damaged reaction centres ($\mathbb{R}_{dam} > 0$) incapable of photochemistry ($k_p = 0$) but dissipating a greater amount of energy as heat than closed reaction centres (addition of the k_I constant).

2.9.4 Reaction Center Quenching

A fourth type of non-photochemical quenching, reaction centre quenching, has been much less studied but is expected to have significant effects on fluorescence (see review by Ivanov et al. 2008a; 2008b). Current understanding suggests that this type of quenching would involve a reversible quenching mechanisms linked to an increased probability of recombination of the radical pair $P680^+Q_a^-$ and would be link to the increased reduction of the plastoquinone pool. This quenching is not linked to inhibition and would have a photoprotective role. It could be particularly important in prokaryotic phototrophes. The timescale for induction and relaxation are rapid of the order of seconds to minutes.

2.10 Transient Changes in Fluorescence

When a photosynthetic organism is suddenly transferred from dark to light, the flux of *in vivo* fluorescence (and ϕ_F) experiences several typical transient changes (e.g. Strasser et al. 1995). The number and magnitude of these transients depend on the energy and duration of light exposure. Each of these transients reflects a specific process of photosynthesis. Figure 5 shows three idealized examples of so-called "induction curves" obtained on samples transferred from dark to continuous high (saturating), moderate (non-saturating) and low irradiance. The different parts of the curve obtained under high light are likely (but see Lazár 2006 for other potential origins) related to the following processes (more details are available in Govindjee and Satoh 1986; Geider and Osborne 1992; Govindjee 1995; Stirbet et al. 1998; Lazár 1999; Boisvert et al. 2006, and Chapter 1 – Cosgrove and Borowitzka).

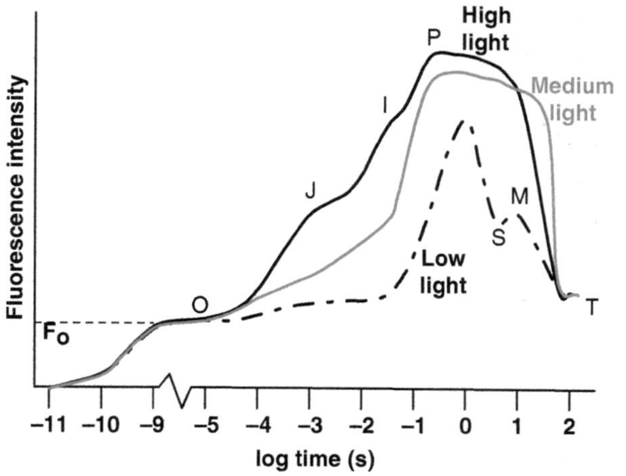

Fig. 5 Idealized light induction curves for three actinic light levels. Labels are according to standard nomenclature. Curves are redrawn from Strasser et al. (1995), extended below 10^{-4} s according to simulations by Lazar (2003)

O When the light is turned on, F shows a quasi-instantaneous (~3 ns) rise to the O level. This signal is generally referred to as "minimum fluorescence" or "initial fluorescence" and denoted F_o (and correspondingly ϕ_{Fo}). This fluorescence occurs before the light closes any reaction centre. At this point, Q_A is maximally oxidized ($\mathbb{R}_O \approx 1$) and the (potential) quantum yield of charge separation is maximal ($\phi_p = \phi_{pO}$). Most of the reaction centres are in the following un reduced state[14]: Q_AQ_B.

O-J F then rises to a first inflection point (indicated by the letter J). The O-J rise results from the progressive reduction of Q_A and the associated closure of the reaction centres (i.e. a decrease in $\mathbb{R}_O$). At the inflexion point, the maximum number of reaction centres with reduced Q_A but oxidized Q_B (e.g. Stirbet et al. 1998) are found. This section is called the "photochemical phase" as it is not slowed at low (physiological) temperatures. The inflexion occurs as the Q_A pool is being reoxidized by transferring electrons to Q_B.

[14] In this section we will describe the state of a reaction centre with respect to the reduction state of the two quinones Q_A and Q_B. The following successive states are possible from the dark-regulated (fully oxidized) state: no charge separation Q_AQ_B; one charge separation $Q_A^-Q_B$; $Q_AQ_B^-$; two charge separations: $Q_A^-Q_B^-$, $Q_AQ_BH_2$, three charge separations $Q_A^-Q_BH_2$. The inflexion points are, here, associated with the transfer of the electron from the state where most reaction centres are in the Q_A reduced the state to the state where the Q_B is reduced after one and two charge separations (the J and I inflexion respectively).

J-I F then rises to a second inflexion (I). This increase in F results from the completion of Q_A reduction to the maximum level under these light conditions (i.e. minimum in $\mathbb{R}_O$. At the inflexion point, the reaction centres are mostly in the $Q_A^-Q_BH_2$ state.

I-P This rise is likely associated with the decrease of non-photochemical quenching associated to oxidized plastoquinones (Vermotte et al. 1979).

P-T F then experiences a continual decrease associated with the increase in non-photochemical quenching (if the light is truly saturating there is no significant reoxidation of Q_A). It reaches a steady-state value generally after several minutes.

Under non-saturating irradiance, the level J, is generally not observed (likely because no significant number of reaction centres in the state $Q_A^-Q_B$ accumulate). At low light, a transient minimum is also observed after P, labelled "S" followed by continuous decrease to T. In this case, the transient from P to T is a complex function of non-photochemical quenching, reoxidation of Q_A and the induction of sinks for electron downstream of PSI. While many models are developed to study the high light case, low light condition is presently not modelled due to the complexities of accounting for factors outside of PSII.

3 Protocols for Measurement of *In Vivo* Phytoplankton Fluorescence, and the Use of Chl *a* Fluorescence to Study Phytoplankton Ecophysiology

A number of chl *a* fluorescence measurement protocols are used for different purposes. These protocols vary depending on the objective of the measurement and the spatial and temporal requirements. Furthermore, often the platform (e.g. gliders, ships of opportunity, buoy, etc.) on which the fluorometers are deployed imposes some constraints. This section describes, for several of these approaches, the basic principles, and the measurement protocols. It also discusses the underlying assumptions and provides commented examples.

3.1 *The Determination of Biomass In Vivo*

This approach is mostly used to measure the concentration of chl *a* in water from its fluorescence emission.

3.1.1 Basic Principle

The assumptions that justifies this approach are straightforward: if $\overset{o}{E}(PAR)$ is from a constant light source (or flashes of equal intensity) and the product ${}^{PS}\bar{a}^{*}_{PSII}\phi_F\Phi_f Q^{*}_{a}$ in Eq. 6 is assumed constant, then F_{PSII} is proportional to [*chla*].

3.1.2 Instruments and Protocols

Commercial fluorometers built for both benchtop (with flow through accessories) and in situ deployments are available. Figure 6 shows the optical setup of a benchtop field fluorometer commercialized by Turner Designs Inc. (Sunnyvale, California, USA). The main features are:

- A low-pressure mercury vapour lamp (4 Watts) that produces continuous light of moderate energy
- A blue (340–500 nm) filter for the excitation of the seawater sample
- A red (680 nm) band filter for the collection of the fluorescence signal
- A photomultiplier tube as detector
- Dual-beam optics to correct for drifts in the lamp output and/or detector sensitivity (represented by the light pipe in Fig. 6); the same detector measures alternatively the signal originating from the lamp and from the sample

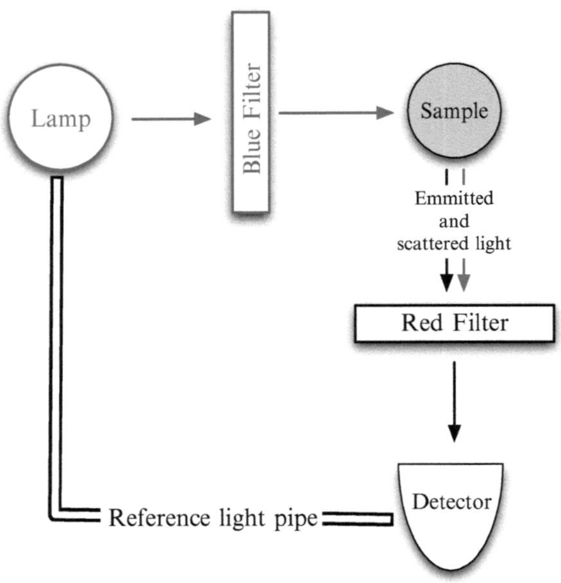

Fig. 6 The optical setup of a typical bench top fluorometer (illustration based on Turner Designs Model 10)

As in most systems measuring fluorescence, to reduce the influence of scattering and straylight,[15] the excitation and detection optics are at a right angle from each other. Such systems can be used for discrete sample analysis or for continuous-flow monitoring.

The influence of the fluorescence from colored dissolved matter or other artifacts can be corrected for by using a blank (see Cullen and Davis 2003). The appropriate blank is an *in situ* water sample filtered just before analysis through a glass fibre filter (e.g. Whatman GF/F) or other kinds of filters with pore size smaller than 0.45 µm (e.g. Whatman Nuclepore® 0.2 µm). When preparing the blank, it is important to verify that the filters used do not leach any fluorescing substance during filtration. This can be done by measuring the signal of pure water before and after passing through the filter, or by comparing the signal from water filtered through different types of filter (e.g. glass fibre filter vs. polycarbonate membrane).

Because the assumptions of constant ${}^{PS}\bar{a}^{*}_{PSII}\phi_F\Phi_f Q^{*}_{a}$ is not generally not met,[16] such *in vivo* fluorescence measurements provide qualitative information about phytoplankton biomass. Nevertheless, a calibration procedure allows converting the fluorometer raw output into rough estimates of [*chla*] values. The calibration procedure simply consists in comparing the fluorometer output with chemical determinations of [*chla*] on common samples. Because the assumptions depend on the species present and their physiological status (see section 3.2.3) when measurements are made in the natural environment the calibration is preferably done using samples from the study area rather than phytoplankton cultures.

The optical setup of a typical underwater fluorometer is similar to that of the flow-through fluorometer described above. Usually light emitting diodes or xenon flash lamps are used for the excitation light. In addition,

[15] Light measured at the detector but not originating from the process that is being measured. In fluorometers, this can originate from, for example, scattered light in the measuring volume that is not perfectly filtered by the emission filter.

[16] When phytoplankton samples are extracted in an organic solvent, these assumptions are generally met such that the method of chlorophyll determination in vitro is much more accurate once the influence of other fluorescing substances can be excluded (mostly chlorophyll b and pheopigments, Welschmeyer 1994).

depending on design and light source the reference path is not always present. This type of profiler can be operated from the surface down to a few hundreds or thousands of meters, depending on the model, and can usually be interfaced with a conductivity-temperature-depth (CTD) sensor.

3.1.3 Validity of the Underlying Assumptions

As mentioned above, this approach relies upon the assumption that the product ${}^{PS}\overline{a}^{*}_{PSII}\phi_F\Phi_f Q^{*}_{a}$ is constant.

The first source of variability is thus the cellular optical properties ${}^{PS}\overline{a}^{*}_{PSII}$ and Q^{*}_{a}. Figure 7 shows frequency distributions for ${}^{PS}\overline{a}^{*}_{PSII}$ and Q^{*}_{a} observed in open ocean waters along a transect in the South Pacific (BIOSOPE cruise). Both ${}^{PS}\overline{a}^{*}_{PSII}$ and Q^{*}_{a} show a factor of 2 to 3 variability. These variations result from changes in pigment composition and pigment packaging (Morel and Bricaud 1986; Bricaud et al. 2004); in the case of Q^{*}_{a} only chlorophyll *a* and divinyl chlorophyll *a* absorb significantly in the emission band (see Bricaud et al. 2004).

To the first order, ${}^{PS}\overline{a}^{*}_{PSII}$ and Q^{*}_{a} covary with [*chla*]. These statistical relationships are well represented by power laws, but significant dispersion occurs around the mean relationships. So, while at large scale, the variations in ${}^{PS}\overline{a}^{*}_{PSII}$ and Q^{*}_{a} may simply make the relationship between F and [*chla*] non-linear, under the assumption of constant $\mathring{E}(PAR)$ (Ostrowska et al. 2000a; 2000b) the large dispersion for a given [*chla*] implies that this relationship may change significantly regionally and at different times of the year with species successions. In the marine environment, the variability in ${}^{PS}\overline{a}^{*}_{PSII}$ has not been thoroughly investigated, but is likely similar or slightly less than the irradiance weighted chlorophyll-specific absorption of all phytoplankton pigments ($\overline{a}^{*}_{\varphi}\, m^{-1}$) which varies by roughly one order of magnitude (Bricaud et al. 1998, the BIOSOPE dataset likely underestimate this variability). The range of variations in Q^{*}_{a} is between ~0.2–1 at 685 nm (Babin et al. 1996b).

To ascertain whether variations in ϕ_F affect F vs. [*chla*], one can look at time series sufficiently short so it can be assumed that ${}^{PS}\overline{a}^{*}_{PSII}$ and Q^{*}_{a} do not vary significantly.

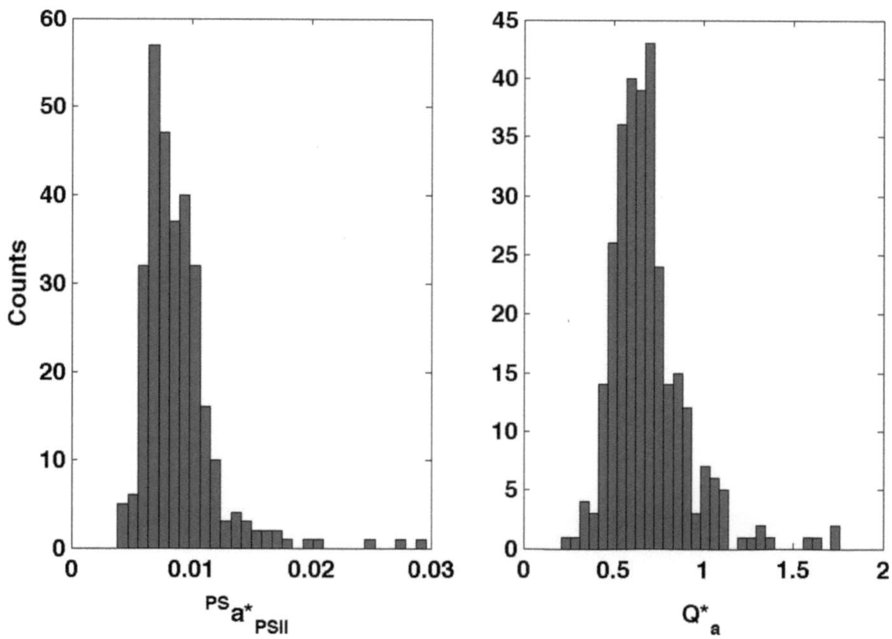

Fig. 7 Frequency distributions of ${}^{PS}\overline{a}^{*}_{PSII}$ and $Q^{*}_{a}(685)$ obtained for the BIOSOPE transect (see Claustre et al. 2008) which covered a wide range of eutrophic regimes in tropical and subtropical waters. The total photosynthetic absorption was calculated according to Babin et al. (1996) and divided by two to obtain the fraction from PSII. The irradiance used to weight the absorption was from a LED (centred at 452 nm) used in a FIRE fluorometer (Satlantic Inc. Halifax) on the cruise. The $Q^{*}_{a}(685)$ was calculated according to Eq. 7 where a_{sol} is computed considering the sum of absorption by chlorophyll *a* and divinyl chlorophyll *a* using the spectrum given in Bricaud et al. (2004). Values greater than 1 are theoretically impossible and reflect measurement errors

Using a bench top fluorometer, Kiefer (1973a, b) observed diel variations in F that spanned over more than a factor of 4. Loftus and Seliger (1975) and Abbott et al. (1982) found that F varies inversely with ambient irradiance at very short time scales (e.g. minutes) as a result of, for instance, cloud passage (Abbott et al. 1982). These typical variations in ϕ_F result mostly from non-photochemical quenching induced by changes in the ambient light. Measurements at night will reduce the influence of non-photochemical quenching, but they may be affected by other physiological processes.

The intensity and length of the excitation protocol and its impact on ϕ_F is another important aspect of *in vivo* chl *a* fluorometry (Cullen et al. 1988; Neale et al. 1989). Figure 8 shows a diagram of energy vs. duration for different excitation sources and protocols used in fluorometers. These source measure at different position along the chlorophyll fluorescence induction curve (see Section 2.10 and Strasser et al. 1995). In the case of the bench top fluorometer commercialized by Turner Designs Inc., the weakness of the source is such that fluorescence is generally measured with most reaction centers open ($\mathbb{R}_O \approx 1$) (Neale et al. 1989, but see section 3.6.2, "Use of DCMU").

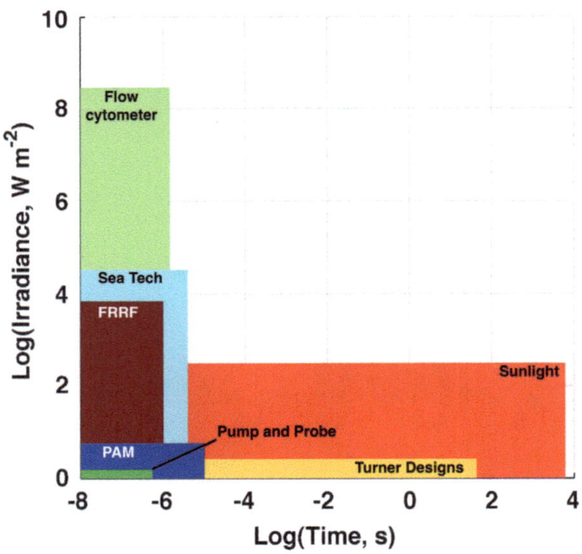

Fig. 8 Diagram of energy vs. duration for various excitation sources (adapted from Cullen et al. 1988). The energy and duration are for the exciting light in each protocol. Note that the Sea Tech fluorometer is an analog to the Aquatracka Mk III

Another critical aspect of *in vivo* fluorometry is the spectrum of the excitation source. It must cover a spectral range in the blue and green domains large enough to cover the PSII absorption bands of most phytoplankton groups. This is especially important for phycobilin-containing groups because the latter pigment absorbs in the green, rather than in the blue as chlorophylls and carotenoids do (see e.g. Beutler et al. 2003) and only a small fraction of the total cellular blue-absorbing pigments are connected to the fluorescing PSII.

Fluorescence from other pigments and from photosystem I (generally weak) will also contribute to different extent to the signal measured. The ratio of photosystem I to photosystem II (particularly high in cyanobacteria, see Mauzerall and Greenbaum 1989) and the presence of phycobilin which can contribute to the emission in the red will also influence the signal.

These important sources of variability thus stress the qualitative nature of even a calibrated instrument.[17]

3.1.4 Examples

To illustrate the kind of observations that can be made with fluorometers as well as the difficulties generally associated with their interpretation, two examples are presented. In the first, underway sub-surface measurements were carried out by Dandonneau and Neveux (1997) using a bench top fluorometer (Turner Model 112 equipped with a flow-through cuvette) onboard a ship of opportunity steaming from New Caledonia in the Pacific to Le Havre in France (through the Panama Canal). The observed fluorescence trace shows large regional variations in F related to changes in phytoplankton biomass (Fig. 9). However, it also shows more than 4-fold diel variations in F resulting from non-photochemical quenching. This pattern is especially clear in the Equatorial Pacific. On the bottom panel of Fig. 9, the daily F traces observed in the Equatorial Pacific superimpose nicely over each other

[17] We note that similar sources of variability are present for fluorometers designed to estimate the biomass of cyanobacteria (whether through phycobilosomes excitation or emission) in particular the presence of NPQ (Karapetyan 2007).

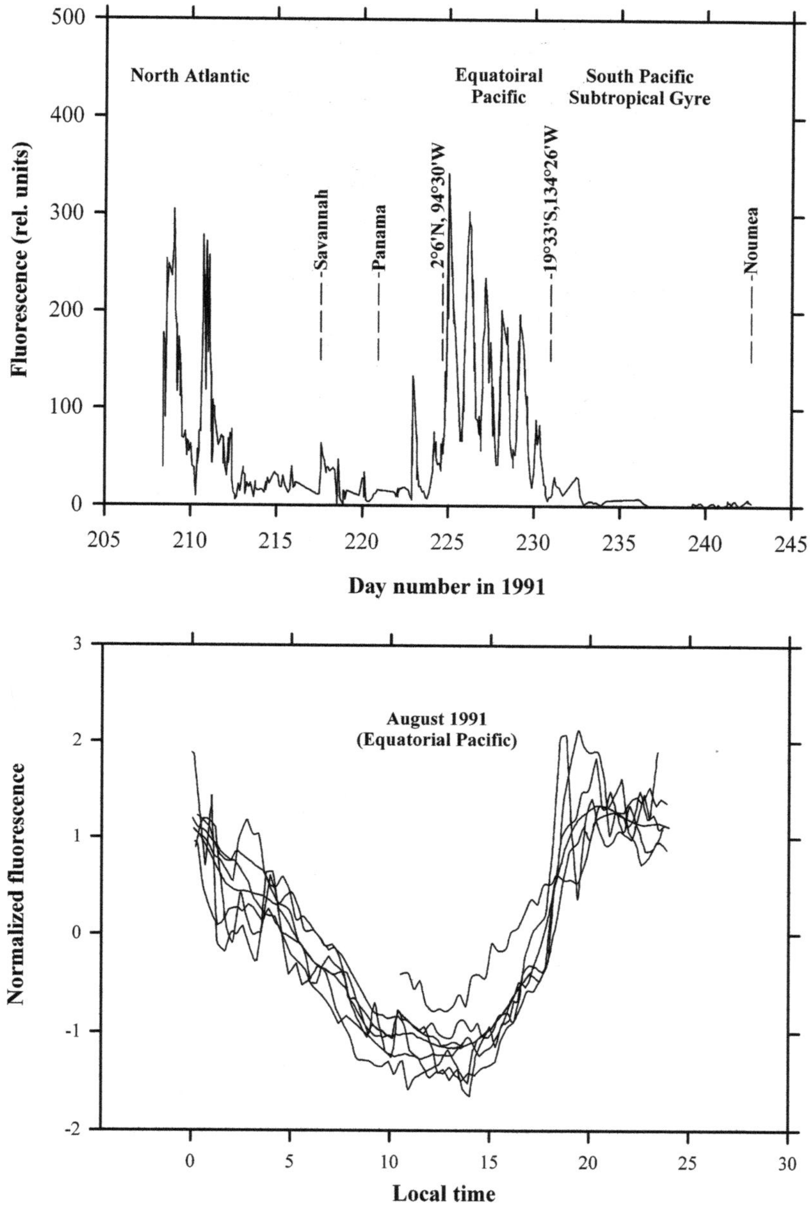

Fig. 9 Changes in surface *F* measured using a bench top fluorometer with a flow-through sampling system along a track between New Caledonia and Le Havre (France) (Redrawn from Dandonneau et al. 1997)

after normalization. They all exhibit a minimum around noon. Obviously, under such strong non-photochemical quenching, it can be difficult to distinguish spatial patterns from diel variation when their scales coincide.

In the second example, Claustre et al. (1999) examined the temporal variations in the vertical profile of fluorescence over five consecutive days in the subtropical Pacific (Fig. 10). These profiles were characterized by a deep chlorophyll maximum that varied in depth between 50 and 80 m over time because of internal waves, and by strong surface minima (with a decreasing impact down to ~50 m) around noon most probably resulting from NPQ. Obviously, the day and night fluorescence values were significantly (up to a factor of ~2) different close to the surface due to NPQ. Nevertheless, the position of the deep chlorophyll maximum could be monitored at high temporal and vertical resolution.

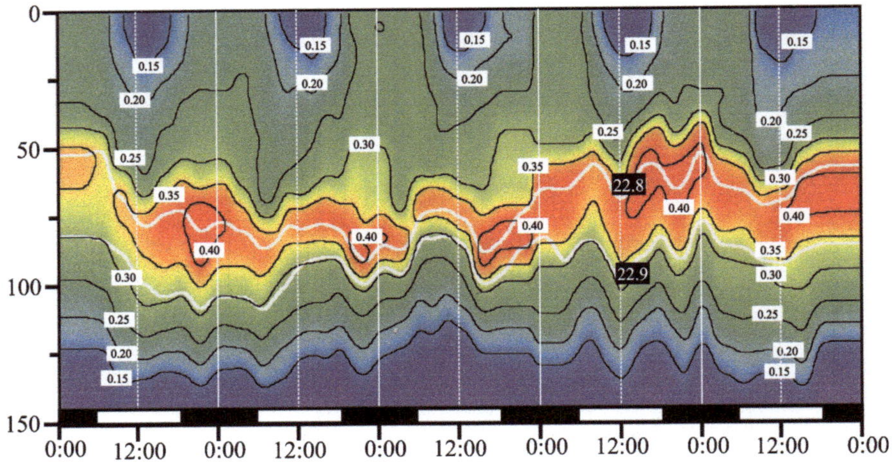

Fig. 10 Temporal changes over five days in the vertical profile of chl *a* fluorescence measured with an underwater fluorometer at 5°S 150°W in the Pacific ocean (redrawn from Claustre et al. 1999). Black contours indicate the output from the fluorometer. White contours represent the density (kg m^{-3}). The depth of the fluorescence maximum follows the variations in the depth of the density contours, indicating the presence of internal waves

These two examples illustrate the fact that horizontal and vertical patterns in phytoplankton biomass estimated from chl *a* fluorescence can be strongly affected by NPQ. Therefore, they must be interpreted with caution. Simultaneous observations of irradiance may help separate the fluorescence quenching effect from the biomass effect (e.g. Cullen and Lewis 1995; Behrenfeld and Boss 2006). In some cases the high spatial resolution obtained from these measurements can compensate for their inaccuracy as long as they are interpreted qualitatively.

It is worth noting that the calibration of fluorometers is especially problematic in the case of sustained observation on moorings over long periods of time, or on autonomous moving platforms (e.g. gliders, floats) over large distances. In this case, the experimenter must decide if it is better to (1) calibrate a fluorometer at time zero (for a mooring) or at initial station (as in a glider) and then interpret several months or hundreds of kilometres of variability to that single point or (2) to calibrate to a culture, a pigment extract or another neutral fluorophore, the conditions of which are controlled and repeatable? In the former case, the reference/calibration value will inevitably change between deployments and may not be valid for the whole time/transect. In the latter case, all deployments will be comparable, but absolute values will be approximate at best.

3.2 Spectrofluorometry

The objective of spectrofluorometry is generally to obtain information regarding the pigments that lead to fluorescence; often the interest is their absorption characteristics. It can also be used to study the emission of fluorescence.

3.2.1 Basic Principle

When fluorescence is measured at a given emission wavelength λ_{em} and excited using a monochromatic beam of irradiance $\mathring{E}(\lambda)$ and λ is varied throughout the visible range, we have from Eq. 2 that the changes in $F_{PSII}/\mathring{E}(\lambda)$ are proportional to spectral changes in ${}^{PS}a_{PSII}(\lambda)$. This proportionality rests on the assumption that the term $\eta_{PSII}(\lambda \rightarrow \lambda_{em})$ remains constant with wavelength.

The $F_{PSII}(\lambda)$ spectrum so obtained is referred to as an *excitation spectrum* (Fig. 1c dashed line). It is often used to infer the absorption spectrum of photosynthetic pigments (e.g. Mitchell and Kiefer 1988; Sakshaug et al. 1991; Sosik and Mitchell 1995; Johnsen et al. 1997). When applied on natural water samples it has the advantage of providing a signal specific to phytoplankton even in the presence of other absorbing substances (e.g. Desiderio et al. 1997).

Spectrofluorometry is also used to determine the spectral shape of fluorescence emission, generally referred to as the *emission spectrum* (Fig. 1a, b, c). In that case, the excitation wavelength is fixed ($\mathring{E}(\lambda)$ is constant), and the wavelength of the emission scanned to get $F_{PSII}(\lambda_{em})$. The emission spectrum allows detection of the presence of other fluorescing pigments in addition to chl *a* (e.g. Cowles et al. 1993; Chekalyuk and Hafez 2008). In vivo, phytoplankton pigments other than chlorophyll *a* with significant emission are limited to the phycobilins, mostly phycoerythrin, phycocyanin, and allophycocyanin that absorb light around 550, 620, 650 nm, and fluoresce around 575, 650, 660 nm, respectively. Cyanobacteria and cryptophytes are the main taxa containing phycobilins.

3.2.2 Instruments and Protocols

The optical setup of a standard commercial bench top spectrofluorometer is sketched in Fig. 11. It is generally composed of an exciting source coupled to a monochromator, a sample compartment for standard 1-cm glass cuvettes, and a monochromator coupled to a detector that collects the fluoresced light. A fraction of the exciting light beam is diverted to a reference detector. This reference beam allows the measurement of $E(\lambda)$ in relative units when a quantum counter is placed in front of the reference detector. A quantum counter is a dye solution that has the following properties : (i) it absorbs nearly all light in the spectral range of interest, (ii) it has a constant fluorescence quantum yield over the spectral range of interest, and (iii) it is stable over time. Because the reference detector measure the light fluoresced from the quantum counter at a fixed wavelength, the measured signal does not depend on spectral variations in the sensitivity of the detector. Moreover, because the quantum yield of the quantum counter is spectrally constant (i.e. it does not depend on the energy of photons), it provides a measure of $E(\lambda)$ in relative quantum units (see Hofstraat et al. 1992 for more details), which is appropriate for the normalization of chl *a* fluorescence. Common solutions used as quantum counters include Rhodamine-B, Basic Blue-3 and Oxazine-170. The latter two are especially convenient for phytoplankton studies because they allow measurement over a wider spectral range (255–700 nm) compared with Rhodamine-B (UV-600 nm) (Kopf and Heinze 1984). Note that to measure up to 700 nm, chl *a* fluorescence must be detected in the near-IR maximum around 730 nm (Neori et al. 1988). This has the consequence of increasing the influence of PSI (Franck et al. 2002). To obtain an emission spectrum spectrally calibrated in relative units, a standard lamp is generally used to derive the sensitivity spectrum of the emission detector (Neori et al. 1988).

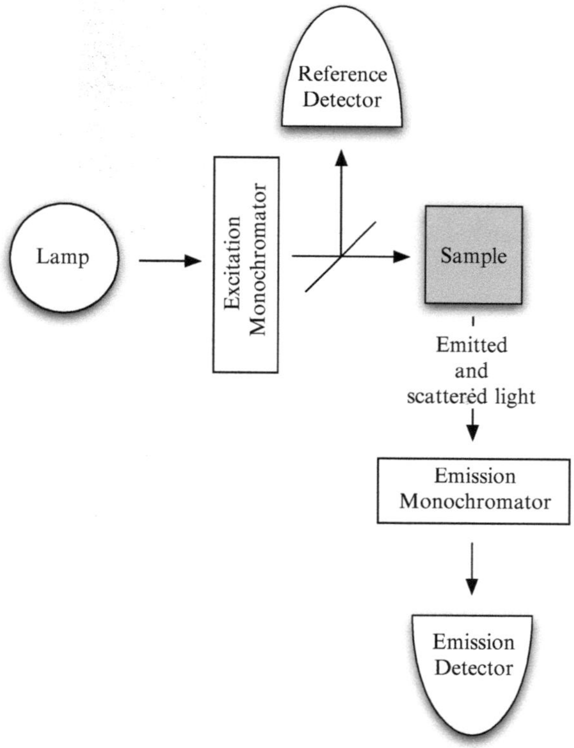

Fig. 11 The optical setup of a typical bench top spectrofluorometer. To measure excitation spectra, the emission monochromator is set at a constant waveband and the excitation monochromator is used to scan wavelengths. For emission spectra the roles of the monochromators are interchanged

3.2.3 Validity of the Underlying Assumptions

When using $F_{PSII}(\lambda)$ to estimate the absorption spectrum of photosynthetic pigments, it is assumed that the $\eta_{PSII}(\lambda \rightarrow \lambda_{em})$ is constant with respect to the excitation wavelength. As mentioned earlier this term represents the efficiency with which photons absorbed at λ will be reemitted at λ_{em}. It thus accounts for:

1. the quenching of fluorescence by photochemical and non-photochemical quenching;
2. the efficiency of transfer of energy from a photosynthetic pigment to the fluorescing chl a; and
3. the wavelength dependence of the emission

To reduce the variability associated with point 1, the measurement is made in the dark-regulated state to remove most of the short timescale non-photochemical quenching while the longer timescale change in q_I can be assumed constant during the time of the wavelength scan. Further manipulations are necessary to avoid the effect of photochemical quenching (different fractions of open reaction centres at different λ). This variability arises because the amount of excitons reaching the reaction centres vary with wavelength due to the spectral nature of $\mathring{E}(\lambda)$ and ${}^{PS}a_{PSII}(\lambda)$. To avoid such changes, a common approach consists of poisoning the sample with 3'-(3,4-dichlrophenyl)-1',1'-dimethyl urea (DCMU). DCMU prevents the re-oxidation of Q_A and, thereby, maintains all RCII closed. That way $\mathbb{R}_O = 0$ and F_m is measured (Neori et al. 1986). DCMU can be dissolved in methanol or ethanol and then introduced in the sample at a final concentration of 50–100 μM (from a concentrated stock solution to avoid the effect of the solvent on the cells). Variability associated with point 2 is generally not assessed and the transfer efficiency is assumed constant. Variability associate with point 3 arise if different excitation wavelength lead to different emission spectrum (this is also generally assumed constant but can be verified easily by creating an excitation-emission matrix).

There are two problems when inferring the absorption coefficient for photosynthetic pigments from the fluorescence excitation spectrum. The first one is that excitation spectra need to be expressed in absorption or specific absorption units [m^{-1} or m^2 (mg chl *a*)$^{-1}$, respectively]. To achieve this, Sakshaug et al. (1991), proposed to scale the excitation spectrum to measured a_ϕ at 675 nm. The underlying assumption is that $a_\phi(\lambda)$ in the spectral region of the chl *a* red absorption band represents only photosynthetic pigments. The second problem in using the excitation spectrum to infer photosynthetic pigment absorption is that the former reflects absorption mostly by pigments associated with PSII. As a result, the excitation spectrum may poorly represents absorption by all photosynthetic pigments (see the example below and Lutz et al. 1998, 2001) especially in the case of phycobilin-containing algae.

3.2.4 Examples

Figure 1c shows absorption and excitation spectra measured on *Cryptomonas sp.* by Sciandra et al. (2000). The excitation spectrum was scaled to absorption at 675 nm. In the blue region (up to 520 nm), $a_\phi(\lambda)$ is larger than absorption inferred from fluorescence probably because of the presence of non-photosynthetic carotenoids (α-carotene and/or alloxanthin). In the green region (520–600 nm), however, absorption inferred from fluorescence is much larger than $a_\phi(\lambda)$. This is certainly because *Cryptomonas sp.* contains large amounts of phycoerythrin only associated with PSII (see Prezelin and Boczar 1986). This example illustrates the difficulty in relying upon excitation spectra to infer absorption by all photosynthetic pigments, especially when dealing with phycobilin-containing species.

Figure 1d shows the emission spectrum of *Cryptomonas sp.* It illustrates very clearly the distinct fluorescence bands of phycoerythrin and chl *a*. The specific fluorescence emission properties of phycobilin-containing phytoplankton are also commonly exploited in flow cytometry, most often to distinguish cyanobacteria from other picoplankton species (see the review by Olson et al. 1991 and see Chapter 8 – Sosik et al.).

The use of excitation and emission fluorescence spectra to discriminate phytoplankton groups in natural waters has been first proposed by Yentsch and Yenstch (1979; see also Yentsch and Phinney 1985). Recently, the feasibility of this approach has been demonstrated by Beutler et al. (2002, 2004) and MacIntyre et al. (Chapter 7 this volume).

3.3 Sun-induced Chlorophyll Fluorescence

Spectra of downward and upward irradiance measured at depth in natural waters typically feature a peak around 685 nm (Fig. 12a). This peak corresponds to the *in vivo* emission of chl *a* fluorescence by phytoplankton (Morel and Prieur 1977; Neville and Gower 1977; Gordon 1979). It is also present in radiance and reflectance spectra. As depth increases, the amplitude of the fluorescence peak in the upward light increases relative to the background (Fig. 12b). This reflects the

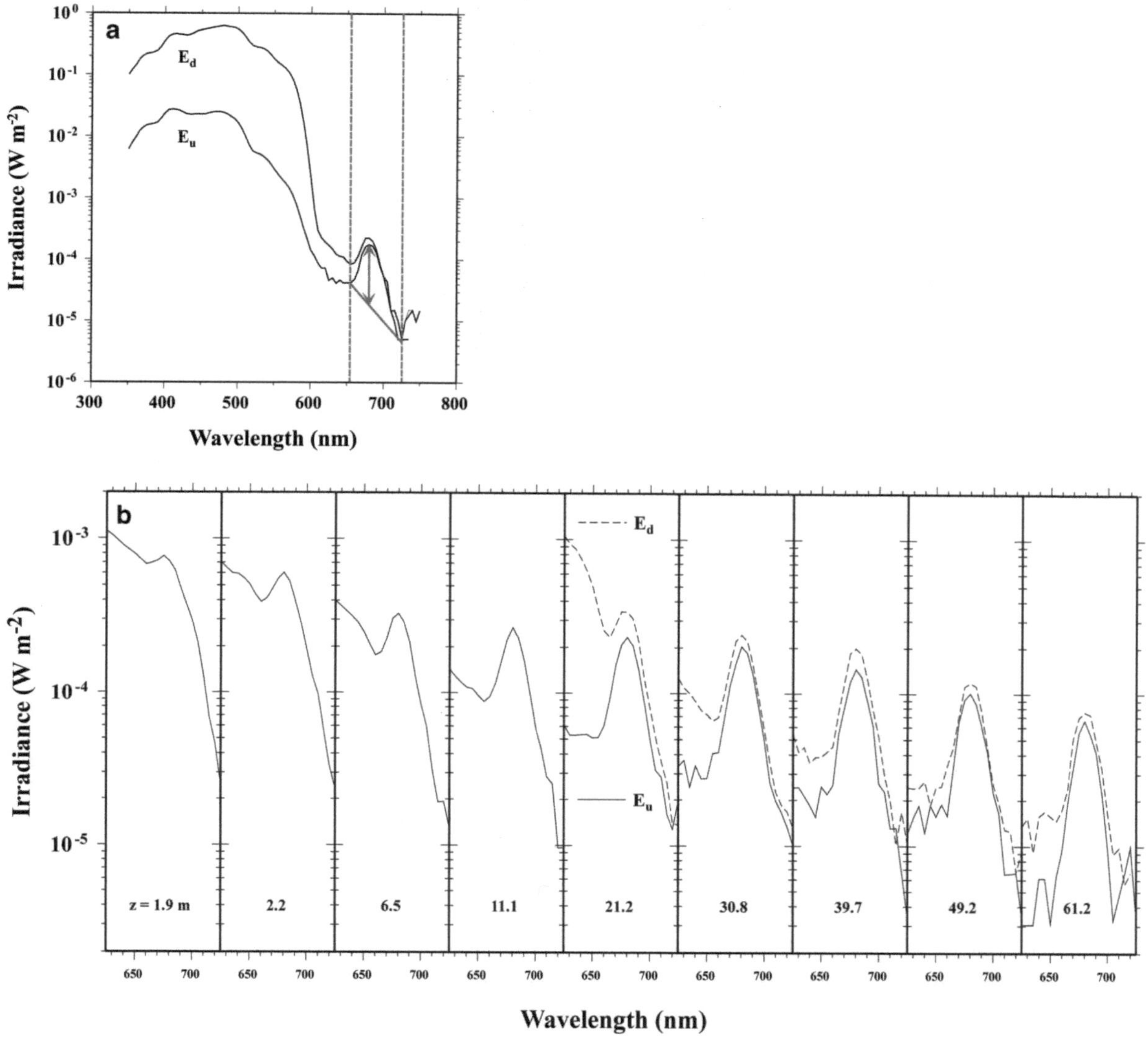

Fig. 12 Sun-induced fluorescence. (**a**) Examples of downward and upward irradiance spectra measured at 31.6 m in the sub-equatorial Pacific, with [*chla*] = 0.09 mg m^{-3} (reproduced from Maritorena et al. 2000). The solid red line depicts the baseline defined by $E_u(\lambda)$ at two adjacent wavelengths (vertical dashed lines), and the double-head arrow depicts the fluorescence line height (see text). (**b**) Examples of downward and upward irradiance spectra measured at different depths in the equatorial Pacific. The [*chla*] was nearly constant at 0.25 ± 0.02 mg m^{-3} over the whole depth range (reproduced from Maritorena et al. 2000)

relative amount of light originating from two different sources. Close to the surface, upward irradiance in the red region results from (1) backscattering of solar light by the water particulate matter contained therein, (2) chl *a* fluorescence and (3) Raman Scattering. With the first two being dominant. Deeper, because seawater itself has a high absorption coefficient in the red, the backscattered sunlight is much reduced, while chl *a* fluorescence is still present as a result of the much lower attenuation of light in the blue part of the spectrum where fluorescence is mostly stimulated. Despite the rapid extinction of the red sunlight, another source, Raman scattering by pure seawater, becomes important at depth. Raman scattering is an inelastic scattering process by which photons of shorter wavelength are reemitted at longer wavelength. The emission at 685 nm is the result of the excitation at 555 nm (green) where vertical attenuation is weaker (in most natural waters) than near 685 nm (Maritorena et al. 2000). Raman scattering accounts for up to 25–35% of total upward irradiance between 655 and 710 nm at depth (Maritorena et al. 2000).

The variations in the *in vivo* emission of chl *a* fluorescence by phytoplankton as observed in the *in situ* radiant field can be described by the following expression:

$$dL_F(z) = \frac{1}{4\pi}\overset{o}{E}(\mathrm{PAR}, z)[chl\,a]\bar{a}^*_\phi\,\phi_F^{app}\;Q_a^*\,e^{-\left[a_{tot}(\lambda_{em})+K(PAR)\right]\ell}\,d\ell \quad (19)$$

where $dL_F(z)$ is the part of radiance due to chl *a in vivo* fluorescence measured at nadir and depth *z* and originating from a thin layer of thickness $d\ell$ at distance ℓ below the sensor, $a_{tot}(\lambda_{em})$ is the total absorption coefficient of seawater over the wavelengths of chl *a in vivo* fluorescence, and *K(PAR)* is the vertical attenuation coefficient for scalar irradiance over the 400–700 nm spectral range. When integrating Eq. 19 over ℓ from 0 to ∞, the following equation is obtained:

$$L_F(z) = \frac{1}{4\pi}\overset{o}{E}(\mathrm{PAR}, z)[chl\,a]\bar{a}^*_\phi\phi_F^{app}\;Q_a^*\frac{1}{a_{tot}(\lambda_{em})+K(PAR)} \quad (20)$$

Here, it is assumed that [*chla*], $\bar{a}^*_\phi$, Q^*_a, and ϕ_F^{app} are vertically constant over the depth range seen by the sensor. This assumption is often valid because 90% of the signal comes from within 2 m below the sensor due to strong absorption by water in this waveband. But strong vertical variations in both [*chla*] and ϕ_F^{app} can be observed, especially close to sea surface in the case of ϕ_F^{app} as a result of non-photochemical quenching. Note that in Eqs. 19 and 20, we use a notation that is slightly different from that in Eq. 6 where the quantum yield is now the apparent quantum yield and accounts for all pigments in the cell. Similarly, the chlorophyll specific absorption accounts for all pigments. This formulation is consistent with past studies using sun-induced fluorescence and the limitation often imposed by the measurement (generally only $\bar{a}^*_\phi$ is accessible); the relationship between the two formulations follows from: $\bar{a}^*_\phi\phi_F^{app} = {}^{PS}\bar{a}^*_{PSII}\phi_F$. The total fluorescence emitted by PSII per unit volume from the thin layer at depth *z* (from Eq. 19) is equal to

$$F_{PSII} = 4\pi\frac{dL_F}{d\ell} \quad (21)$$

If $\bar{a}^*_\phi$, Q^*_a, and ϕ_F^{app} are spatially constant, and $a_{tot}(\lambda_{em})$ and *K(PAR)* are known [measured or expressed as a function of [*chla*] as is generally the case in open ocean (Case 1) waters], then :

$$[chl\,a] = f\left(\frac{L_F(z)}{\overset{o}{E}(\mathrm{PAR}, z)}\right) \quad (22)$$

L_F can be measured using underwater spectrometers.

The PNF-300 underwater spectrometer, commercialized by Biospherical Instruments Inc. (San Diego, CA, USA), was specifically designed for the measurement of so-called "natural" fluorescence (Kiefer et al. 1989). Actually, it measures vertical profiles of nadir radiance around 685 nm [L_u(685,z)] and of *PAR*. When using this instrument, it is assumed that below a few meters depth, $L_F \sim L_u(685)$. L_F can also be measured using the underwater spectroradiometers commercialized by different companies. The advantage of spectroradiometers over single-wavelength radiometers is that L_F (or irradiance and reflectance equivalents) can be extracted from the $L_u(\lambda)$ spectrum using a baseline (Maritorena et al. 2000) or a modelling (Morrison 2003; Huot et al. 2007) approach. The baseline approach (e.g. Neville and Gower 1977) is illustrated in Fig. 12. It simply consists in subtracting from $L_u(685)$ a background signal determined usually by linear interpolation between $L_u(\lambda)$ at two wavelengths on each side of the fluorescence peak. The resulting fluorescence signal is often referred to as the "fluorescence line height" (FLH). It is now applied to reflectance spectra measured from space using the MODIS (NASA) and MERIS (ESA) sensors (e.g. Gower et al. 2004). Fluorescence from space is attractive because, in many coastal, so-called "Case 2" waters, it provides a signal that is more specific to chl *a* than the traditional blue-to-green reflectance ratio. For Case 1 or open ocean waters waters, the good accuracy of standard algorithms using the phytoplankton absorption characteristics and the relatively high detection limit (between 0.07 and 0.5 mg chl m^{-3}, Babin et al. 1996a, with the lower estimates being somewhat optimistic; Letelier and Abbott 1996) of the fluorescence FLH reduces its scope.

3.3.1 Validity of the Underlying Assumptions

The major difficulties in using FLH to estimate chl *a* concentration are the following:

1. ϕ_F^{app} is highly variable in the natural environment. According to the observations of Maritorena et al.

(2000), Morrison (2003) and Schallenberg et al. (2008) in the Pacific, Atlantic and Bering Sea respectively, ϕ_F^{app} varies from ca. 0.005 to 0.05 mol emitted photons (mol absorbed photons)$^{-1}$ [with some values up to ~0.12 mol emitted photons (mol absorbed photons)$^{-1}$ observed]. The small values are generally observed close to surface when non-photochemical quenching is maximal. Therefore, the assumption of constant ϕ_F^{app} is obviously problematic.

2. It is well known that *K(PAR)* also varies significantly at sea. However, it can be easily measured *in situ*. For remote sensing applications, different approaches can be adopted to retrieve it, one of the most commonly used is based on the good relationship observed between $K_d(PAR)$[18] and [*chla*] (Morel 1988). Because of this good relationship, FLH remains well correlated with [*chla*] despite variability in $K_d(PAR)$. In Case 2 waters, however, $K_d(PAR)$ is poorly correlated with [*chla*]. Consequently, the large variations in *PAR*, and the changes in the spectral shape of $E(\lambda)$ that affect $\bar{a}_\phi^*$ (see Eq. 4, an equivalent equation can be written for $\bar{a}_\phi^*$), cannot be accounted for (Fisher and Kronfeld 1990). So, the variability in the apparent optical properties of Case 2 waters is such that the interpretation of FLH variations is more problematic. Furthermore, in Eqs. 19 and 20 *K(PAR)* is used as a surrogate for the spectral attenuation coefficient of the exciting light, this can be a good approximation in many cases. However, when *K(PAR)* is high relative to $a_{tot}(\lambda_{em})$, it is better to account for the spectral nature of light absorption by phytoplankton (see, Huot et al. 2005, 2007).

3. It has been shown by Gower et al. (1999) that the chl *a* fluorescence peak in reflectance spectra moves from 685 to more than 700 nm as [*chla*] and/or turbidity increase. This apparent shift in the fluorescence peak position results from the occurrence, in high-chl *a* and turbid waters, of an elastic scattering peak around 700 nm. Scattering by phytoplankton and suspended particles, which is more or less spectrally neutral over the limited spectral range between 680 and 700 nm, forms a peak because backscattered light is strongly absorbed below 700 nm by chl *a*, and highly absorbed above 700 nm by pure water. The problem may strongly compromise the baseline approach used to derive FLH (see Gower et al. 1999). To circumvent that problem, the solution may be a full inversion of the reflectance spectrum into inherent optical properties and fluorescence (Roesler and Perry 1995; Morrison 2003; Huot et al. 2007). Such an inversion is not, however, easily carried out on multispectral (instead of hyperspectral) data as available on satellite sensors.

3.3.2 Examples

Only a few studies have included true attempts to estimate [*chla*] using sun-induced chl *a* fluorescence. For instance, Kiefer et al. (1989) and Chamberlin et al. (1990) used the PNF-300 fluorometer (or a precursor of it) to measure $L_u(683)$ and PAR profiles. When comparing estimated with measured [*chla*] on limited data sets from Case 1 waters, they obtained a R^2 of around 0.9. This result is apparently very good. But it can be expected that when applied to a larger data set that covers several seasons and locations, this approach will provide poorer results. Further problems will also arise in coastal waters. Nevertheless, the FLH deserves further studies as it is a promising approach and, perhaps, the most interesting approach for remote sensing of [*chla*] in coastal waters. Recently, Gower et al. (2004) published a study suggesting (their Fig. 6) that [*chla*] can be retrieved within a factor of ca. 4 from the fluorescence signal measured by the satellite sensor MODIS. Huot et al. (2005) proposed algorithms to use the FLH signal measured by MODIS and applied them over the Pacific and Indian Ocean. In this study the chlorophyll retrieved using fluorescence (but also using information from the blue and green bands) was well correlated with the chlorophyll estimated solely from the blue and green bands. In coastal water, Huot et al. (2007) carried out a complete inversion of the reflectance and diffuse attenuation coefficient and could estimate the [*chla*] concentration in waters where ocean color methods based on phytoplankton absorption failed. Using aircraft overflights, with limited attempts for correcting the signal for the optical properties of the water, good correlation between FLH and in situ chlorophyll sampled at a few stations were found in two studies (Neville and Gower 1977; Sathyendranath et al. 2004). Good correlation were

[18] K_d(PAR) is the vertical diffuse attenuation coefficient for the downwelling planar irradiance. It is a good proxy of K(PAR) and much better described in the literature.

also found by Timmermans et al. (2008) between MODIS FLH and in situ match ups.

3.4 Flow Cytometry

Flow cytometry is used to measure optical properties of individual particles by shining a thin beam of light (generally a laser) on a fine stream of seawater. The stream is sufficiently fine that only one living particle (in the size range of interest) at a time intercepts the beam of light. Flow cytometry is generally used to study two properties: the fluorescence emission and the scattering of particles (at different angles). The latter is generally used as a proxy for the size of particles. The fluorescence emission allows one to gain information on the type of particles to be obtained as well as in the case of phytoplankton its physiology (even more so when fluorescent stains are used in conjunction). We only briefly touch upon the main uses of fluorescence here (see Sosik et al. (Chapter 8 this volume) in this book for a more thorough discussion of this method).

Most flow cytometers configured for work in the aquatic environment measure fluorescence in two spectral bands, one for the detection of chl *a* fluorescence and another for the detection of phycobilin fluorescence. The fluorescence measurement is thus used to separate phytoplankton from the rest of the particle population as well as separate phycobilin containing plankton (mostly cyanobacteria) from other cells. Separation of particle type along with the sizing capacity of the scattering measurement allows group specific studies to be carried out and to obtain information about cell specific properties.

It has among other uses been utilized to determine growth rates and division time of phytoplankton population in the sea (Vaulot and Marie 1999; Sosik et al. 2003). Flow cytometry has also been adapted to variable fluorescence technique (see section "Variable fluorescence") to obtain physiological information on individual cells (Olson et al. 1996).

3.5 Laser Excitation and LIDAR Fluorometry

Bulk fluorescence can be excited with lasers. The intensity, monochromaticiy and low divergence of laser beams make them particularly suited for remote sensing of fluorescing pigments in LIDAR (abbreviation for Light Detection and Ranging) systems (often more specifically referred to as laser fluorosensors). Because only a few groups have the means and know-how to utilize these systems, the technique has seen relatively little use compared to the other approaches described herein, we thus describe it only briefly here; it is nevertheless an attractive approach for rapid mapping of phytoplankton biomass fields (Hoge and Swift 1981). Lasers have also been used in benchtop applications (Exton et al. 1983) and, with appropriate inversion techniques, allows semi-quantitative separation of different fluorescing pigments based on their emission spectra (Chekalyuk and Hafez 2008), but this will not be discussed here.

The introduction of LIDAR technology for the measurement of phytoplankton concentration in water was made by Kim (1973). The LIDAR, essentially a RADAR for visible radiation, shines a laser beam into the medium and measures the returned signal. Just like in the case of Sun-induced fluorescence, the returned signal can arise from (1) backscattering by particles in seawater and the water itself, (2) from Raman scattering, and (3) from fluorescing substances. The distinct spectral signature of the different emissions allows relatively easy separation of these sources. The theory of the LIDAR is analogous to that of Sun-induced fluorescence (Browell 1977); one must account for both the attenuation of downwelling irradiance and of upwelling radiance to correct the signal and obtain a meaningful measure of biomass. Furthermore, as in any fluorescence measurement that utilizes in vivo fluorescence to estimate biomass it is influenced by the fluorescence quantum yield, it must thus be considered a semi-quantitative method.

Because of the monochromatic light used, an important aspect of LIDAR fluorometry is the possibility of using the Raman emission to partly correct the signal for the attenuation of light (Bristow et al. 1981). Indeed, because Raman emission originates only from water molecules, its emission for a given excitation irradiance is constant across any water type. However, the attenuation of the incident laser irradiance and upwelling Raman emission are the same as those of fluoresced photons if wavelengths are the same. Practically, with commonly used laser, Raman emission occurs at a different wavelength (this is also necessary to separate the signal). When the excitation is carried out at 532 nm (common frequency:doubled Nd:Yag laser), the emission is near 650 nm and it was found theoretically to provide a satisfactory correction (Poole and Esaias 1982) as the fluorescence and Raman emission are originating from similar integration depth (due to similar, and high attenuation of the upwelling light).

The laser fluorosensing technology has been utilized in conjunction with variable fluorescence techniques to obtain mapping of physiological parameters of phytoplankton (e.g. Chekalyuk et al. 2000).

3.6 Variable Fluorescence

This technique uses the fluorescence characteristics of the different states (i.e. open, closed, damaged, with and without NPQ) of a reaction centre to obtain information about the efficiency of different processes within the pool of reaction centres as well as on the fraction of reaction centres in these different states. More specifically five basic measurements can be distinguished[19,20]:

1. F_o – The fluorescence in the dark regulated state when all undamaged reaction centres are in the open state and the fast components of non-photochemical quenching are relaxed (this is the O level in the induction curve).
2. F_m – The fluorescence in the dark regulated state when all undamaged reaction centres are in the closed state and the fast components of non-photochemical quenching are relaxed (equivalent to the P level in a high-light induction curve).
3. F_o' – The fluorescence in the light regulated state when all undamaged reaction centres are in the open state.
4. F' – The fluorescence in the light regulated state (generally at steady state) when the reaction centres are in a mix of closed, opened and damaged states.
5. F_m' – The fluorescence in the light regulated state when all undamaged reaction centres are in the closed state.

An hypothetical representation (as would be measured with a PAM fluorometer – see section 3.6.2, "Pulse amplitude modulation") of these different parameters are presented in Fig. 13b. The instruments and protocols presented in upcoming section "Instruments and protocols" differ in their ability to observe these different states and by the approach they utilize to manipulate the reaction centres to reach these states.

[19] Here instead of defining these five measurements with respect to a protocol, we define them with respect to their theoretical meaning.

[20] The fluorescence parameters in the recovery phase following illumination by an actinic light are usually denoted with double primes (e.g. F_m'') this will not be discussed herein.

3.6.1 Basic Principle

The maximum (ϕ_{Fm}) and minimum (ϕ_{Fo}) quantum yields of fluorescence for undamaged ($\mathbb{R}_{dam} = 0$), dark regulated state ($k_{qe} = 0$) reaction centres are expressed as:

$$\phi_{Fm} = \frac{k_F}{k_F + k_D} \tag{23}$$

$$\phi_{Fo} = \frac{k_F}{k_F + k_D + k_P} \tag{24}$$

When replacing ϕ_F by ϕ_{Fm} and ϕ_{Fo} in Eq. 5 to obtain the expressions for F_m and F_o, respectively, the following equation can be derived:

$$\frac{F_v}{F_m} = \frac{\phi_{Fm} - \phi_{Fo}}{\phi_{Fm}} = \frac{k_P}{k_D + k_F + k_P} = \phi_{PO}^{\max} \tag{25}$$

where F_v is variable fluorescence defined as $F_m - F_o$, and $\phi_{PO}^{\max}$ is the maximum quantum yield for energy conversion in RCII (see Eq. 14). Two fluorescence measurements (F_m and F_o) in the dark regulated state thus provide direct information about the efficiency of photosynthetic processes at PSII.

In the presence of damage, a reduction of F_v/F_m is observed. Using the same approach as above we find

$$\frac{F_v}{F_m} = \phi_{PO}^{dam} = \phi_{PO}^{\max}\left[\frac{1-\mathbb{R}_{dam}}{1-\mathbb{R}_{dam}\left(1-[k_F+k_D]/[k_F+k_D+k_I]\right)}\right] \tag{26}$$

where ϕ_{PO}^{dam} is the maximum quantum yield in the presence of damage. We thus find that F_v/F_m is, according to our model, a non-linear function of the fraction of damaged reaction centres[21] and thus does not strictly provide an estimate of the quantum yield of charge separation of open reaction centres in the presence of damage (consistent with observation in higher plants Park et al. 1995).

[21] In the hypothetical case of non-quenching damaged reaction centers ($k_I = 0$), F_v/F_m becomes a linear function of the fraction of damaged reaction centre.

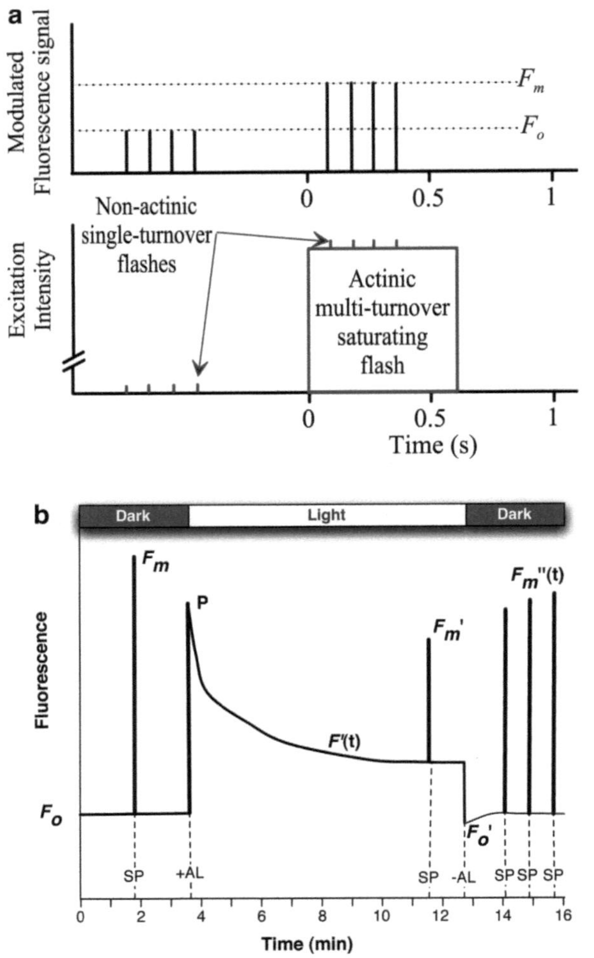

Fig. 13 PAM protocol. (**a**) Measurement protocol applied to measure F_o and F_m with the PAM fluorometer. *Top panel*, the fluorescence signal resulting from modulated dim flashes, as filtered out by the electronics of the PAM from the total fluorescence signal which results from both the dim flashes and actinic light. *Bottom panel*, the excitation intensity, showing the probe flash and the actinic flash; note the break in the y-axis, used to represent the very high actinic intensity relative to the probe flash. The modulated electronics measures only the fluorescence emitted from the probe flash which changes by a factor of ~2-3 due to changes in photochemical quenching, while the total fluorescence emitted is proportional to the excitation intensity which changes by several orders of magnitude. (**b**) Measurement of basic fluorescence parameters in the light and dark regulated states. In the first dark period (0–3.5 min) the measured fluorescence level with a dim exciting light consists of F_o. Upon closure of the reaction center (using a saturation pulse; noted SP), the fluorescence measured (still with a weak light) reaches F_m. Upon turning the actinic light on (note +AL) the fluorescence reaches a maximum denoted P, the fluorescence, now noted F'(t) then decreases to a steady state value. At this steady state value, a saturation pulse is applied and the fluorescence reaches the level F'_m. Upon turning the actinic lights off (-AL), the fluorescence reaches the F'_o level. While the sample is in the dark, application of saturating pulse at constant interval allows the measurement of F''_m. During this treatment, the fluorescence is always measured with the same dim light source. See text in "Pulse amplitude modulation"

In the undamaged case, in the light regulated state the following equations can also be obtained from a similar approach:

$$q_p = \frac{F'_m - F'}{F' - F'_o} = \mathbb{R}_O \tag{27}$$

This parameter is called q_p as it is related to photochemical quenching and is equal to the fraction of closed reaction centres (for unshared antenna). By fitting Eq. 17 (or another induction curve model e.g. Trissl and Lavergne 1995) to experimental $\mathbb{R}_O$ vs. ε data obtained using Eq. 27, one can derive $\bar{\sigma}'_{PSII}$. Another useful relationship is

$$\phi'_p = \frac{F'_m - F'}{F'_m} = \mathbb{R}_O \frac{k_p}{k_F + k_P + k_D + k_{qe}} = \mathbb{R}_O \phi'_{PO} \tag{28}$$

This parameter first derived by Genty et al. (1989) is often referred to as the "Genty yield". It provides the realized quantum yield of charge separation (see Eq. 15).

When damage is taken into account, the relationship becomes:

$$\phi_p^{dam'} = \frac{F'_m - F'}{F'_m}$$

$$= \mathbb{R}_O \phi'_{PO} \left[\frac{k_i + k_D + k_F + k_{qe}}{k_I \left(1 - \mathbb{R}_{dam}\right) + k_D + k_F + k_{qe}} \right] \tag{29}$$

where $\mathbb{R}_O$ and $\mathbb{R}_{dam}$ are the fractions of open and damaged reaction centres in the F' state. Again, it is found that, when damage is present, this measurement does not provide a direct estimate of the realized quantum yield of photochemistry.

These relationships hold only for the puddle model. When antenna are shared some of these equations require modification for proper interpretation (e.g. Havaux et al. 1991; Lavergne and Trissl 1995; Kramer et al. 2004). Nevertheless, they provide a good basis for the interpretation of current protocols as they represent one of the simplest scenarios[22]: if current protocols do not capture the complexity of these relationships, it is unlikely that they will be able to deal with different levels of antenna sharing. Many more fluorescence parameters have been published (see for examples Cosgrove and Borowitzka (Chapter 1) and Rohácek and Barták 1999 for a more complete

[22] The fully connected case also reduces to relatively simple expressions, while intermediate levels of connectivity, probably more true of the reality, are more complex.

list of parameters used), especially for examining non-photochemical quenching; these are not discussed here.

There are now a number of measurement protocols and commercial instruments that allow the experimental determination of these parameters on natural phytoplankton samples. In limnology and oceanography, ϕ_{PO}^{max} has been used as an index of photoinhibition (i.e. photodamage; see Eq. 26 for rationale) and nutrient stress (suggested to increase the fraction of damaged reaction centres due to a reduction in the repair capacity, see Marshall et al. 2000 and references therein). Changes in $\overline{\sigma}_{PSII}$ are expected for certain type of photoacclimation (e.g. Prézelin 1981; Suggett et al. 2004; Moore et al. 2006) as well as with nutrient stress (Suggett et al. 2009). Together they are also used to estimate primary production (Kolber and Falkowski 1993; Suggett et al. 2001; see also Suggett et al. (Chapter 6)).

3.6.2 Instruments and Protocols

Use of DCMU

As mentioned above, DCMU prevents the re-oxidation of Q_A and, thereby, maintains all RCII closed under incident irradiance, in which case $\mathbb{R}_O = 0$ and F_m or F_m' is measured depending on whether the cell is in the dark or light regulated state. In the late 1970s and early 1980s, several groups started to use DCMU on natural phytoplankton samples in order to derive variable fluorescence (e.g. Cullen and Renger 1979; Roy and Legendre 1979; Vincent 1981). The protocol is generally as follows. A first measurement is made after leaving the sample for ~30 min in the dark in order to relax energy-dependent and state-transition NPQ (q_E and q_T). This measurement is carried out under sufficiently low excitation irradiance to avoid measuring fluorescence from closed[23] reaction centres and represent F_o. A second measurement is then made after the addition of DCMU to obtain F_m.

In the case of low-light acclimated phytoplankton cells, it is safer to verify that F_o is actually measured. This can be done by conducting a few measurements of F_o and F_m on subsamples while increasingly attenuating the excitation light source of the fluorometer using neutral filters. The maximum light intensity providing the correct estimate is found when the calculated F_v/F_o stop increasing by further attenuating the fluorometer excitation light source.

Pulse Amplitude Modulation

The pulse amplitude modulation (PAM) fluorometer was introduced by Schreiber et al. (1986). The measurement protocol typically followed to determine F_o and F_m using the PAM fluorometer is sketched in Fig. 13a and leads to a fluorescence trace as shown in Fig. 13b. Prior to the measurement, the sample is generally dark regulated for ~30 min to relax q_E and q_T. The sample is then exposed to a short series of weak and short (~10 μs) flashes. These flashes often referred to as "probe flashes", do not induce any photochemical quenching or NPQ and, therefore, provide a measurement of F_o. The sample is then exposed to a saturating actinic flash lasting around 600 ms and thus leading to multiple photosynthetic events (or turnovers), that is multiple successive charge separations at each RCII. During the actinic flash, the sample is, simultaneously, exposed to probe flashes. The modulation electronics of the PAM fluorometer is such that it only records the part of the fluorescence induced from the probe flash. This measurement corresponds to F_m. Therefore the PAM measures only the changes in the quantum yield of fluorescence and not the absolute change in fluorescence. The same measurements can also be carried out in the light (see Fig. 13b) providing a continuous measurement of F' and, upon the use of an actinic flash, F_m'. Using these five measurements a series of parameters can be derived to examine the different processes within PSII (see for e.g. Eqs. 25 and 28 and Rohácek and Barták 1999)

Pump-and-Probe

Pump-and-probe (P&P) fluorometry was introduced by Mauzerall (1972), and then intensively used to study phytoplankton photosynthesis in the laboratory and at sea by Falkowski and colleagues (e.g., Falkowski et al. 1986; Kolber et al. 1990).

The most significant differences between the P&P and the PAM protocols are that P&P: (1) uses of a

[23] It takes at least two photons per reaction centre to measure the fluorescence of a closed reaction center; the first photon closes the reaction centre and the second measures or "feels" its closed state. At sufficiently low irradiance the second photon arrives after the reaction centre has reopened.

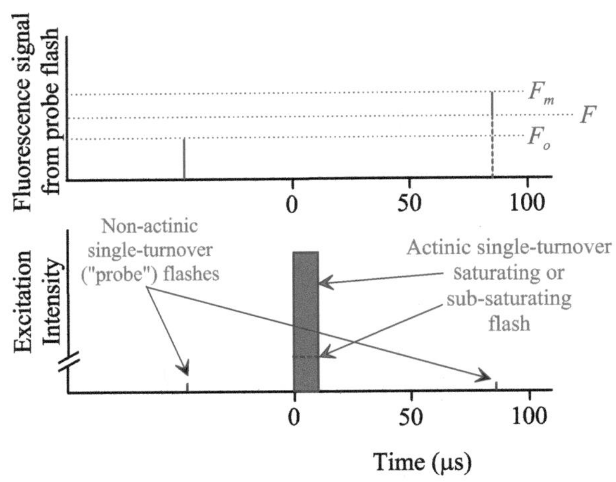

Fig. 14 Diagram describing the pump-and-probe measurement protocol to measure F_o, F and F_m. See text in section "Pump-and-probe"

single-turnover actinic flash, and (2) applies actinic flashes of variable energies (Fig. 14). First, the measurement of F_o is made as described above for the PAM fluorometer. Then the sample is exposed to a short (~10 μs) saturating actinic flash (so-called "pump" flash), and a probe flash is applied ca. 80 μs later to measure F_m. The 80-μs delay is to allow the relaxation of a short lived quenched state resulting from the pump flash (see Falkowski and Kiefer 1985 and references therein). Beyond that delay, Q_A begins to be reoxidized by the transfer of an electron to Q_B. The pump flash produces a single turnover because it is shorter than 80 μs thus very little reoxidation of Q_A takes place. By varying the energy of the pump flash, F can be measured at intermediate levels between F_o and F_m therefore $\bar{\sigma}_{PSII}$ can be derived using the P&P protocol (see Eqs. 17 and 27).

Fast Repetition Rate

Data from the fast repetition rate (FRR) fluorometry protocol were first published in 1994 (e.g. Greene et al. 1994; Kolber et al. 1994) but the technique was described in details by Kolber et al. (1998). This protocol was developed as an alternative to P&P for fast measurements of the five basic fluorescence components (F_o, F_m, F', F'_m, F'_o) as well as $\bar{\sigma}_{PSII}$ and an estimate of the connectivity between antennas. It is especially appropriate for high frequency measurements *in situ*.

The FRR protocol is sketched in Fig. 15a. The sample is exposed to a series (e.g. 100) of short (~1 μs) and intense[24] (see Fig. 8) consecutive flashes (often referred to as flashlets), at a frequency of about 0.35 MHz, over an interval of about 280 μs. These flashes serve both to: (1) progressively populate the RCII (i.e. reduce Q_A), and (2) probe the fluorescence signal. After this first 280-μs period, the sample is exposed to a series (e.g. 20) of flashes at lower frequency (~ 20 kHz) to determine the kinetics of Q_A reoxidation by the plastoquinone pool. Because little work has been carried out on these kinetics, we will not discuss them further herein. An example of (hypothetical) FRRF measurement is shown in Fig. 15b. If the measurement is made in the dark regulated state, F_o, F_m and $\bar{\sigma}_{PSII}$ are obtained by fitting a model described in Trissl and Lavergne (1995) and adapted by Kolber et al. (1998) to the experimental data. If the measurement is made in the light, F' and F'_m are obtained; the value of $\bar{\sigma}'_{PSII}$ is in practice difficult to obtain except under low incident light.

Generally, *in situ*, concurrent measurements are made on water flowing through a dark chamber and a chamber exposed to ambient irradiance. In both cases, there is no time for the algae to dark-regulate. Therefore, F' and F'_m are measured in the light chamber and F'_o and F'_m are measured in the dark chamber. These measurements, of course, include all non-photochemical quenching processes.

A similar protocol called "pump during probe" was used by Olson and Sosik (1996) which uses a continuous flash to excite the fluorescence instead of a series of short flashes as in the FRR protocol and allows the same parameters to be derived. This protocol has been adapted to single cell analysis for flow cytometry and microscopy (Olson et al. 1996).

3.6.3 Validity of the Underlying Assumptions

The FRRF protocol relies on fitting an appropriate model to the fluorescence induction curve (e.g. data Fig. 15). Equation 17 is one such model and is valid under the assumption that no transfer of excitons

[24] The advantage of the high intensity of the flash is that fluorescence can be measured in low chlorophyll concentration as the fluorescence is proportional to the excitation light (see Eq. 2) while the short duration makes the flashlet subsaturating.

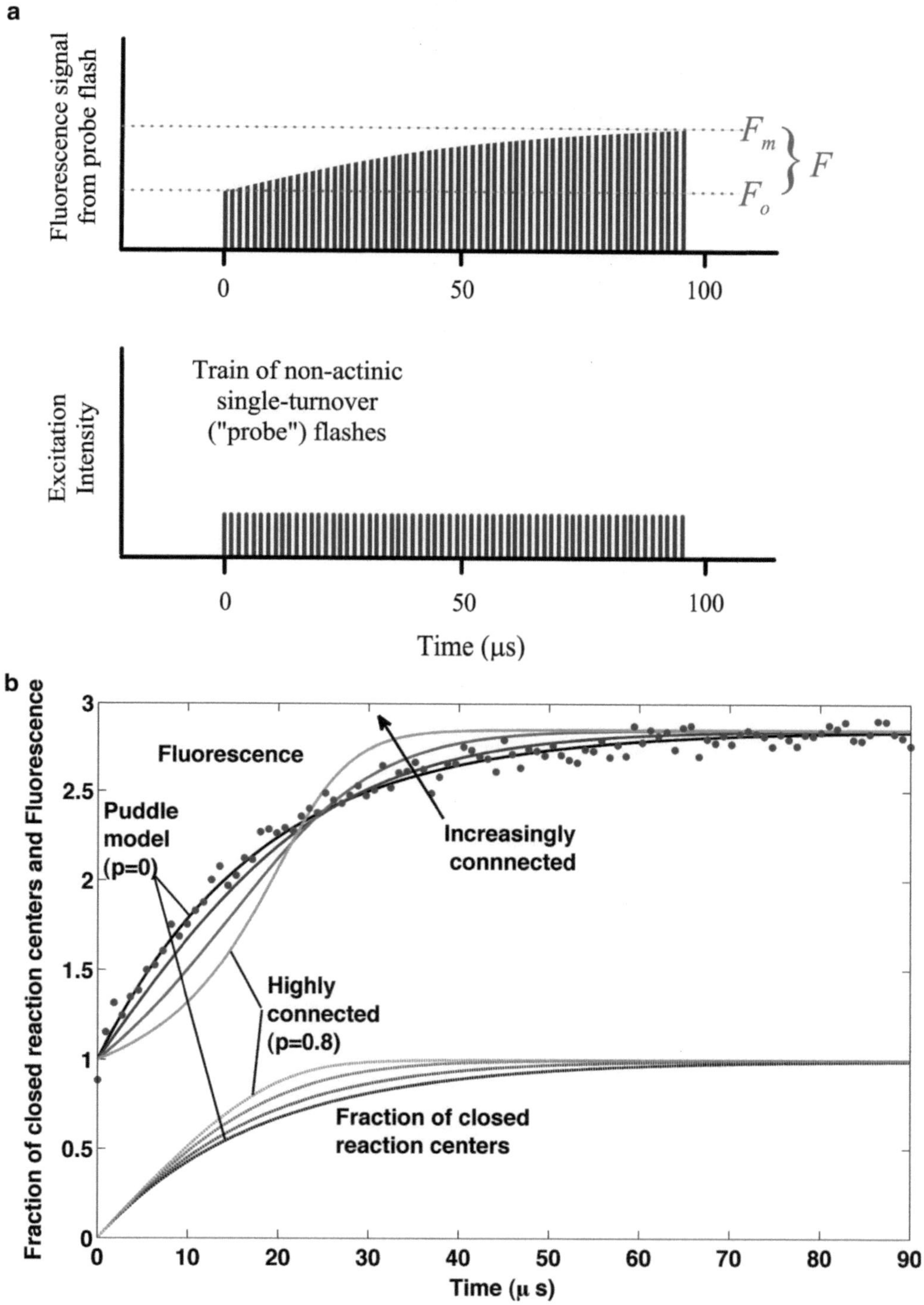

Fig. 15 Fast repetition rate fluorescence protocol. a) Diagram describing the FRR measurement protocol. See text in section "Fast Repetition Rate". b) Fluorescence induction and fraction of closed reaction centers as a function of time based on Eqs. 1 and 3 in Kolber et al. (1998). The variable p represents the interconnectivity parameter and is allowed to vary between 0 and 1 with the latter representing one connected matrix of pigments serving all reaction centers. The data point simulates the fluorescence signal as would be measured for each probe flash for a puddle model (modelled signal with added random error). The abscissa indicates the time elapsed after the first flash

occurs from closed to open RCII.[25] When such transfers occur[26] (so called "interconnectivity"), F_o and F_m do not change, but F becomes increasingly sigmoidal as connectivity increases (see Fig. 15b). Consequently, Eq. 17 cannot be fit to the experimental data to derive $\bar{\sigma}_{PSII}$. The model used by Kolber et al. (1998) to fit the FRR data includes an interconnectivity parameter, p, that theoretically varies between 0 (no interconnectivity) and 1 (full interconnectivity). An adequate retrieval of this parameter when fitting Kolber et al.'s model to experimental FRR data is important, as it affects the retrieval of all other parameters (F, F_o, F_m and $\bar{\sigma}_{PSII}$). When the saturation curve derived by FRR fluorometry (Fig. 15b) is noisy, as is often the case in low-chl *a* environments, the retrieval of p becomes difficult. To circumvent this problem, Suggett et al. (2001) chose to fix this parameter at a value of 0.3. A sensitivity analysis where different fixed p (0 to 0.6) values have been adopted for processing actual FRRF data from the Mediterranean Sea, suggests that the use of different fixed p values leads to variations in F_v/F_m of 20% on average (M. Babin, 2002). Because the actual variability of p in the natural environment is unknown, it is presently unclear whether or not the use of a fixed p leads to significant error in the retrieval of the other fluorescence parameters. When biomass is high as in phytoplankton culture, p is retrieved with good confidence, and F_v/F_m compares very well with that measured using pump-and-probe fluorometry (Bruyant et al. 2005).

Laney (2003; see also Laney and Letelier 2008 for a more theoretical framework) published a detailed study on the different sources of experimental and computational errors associated with the data provided by the FASTtracka FRR fluorometer commercialized by Chelsea Technologies Group. He emphasized two major problems: (1) the instrument response function[27] (IRF) varies from one gain to another, and (2) the data processing software provided by the manufacturer produced systematic biases in the retrieval of the fluorescence parameters (F_v/F_m, $\bar{\sigma}_{PSII}$, p and τ_Q). To solve the first problem, the IRF must be characterized using a dye solution separately for each of the five built-in gains. To resolve the second issue, Laney (2003) developed and successfully tested a new processing software which has been made publicly available.

While phaeophytin is typically present at low levels compared to chl *a* in open ocean waters (Trees et al. 2000), this is not the case for coastal waters. Phaeophytin contributes to fluorescence detected around 685 nm. More specifically, it contributes to F_o and F_m, but not to F_v as it is not involved in photosynthesis. As a result, the presence of phaeophytin in seawater leads to an underestimate of $\phi_{pO}^{\max}$ (see Eq. 25). Fuchs et al. (2002) have shown than when the phaeophytin to total pigment ratio is 0.7, a realistic value in coastal waters, $\phi_{pO}^{\max}$ can be underestimated by 25%. Fuchs et al. (2002) were able to predict the impact of phaeophytin on $\phi_{pO}^{\max}$ based on seawater pigment composition. They also found that the measurement of $\bar{\sigma}_{PSII}$ is not affected by the presence of phaeophytin. A similar effect is found for CDOM fluorescence (Chekalyuk and Hafez 2008).

3.6.4 Examples

Figure 16 shows the changes in variable fluorescence (in this case expressed as F_v/F_o, and denoted $\Delta\phi_{sat}$) over a vertical section located through an eddy in the subtropical Pacific. This figure also shows the distribution of chl *a* concentration, temperature and nitrate concentration. These results were published by Falkowski et al. (1991) who argued that the highest variable fluorescence values observed close to the surface around 25 km result from the eddy pumping of nitrate to the surface. Higher chl *a* concentration and lower temperature were also observed in that area.

The FRRF has also been central to verifying the so-called iron hypothesis which states that in large areas of the ocean phytoplankton is limited by iron. Behrenfeld et al. (1996) interpreted changes in

[25] This corresponds to the "puddle" model.

[26] This corresponds to the connected model and when fully connected often referred to the "lake" model.

[27] The Fasttracka fluorometer shows a systematic artefactual transient in the measured fluorescence signal over the series of ~100 flashes used to close reaction centres. This artefact can be observed using a fluorophore solution that shows no variable fluorescence (e.g. extracted chl *a*, rhodamine B).

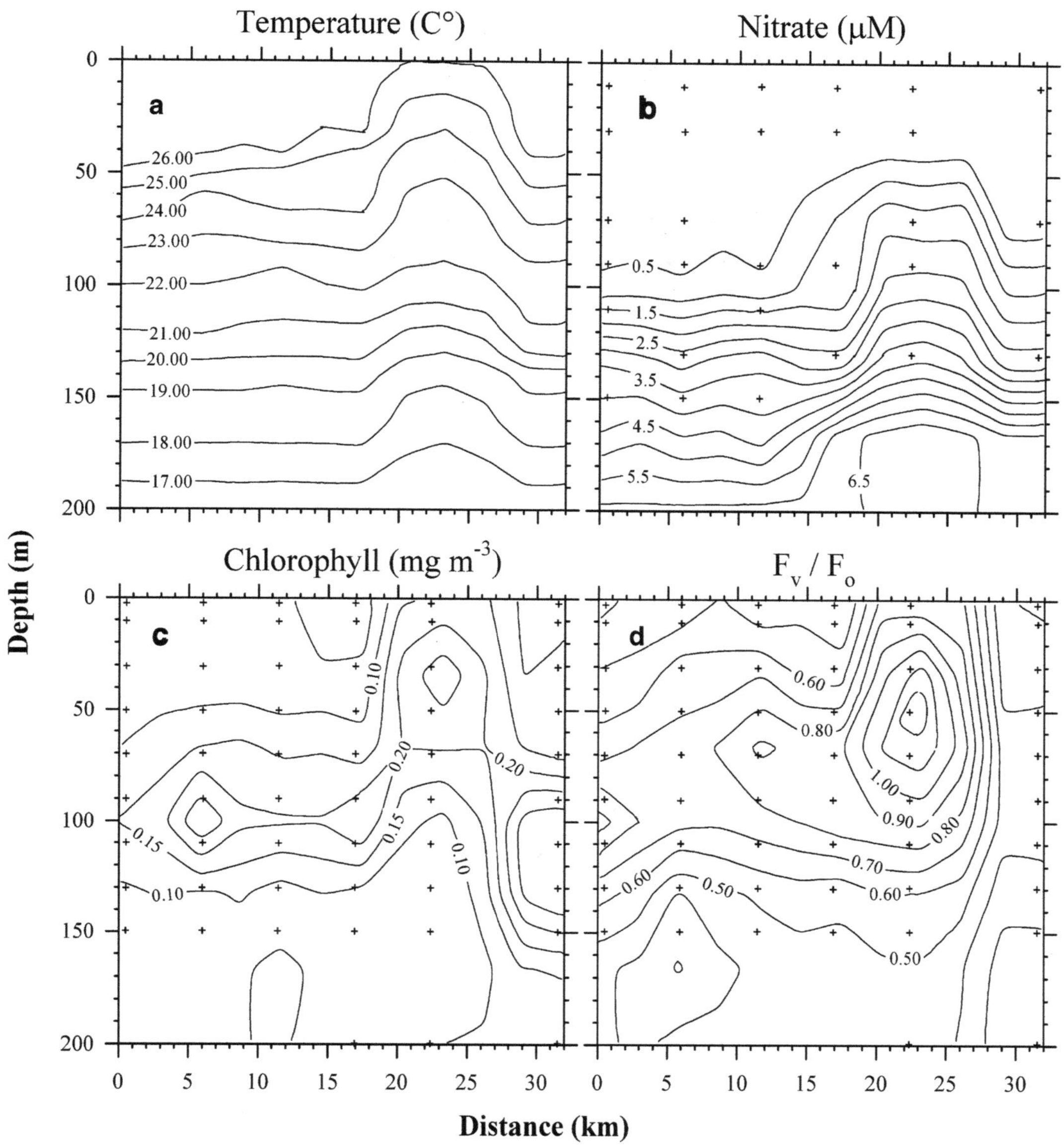

Fig. 16 Changes in temperature, nitrate concentration, chl *a* concentration, and variable fluorescence (expressed here as F_v/F_o) over a vertical section across an upwelling eddy in the subtropical Pacific (Reproduced from Falkowski et al. 1991)

variable fluorescence observed in the Pacific and Southern Ocean as an indication of iron limitation. More recently, a study by Behrenfeld et al. (2006) using special diel characteristic of variable fluorescence (Behrenfeld and Kolber 1999) present in iron limited waters, mapped regions with different levels of iron stress.

The relationship between variable fluorescence and nitrate concentration has been documented by several laboratory and field studies (Kolber et al. 1988; Kolber et al. 1990; Babin et al. 1996a) while others could not find any trend (Olaizola et al. 1996). In a laboratory study using a species of diatoms, Parkhill et al. (2001) have shown that under nitrogen limited balanced growth[28] F_v/F_m remains maximal (see also Cullen et al. 1992; Suggett et al. 2009). Parkhill et al. (2001) suggested that the decrease in variable fluorescence under nitrogen stress may occur only under unbalanced growth (which they refer to as nitrogen starvation, see also Cullen et al. 1992). If this observation can be extended to other nutrients, species, and in situ growth conditions, the new paradigm would thus suggest that F_v/F_m is an indicator of nutrient starvation rather than limitation (see Shelly et al. (Chapter 11) for more details on the use of fluorescence as a measure of nutrient stress). More recently, Suggett et al. (2009) have shown strong species-specific characteristics of variable fluorescence parameters (mostly F_v/F_m and $\bar{\sigma}_{PSII}$) and warned against a strictly physiological interpretation of these parameters at sea.

4 The Use of Chlorophyll Fluorescence to Estimate Primary Production

4.1 *Variable Fluorescence*

4.1.1 If $\bar{\sigma}_{PSII}$ is Available (FRRF and Pump and Probe Protocol)

Kolber and Falkowski (1993) proposed a model for estimating primary production (i.e. the instantaneous gross rate of carbon fixation by phytoplankton) from variable fluorescence information obtained using *in situ* pump and probe fluorometry. This protocol is also applicable to FRR fluorometry. The model is based on the following rationale: the rate carbon fixation (P_c) can be estimated if (1) the rate of charge separation (P_e) can be obtained from fluorescence; (2) the ratio of oxygen evolution to charge separation (γ_e^{O2}) is known; and (3) the ratio of oxygen evolution to carbon fixation is known (PQ, so-called photosynthetic quotient):

$$P_c \propto P_e \gamma_e^{O2} / PQ \tag{30}$$

The original model for P_e[29] was revisited by the same group in 2000 (Gorbunov et al. 2000; 2001) and we use this version here.[30] It is actually based on Eq. 16 (last line; note that the second to last line could also be used, but the determination of $\bar{\sigma}'_{PSII}$ under high light is practically not possible, see chapter by Suggett et al. (Chapter 6) where $\mathbb{R}_O \phi'_{pO}$ and $\phi_{pO}^{\max}$ are replaced by their fluorescence-derived equivalents from Eqs. 25 and 28. It is thus expressed as[31]

$$P_e = \bar{\sigma}_{PSII} \frac{(F'_m - F')/F'_m}{(F'_m - F_o)/F_m} N_{PSII} \, \mathring{E}(PAR) \tag{31}$$

The parameter γ_e^{O2} is usually taken as equal to 0.25 O_2 electron^{-1} while a value near 1.2 is usually used for PQ. The value of N_{PSII} must also be estimated. This is usually done based on oxygen flash yields by calculating the ratio of oxygen evolved per single turnover flash. The number of oxygen evolved is multiplied by four thus providing the number of active (oxygen evolving) reaction centres. In practice, it is a difficult measurement that requires large concentration of chlorophyll and it is thus rarely measured in the field. A canonical value of one reaction centre (or electron) per 500 chl molecules can be used. This ratio (1/500) is referred to as n_{PSII} and, therefore, $N_{PSII} = [chla] n_{PSII} = [chla]/500$. This conversion is adopted for practical reasons as $[chla]$ is much more readily measured; note that the

[28] Parkhill et al. (2001) make the following distinctions regarding the growth conditions and nutrient status of phytoplankton. When nutrients are available in concentrations that do not limit the growth rate and other environmental factors are constant, the physiological condition is said *nutrient replete* and phytoplankton assume *balanced growth*. That is, over a daily cycle the growth rate will be the same if measured by the concentrations of different cellular components (e.g. DNA, chlorophyll, carbon). When a nutrient is limiting growth, the physiological condition of phytoplankton is said to be *nutrient stressed*. This refers to two nutritional states: *nutrient limitation* and *nutrient starvation*. In the former, a nutrient is in short supply, but the fluxes are steady and sufficient to allow the phytoplankton to assume a balanced, albeit reduced, growth rate. Under nutrient starvation, the availability of the nutrient decreases with time relative to the demand so that phytoplankton cannot acclimate physiologically and their *growth is unbalanced*. Beyond nutrient stress, unbalanced growth is also generally assumed when environmental conditions change.

[29] This protocol is in practice somewhat difficult to apply because some parameters cannot be estimated easily simply with fluorometry. The main advantage of the more recent version is that it does not require an estimate of the turnover time for electron transport through the photosynthetic chain and carbon fixation.

[30] One recent alternative protocol was proposed by Smyth et al. (2004) using profiles of fluorescence properties.

[31] We note that Suggett et al. (2003) interpret the formalism of Gorbunov et al. (2000) differently when forming the quantum yield of photochemistry.

oxygen evolution technique provides the number of active centres while N_{PSII} is defined here as the total number of reaction centres. Thus care must be taken to use a sample with an insignificant fraction of damaged reaction centres to obtain n_{PSII} within the formalism used here.

The values of $\bar{\sigma}_{PSII}$ and fluorescence parameters in Eq. 31 are determined *in situ* from FRRF observations. The [*chla*] and $\overset{o}{E}(PAR)$ can be easily measured, or estimated using other *in situ* sensors. The other quantities, PQ, γ_e^{O2} and N_{PSII} have to be assumed constant (see for instance Suggett et al. 2001) as their measurement on natural phytoplankton is difficult.

Kolber and Falkowski (1993) compared primary production estimated from FRR fluorometry using their model, with primary production measured using the commonly used radiocarbon technique. The intercept and slope of the linear regression were close to 0 and 1, respectively, and the value of R^2 was 0.67. Similar results were obtained by Suggett et al. (2001).

As recognized by both Kolber and Falkowski (1993) and Suggett et al. (2001), the most significant weaknesses of this approach are the following:

- FRR fluorometry in fact measures the photosynthetic activity at the scale of photosynthetic units. To obtain bulk measurements one must multiply by the number of PSII. Because at sea the pool size of photosynthesizing matter is in fact quantified in terms of bulk chl *a* concentration, n_{PSII} is used to convert from a single photosynthetic unit to a whole phytoplankton community in sea water. But n_{PSII} is known to vary over at least a factor of 5 (e.g. Dubinsky et al. 1986; Suggett et al. 2004) among phytoplankton species and in response to changes in environmental factors (mostly light and nutrients), and it is currently not possible to measure n_{PSII} on most natural samples because of the low sensitivity of available measurement protocols. To estimate primary production from FRR fluorometry measurements, constant n_{PSII} values have thus been used (e.g. Kolber and Falkowski 1993; Suggett et al. 2001). A relative error in the n_{PSII} value used leads to the same relative errors in the estimates of primary production.
- In order to actually reflect the quantum yield of functional fraction of RCIIs, ϕ_{pO}^{max} must be measured after 20 to 30 min of dark adaptation to relax q_E and q_T (i.e. $[F_m - F_o]/F_m$ must be measured). In practice, this cannot be achieved *in situ* using the FASTtracka fluorometer and $[F'_m - F'_o]/F'_m$ is in fact measured. As a result, the ϕ_{pO}^{max} measured *in situ* in the dark chamber of the FASTtracka is smaller than it should be. To avoid this problem, $\bar{\sigma}'_{PSII}$ can sometimes be used instead (Suggett, this volume).
- The photosynthetic quotient, *PQ*, is known to vary between ca. 1 and 1.4 mol C (mol O_2)$^{-1}$ mostly as a function of the type of nitrogen available in seawater (Laws 1991). It is lowest when ammonium dominates, and highest when nitrate is the main source because the cell must reduce nitrate to ammonium with electrons produced by the light reactions of photosynthesis before it is utilized in cellular metabolism. Compared to other sources of error mentioned above the error on *PQ* is likely small. Its value can be adjusted based on climatological knowledge of the sampled ocean system.
- The presence of cyanobacteria is also problematic. The most commonly used FRR fluorometer at sea (FASTracka, Chelsea, U.K.) has only blue-LED excitation that does not excite the cyanobacteria PSII antenna efficiently (Raateoja et al. 2004). Under these conditions, the F_m or F'_m values are not reached during excitation and ϕ_{pO}^{max} is underestimated. The contribution of phycobilin fluorescence to F_m and F_o is also a source of biases (Campbell et al. 1998).

There are a number of other potential problems with the assumptions made in this model such as the spectral nature of $\bar{\sigma}'_{PSII}$, the energy losses between RCII and the dark reactions of photosynthesis (e.g. Noctor and Foyer 2000; Behrenfeld et al. 2004). Furthermore, accurate accounting for damaged reaction centres remains an unsettled issue. Equations 26 and 29 show that within the simplified puddle model, F_v/F_m or $[F'_m - F']/F'_m$ do not provide an adequate measures of these damaged reaction centres. A similar development to Eq. 26 can be made which suggests that the use of F_v/F_O (see for example Babin et al. 1996a) provides a linear representation of the effect of damaged reaction centres under the condition where $k_p = k_I$. While the later possibility remains to be verified and could provide an estimate of damaged reaction centers under certain conditions, it is in any case unlikely that a proper accounting for these damaged reaction centres can be made in profile mode since the dark regulated state has to be reached to obtain unambiguous information about them.

4.1.2 When $\bar{\sigma}_{PSII}$ is not Available (PAM Protocol)

Instruments where the measuring light is modulated can measure the fluorescence under ambient irradiance to obtain F' and F_m' and thus allowing one to calculate ϕ_p' (Eq. 28) and obtain the rate of charge separation using Eq. 16 (first line is generally used with the PAM, with various proxies to estimate ${}^{PS}\bar{a}_{PSII}$).

A protocol often used with the PAM instrument is that of the F *vs.* E curve (or rapid light curve White and Critchley 1999) to estimate the rate of charge separation and other parameters as a function of light level. Typically in this protocol, the sample will be dark regulated and F_o and F_m measured in the dark. Then the sample will be illuminated in steps of increasing irradiance. These steps last generally a few minutes to allow the steady state to be reached. During the light steps, F_m', and F' are measured. A recovery phase after the last step is often used to look at the kinetics of non-photochemical quenching relaxation (the fluorescence parameters during this phase often indicated with two primes, e.g. F_m''). Figure 13 shows the trace obtained with such protocols, but for only one light level.

The PAM protocol is generally used to obtain the rate of charge separation (called electron transport rate and abbreviated ETR in the PAM literature, i.e. $ETR = P_e = \phi_p' \, {}^{PS}\bar{a}_{PSII} \, \overset{o}{E}(PAR)$, where ϕ_p' is obtained from Eq. 28); rarely are estimates of carbon fixation attempted. When they are, similar assumptions to those made with the FRRF protocol regarding PQ and γ_e^{O2} are required (e.g. Kromkamp and Forster 2003). Because the commercially available in situ PAM fluorometers are generally not sufficiently sensitive to measure phytoplankton fluorescence, the measurements are usually made from samples collected and brought back to the lab or from cultures.

4.2 Sun-induced Chlorophyll Fluorescence

Kiefer et al. (1989) proposed a simple model that allows estimation of primary production from Sun-induced chl *a* fluorescence (P_C, mol carbon m^{-3} s^{-1}). It can be summarized by the following equations:

$$P_C = \overset{o}{E}(PAR)\,[chla]\,\bar{a}_\phi^*\,\phi_c \tag{32}$$

where ϕ_C is the quantum yield of carbon fixation (moles of carbon fixed per mole of absorbed photon by all cellular pigments). By combining Eqs. 32, 19 and 21 (setting $\ell = 0$), the following equation is obtained:

$$P_C = F_{PSII}\,\frac{\phi_c}{Q_a(\gamma_{em})\,\phi_F^{app}} \tag{33}$$

The ratio ϕ_C / ϕ_F^{app} is known to be highly variable because of photochemical and non-photochemical quenching (Kiefer and Reynolds 1992; Westberry and Siegel 2003) as well as variability in PQ which are in turn dependent on species and physiological status.

To address this problem, Chamberlin et al. (1990) presented an empirical relationship between ϕ_C / ϕ_F^{app} and *PAR*. With the above model (Eq. 33) and this empirical relationship, they obtained an R^2 of 0.9 when comparing fluorescence-based primary production with measured primary production in different ocean systems. Stegmann et al. (1992) obtained similar results in the Equatorial Pacific. Westberry and Siegel (2003) obtained much poorer results on a time series in the Sargasso Sea.

The sun-induced chl *a* fluorescence approach for estimating primary production suffers from some of the difficulties experienced by variable-fluorescence. Furthermore, while the variable fluorescence approach is capable of dealing with different dependence of the quantum yield of photochemistry on irradiance as it is reflected by the fluorescence characteristics of the cell, the approach proposed by Chamberlin et al. (1990) assumes a single relationship between fluorescence and the quantum yield of carbon fixation. However, the sun-induced fluorescence approach is not sensitive to uncertainties in n_{PSII}. When comparing these two fluorescence-based methods for estimating primary production, Kiefer and Reynolds (1992) found R^2 values of 0.74 and 0.86 with FLH and FRR approaches, respectively, when comparing with radiocarbon-based measurements of primary production. This unique comparison, which is certainly not sufficient, suggests that the FRR-fluorometry approach is a slightly better predictor of radioisotope-estimated production.

5 Conclusion

The use of chlorophyll fluorescence can be an extremely powerful tool to investigate the ecophysiology of phytoplankton and monitor its biomass. However, its underlying principles must be well understood for accurately interpreting its measurement. No user should buy a fluorometer out of the box and assume that the manufacturer protocol will provide them with the parameter they seek under all conditions. Instead, the measurement must always be interpreted within a framework similar to the one developed herein. The main points are as follows:

- Chlorophyll *a* is not the only fluorescing molecule in the aquatic environment.
- In the cell, most – but not all – fluorescence originate from photosystem II.
- The flux of fluorescence is proportional to the exciting light, the biomass of phytoplankton (expressed by its absorption coefficient) and the quantum yield of fluorescence.
- The quantum yield of chlorophyll a fluorescence in vivo is highly variable due to a variety of environmental cues that dictates the physiological responses and lead to changes in photochemical and non-photochemical quenching.
- Variable fluorescence provides a means to probe photosystem II photosynthetic efficiency under dark and light regulated conditions. Important assumptions are made when it is used to infer carbon based primary production and as such fluorometry should not be used as a replacement but as a complement when accurate estimates of carbon fixation are required.

It must also be kept in mind that we have presented a framework to support our discussion that is heavily based on model representations of the optics and photosynthetic processes. These models are just that: an attempt at representing the observations. These models are constantly evolving as our observations allow a better understanding. As such, our attempt, in this chapter at synthesizing the different approaches and techniques as well as the conclusions we draw from them must be critically examined in the future against new observations and models.

With these caveats in mind we believe there is a bright future for the continuing role of chlorophyll fluorescence in supporting discoveries in aquatic sciences and the monitoring of our aquatic environments for decades to come.

6 List of Symbols

Symbol[1]	Definition	Units
a_ϕ	Absorption coefficient of phytoplankton	m^{-1}
$\overline{a}_\phi^*$	Chlorophyll-specific, irradiance-weighted absorption coefficient of phytoplankton	$m^2\ mg^{-1}$
${}^{PS}a_{PSII}$ ${}^{PS}\overline{a}_{PSII}$	Absorption coefficient and irradiance-weighted (Eq. 4) absorption coefficient of photosynthetic pigments associated with photosystem II	m^{-1}
${}^{PS}\overline{a}_{PSII}^*$	Chlorophyll-specific, irradiance weighted (Eq. 4) absorption coefficient of photosynthetic pigments associated with photosystem II	$m^2\ mg^{-1}$
a_{tot}	Absorption coefficient of seawater and all its constituents	m^{-1}
a_{sol}	Unpackaged phytoplankton absorption	m^{-1}
$[chla]$	In-water concentration of chlorophyll *a*	$mg\ m^{-3}$
$\overset{o}{E}$	Scalar irradiance	$\mu mol\ photons\ m^{-2}\ s^{-1}$
F_{tot}, F_{PSII}, F_{PSI}, F_{phyco}, F_{phaeo}, F_{CDOM}	Flux of *in vivo* fluorescence emitted by an elementary volume by all constituents (tot), Photosystem II (PSII), Photosystem I (PSI), phycobilins (phyco), phaeopigments (phaeo), and Colored Dissolved Organic Matter (CDOM)	$\mu mol\ photons\ m^{-3}\ s^{-1}$
F'	Flux of *in vivo* fluorescence from PSII emitted by an elementary volume, under the light regulated condition (often under steady state conditions)	$\mu mol\ photons\ m^{-3}\ s^{-1}$
F_m, F_m'	Maximum flux of *in vivo* fluorescence measured when all reaction centres are closed in the dark and light regulated state	$\mu mol\ photons\ m^{-3}\ s^{-1}$
F_O F_O'	Minimum flux of *in vivo* fluorescence measured when all reaction centres are open in the dark and light regulated state	$\mu mol\ photons\ m^{-3}\ s^{-1}$

(continued)

(continued)

Symbol[1]	Definition	Units
F_v	Variable fluorescence ($F_v = F_m - F_o$)	µmol photons m^{-3} s^{-1}
F_{Raman}	Flux of Raman emission emitted by an elementary volume of seawater.	µmol photons m^{-3} s^{-1}
k_x	Rate constant of chl *a* de-excitation processes, the subscript *x* is D for basal thermal decay in vivo, qe for thermal decay due to energy dependent non-photochemical quenching, *F* for fluorescence, H for heat (in solution), I for thermal decay due to photoinhibitory quenching and *P* for photochemistry (or charge separation)	s^{-1}
K, K_d	Vertical attenuation coefficient for scalar and planar irradiance	m^{-1}
L_F	Part of radiance due to *in vivo* chl *a* fluorescence	µmol photons m^{-2} s^{-1} sr^{-1}
L_u	Upwelling radiance	µmol photons m^{-2} s^{-1} sr^{-1}
N_{PSII}	Number of photosystem II per unit volume of water	PSII (or electrons) m^{-3}
n_{PSII}	Number of oxygen-evolving RCII per unit chl *a*	O_2 (mg chl *a*)$^{-1}$ or electrons (mg chl *a*)$^{-1}$
P_e	Rate of charge separation per unit volume of water	electrons m^{-3} s^{-1}
PQ	Photosynthetic quotient	mol O_2 (mol C)$^{-1}$
Q_a^*	Fluorescence intracellular reabsorption factor	dimensionless
q_p	Photochemical quenching parameter	dimensionless
$\mathbb{R}_x$	Fraction of reaction centre is different states given by the subscript: *open :* open; *closed*: closed; *dam*:damaged.	
ε	Photons per unit area during a flash	photons m^{-2}
η_{PSII}	Spectral fluorescence quantum efficiency function	mol emitted photons (mol absorbed photon)$^{-1}$ nm^{-1}
ϕ_C	Quantum yield of carbon fixation in phytoplankton when the absorption of all pigments is considered.	mol carbon (mol absorbed photon)$^{-1}$
ϕ_F	Quantum yield of fluorescence of PSII chlorophyll a when only the absorption of photosynthetic pigments in PSII is considered in the dark and light regulated state.	mol emitted photons (mol absorbed photon)$^{-1}$
ϕ_F^{app}	Apparent quantum yield of fluorescence of PSII chlorophyll a when the absorption of all pigments are considered.	mol emitted photons (mol absorbed photon)$^{-1}$
ϕ_F^{sol}	Quantum yield of fluorescence of chlorophyll a in solution.	mol emitted photons (mol absorbed photon)$^{-1}$
ϕ_{Fm}	Maximum quantum yield of PSII fluorescence when only the absorption of photosynthetic pigments in PSII is considered	mol emitted photons (mol absorbed photon)$^{-1}$
ϕ_{Fo}	Minimum quantum yield of fluorescence when only the absorption of photosynthetic pigments in PSII is considered	mol emitted photons (mol absorbed photon)$^{-1}$
Φ_f	Fluorescence emission spectrum of chlorophyll a within the cell normalized to have an area under the curve of 1.	nm^{-1}
γ_e^{O2}	Ratio of oxygen evolution to charge separation	mol of O_2 (mol of electron)$^{-1}$
λ	Wavelength	nm
λ_{em}	Wavelength of fluorescence emission	nm
$\bar{\sigma}_{PSII}^{opt}$	Irradiance-weighted optical absorption cross section of photosystem II	m^2 $photon^{-1}$
$\bar{\sigma}_{PSII}$	Irradiance-weighted effective absorption cross section of photosystem II	m^2 $photon^{-1}$
ϕ_p'	Realized quantum yield of charge separation in the light regulated state.	mol photons used for charge separation (mol absorbed photon)$^{-1}$
ϕ_{pO}^{max} ϕ_{pO}'	Quantum yield of charge separation at open reaction centres in the dark and light regulated state.	mol photons used for charge separation (mol absorbed photon)$^{-1}$

Acknowledgements We are very thankful to John Cullen for sharing with us his insights on chlorophyll fluorescence over the years. This work was funded by a CNES grant to M. Babin and a CNES fellowship to Y. Huot.

References

Abbott MR, Richerson PJ, Powell TM (1982) In situ response of phytoplankton fluorescence to rapid variations in light. Limnol Oceanogr 27:218–225

Ahn TK et al (2008) Architecture of a charge-transfer state regulating light harvesting in a plant antenna protein. Science 320:794–797

Aro E-M, Ohad I (2003) Redox regulation of thylakoid protein phosphorylation. Antioxidants and Redox Signaling 5:55–67

Aro E-M, Soursa M, Rokka A, Allahverdiyeva Y, Paakkarinen V, Saleem A, battchikova N, Rintamäki E (2005) Dynamics of photosystem II: a proteomics approach to thylakoid protein complexes. J Exp Bot 56:347–356

Aro E-M, Virgin I, Andersson B (1993) Photoinhibition of photosystem II. Inactivation, protein damage and turnover. Biochim Biophys Acta 1143:113–134

Babin M (2008) Phytoplankton fluorescence: theory, current literature and in situ measurement. In: Babin M, Roesler CS, Cullen JJ (eds) Real-time coastal observing systems for marine ecosystem dynamics and harmful algal blooms. UNESCO Publishing, Paris. p 860

Babin M, Morel A, Claustre H, Bricaud A, Kolber Z, Falkowski PG (1996a) Nitrogen- and irradiance-dependent variations of the maximum quantum yield of carbon fixation in eutrophic, mesotrophic and oligotrophic marine systems. Deep Sea Res I 43:1241–1272

Babin M, Morel A, Gentili B (1996b) Remote sensing of sea surface sun-induced chlorophyll fluorescence: consequences of natural variations in the optical characteristics of phytoplankton and the quantum yield of chlorophyll a fluorescence. Int J Remote Sens 17:2417–2448

Baker NR (2008) Chlorophyll fluorescence: a probe of photosynthesis in vivo. Annu Rev PlantBiol 59:89–113

Barber J (2006) Photosystem II: an enzyme of global significance. Biochem Soc Trans 34:619–631

Bassi R, Pineau B, Dainese P, Marquardt J (1993) Carotenoid-binding proteins of photosystem II. Eur J Biochem 212:297–303

Behrenfeld MJ, Bale AJ, Kolber ZS, Aiken J, Falkowski PG (1996) Confirmation of iron limitation of phytoplankton photosynthesis in the equatorial Pacific Ocean. Nature 383:508–511

Behrenfeld MJ, Boss E (2006) Beam attenuation and chlorophyll concentration as alternative optical indices of phytoplankton biomass. J Mar Res 64:431–451

Behrenfeld MJ, Kolber Z (1999) Widespread iron limitation of phytoplankton in the South Pacific Ocean. Science 283:840–843

Behrenfeld MJ, Prasil O, Babin M, Bruyant F (2004) In search of a physiological basis for covariations in light-limited and light-saturated photosynthesis. J Phycol 40:4–25

Behrenfeld MJ, Worthington K, Sherrell RM, Chavez FP, Strutton P, McPhaden M, Shea DM (2006) Controls on tropical Pacific Ocean productivity revealed through nutrient stress diagnostics. Nature 442:1025–1028

Beutler M, Wiltshire KH, Arp M, Kruse J, Moldaenke C, Hansen UP (2003) A reduced model of the fluorescence from cyanobacterial photosynthetic apparatus designed for the in situ detection of cyanobacteria. Biochim Biophys Acta 1604:33–46

Beutler M, Wiltshire KH, Meyer B, Moldaenke C, Lüring C, Meyerhöfer M, Hansen UP, Dau H (2002) A fluorometric method for the differentiation of algal populations in vivo and in situ. Photosynth Res 72:39–53

Beutler M, Wiltshire KH, Reineke C, Hansen UP (2004) Algorithms and practical fluorescence models of the photosynthetic apparatus of red cyanobacteria and Cryptophyta designed for the fluorescence detection of red cyanobacteria and cryptophytes. Aquat Microb Ecol 35:115–129

Bidigare RR, Ondrusek ME, Morrow JH, Kiefer DA (1990) In vivo absorption properties of algal pigments. In: Spinrad RW (ed) Ocean optics X. Proc SPIE 1302:290–302

Blankenship RE (2002) Molecular mechanisms of photosynthesis. Blackwell Science, Oxford

Boisvert S, Joly D, Carpentier R (2006) Quantitative analysis of the experimental O-J-I-P chlorophyll fluorescence induction kinetics. FEBS J 273:4770–4777

Bricaud A, Claustre H, Ras J, Oubelkheir K (2004) Natural variability of phytoplankton absorption in oceanic waters: influence of the size structure of algal populations. J Geophys Res-Oceans 109:C11010. doi:11010.11029/12004JC002419

Bricaud A, Morel A, Babin M, Allali K, Claustre H (1998) Variations of light absorption by suspended particles with the chlorophyll a concentration in oceanic (Case 1) waters: analysis and implications for bio-optical models. J Geophys Res-Oceans 103:31033–31044

Bristow M, Nielsen D, Bundy D, Furtek R (1981) Use of water Raman emission to correct airborne laser fluorosensor data for effects of water optical attenuation. Appl Optics 20: 2889–2906

Browell EV (1977) Analysis of laser fluorosensor systems for remote algae detection and quantification. In: NASA technical note. National Aeronoutics and Space Administration, Washington, DC p 39

Bruyant F, babin M, genty B, Prasil O, Behrenfeld MJ, Claustre H, Bricaud A, Garczarek L, Holtzendorff J, Koblizek M, Dousova H, Partensky F (2005) Diel variations in the photosynthetic parameters of *Prochlorococcus* strain PCC 9511: Combined effects of light and cell cycle. Limnol Oceanogr 50:850–863

Bryant DA, Frigaard N-U (2006) Prokaryotic photosynthesis and phototrophy illuminated. Trends Microbiol 14: 488–496

Butler WL (1978) Energy distribution in the photochemical apparatus of photosynthesis. Annu Rev Plant Physiol 29:345–378

Campbell D, Vaughan H, Clarke AK, Gustafsson P, Öquist G (1998) Chlorophyll fluorescence analysis of cyanobacterial photosynthesis and acclimation. Microbiol Mol Biol Rev 62:667–683

Chamberlin WS, Booth CR, Kiefer DA, Morrow JH, Murphy RC (1990) Evidence for a simple relationship between natural

fluorescence, photosynthesis and chlorophyll in the sea. Deep Sea Res 37:951–973
Chekalyuk AM, Hafez M (2008) Advanced laser fluorometry of natural aquatic environments. Limnol Oceanogr Meth 6:591–609
Chekalyuk AM, Hoge FE, Wright CW, Swift RN, Yungel JK (2000) Airborne test of laser pump-and-probe technique for assessment of phytoplankton photochemical characteristics. Photosynth Res 66:45–56
Chow WS (2001) The photoinactivation of photosystem II in leaves: a personal perspective. J Photosci 2001:43–53
Claustre H, Morel A, Babin M, Cailliau C, Marie D, Marty J-C, Taillez D, Vaulot D (1999) Variability in particle attenuation and chlorophyll fluorescence in the tropical Pacific: Scales, patterns, and biogeochemical implications. J Geophys Res 104:3401–3422
Claustre H, Sciandra A, Vaulot D (2008) Introduction to the special section bio-optical and biogeochemical conditions in the South East Pacific in late 2004: the BIOSOPE program. Biogeosciences 5:679–691
Clayton RK (1980) Photosynthesis. Cambridge University Press, Cambridge
Collins DJ, Kiefer DA, SooHoo JB, McDermid IS (1985) The role of reabsorption in the spectral distribution of phytoplankton fluorescence emission. Deep Sea Res 32:983–1003
Cowles TJ, Desiderio RA, Neuer S (1993) In situ characterization of phytoplankton from vertical profiles of fluorescence emission spectra. Mar Biol 115:217–222
Cullen JJ, Davis RF (2003) The blank can make a big difference in oceanographic measurements. Limnol Oceanogr Bull 12:21–35
Cullen JJ, Lewis MR (1995) Biological processes and optical measurements near the sea-surface: some issues relevant to remote sensing. J Geophys Res 100:13255–13266
Cullen JJ, Renger EH (1979) Continuous measurement of the DCMU-induced fluorescence response of natural phytoplankton populations. Mar Biol 53:13–20
Cullen JJ, Xiaolong Y, MacIntyre HL (1992) Nutrient limitation of marine photosynthesis. In: Falkowski PG (ed) Primary productivity and biogeochemical cycles in the sea. Plenum Press, NY, pp 69–88
Cullen JJ, Yentsch CS, Cucci TL, MacIntyre HL (1988) Autofluorescence and other optical properties as tools in biological oceanography. Proc SPIE, Int Soci Opt Eng 925: 149–156
Dandonneau Y, Neveux J (1997) Diel variations of the *in vivo* fluorescence in the eastern equatorial Pacific: an unvarying pattern. Deep Sea Res II 44:1869–1880
Demmig-Adams B, Adams WW III (1992) Photoprotection and other responses of plants to high light stress. Annu Rev Plant Physiol Plant Mol Biol 43:599–626
Desiderio RA, Moore C, Lantz C, Cowles TJ (1997) Multiple excitation fluorometer for in situ oceanographic applications. Appl Optics 36:1289–1296
Dubinsky Z, Falkowski PG, Wyman K (1986) Light Harvesting and utilization by phytoplankton. Plant Cell Physiol 27:1335–1349
Exton RJ, Houghton WM, Esaias W, Harriss RC, Farmer FH, White HH (1983) Laboratory analysis of techniques for remote sensing of estuarine parameters using laser excitation. Appl Optics 22:54–64
Falkowski PG, Fujita Y, Ley A, Mauzerall D (1986) Evidence for cyclic electron flow around photosystem-II in *Chlorella pyrenoidosa*. Plant Physiol 81:310–312
Falkowski PG, Greene RM, Kolber Z (1995) Light utilization and photoinhibition of photosynthesis in marine phytoplankton. In: Baker NR, Bowyer JR (eds) Photoinhibition of photosynthesis from molecular mechanisms to the field Environmental plant biology. Bios Scientific Publishers, Oxford, pp 407–432
Falkowski PG, Kiefer DA (1985) Chlorophyll a fluorescence in phytoplankton: relationship to photosynthesis and biomass. J Plankton Res 7:715–731
Falkowski PG, LaRoche J (1991) Acclimation to spectral irradiance in algae. J Phycol 27:8–14
Falkowski PG, Raven JA (2007) Aquatic photosynthesis. Princeton University Press, Princeton, NJ
Falkowski PG, Ziemann D, Kolber Z, Bienfang PK (1991) Role of eddy pumping in enhancing primary production in the ocean. Nature 352:55–58
Finazzi G, Rapaport F, Furia A, Fleischmann M, Rochaix J-D, Zito F, Forti G (2002) Involvement of state transitions in the switch between linear and cyclic electron flow in *Chlamydomonas reihardtii*. EMBO Reports 3:280–285
Fisher J, Kronfeld U (1990) Sun-stimulated chlorophyll fluorescence 1: influence of oceanic properties. Int J Remote Sens 11:2125–2147
Franck F, Juneau P, Popovic R (2002) Resolution of the photosystem I and photosystem II contributions to chlorophyll fluorescence of intact leaves at room temperature. Biochim Biophys Acta 1556:239–246
Fuchs E, Zimmerman RC, Jaffe JS (2002) The effect of elevated levels of phaeophytin in natural water on variable fluorescence measured from phytoplankton. J Plankton Res 24:1221–1229
Geider RJ, Osborne BA (1992) Algal photosynthesis. Chapman & Hall, New York
Genty B, Briantais J-M, Baker NR (1989) The relationship between the quantum yield of photosynthetic electron transport and quenching of chlorophyll fluorescence. Biochim Biophys Acta 990:87–92
Gilmore AM, Govindjee (1999) How higher plants respond to excess light: energy dissipation in photosystem II. In: Singhal GS, Renger R, Sopory SK, Irrgang K-D, Govindjee (eds) Concepts in photobiology: photosynthesis and photomorphogenesis. Narosa-Publishing, New Delhi, pp 513–548
Gorbunov MY, Falkowski PG, Kolber ZS (2000) Measurement of photosynthetic parameters in benthic organisms *in situ* using a SCUBA-based repetition rate fluorometer. Limnol Oceanogr 45:242–245
Gorbunov MY, Kolber ZS, Falkowski PG (1999) Measuring phtosynthetic parameters in individual algal cells by Fast Repetition Rate fluorometry. Photosynth Res 62:141–153
Gorbunov MY, Kolber ZS, Lesser MP, Falkowski PG (2001) Photosynthesis and photoprotection in symbiotic corals. Limnol Oceanogr 46:75–85
Gordon HR (1979) Diffuse reflectance of the ocean: the theory of its augmentation by chlorophyll *a* fluorescence. Appl Optics 21:2489–2492
Govindjee, Satoh K (1986) Fluorescence properties of chlorophyll b- and chlorophyll c-containing algae. In: Govindjee, Amesz J, Fork DC (eds) Light emission by plant and bacteria. Academic Press, Orlando, pp 497–537

Govindjee (1995) Sixty-three years since Kautsky: chlorophyll *a* fluorescence. Aust J Plant Physiol 22:131–160

Gower JFR, Brown L, Borstad GA (2004) Observations of chlorophyll fluorescence in west coast waters of Canada using the MODIS satellite sensor. Can J Remote Sens 30:17–25

Gower JFR, Doerffer R, Borstad GA (1999) Interpretation of the 685 nm peak in water-leaving radiance spectra in terms of fluorescence, absorption and scattering, and its observation by MERIS. Int J Remote Sens 20:1771–1786

Greene RM, Kolber ZS, Swift DG, Tindale NW, Falkowski PG (1994) Physiological limitation of phytoplankton photosynthesis in the eastern equatorial Pacific determined from variability in the quantum yield of fluorescence. Limnol Oceanogr 39:1061–1074

Hall DO, Rao KK (1999) Photosynthesis, 6th edn. Cambridge University Press, Cambridge

Han BP (2001) Photosynthesis-irradiance response at physiological level: A mechanistic model. J Theoret Biol 213:121–127

Han BP (2002) A mechanistic model of algal photoinhibition induced by photodamage to Photosystem-II. J Theoret Biol 214:519–527

Hankamer B, Barber J, Boekema EJ (1997) Structure and membrane organization of photosystem II in green plants. Annu Rev Plant Physiol Plant Mol Biol 48:641–671

Havaux M, Strasser RJ, Greppin H (1991) A theoretical and experimental analysis of the q_p and q_n coefficients of chlorophyll fluorescence quenching and their relation to photochemical and nonphotochemical events. Photosynth Res 27:41–55

Hofstraat JW, Rubelowsky K, Slutter S (1992) Corrected fluorescence excitation and emission spectra of phytoplankton: toward a more uniform approach to fluorescence measurements. J Plankton Res 14:625–636

Hoge FE, Swift RN (1981) Airborne simultaneous spectroscopic detection of laser-induced water Raman backscatter and fluorescence from chlorophyll *a* and other naturally occurring pigments. Appl Optics 20:3197–3205

Horton P, Ruban AV, Walters RG (1996) Regulation of light harvesting in green plants. Annu Rev Plant Physiol Plant Mol Biol 47:655–684

Huot Y, Brown CA, Cullen JJ (2005) New algorithms for MODIS sun-induced chlorophyll fluorescence and a comparison with present data products. Limnol Oceanogr Meth 3:108–130

Huot Y, Brown CA, Cullen JJ (2007) Retrieval of phytoplankton biomass from simultaneous inversion of reflectance, the diffuse attenuation coefficient, and Sun-induced fluorescence in coastal waters. J Geophys Res 112. doi:10.1029/2006JC003794

Ivanov AG, Hurry V, Prafullanchandra VS, Öquist G, Huner NPA (2008a) Reaction centre quenching of excess light energy and photoprotection of photosystem II. J Plant Biol 51:85–96

Ivanov AG, Prafullanchandra VS, Hurry V, Öquist G, Huner NPA (2008b) Photosystem II reaction centre quenching: mechanisms and physiological role. Photosynth Res 98:565–574

Jeffrey SW, Vesk M (1997) Introduction to marine phytoplankton and their pigment signature. In: Jeffrey SW, Mantoura RFC, Wright SW (eds) Phytoplankton pigments in oceanography. UNESCO Publishing, Rome

Johnsen G, Prézelin BB, Jovine RVM (1997) Fluorescence excitation spectra and light utilization in two red tide dinoflagellates. Limnol Oceanogr 42:1166–1177

Karapetyan NV (2007) Non-photochemical quenching of fluorescence in cyanobacteria. Biochemistry (Moscow) 72:1127–1135

Kargul J, Barber J (2008) Photosynthetic acclimation: Structural reorganisation of light harvesting antenna - role of redox-dependent phophorylation of major and minor chlorophyll a/b binding proteins. FEBS J 275:1056–1068

Kiefer DA (1973a) Chlorophyll *a* fluorescence in marine centric diatoms: responses of chloroplasts to light and nutrient stress. Mar Biol 23:39–46

Kiefer DA (1973b) Fluorescence properties of natural phytoplankton populations. Mar Biol 22:263–269

Kiefer DA, Chamberlin WS, Booth CR (1989) Natural fluorescence of chlorophyll a: relationship to photosynthesis and chlorophyll concentration in the western South Pacific gyre. Limnol Oceanogr 34:868–881

Kiefer DA, Reynolds RA (1992) Advances in understanding phytoplankton fluorescence and photosynthesis. In: Falkowski PG (ed) Primary productivity and biogeochemical cycles in the sea. Plenum Press, New York, pp 155–174

Kim HH (1973) New algae mapping technique by the use of an airborne laser fluorosensor. Appl Optics 12:1454–1459

Kolber Z, Falkowski PG (1993) Use of active fluorescence to estimate phytoplankton photosynthesis in situ. Limnol Oceanogr 38:1646–1665

Kolber Z, Wyman KD, Falkowski PG (1990) Natural variability in photosynthetic energy-conversion efficiency – A field-study in the Gulf of Maine. Limnol Oceanogr 35:72–79

Kolber Z, Zehr JR, Falkowski PG (1988) Effects of growth irradiance and nitrogen limitation on photosynthetic energy conversion in photosystem II. Plant Physiol 88:923–929

Kolber ZS, Barber RT, Coale KH, Fitzwater SE, Greene RM, Johnson S, Lindley KS, Falkowski PG (1994) Iron limitation of phytoplankton photosynthesis in the equatorial Pacific Ocean. Nature 371:145–149

Kolber ZS, Prásil O, Falkowski PG (1998) Measurements of variable chlorophyll fluorescence using fast repetition rate techniques: defining methodology and experimental protocols. Biochim Biophys Acta 1367:88–106

Kopf U, Heinze J (1984) 2, 7-Bis(diethylamino)phenazoxonium chloride as a quantum counter for emission measurements between 240 and 700 nm. Anal Chem 56:1931–1935

Kramer DM, Johnson G, Kiirats O, Edwards GE (2004) New fluorescence parameters for the determination of Q(A) redox state and excitation energy fluxes. Photosynth Res 79:209–218

Krause GH, Jahns P (2004) Non-photochemical energy dissipation determined by chlorophyll fluorescence quenching: characterization and function. In: Papageorgiou GC, Govindjee (eds) Chlorophyll *a* fluorescence: a signature of photosynthesis. Springer, Berlin, pp 463–495

Krause GH, Weis E (1991) Chlorophyll fluorescence: the basics. Annu Rev Plant Physiol 42:313–349

Kromkamp JC, Forster RM (2003) The use of variable fluorescence measurements in aquatic ecosystems: differences between multiple and single turnover measuring protocols and suggested terminology. Eur J Phycol 38:103–112

Laney SR (2003) Assessing the error in photosynthetic properties determined by fast repetition rate fluorometry. Limnol Oceanogr 48:2234–2242

Laney SR, Letelier RM (2008) Artifacts in measurements of chlorophyll fluorescence transients, with specific application to fast repetition rate fluorometry. Limnol Oceanogr Meth 6:40–50

Lavaud J, Rousseau B, Etienne A-L (2004) General features of photoprotection by energy dissipation in planktonic diatoms (Bacillariophyceae). J Phycol 40:130–137

Lavergne J, Trissl H-W (1995) Theory of fluorescence induction in photosystem II: Derivation of analytical expressions in a model including exciton-radial-pair equilibrium and restricted energy transfer between photosynthetic units. Biophys J 68:2474–2492

Laws EA (1991) Photosynthetic quotients, new production and net community production in the open ocean. Deep Sea Res Part A 38:143–167

Lazar D (2003) Chlorophyll *a* fluorescence rise induced by high light illumination of dark-adapted plant tissue studied by means of a model of photosystem II and considering photosystem II heterogeneity. J Theoret Biol 220:469–503

Lazár D (1999) Chlorophyll *a* fluorescence induction. Biochim Biophys Acta 1412:1–28

Lazár D (2006) The polyphasic chlorophyll *a* fluorecence rise measured under high intensity of exciting light. Funct Plant Biol 33:9–30

Letelier R, Abbott MR (1996) An analysis of chlorophyll fluorescence algorithms for the Moderate Resolution Imaging Spectrometer (MODIS). Remote Sens Environ 58:215–223

Liu G, Janowitz GS, Kamykowski D (2001) A biophysical model of population dynamics of the autotrophic dinoflagellate *Gymnodinium breve*. Mar Ecol Prog Ser 210:101–124

Loftus ME, Seliger HH (1975) Some limitations of the *in vivo* fluorescence technique. Chesapeake Sci 16:79–92

Lorenzen CJ (1966) A method for the continuous measurement of *in vivo* chlorophyll concentration. Deep Sea Res 13:223–227

Lutz VA, Sathyendranath S, Head EJH, Li WKW (1998) Differences between in vivo absorption and fluorescence excitation spectra in natural samples of phytoplankton. J Phycol 34:214–227

Lutz VA, Sathyendranath S, Head EJH, Li WKW (2001) Changes in the *in vivo* absorption and fluorescence excitation spectra with growth irradiance in three species of phytoplankton. J Plankton Res 23:555–569

MacIntyre HL, Kana TM, Anning T, Geider RJ (2002) Photoacclimation of photosynthesis irradiance response curves and photosynthetic pigments in microalgae and cyanobacteria. J Phycol 38:17–38

Maritorena S, Morel A, Gentili B (2000) Determination of the fluorescence quantum yield by oceanic phytoplankton in their natural habitat. Appl Optics 39:6725–6737

Marshall HL, Geider RJ, Flynn KJ (2000) A mechanistic model of photoinhibition. New Phytol 145:347–359

Mauzerall DC (1972) Light-induced changes in *Chlorella*, and the primary photoreaction for the production of oxygen. Proc Nat Acad Sci USA 69:1358–1362

Mauzerall DC, Greenbaum NL (1989) The absolute size of a photosynthetic unit. Bioch Biophys Acta 974:119–140

Melis A (1999) Photosystem-II damage and repair cycle in chloroplasts: what modulates the rate of photodamage in vivo. Trends Plant Sci 4:130–135

Miller WL, Moran MA, Sheldon WM, Zepp RG, Opsahl S (2002) Determination of apparent quantum yield spectra for the formation of biologically labile photoproducts. Limnol Oceanogr 47:343–352

Millie DF, Schofield OME, Kirkpatrick GJ, Johnsen G, Evens TJ (2002) Using absorbance and fluorescence spectra to discriminate microalgae. Eur J Phycol 37:313–322

Mitchell BG, Kiefer DA (1988) Chlorophyll *a* specific absorption and fluorescence excitation spectra for light-limited phytoplankton. Deep Sea Res 35:639–663

Mobley CD (1994) Light and water radiative transfer in natural waters, 1st edn. Academic Press, San Diego

Moore CM, Suggett DM, Hickman AE, Kim Y-N, Tweddle JF, Sharples J, Geider RJ, Holligan PM (2006) Phytoplankton photoacclimation and photoadaptation in response to environmental gradients in a shelf sea. Limnol Oceanogr 51: 936–949

Morel A (1988) Optical modeling of the upper ocean in relation to its biogenous matter content (case 1 waters). J Geophys Res-Oceans 93:10749–10768

Morel A, Bricaud A (1981) Theoretical results concerning light absorption in a discrete medium, and application to specific absorption of phytoplankton. Deep-Sea Res 28:1375–1393

Morel A, Bricaud A (1986) Inherent properties of algal cells including picoplankton: Theoretical and experimental results. In: Platt T, Li WKW (eds) Photosynthetic picoplankton. Can Bull Fish Aquat Sci 214:521–559

Morel A, Prieur L (1977) Analysis of variations in ocean color. Limnol Oceanogr 22:709–722

Morrison JR (2003) In situ determination of the quantum yield of phytoplankton chlorophyll *a* fluorescence: a simple algorithm, observations, and a model. Limnol Oceanogr 48:618–631

Neale PJ, Cullen JJ, Yentsch CM (1989) Bio-optical inferences from chlorophyll a fluorescence: what kind of fluorescence is measured in flow cytometry? Limnol Oceanogr 34:1739–1748

Nelson N, Yocum CF (2006) Structure and function of photosystems I and II. Annu Rev Plant Biol 57:521–565

Neori A, Vernet M, Holm-Hansen O, Haxo FT (1986) Relationship between action spectra for chlorophyll *a* fluorescence and photosynthetic O_2 evolution in algae. J Plankton Res 8:537–548

Neori A, Vernet M, Holm-Hansen O, Haxo FT (1988) Comparison of chlorophyll far-red and red fluorescence excitation spectra with photosynthetic oxygen action spectra for photosystem II in algae. Mar Ecol Prog Ser 44:297–302

Neville RA, Gower JFR (1977) Passive remote sensing of phytoplankton via chlorophyll fluorescence. J Geophys Res 82:3487–3493

Niyogi KK (1999) Photoprotection revisited: Genetic and molecular approaches. Annu Rev Plant Physiol Plant Mol Biol 50:333–359

Noctor G, Foyer CH (2000) Homeostasis of adenylate status during photosynthesis in a fluctuating environment. J Exp Bot 51:347–356

Olaizola M, Geider RJ, Harrison WG, Graziano LM, Ferrari GM, Schlittenhardt PM (1996) Synoptic study of variations in the fluorescence-based maximum quantum efficiency of photosynthesis across the North Atlantic Ocean. Limnol Oceanogr 41:755–765

Olaizola M, Yamamoto HY (1994) Short-term responses of the diadinoxanthin cycle and fluorescence yield in *Chaetoceros muelleri*. J Phycol 30:606–612

Olson RJ, Chekalyuk AM, Sosik HM (1996) Phytoplankton photosynthetic characteristics from fluorescence induction assays of individual cells. Limnol Oceanogr 41:1253–1263

Olson RJ, Sosik HM, Chekalyuk AM (1999) Photosynthetic characteristics of marine phytoplankton from pump-during-probe fluorometry of individual cells at sea. Cytometry 37:1–13

Olson RJ, Zettler ER, Chisholm SW, Dusenberry JA (1991) Flow cytometry in oceanography. In: Demers S (ed) Particle analysis in oceanography. Springer, Berlin, pp 351–399

Ostrowska M, Darecki M, Wozniak B (1997) An attempt to use measurements of sun-induced chlorophyll fluorescence to estimate chlorophyll *a* concentration in the Baltic Sea. Proc SPIE, Int Soci Optical Eng 3222:528–537

Ostrowska M, Majchrowski R, Matorin DN, Wozniak B (2000a) Variability of the specific fluorescence of chlorophyll in the ocean. Part 1. Theory of classical *in situ* chlorophyll fluorometry. Oceanologia 42:203–219

Ostrowska M, Matorin DN, Ficek D (2000b) Variability of the specific fluorescence of chlorophyll in the ocean. Part 2. Fluorometric method of chlorophyll *a* determination. Oceanologia 42:221–229

Oxborough K, Baker NR (2000) An evaluation of the potential triggers of photoinactivation of photosystem II in the context of a Stern-Volmer model for downregulation and the reversible radical pair equilibrium. Phil Trans Roy Soc London 355:1489–1498

Owens, TG (1991) Energy Transformation and Fluorescence in Photosynthesis. In Particle Analysis in Oceanography. Edited by Serge Demers. NATO ASI Series. Springer-Verlag

Park Y-I, Chow WS, Anderson JM (1995) Light inactivation of functional photosystem II in leaves of peas grown in moderate light depends on photon exposure. Planta 196:401–411

Parkhill J-P, Maillet G, Cullen JJ (2001) Fluorescence-based maximal quantum yield for PSII as a diagnostic of nutrient stress. J Phycol 37:517–529

Poole LR, Esaias WE (1982) Water Raman normalization of airborn laser fluorosensor measurements: a computer model study. Appl Optics 21:1982

Prézelin BB (1981) Light reactions in photosynthesis. In: Platt T (ed) Physiological bases of phytoplankton ecology. Can Bull Fish Aquat Sciences 210:1–43

Prezelin BB, Boczar BA (1986) Molecular bases of cell absorption and fluorescence in phytoplankton: potential applications to studies in optical oceanography. Prog Phycol Res 4:349–464

Raateoja M, Seppälä J, Ylöstalo P (2004) Fast repetition rate fluorometry is not applicable to studies of filamentous cyanobacteria from the Baltic Sea. Limnol Oceanogr 49:1006–1012

Raven JA, Geider RJ (2003) Adaptation, acclimation and regulation in algal phtosynthesis. In: Larkum AWD, Douglas SE, Raven JA (eds) Photosynthesis in algae. Kluwer, Dordrecht, pp 385–412

Roesler CS, Perry MJ (1995) In situ phytoplankton absorption, fluorescence emission, and particulate backscattering spectra determined from reflectance. J Geophys Res 100:13279–13294

Rohácek K (2002) Chlorophyll fluorescence parameters: the definitions, photosynthetic meaning, and mutual relationships. Photosynthetica 40:13–29

Rohácek K, Barták M (1999) Technique of the modulated chlorophyll fluorescence: basic concepts, useful parameters, and some applications. Photosynthetica 37:339–363

Roy S, Legendre L (1979) DCMU-enhanced fluorescence as an index of photosynthetic activity of phytoplankton. Mar Biol 55:93–101

Ruban AV Berera R, Ilioaia C, van Stokkum IHM, Kennis JTM, Pascal AA, van Amerongen H, Robert B, Horton P, van Grondelle R (2007) Identification of a mechanism of photoprotective energy dissipation in higher plants. Nature 450:575–578

Ruban AV, Lavaud J, Rousseau B, Guglielmi G, Horton P, Etienne A-L (2004) The super-excess energy dissipation in diatom algae: comparative analysis with higher plants. Photosynth Res 82:165–175

Sakshaug E, Johnsen G, Andresen K, Vernet M (1991) Modeling of light-dependent algal photosynthesis and growth: experiments with the Barents Sea diatoms *Thalassiosira nordenskioeldii* and *Chaetoceros furcellatus*. Deep Sea Res A 38:415–430

Sathyendranath S, Platt T, Irwin B, Horne E, Borstad G, Stuart V, Payzant L, Maass H, Kepkay P, Li P, Spry J, Gower J (2004) A multispectral remote sensing study of coastal waters off Vancouver Island. Int J Remote Sens 25:893–919

Schallenberg C, Lewis MR, Kelley DE, Cullen JJ (2008) The inferred influence of nutrient availability on the relationship between sun-induced fluorescence and incident irradiance in the Bering Sea. J Geophys Res 113:C07046. doi:07010.01029/02007JC004355

Schatz GH, Brock H, Holzwarth AR (1988) Kinetic and energetic model for the primary processes in photosystem II. Biophys J 54:397–405

Schreiber U, Schliwa U, Bilger W (1986) Continuous recording of photochemical and non-photochemical chlorophyll fluorescence quenching with a new type of modulation fluorometer. Photosynth Res 10:51–62

Sciandra A, Lazzara L, Claustre H, Babin M (2000) Responses of growth rate, pigment composition and optical properties of *Cryptomonas* sp. to light and nitrogen stresses. Mar Ecol Prog Ser 201:107–120

Smyth TJ, Pemberton KL, Aiken J, Geider RJ (2004) A methodology to determine primary production and phytoplankton photosynthetic parameters from fast repetition rate fluorometry. J Plankton Res 26:1337–1350

SooHoo JB, Kiefer DA (1982) Vertical distribution of phaeopigments — II. Rates of production and kinetics of photooxidation. Deep-Sea Res 29:1553–1563

Sosik HM, Mitchell BG (1995) Light absorption by phytoplankton, photosynthetic pigments and detritus in the California Current System. Deep Sea Res I 42:1717–1748

Sosik HM, Olson JS, Neubert MG, Shalapyonok A, Solow AR (2003) Growth rates of coastal phytoplankton from time-series measurements with a submersible flow cytometer. Limnol Oceanog 48:1756–1765

Stegmann PM, Lewis MR, Davis CO, Cullen JJ (1992) Primary production estimates from recordings of solar-stimulated fluorescence in the Equatorial Pacific at 150°W. J Geophys Res 97:627–638

Stirbet A, Govindjee SBJ, Strasser RJ (1998) Chlorophyll a fluorescence induction in higher plants: modelling and numerical simulation. J Theoret Biol 193:131–151

Strasser RJ, Srivastava A, Govindgee (1995) Polyphasic chlorophyll *a* fluorescence transient in plants and cyanobacteria. Photochem Photobiol 61:32–42

Suggett DJ, Kraay G, Holligan P, Davey M, Aiken J, Geider RJ (2001) Assessment of photosynthesis in a spring cyanobacterial bloom by use of a fast repetition rate fluorometer. Limnol Oceanogr 46:802–810

Suggett DJ, MacIntyre HL, Geider RJ (2004) Evaluation of biophysical and optical determination of light absorption by photosystem II in phytoplankton. Limnol Oceanogr Meth 2:316–332

Suggett DJ, Moore CM, Hickman AE, Geider RJ (2009) Interpretation of fast repetition rate (FRR) fluorescence: signature of phytoplankton community structure versus physiological state. Mar Ecol Prog Ser 376:1–19

Suggett DJ, Oxborough K, Baker NR, MacIntyre HL, Kana TM, Geider RJ (2003) Fast repetition rate and pulse amplitude modulation chlorophyll *a* fluorescence measurements for assessment of photosynthetic electron transport in marine phytoplankton. Eur J Phycol 38:371–384

Timmermans KR, van der Woerd HJ, Wernand MR, Sligting M, Uitz J, de Baar HJW (2008) In situ and remote-sensed chlorophyll fluorescence as indicator of the physiological state of phytoplankton near the Isles Kerguelen (Souther Ocean). Polar Biol 31:617–628

Ting CS, Rocap G, King J, Chisholm SW (2002) Cyanobacterial photosynthesis in the oceans: the origins and significance of divergent light-harvesting strategies. Trends Microbiol 10:134–142

Trees CC, Clark DK, Bidigare RR, Ondrusek ME, Mueller JL (2000) Accessory pigments versus chlorophyll *a* concentrations within the euphotic zone: A ubiquitous relationship. Limnol Oceanogr 45:1130–1143

Trissl H-W, Lavergne J (1995) Fluorescence Induction from photosystem II: Analytical equations for the yields of photochemistry and fluorescence derived from analysis of a model including exciton-radical pair equilibrium and restricted energy transfer between photosynthetic units. Aust J Plant Physiol 22:183–193

Vaulot D, Marie D (1999) Diel variability of photosynthetic picolankton in the equatorial Pacific. J Geophys Res 104:3297–3310

Vermotte C, Etienne A-L, Briantais J-M (1979) Quenching of the system II chlorophyll fluorescence by the plastoquinone pool. Biochim Biophys Acta 545:519–527

Vincent WF (1981) Photosynthetic capacity measured by DCMU-induced chlorophyll fluorescence in an oligotrophic lake. Freshw Biol 11:61–78

Welschmeyer NA (1994) Fluorometric analysis of chlrophyll *a* in the presence of chlorophyll *b* and pheopigments. Limnol Oceanogr 39:1985–1992

Westberry TK, Siegel DA (2003) Phytoplankton natural fluorescence variability in the Sargasso Sea. Deep Sea Res I 50:417–434

White AJ, Critchley C (1999) Rapid light curves: a new fluorescence method to assess the state of the photosynthetic apparatus. Photosynth Res 59:63–72

Whitmarsh J, Govindjee (1999) The photosynthetic process. In: Singhal GS, Renger R, Sopory SK, Irrgang K-D, Govindjee (eds) Concepts in photobiology: photosynthesis and photomorphogenesis. Narosa-Publishing, New Delhi, pp 11–51

Yentsch CM, Horan PK, Muirhead K, Dortch Q, Haugen E, Legendre L, Muphy LS, Perry MJ, Phinney DA, pomponi SA, Spinrad RW, Wood M, Yentsch CS Zahoranec BJ (1983) Flow cytometry and cell sorting: a technique for analysis and sorting of aquatic particles. Limnol Oceanogr 28: 1275–1280

Yentsch CS, Menzel DW (1963) A method for the determination of phytoplankton chlorophyll and phaeophytin by fluorescence. Deep Sea Res 10:221–231

Yentsch CS, Phinney DA (1985) Spectral fluorescence: a taxonomic tool for studying the structure of phytoplankton populations. J Plankton Res 7:617–632

Yentsch CS, Yentsch CA (1979) Fluorescence spectral signatures: the characterization of phytoplankton populations by the use of excitation and emission. J Mar Res 37:471–483

Zonneveld C (1997) Modelling the effects of photoadaptation on the photosynthesis-irradiance curve. J Theoret Biol 186:381–388

Zouni A, Witt HT, Fromme P, Krauss N, Saenger W, Orth P (2001) Crystal structure of photosystem II from *Synechococcus elongatus* a 3.8Å resolution. Nature 409:739–743

Chapter 4
Fluorescence as a Tool to Understand Changes in Photosynthetic Electron Flow Regulation

Peter J. Ralph, Christian Wilhelm, Johann Lavaud, Torsten Jakob, Katherina Petrou, and Sven A. Kranz

1 Introduction

The physiological state of a chloroplast is strongly influenced by both biotic and abiotic conditions. Unfavourable growth conditions lead to photosynthetic stress. Chlorophyll *a* fluorescence is a widely used probe of photosynthetic activity (specifically PSII), and therefore stress which specifically targets the electron transport pathway and associated alternative electron cycling pathways. By manipulating the processes that control photosynthesis, affecting the chlorophyll *a* fluorescence, yields detailed insight into the biochemical pathways. Light that is captured by a chlorophyll molecule can be utilised in three competing processes; electron transport, energy dissipation (via heat) and chlorophyll *a* fluorescence emission. Electrons produced by water-splitting are not always used in carbon fixation; if the incident irradiance generates more electrons than the dark reactions can use in carbon fixation, damage will occur to the photosynthetic apparatus. If carbon fixation is inhibited by temperature or reduced inorganic carbon (Ci), ATP or NADPH availability, then the photosystem dynamically adjusts and uses alternate sinks for electrons, such as molecular oxygen (water-water cycle or Mehler ascorbate peroxidase reaction). The process of stress acclimation leads to a number of photoprotective pathways and we describe how inhibitors can be used to identify these particular processes. In this chapter, we describe the processes controlling electron transport as influenced by light-induced stress.

P.J. Ralph (✉) and K. Petrou
Plant Function Biology & Climate Change Cluster,
University of Technology, PO Box 123, Broadway, Sydney
NSW, Austraila
e-mail: Peter.Ralph@uts.edu.au

C. Wilhelm and T. Jakob
Department of Plant Physiology, University of Leipzig,
D-04103, Leipzig, Johannisallee 23,
Germany

J. Lavaud
CNRS UMR6250 'LIENSs', Institute of Coastal and
Environmental Research (ILE), University of La Rochelle,
France

S.A. Kranz
Alfred Wegener Institute for Polar and Marine Research,
Am Handelshafen 12, 27570, Bremerhaven,

2 Electron Usage in Photosynthesis

Photosynthesis drives the light reactions which ultimately lead to carbon fixation; however predicting photosynthetic rates from fluorescence is a complex issue. As outlined in other chapters of this book (Chapter 3 and Chapter 6) different fluorescence tools are available to measure the electron flow through Photosystem II (PSII). The quantum yield of PSII can be multiplied by the amount of absorbed quanta which can be obtained from the incident light and either the PSII absorption cross section or the spectral overlap between the light spectrum and the *in-situ* absorption spectra. From these data, the electron transport rate per chlorophyll molecule over time can be assessed for an entire day to determine the daily primary production (Wagner et al. 2005). However, growth and photosynthesis are rarely equivalent. Electrons transported by PSII can follow several competing pathways: the majority of the electrons are normally used to reduce CO_2 to carbohydrates, allowing the synthesis of other cellular macromolecules like proteins, lipids or nucleotides, but some of them might be lost by alternate

D.J. Suggett et al. (eds.), *Chlorophyll* a *Fluorescence in Aquatic Sciences: Methods and Applications*, Developments in Applied Phycology 4, DOI 10.1007/978-90-481-9268-7_4, © Springer Science+Business Media B.V. 2011

cellular processes (see alternate electron cycling) or dissipated (non-photochemical quenching). Therefore, the ratio of electrons per carbon incorporated into the biomass may vary tremendously, either by losses or by the synthesis of highly reduced biomolecules like proteins or lipids.

2.1 Alternative Electron Cycling (AEC)

In principle, PSII electron transport rates should match the gross rates of oxygen evolution. Studies have tried to verify this assumption with divergent results (Falkowski et al. 1986; Kolber et al. 1998; Suggett et al. 2001; Jakob et al. 2005). It was shown that linearity between PSII electron transport and oxygen evolution can be found, but non-linear behaviour was also observed, especially under conditions when photosynthesis was over-saturated (excess irradiance). These experiments indicated that PSII electron transport might over-estimate the primary production under some conditions, because oxygen evolution rates were found to be lower than PSII electron flow (Gilbert et al. 2000). Several explanations for this disparity have been suggested:

(a) Cyclic electron flow around PSII (Prasil et al. 1996; Lavaud et al. 2002)
(b) Water-water cycle (Asada 1999) where oxygen uptake on the acceptor side of PSI leads to superoxide which is then dismutated to H_2O_2 and then detoxified to water and
(c) Cyclic electron flow around PSI (Bendall and Manesse 1996)

These processes can be summarised as alternative electron cycling (AEC) which are not energetic losses (such as non-photochemical quenching: NPQ), because at least the water-water cycle and the cyclic flow around PSI generate a proton gradient which can be used for additional ATP synthesis. Therefore, it is suggested that alternative electron cycling is a normal stress response and might be of less importance under balanced growth conditions. Recently, Wagner et al. (2005) described an experimental setup to estimate the alternative electron cycling activity by comparing the electron flow through PSII with oxygen evolution relative to the amount of absorbed quanta. The result is shown in Fig. 1.

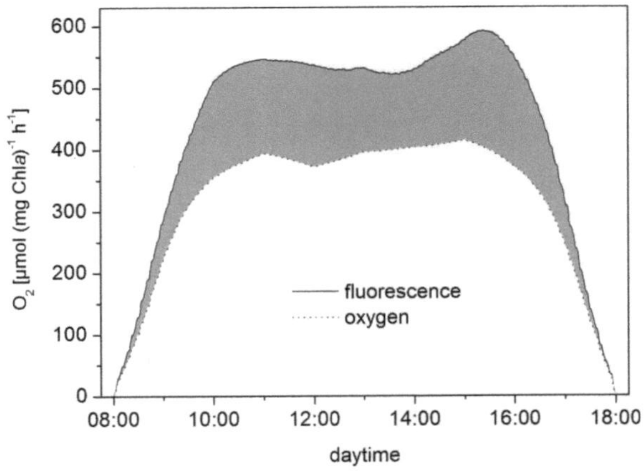

Fig. 1 Modelling of fluorescence and oxygen-based photosynthesis rates in *Phaeodactylum tricornutum* grown in a turbidostat under sine light conditions (10 h light period). Photosynthesis-irradiance curves were measured hourly and fitted using the dynamic model of Eilers and Peters (1988). With the derived fitting parameters, oxygen and fluorescence-based electron transport rates can be calculated for any given light intensity during the daily course of the light climate. The difference between fluorescence-based electron transport rates and oxygen-based photosynthesis rates (grey area between the curves) is linked to the proportion of alternative electron cycling

Obviously, at low light intensities in the morning and in the late afternoon, the fluorescence-based electron transport rates closely match the oxygen evolution rates as measured by a Clark-type electrode, whereas at high light intensities the "alternative electron cycling" can account for up to 40% of the fluorescence-based electron transport. This mismatch is not due to inappropriate measuring techniques, but to the physiological variability between linear and alternative electron pathways across the photosynthetic membrane. Interestingly, the ratio of linear to alternative electron cycling is not only light-dependent, but can be linked to species-specific physiological regulation, as shown in Fig. 2.

When light pulse frequency is manipulated, the relationship between electron transport and oxygen evolution is further altered. In the sine light (SL) climate which simulates a sunny day, the ratio PAM/oxygen is always higher than in the exponentially fluctuating light (FL) climate (Fig. 2), where the light intensity oscillates with a frequency of half an hour between the maximum value and zero. The green

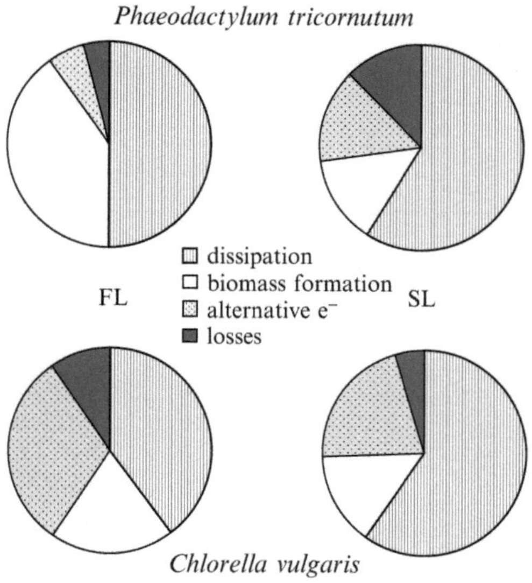

Fig. 2 The fate of absorbed photons in the comparison of *Phaeodactylum tricornutum* and *Chlorella vulgaris* grown in a turbidostat under dynamic light conditions (fluctuating light, FL; sine light, SL). Data are given as percentage of absorbed quanta (Q_{phar}). Energy losses by dissipation include the conversion of absorbed light into heat and fluorescence and were derived from $[(1-q_P)\ Q_{phar}]$, where q_P is the photosynthetic quantum efficiency at Photosystem II measured by PAM fluorescence. The amount of quanta lost by alternative electron sinks was calculated from the difference of fluorescence and oxygen-based photosynthesis rates and the assumption of a quantum efficiency of 0.125 [mol O_2 (mol quanta)$^{-1}$]. The amount of quanta used for biomass formation was derived from Φ_μ according to Jakob et al. (2007). Energy losses which are not directly quantifiable as absorbed quanta, like mitochondrial respiration, have been depicted as 'losses'

alga, *Chlorella vulgaris,* performs much more alternative electron transport than the diatom, *Phaeodactylum tricornutum,* under both light conditions (FL and SL). In the sine light where the cells are exposed to a photon flux which exceeds the capacity of the Calvin cycle, the alternative electron cycling is highest indicating, that it can act as a photoprotective mechanism which compliments other photoprotective processes. Given that alternative electron cycling in the green alga was higher than in the diatom, this corresponds with the observation that in diatoms the energy dissipation capacity of the diadinoxanthin/diatoxanthin xanthophyll cycle is more active than in green algae or higher plants (Ruban et al. 2004; Goss et al. 2006) and will be discussed later in this chapter. It should be noted that green algae and higher plants have a different suite of xanthophyll pigments to diatoms and dinoflagellates.

2.2 Electron Usage to Produce New Biomass

Under continuous light, the oxygen production or the uptake of inorganic carbon shows a clear linear relationship with biomass production (Toepel et al. 2004) indicating that 55–60 μmol oxygen released is equivalent to 1 mg dry weight. However, the ratio of oxygen released to carbon incorporated is highly variable, for several reasons. Firstly, the reduction in biomass strongly depends on the species and the environmental conditions (Kroon and Thoms 2006). For example, under N or P limitation, the relative proportion of carbon incorporated into carbohydrates is strongly increased and therefore the reduction in biomass is relatively low. The energetic cost of converting the products of the Calvin cycle into lipid or proteins are incorporated by higher rates of mitochondrial respiration. Therefore, it can not be expected that the ratio of oxygen production in the light, per oxygen molecule consumed in the dark to be constant. This ratio is modulated, not only by the availability of nutrients and the reduction in biomass, but also by the turn-over rates of proteins and lipids. It is well documented that cells growing under high-light have significantly higher mitochondrial respiration rates, as well as under nutrient replete conditions and optimal temperature (Wilhelm and Wild 1984). Table 1 shows that the ratio photosynthesis/respiration varies not only in response to the light climate (sine versus fluctuating light) but also with the C/N ratio. Therefore, the ratio of photosynthetic electrons to carbon incorporated into the newly formed cells has to be variable. However, such parameters have not been measured under an adequate range of conditions or with sufficient species to make broad speculation.

The conversion of electron transport rates into actual new biomass requires accurate estimates for the ratio of electrons per carbon in the macromolecules of the new biomass. In future, the FTIR spectroscopy (Stehfest et al. 2005) might become a tool to measure this parameter, and is also possible with single cells.

Table 1 Comparison of *Phaeodactylum tricornutum* and *Chlorella vulgaris* with respect to the activity of alternative electron transport (expressed as the ratio of fluorescence-based to oxygen-based photosynthesis rates; P_F/P_O), to C/N ratios (given as mol mol^{-1}), and the activity of mitochondrial respiration (expressed as the ratio of respiration rate to net photosynthesis rate; R/P_{net}). Algal cultures were grown in a turbidostat under dynamic light conditions (10 h and 12 h light periods) which have been applied either as a non-fluctuating sine light climate or as oscillating light (osc. Light). In addition, *P. tricornutum* was exposed to nitrate-limited conditions (N-limited) (Data are adapted from Wagner et al. 2006 and Jakob et al. 2007)

Species	Growth condition	P_F/P_O	C/N	R/P_{net}
P. tricornutum	Replete – sine light (10 h)	1.4	7.7	0.8
P. tricornutum	Replete – osc. light (10 h)	1.3	7.9	1.0
P. tricornutum	Replete – sine light (12 h)	1.6	6.6	0.4
P. tricornutum	Replete – osc. light (12 h)	1.1	6.7	1.4
P .tricornutum	N-limited – sine light (10 h)	1.6	14.5	0.7
P. tricornutum	N-limited – osc. light (10 h)	1.2	10.8	1.5
C. vulgaris	Replete – sine light (12 h)	2.1	6.8	0.4
C. vulgaris	Replete – osc. light (12 h)	2.0	6.8	0.9

This opens the perspective to improve the robustness of estimates for primary production by using advanced fluorescence techniques.

3 Effect of Light Stress on Fluorescence Signatures and their Interpretation

When captured light energy cannot be completely utilized for metabolic processes, the excess energy accumulates within the photosynthetic apparatus (Nixon and Mullineaux 2001). This typically occurs when the light intensity is too high (over minimum saturating irradiance: E_k). However, this also occurs when the cells are suddenly switched from a dark/low light environment or to a higher irradiance (not necessarily over E_k) depending on the physiological state of the cells and their response to other environmental cues. Accumulation of excess energy within the photosynthetic apparatus can be harmful for photosynthesis, and especially for the activity of PSII, because the over-reduction of the primary electron acceptor (Q_A) generates free radicals which leads to oxidative stress (Ledford and Niyogi 2005); stress which will ultimately cause a decrease in the photosynthetic rate (i.e. photoinhibition). Photosynthetic organisms have developed a number of fast photoprotective (or photoacclimative) processes to minimize the level of oxidative stress, especially linked to the dissipation of the excess absorbed energy (Niyogi 2000). Non-photochemical quenching (NPQ) is believed to be one of the most important of these mechanisms for the fast regulation of photosynthesis in higher plants as well as in algae (Szabo et al. 2005; Demmig-Adams and Adams 2006; Lavaud 2007). It should be noted that NPQ is not a form of AEC, but rather it is especially efficient for organisms growing in a fluctuating light environment where it helps to balance the absorption of light energy with its use, and ultimately plays a role in the maintenance of their fitness (Külheim et al. 2002; Ralph et al. 2002; Demmig-Adams and Adams 2006; Lavaud 2007).

Non-photochemical quenching (NPQ) originates in the light-harvesting antenna (LHC) of PSII. When the available excitation energy exceeds the photochemical capacity, it can then be dissipated as heat (or reallocated) before it reaches the PSII reaction center. This process arise from reactions not directly related to photochemistry, which have been defined as 'non-photochemical quenching' to be distinguished from the processes dealing with the 'photochemical quenching' (qP) which is directly related to photochemistry and the linear transport of electrons (Fig. 3a) (Maxwell and Johnson 2000; Baker 2008). In that framework, the redox state of quinones (Q_A and Q_B) and plastoquinones can strongly influence the emission of fluorescence in parallel to NPQ under high light conditions (Perkins et al. 2006). NPQ reduces the lifetime of excited chlorophylls ($^1Chl^*$) and thereby the quantum yield of Chl *a* fluorescence, which is seen by a decrease of F_m to F'_m level (see Fig. 3a). For that reason, it is calculated as $(F_m–F'_m)/F'_m$ (or $F_o–F'_o/F'_o$; Lavaud et al. 2002). In higher plants, green algae and dinoflagellates, where the NPQ mechanism has been investigated, it consists of three components (Fig. 3a) (Stroch et al. 2004;

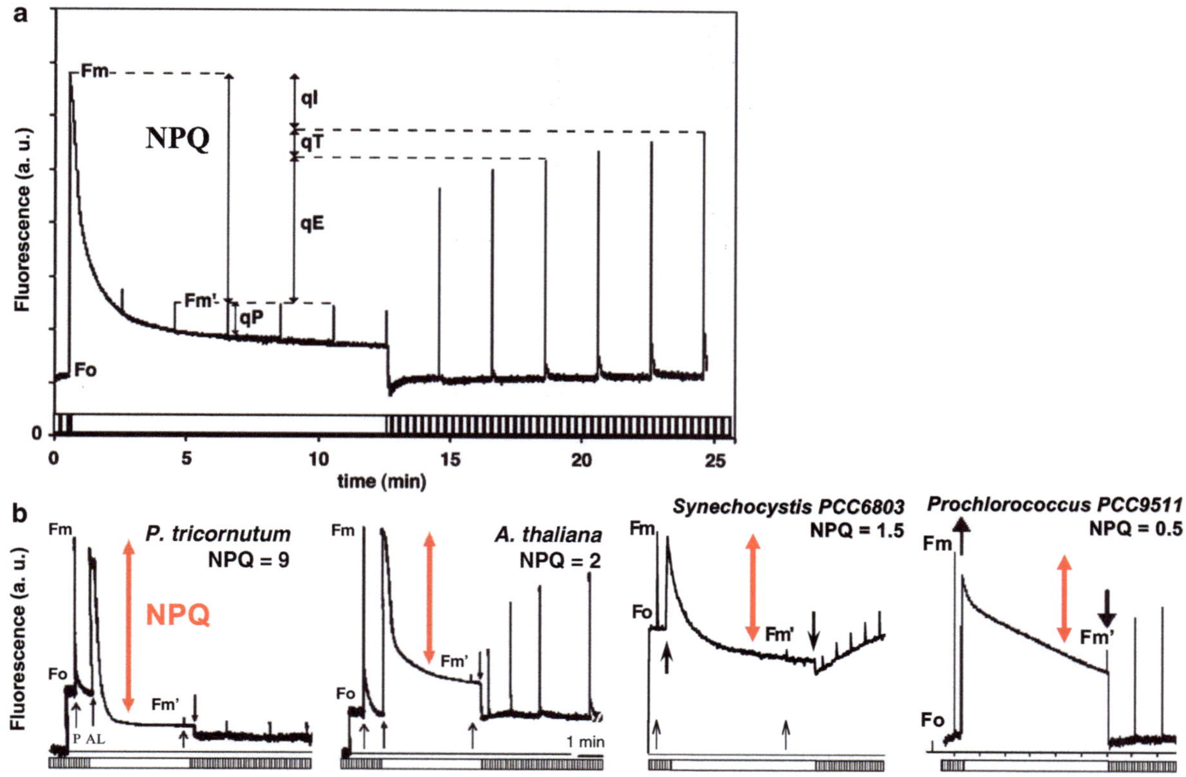

Fig. 3 (**a**) Chlorophyll *a* (Chl *a*) fluorescence signal as measured with a PAM fluorometer on an *Arabidopsis thalania* leaf. After dark-adaptation, in the presence of the detector beam (dashed bottom bar), the minimal fluorescence level (F_o) is measured. When a saturating light pulse (P) is given, the photosynthetic light reactions become saturated and fluorescence reaches a maximum level (F_m). Upon continuous actinic light (AL On, white bottom bar) with moderately excess light (750 μmol photons m^{-2} s^{-1}; growth light was 130 μmol photons m^{-2} s^{-1}), a combination of qP and NPQ lowers the fluorescence yield. NPQ (qE + qT + qI) is the difference between F_m and the measured maximal fluorescence after a saturating light pulse during illumination (F'_m): NPQ = (F_m-F'_m)/F'_m. After switching off the actinic light (AL Off), the quenching on the F_o level can be observed (F'_o). Also, the recovery of F'_m within a few minutes reflects relaxation of the qE component of NPQ. qT takes a longer time to relax while qI is a sustained quenching. Adapted from Müller et al. (2001). (**b**) Characteristic Chl *a* fluorescence signals as measured with a PAM fluorometer in cells of the diatom *Phaeodactylum tricornutum*, leaf of the higher plant *A. thalania*, cells of the cyanobacterium *Synechocystis* PCC6803, and cells of the Prochlorophyte *Prochlorococcus* PCC9511, illumination was: 5 min-2000 μmol photons m^{-2} s^{-1}. The time scale is given on the *A. thaliana* trace. Adapted from Ruban et al. (2004), Cadoret et al. (2004), Bailey et al. (2005). For (**a**) and (**b**): F_o, minimum fluorescence level in the dark; F'_o, minimum fluorescence level after light exposure (detector beam only for both); F_m, maximum fluorescence level in the dark; F'_m, maximum fluorescence level at light; AL, actinic continuous light (bold arrow up/down: AL on/off); P, over-saturating pulses (600–800 ms duration, thin arrows: pulse fire). Bars: dashed, detector beam only; white; detector beam+AL on

Hill et al. 2005; Szabo et al. 2005): the energy-dependent quenching (qE) which is regulated by the built-up of a trans-thylakoid proton gradient (ΔpH) and the operation of the xanthophyll cycle (XC); state-transition quenching (qT); which relies on the redistribution of excitation energy between photosystems by physical modulation of the cross-section of light-harvesting antennas (Ruban and Johnson 2009) and the sustained quenching which is heterogeneous (Demmig-Adams and Adams 2006) which partially depends on xanthophylls (Garcia-Mendoza and Colombo-Pallotta 2007) as well as on photoinactivation/photoinhibition (qI) of PSII (Stroch et al. 2004). Quantification of these three components is either based on their relaxation kinetics in the dark (Müller et al. 2001) or requires photosynthetic inhibitors (Horton and Hague 1988). The characteristics of their relaxation kinetics can vary according to environmental stresses and between groups of organisms. Such that, qE relaxes very rapidly (within tens of seconds after the offset of light), qT takes several minutes (shorter for cyanobacteria and rhodophytes), while qI is sustained and can last for hours even days under certain extreme environmental conditions (Demmig-Adams and Adams 2006; Garcia-Mendoza and Colombo-Pallotta 2007). Furthermore, in diatoms qE usually relaxes very slowly in comparison to higher plants (compare the two organisms in Fig. 3b) so that it could be confounded with qI due to overlaps with time. In general, with non-stressed leaves, qE is the major component under moderate to saturating irradiance. qI can become prominent under over-saturating irradiances and possibly in combination with other stresses (nutrient/water deficiency, temperature and salinity) (Demmig-Adams and Adams 2006). In this context, qT is not as relevant since it generally only makes a small contribution to overall relaxation of fluorescence (see Fig. 3a) (Nixon and Mullineaux 2001). qT is usually significant only under low light levels (Mullineaux and Emlyn-Jones 2005) while some dinoflagellates increase qT under thermal and light stress (Hill et al. 2005). The amplitude and kinetics of the whole NPQ process and the importance of each component (Fig. 3b) can be extremely divergent between taxa, especially among microalgal groups (Casper-Lindley and Bjorkman 1998; Hill et al. 2005; Juneau and Harrison 2005), and even between species within a taxonomic group (Lavaud et al. 2004; Lavaud et al. 2007). For example, qE shows high amplitude and fast onset in diatoms and brown macroalgae, while being of minor importance in most of the green microalgae (Finazzi et al. 2006) and cyanobacteria (Kirilovsky 2007). Nevertheless, within the diatoms (see Chapter 7) as well as higher plants (Johnson et al. 1993) there are clear differences in qE amplitude that have been highlighted. Whereas, qT is currently unknown in diatoms (Owens 1986) and brown macroalgae (Fork et al. 1991), and of only moderate importance in higher plants and dinoflagellates (Hill et al. 2005), yet highly developed in some green microalgae and cyanobacteria (Finazzi 2005; Mullineaux and Emlyn-Jones 2005).

Amongst the three components of NPQ, qE has a major influence on the Chl *a* fluorescence signal under normal growth conditions (Logan et al. 2007; see also Chapter 7). The interpretation of qE is possibly the most complex of the NPQ components, as it is linked to faster regulation of photosynthesis than qT and qI with most organisms, especially under naturally fluctuating environment (Lavaud 2007). Frenkel et al. (2007) demonstrated that qE is critical for maintaining the fitness of plants under natural temperate-light conditions, rather than qT. Also, in cyanobacteria and green microalgae, qT has no significant physiological importance in photoprotection towards high-light stress, yet is more relevant in low light conditions (Mullineaux and Emlyn-Jones 2005; Ruban and Johnson 2009) and for acclimation to different light quality (Pfannschmidt 2005). The qT mechanism has been documented, as well as its impact on the fluorescence signal, especially in cyanobacteria and green microalgae (see Campbell et al. 1998; Nixon and Mullineaux 2001; Finazzi 2005; Mullineaux and Emlyn-Jones 2005). Even though qI has been well documented in some species of higher plants growing in extreme environments (Demmig-Adams and Adams 2006), its occurrence and control mechanism remains unknown in some of the algal groups. Also, the part of qI which depends on xanthophylls is also linked to the qE process (Demmig-Adams and Adams 2006), although clear mechanistic differences have only been recently demonstrated (Dall'Osto et al. 2005).

The qE mechanism has been described in a molecular context for higher plants and green microalgae (Standfuss et al. 2005; Cogdell 2006; Ruban et al. 2007). The machinery triggering and controlling qE amplitude and kinetics is now quite well known for groups of algae like the diatoms and brown macroalgae (Goss et al. 2006; Lavaud 2007), as well as in the cyanobacteria and prochlorophytes (Bailey et al. 2005; Kirilovsky 2007). The NPQ process is based on a feed-back reaction from the linear electron transport through the build-up of a trans-thylakoid ΔpH and subsequent acidification of the thylakoid lumen (see Nixon and Mullineaux 2001). Consequently, the activity of the ATP synthase (Dal Bosso et al. 2004) the cytochrome b_6f (Munekage et al. 2001), or the cyclic electron flow around PSI (Miyake et al. 2005) can indirectly influence qE. Hence, in a simple direct relationship, the higher the irradiance, the higher the electron transport rate, the higher the accumulation of protons in the lumen, the higher qE. In some organisms such as diatoms, it appears there is a relative independence of the PSII redox-state from the proton-motive electron transfer and subsequent NPQ (Ruban et al. 2004; Lavaud et al. 2007). To summarize NPQ responses, the lumen acidification triggers two events (Fig. 4): (1) the protonation of specific sites of the LHC antenna, and (2) the activation of an enzyme, a de-epoxidase, which drives the conversion of epoxized xanthophyll to a de-epoxidized form. This conversion is reversible as the backward reaction is driven by an epoxidase which also depends on the trans-thylakoid ΔpH. The accumulation of de-epoxidized xanthophylls thus depends on the balance between the activity of both enzymes within the xanthophyll cycle (XC) (see Lavaud 2007 for a detailed description). In higher plants, green microalgae and brown macroalgae, the XC involves the conversion of violaxanthin to zeaxanthin (ZX) via antheraxanthin (AX) (Fig. 4) while the diatoms and dinoflagellates use diadinoxanthin (DD) which is converted to diatoxanthin (DT) under elevated light (see MacIntyre et al. – Chapter 7). Both protonated LHC protein(s) and the presence of DT or ZX/AX in the LHC antenna of PSII are thought to act together as the trigger of the qE. The whole LHC antenna switches into a dissipative mode when excess excitation energy should be converted into heat while Chl *a* fluorescence is quenched (Fig. 4) (Stroch et al. 2004; Szabo et al. 2005). More precisely, protonation would promote and transduce conformational changes ('aggregations') which bring pigments closer together and especially chlorophyll/xanthophyll molecules. In higher plants the 'special' polypeptide which undergoes protonation is PsbS (Niyogi et al. 2005). The function of PsbS is

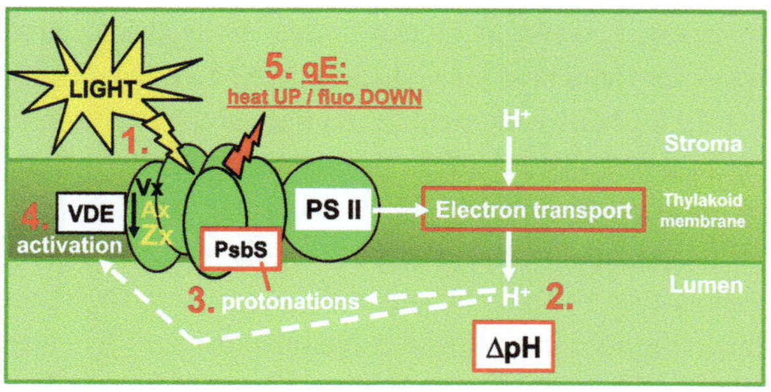

Fig. 4 Simplified model of the qE mechanism in higher plants (see the text for a full description). The numbering refers to the sequence of the qE process steps. AX, antheraxanthin; H^+, protons; PS II, photosystem II; VDE, violaxanthin de-epoxidase; VX, violaxanthin; ZX, zeaxanthin; ΔpH, transthylakoid proton gradient. (Adapted from Lavaud 2007)

essentially to sense the lumen pH, that is linked to several H^+-binding amino acid residues present on the luminal loops of this protein (Dominici et al. 2002; Li et al. 2002). In green microalgae (Peers, G and Niyogi, KK, personal communication, 2008) and diatoms (Zhu and Green 2008), the Li-818 proteins which are up-regulated under high light, could play a similar role as PsbS in qE. Simultaneously, with PsbS protonation, de-epoxidized xanthophylls would also act as 'allosteric regulators' by amplifying the conformational changes within the whole LHC antenna. The physical process by which excitation energy is effectively converted into heat has only recently been understood (Holt et al. 2005; Pascal et al. 2005; Ruban et al. 2007). The qE mechanism is rather similar in other organisms like the diatoms and the brown macroalgae given some peculiarities (see Chapter 7). In other groups like the red algae, cyanobacteria and prochlorophytes, the process is quite different even though it involves xanthophylls and special proteins of the antenna system (Lavaud 2007). Therefore, qE in these taxa is not as controlled as in higher plants or diatoms since these organisms do not display a finely regulated xanthophyll cycle, also cyanobacteria and prochlorophytes show no involvement of a trans-thylakoid ΔpH (Kirilovsky 2007).

Quenching based in the PSII reaction center (as opposed to LHC antenna) has also been observed in higher plants (Bukhov et al. 2001; Stroch et al. 2004) and green microalgae (Finazzi et al. 2004), and possibly in diatoms (Eisenstadt et al. 2008). Appearance of reaction centre quenching depends on the balance between light and carbon fixation fluxes (Finazzi et al. 2004) along with a clear temperature influence (Kornyeyev et al. 2004). This quenching appears to drive both qE and qI components of NPQ with both fast and slow relaxation kinetics, respectively. In contrast to the antenna-based quenching, it cannot cause changes in the F_o level (Maxwell and Johnson 2000). Nevertheless, as well as the antenna-based quenching it requires thylakoid acidification, but it does not require de-epoxidized xanthophylls (Bukhov et al. 2001; Finazzi et al. 2004). The qI part of this reaction center based quenching is associated with a reversible inactivation of a sub-population of the PSII (Finazzi et al. 2004) as well as with PSII photodamage (Kornyeyev et al. 2004).

4 Use of Chemicals for the Differentiation of Photosynthetic Processes

Photosynthesis is a complex interaction of complementary processes such as alternate electron cycling (AEC) and non-photochemical quenching (NPQ). A common method of isolating specific processes is using biochemical inhibitors such as herbicides. Electron transport inhibitors, uncouplers, artificial electron acceptors and donors have all proved to be essential tools in elucidating the function of various components of the photosynthetic electron transport chain, metabolic pathways and photosynthetic regulatory processes. Using herbicides to understand the regulation of photosynthesis and related biochemical pathways requires the basic understanding of how these herbicides interact with the photosynthetic apparatus.

Determining the appropriate concentration of herbicide is very important and often problematic, because depending on the organism, cells can have different cell wall composition, membrane transporters and a variation in the number of reaction centres per cell, thus requiring different concentrations of herbicide (Durnford et al. 1998). Therefore, any concentrations specified herein are only an indication of what has been used based the range of concentrations found in the literature. The most effective and correct way to determine the concentration at which a cellular response occurs is by titration of the herbicide against a known cell density or chlorophyll *a* concentration while measuring the physiological impact (oxygen evolution or chl *a* fluorescence).

4.1 Inhibitors of Linear Electron Transport

DCMU (3′-(3,4-dichlorophenyl)-1′,1′-dimethylurea), also known as Diuron, is the most extensively used inhibitor of photosynthetic electron transport. DCMU inhibits electron transport between PSII and PSI, impacting on the acceptor side of PSII by supplanting a bound plastoquinone from the Q_B^- binding site of PSII (Fig. 5). Binding of this herbicide to the Q_B^- site of PSII, results in the effective blocking of electron flow and leads to the subsequent inhibition of photosynthesis. Blocking of electron flow is a consequence of the herbicide being incapable of receiving electrons, and therefore electrons remain trapped in Q_A, the first quinone acceptor (Kleczkowski 1994). This trapping of electrons prevents the reduction of plastoquinone, by holding the electrons in the D1 dimer, thus affecting the redox state of the PQ pool, which becomes completely oxidised (Durnford et al. 1998).

DCMU causes a rapid increase to maximum fluorescence (F_m), where all PSII reaction centres are closed and the plastoquinone pool fully oxidised (Trebst 2007). DCMU has no impact on the membrane potential of the thylakoid in darkness, yet completely inhibits the light-induced membrane pH gradient. The amount of DCMU required for the inhibition of 50% of PSII reaction centres will vary depending on cell concentration and species. Published concentrations range from 1–20 μM (Falkowski and Raven 2007). However, incremental increases in the amount of DCMU added to cells will result in changes in variable fluorescence and the rate of Q_A re-oxidation (Durnford et al. 1998), which will invariably allow for the determination of the appropriate concentration of DCMU needed to elicit the desired effect. In addition, the light acclimation state of the cells needs to be taken into account, as cells grown at low photon flux densities will have a plastiquinone (PQ) pool that is predominantly oxidized and therefore the addition of DCMU will have very little effect on the redox state of the PQ pool (Durnford et al. 1998) similarly, chlororespiration can alter the PQ redox state, allowing fluorescence yield to occur with saturating DCMU concentrations (Wilhelm and Duval 1990). In the presence of saturating DCMU concentrations, fluorescence yield becomes maximum ($F_o \Rightarrow F_m$ so $F_v/F_m \Rightarrow 0$), as Q_A can no longer pass electrons to PQ, so electron transport stops and the maximum amount of captured energy is dissipated as fluorescence.

Like DCMU, DBMIB (2,5-dibromo-3-methyl-6-isopropyl-p-benzoquinone) is also an inhibitor of electron transport, however it blocks further along the electron transport chain near the Cytochrome b_6f complex (Trebst 2007). DBMIB is thought to interfere at the Reiske iron-sulfur centre (Trebst 2007), thus blocking photosynthetic electron flow through the Cytochrome b_6f complex. DBMIB binds close to the Qo pocket (Cramer et al. 2006), the plastoquinol binding site of the Cytochrome b_6f complex (Fig. 5), inhibiting the reoxidation of PQH_2 thus keeping the PQ pool completely reduced (Trebst 1980).

Some precaution should be taken when using DBMIB, because site of action is concentration dependent, as well as redox sensitive, where DBMIB becomes reduced under light and oxidised in the dark (Bukhov et al. 2003). At low concentrations, DBMIB inhibits electron transport on the reducing side of the PQ, but at

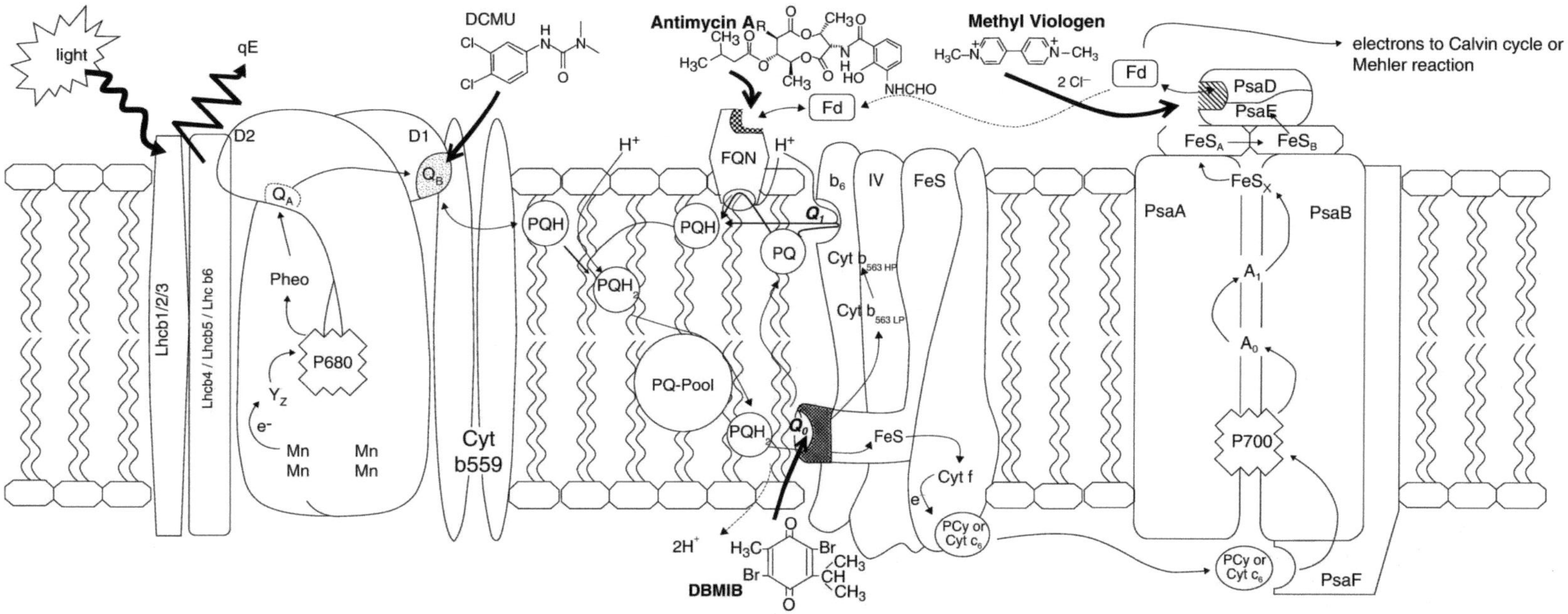

Fig. 5 Electron transport flow through PSII, cytochrome b_6f complex and PSI. The D1, D2 and Cyt b559 proteins are shown. The thin arrows indicate the electron flow pathway through PSII, while the thick arrow indicates the site of inhibition by DCMU. The figure shows the DCMU molecule binding to the Q_B site of PSII, thereby effectively blocking the continuation of electron flow from PSII to the plastoquinone, cytochrome b_6f complex and onto PSI. After Q_B the thin arrows indicate electron flow pathways and proton (H^+) pathways. The thick arrow in the lower middle of the diagram indicates the site of impact by the inhibitors DBMIB and antimycin A. One DBMIB molecule competes with the PQH_2 resulting in the blocking of the release of electrons from the plastoquinone to PSI at the Fe-S complex. Antimycin A (upper middle) inhibits the reduction of ferrodoxin in PSI, intercepting electron transport at the ferrodoxin-plastoquinone reductase, resulting in the blocking of cyclic electron transport. Electrons move through PSI indicated by thin arrows, while the thick arrow (upper right) indicates the impact site of methyl viologen. Methyl viologen interacts at the binding site of ferredoxin, accepting the terminal electron, thus preventing the reduction of ferredoxin and the continued pathway of the electron to carbon fixation or cyclic electron transport

higher concentrations excess DBMIB will inhibit the Q_B site of PSII, located on the oxidising side of the PQ (Moreland 1980; Rich et al. 1991). To prevent fluorescence quenching from oxidised DBMIB, it can be used in conjunction with an excess of sodium ascorbate (Kufryk and Vermaas 2006). For every Cytochrome b_6f complex, one molecule of $DBMIB_{red}$ is needed for complete inhibition of electron transfer through the Cytochrome b_6f complex (Rich et al. 1991). DBMIB can also inhibit mitochondrial electron transport (Durnford et al. 1998). This is a good example of co-inhibition, where some inhibitors have more than one impact site, and therefore interpretation of results must carefully consider the possibility of an alternate component of the cell machinery being affected by the inhibitor. DBMIB reduces minimum fluorescence (F_0) as well as the rise of variable fluorescence (F_v). While $DBMIB_{red}$ quenches chlorophyll *a* fluorescence, it does so less efficiently than the oxidised form ($DBMIB_{ox}$) and both forms alter NPQ estimates (Tyystjarvi et al. 1999).

4.2 Inhibitors of Cyclic Electron Transport

The antibiotic, antimycin A, is an effective inhibitor of one of the alternate electron cycles (AEC), cyclic electron transport around PSI (Tagawa et al. 1963). It has been proposed that inhibition of photosynthetic electron transport by antimycin A is associated with the ferredoxin-plastoquinone reductase (FQR) activity in cyclic electron transport (Simonis and Urbach 1973; Moss and Bendall 1984; Cleland and Bendall 1992). In addition to inhibiting cyclic electron transport, antimycin A is also known to inhibit excess light energy dissipation measured through NPQ (Oxborough and Horton 1987). The decline in qE (energy-dependent quenching) formation in the presence of this antibiotic is due to a change in the redox state of the electron transport chain. However, since antimycin A has no direct impact on linear electron transport rate, the redox change is most likely the result of a change in the redox state of a component located in the cytochrome complex (Oxborough and Horton 1987). Before inhibiting cyclic electron transport, it is important to understand that there are two potential transport pathways that cycle around PSI (Joët et al. 2001; Munekage et al. 2004). The first, cycles electrons from ferrodoxin to the PQ pool and is sensitive to antimycin A, while the second, involves the NDH complex which is insensitive to the antibiotic (Joët et al. 2001). In the case of the NDH cycle, it is not yet fully understood and no known inhibitor has been identified. Published concentrations of antimycin A range from 0.1–50 μM (Falkowski and Raven 2007).

4.3 Inhibitors of Alternative Electron Cycling (AEC)

Distinguishing between different electron pathways is important to describe the discrepancies often seen between oxygen evolution and chlorophyll *a* fluorescence under stressful conditions. Molecular oxygen can be reduced downstream of PSII at various sites, using different forms of AEC. In the case of the Mehler reaction, oxygen is reduced at the acceptor side of PSI (Mehler 1951) where it competes for electrons with both linear and cyclic electron transport pathways (Heber 2002). The Mehler reaction itself cannot be inhibited; however, the addition of potassium cyanide (KCN) can be used to inhibit the formation of H_2O and monodehydroascorbate (MDA) during the ascorbate peroxidase reaction, which is part of the Mehler cycle (Neubauer and Yamamoto 1992). The inhibition of H_2O formation as a result of altered peroxide turnover, impacts on the zeaxanthin-dependent light energy dissipation, by suppressing zeaxanthing formation and consequently NPQ (Neubauer and Yamamoto 1992). In fluorescence, the addition of KCN results in a decline in NPQ as well as a decrease in linear electron flow, shown as a suppression of q_P (Neubauer and Yamamoto 1992). Published KCN concentrations vary from 0.1 mM to 3 mM (Neubauer and Yamamoto 1992, Hormann et al. 1994, Singh et al. 1996).

Another more recently discovered pseudo-cyclic electron transport pathway which cycles around PSII via the plastoquinol terminal oxidase (PTOX), reduces molecular oxygen by utilising electrons from the PQ pool to generate H_2O (Cournac et al. 2000; Peltier and Cournac 2002; Josse et al. 2003). This alternative electron flow around PSII (upstream of PSI and Cytochrome b_6f) is believed to be advantageous in both a high-light environment and under iron limitation (Bailey et al. 2008), as it alleviates PSII excitation pressure by transporting electrons directly to oxygen while simultaneously ensuring that the electrons bypass the iron-demanding cytochrome b_6f and PSI complexes

(Mackey et al. 2008). Propyl-gallate (PGal) is an oxidase inhibitor specific to PTOX (Cournac et al. 2000; Bailey et al. 2008). PGal helps determine the role PTOX plays in alternative electron flow, and establish whether or not electrons are being used to reduce oxygen through PTOX activity (Mackey et al. 2008). The addition of 1 mM PGal results in a decrease in electron flow through PSII (Bailey et al. 2008; Mackey et al. 2008), highlighting the role the oxidase plays in keeping the PSII reaction centres oxidised in cells where Cytochrome b_6f and PSI activity are limiting. As in the case of DBMIB, PGal has more than one impact site in eukaryotes, as it can also lead to the inhibition of mitochondrial electron transport (Bailey et al. 2008).

4.4 Inhibitors of CO_2 Fixation

Iodoacetamide has been used as an inhibitor of carbon fixation (Miller et al. 1988; Miller and Canvin 1989), however when added during steady state photosynthesis, it inhibits CO_2 very slowly and may induce O_2 uptake in the light (Miller and Canvin 1989). An alternative inhibitor of CO_2 fixation is D, L-glyceraldehyde (Stokes and Walker 1972), which at very high concentrations (>25 mM) completely inhibits CO_2 fixation (Shelp and Canvin 1989) and blocks the conversion of triose-P to ribulose-1,5-bisphosphate (Stokes and Walker 1972). However, more recently, glycolaldehyde (GA) has become the preferred inhibitor of CO_2 fixation, as it uses concentrations an order of magnitude lower than those of D, L-glyceraldehyde, while rapidly and effectively inhibiting CO_2 fixation (Sicher 1984) without inhibiting CO_2 or HCO_3^- transport (Miller and Canvin 1989; Rotatore et al. 1992). The addition of GA to cells eliminates the chlorophyll *a* fluorescence quenching that is seen with the addition of inorganic carbon; however, oxygen evolution is greatly impacted by the presence of GA (Miller and Canvin 1989).

4.5 Electron Transport Uncouplers

Uncouplers function by dissociating electron transport from ATP synthesis during photosynthetic phosphorylation (Moreland 1980; McCauley et al. 1987). This is accomplished by dissipating the energised state (H^+) of the membrane (ΔpH) before the energy can be utilised in ADP phosphorylation (Moreland 1980) and thus prevent the formation of the trans-thylakoid ΔpH gradient. In addition to this major effect on the energy budget of the cell, the electron flow continues but the collapsed proton gradient no longer regulates electron transport rate. This type of inhibitor can be useful when examining processes triggered by ΔpH, such as NPQ and in particular qE. Common uncouplers of photophosphorylation include ammonia chloride ((NH_4Cl), carbonyl cyanide 4-trifluoromethyloxyphenylhydazone (FCCP) and nigericin.

Ammonium chloride (NH_4Cl) is a potent uncoupler of electron transport. As described above, it works in the classical way by relaxing the pH gradient across the thylakoid membrane, inhibiting ATP synthesis. The addition of NH_4Cl before the application of saturating light will prevent all quenching of F'_m. In contrast, if the uncoupler is added after fluorescence quenching has already formed (following a series of saturating pulses), it will result in a complete reversal of all F'_m quenching (Delphin et al. 1998).

Carbonyl cyanide p-trifluoromethoxy phenylhydrazone (FCCP) is a powerful uncoupler of photophosphorylation. It acts as an ionophore completely dissipating the pH gradient, while leaving the electron transport system uninhibited (Canaani and Havaux 1990). FCCP prevents the long-term fluorescence induction, meaning that the inhibition of the induction is likely the result of an increase in the dark decay processes (Canaani and Havaux 1990). Typical concentrations of FCCP are 1–10 μM (Shyam et al. 1993; Sigalat et al. 1993; Singh et al. 1996). At low concentrations FCCP quenches PSII fluorescence, indicative of the reoxidation of Q_A^- (McCauley et al. 1987), while it requires much higher concentrations to perform in its function as an uncoupler of oxidative phosphorylation (Canaani and Havaux 1990). When incubated with cells under photoinhibitory light, FCCP accelerates photoinhibition and rapidly quenches fluorescence yield (McCauley et al. 1987; Shyam et al. 1993; Singh et al. 1996).

Another type of uncoupler is the protonophore, such as nigericin which dissipates the proton gradient across the thylakoid membrane. Nigericin relaxes the ΔpH gradient by antiporting H^+ at the expense of K^+ across membranes, resulting in the collapse qE (Pressman et al. 1967). As a result of a breakdown in the pH gradient, the addition of nigericin to illuminated samples, results in an increase in F_m' and strong inhibition of NPQ with a concomitant large increase in steady state fluorescence F_t. The typical concentration range for nigericin is 1–5 mM (Falkowski and Raven 2007).

4.6 Electron Acceptors

Electron acceptors are compounds with very strong reducing capacity, such as methyl viologen (N,N′-Dimethyl-4,4′-bipyridinium dichloride; MV^{2+}) also known as Paraquat. Methyl viologen is an artificial electron acceptor, intercepting electron flow between PSI and the Calvin cycle by competing with ferrodoxin for the binding site at PSI (Fig. 5) (Dan Hess 2000). MV^{2+} is an extremely powerful electron acceptor, due to the nature of the bipyridinium salts, which temporarily become a stable radical with the addition of an electron, neutralising the positive charge of the cation (Moreland 1980; Peon et al. 2001). MV^{2+} oxidizes the primary acceptor (ferrodoxin) of linear electron transport, allowing a ΔpH to become established. However, this temporary neutral radical rapidly reverts back to its ion form, a process that results in the production of superoxide radicals (Hormann et al. 1993; Dan Hess 2000). MV^{2+} can be used to demonstrate damage to the electron transport chain beyond PSI (typically Calvin cycle), where incubation with MV^{2+} will oxidise the electron transport chain and increase Φ_{PSII} by supplementing the slow carbon fixation rate. In the presence of MV^{2+}, non-photochemical quenching (NPQ) is reduced, because the excess electrons that are usually held up by the Calvin cycle, are being accepted by the MV^{2+} allowing for continual rapid electron transport and a reduced need for excess light energy dissipation in the form of NPQ. Published concentrations of MV^{2+} range from 0.05 to 1 mM (Falkowski and Raven 2007).

In conclusion, we have illustrated how photosynthetic electron transport is strongly influenced by a range of internal feedback processes (AEC and NPQ) to ensure maximum efficiency, whilst preventing potential damage from excess excitation energy. Light stimulated processes such as NPQ are closely linked with pigments, however the control mechanisms are species-specific and show wide variability. Chemical inhibitors can be used to isolate specific components of the electron transport chain allowing a mechanistic understanding of the control of these photosynthetic pathways.

Acknowledgement PR thanks the Australian Research Council for financial support. CW and TJ thank the Deutsche Forschungsgemeinschaft for financial support Wi 764/12. JL thanks the Centre National de la Recherche Scientifique (CNRS), the Agence Nationale de la Recherche (ANR program VasiRéMi) and the Deutsche Forschungsgemeinschaft (DFG, grant LA2368/2-1) for their financial support. Ross Hill is thanked for editorial comments.

References

Asada K (1999) The water-water cycle in chloroplasts: scavenging of active oxygens and dissipation of excess photons. Annu Rev Plant Physiol Plant Mol Biol 50:601–639

Baena-Gonzalez E, Allahverdiyeva Y, Svab Z Maliga P, Josse EM, Kuntz M, Maenpää, Aro EM (2003) Deletion of the tobacco plastid psbA gene triggers an upregulation of the thylakoid-associated NAD(P)H dehydrogenase complex and the plastid terminal oxidase (PTOX). Plant J 35:704–716

Baker NR (2008) Chlorophyll fluorescence: a probe of photosynthesis *in vivo*. Annu Rev Plant Biol 59:89–113

Bailey S, Mann NH, Robinson C, Scanlan P (2005) The occurrence of rapidly reversible non-photochemical quenching of chlorophyll a fluorescence in cyanobacteria. FEBS Lett 579:275–280

Bailey S, Melis A, Mackey KRM, Cardol P, van Dijken FGG, Berg G, Arrigo KR, Shrager J, Grossman A (2008) Alternative photosynthetic electron flow to oxygen in marine *Synechococcus*. Biochim Biophys Acta 1777:269–276

Bukhov NG, Heber U, Wiese C, Shuvalov VA (2001) Energy dissipation in photosyntesis: does the quenching of chlorophyll fluorescence originate from antenna complexes of photosystem II or from the reaction center? Planta 212: 749–758

Bukhov NG, Sridharan G, Egorova EA, Carpentier R (2003) Interaction of exogenous quinones with membranes of higher plant chloroplasts: modulation of quinone capacities as photochemical and non-photochemical quenchers of energy in Photosystem II during light-dark transitions. Biochim Biophys Acta - Bioenergetics 1604:115–123

Campbell D, Hurry V, Clarke AK, Gustafsson P, Oquist G (1998) Chlorophyll fluorescence analysis of cyanobacterial photosynthesis and acclimation. Microbiol Mol Biol Rev 62:667–683

Canaani O, Havaux M (1990) Evidence for a biological role in photosynthesis for Cytochrome b-559-a component of Photosystem II reaction center. PNAS 87:9295–9299

Casper-Lindley C, Bjorkman O (1998) Fluorescence quenching in four unicellular algae with different light-harvesting and xanthophyll-cycle pigments. Photosynth Res 56:277–289

Cleland RE, Bendall DS (1992) Photosystem I cyclic electron transport: measurement of ferredoxin-plastoquinone reductase activity. Photosynth Res 34:409–418

Cogdell RJ (2006) The structural basis of non-photochemical quenching revealed? Trends Plant Sci 11:59–60

Cournac L, Redding K, Ravenel J, Rumeau D, Josse EM, Kuntz M, Peltier G (2000) Electron flow between photosystem II and oxygen in chloroplasts of photosystem I-deficient algae is mediated by a quinol oxidase involved in chlororespiration. J Biol Chem 275:17256–17262

Cramer WA, Zhang H, Yan J, Kurisu G, Smith JL (2006) Transmembrane traffic in the Cytochrome b6 f Complex. Annu Rev Biochem 75:769–790

Dal Bosso C, Lezhneva L, Biehl A, Leister D, Strotmann H, Wanner G, Meurer J (2004) Inactivation of the chloroplast ATP synthase g subunit results in high non-photochemical fluorescence quenching and altered nuclear gene expression in *Arabidopsis thaliana*. J Biol Chem 279:1060–1069

Dall'Osto L, Caffarri S, Bassi R (2005) A mechanism of non-photochemical energy dissipation, independent from PsbS, revealed by a conformational change in the antenna protein CP26. Plant Cell 17:1217–1232

Dan Hess F (2000) Review: light-dependent herbicides: an overview. Weed Sci 48:160–170

Delphin E, Duval JC, Etienne AL, Kirilovsky D (1998) ΔpH-dependent photosystem II fluorescence quenching induced by saturating, multiturnover pulses in red algae. Plant Physiol 118:103–113

Demmig-Adams B, Adams WW (2006) Photoprotection in an ecological context: the remarkable complexity of thermal energy dissipation. New Phytol 172:11–21

Dominici P, Caffarri S, Armenante F, Ceoldo S, Crimi M, Bassi R (2002) Biochemical properties of the PsbS subunit of photosystem II either purified from chloroplast or recombinant. J Biol Chem 25:22750–22758

Durnford DG, Prasil O, Escoubas JM, Falkowski P (1998) Assessing the potential for chloroplast redox regulation of nuclear gene expression. Meth Enzymol 297:220–234

Eisenstadt D, Ohad I, Keren N, Kaplan A (2008) Changes in the photosynthetic reaction centre II in the diatom *Phaeodactylum tricornutum* result in non-photochemical fluorescence quenching. Environ Microbiol 10:1997–2007

Falkowski PG, Raven JA (2007) Aquatic photosynthesis, 2nd edition. Princeton University Press, Princeton, p 484

Falkowski PG, Wyman K, Ley AC, Mauzerall DC (1986) Relationship of steady-state photosynthesis to fluorescence in eukaryotic algae. Biochim Biophys Acta 849:183–192

Finazzi G (2005) The central role of the green alga *Chlamydomonas reinhardtii* in revealing the mechanism of state transitions. J Exp Bot 56:383–388

Finazzi G, Johnson GN, Dallosto L, Joliot P, Wollman F-A, Bassi R (2004) A zeaxanthin-independent nonphotochemical quenching mechanism localized in the photosystem II core complex. Proc Nat Acad Sci USA 101:12375–12380

Finazzi G, Johnson GN, Dall'Osto L, Zito F, Bonente G, Bassi R, Wollman F-A (2006) Non-photochemical quenching of chlorophyll fluorescence in *Chlamydomonas reinhardtii*. Biochemistry 45:1490–1498

Fork DC, Herbert SK, Malkin S (1991) Light energy distribution in the brown alga *Macrocystis pyrifera* (giant kelp). Plant Physiol 95:731–739

Frenkel M, Bellafiore S, Rochaix J-D, Jansson S (2007) Hierarchy amongst photosynthetic acclimation responses for plant fitness. Physiol Plant 129:455–459

Garcia-Mendoza E, Colombo-Pallotta MF (2007) The giant kelp *Macrocystis pyrifera* presents a different non-photochemical quenching control than in higher plants. New Phytol 173:526–536

Gilbert M, Richter M, Wilhelm C (2000) Bio-optical modelling of oxygen evolution using *in-vivo* fluorescence: comparison of measured and calculated P-I curves in four representative phytoplankton species. J Plant Physiol 157:307–314

Goss R, Ann Pinto E, Wilhelm C, Richter M (2006) The importance of a highly active and delta-pH-regulated diatoxanthin epoxidase for the regulation of the PS II antenna function in diadinoxanthin cycle containing algae. J Plant Physiol 163:1008–1021

Heber U (2002) Irrungen, Wirrungen? The Mehler reaction in relation to cyclic electron transport in C3 plants. Photosynth Res 73:223–231

Hill R, Frankart C, Ralph PJ (2005) Impact of bleaching conditions on the components of non-photochemical quenching in the zooxanthellae of a coral. J Exp Mar Biol Ecol 322:83–92

Holt NE, Zigmantas D, Valkunas L, Li XP, Niyogi KK, Fleming GR (2005) Carotenoid cation formation and the regulation of photosynthetic light harvesting. Science 307:433–436

Hormann H, Neubauer C, Asada K, Schreiber U (1993) Intact chloroplasts display pH 5 optimum of O_2-reduction in the absence of methyl viologen: indirect evidence for a regulatory role of superoxide protonation. Photosynth Res 37:69–80

Horton P, Hague A (1988) Studies on the induction of chlorophyll fluorescence in isolated barley protoplasts IV. Resolution of non-photochemical quenching. Biochim Biophys Acta 932:107–115

Jakob T, Schreiber U, Kirchesch V, Langner U, Wilhelm C (2005) Estimation of chlorophyll content and daily primary production of the major algal groups by means of multiwavelength-excitation PAM chlorophyll fluorometry: performance and methodological limits. Photosynth Res 83:343–361

Jakob T, Wagner H, Stehfest K, Wilhelm C (2007) Uncoupling of growth rates and biomass production under nitrate limitation in combination with dynamic light conditions in the diatom *Phaeodactylum tricornutum*. J Exp Bot 58: 2101–2113

Joët T, Cournac L, Horváth EM, Medgyesy P, Peltier G (2001) Increased sensitivity of photosynthesis to antimycin A induced by inactivation of the chloroplast *ndhB* gene Evidence for a participation of the NADH-dehydrogenase complex to cyclic electron flow around photosystem I. Plant Physiol 125:1919–1929

Johnson GN, Young AJ, Scholes JD, Horton P (1993) The dissipation of excess excitation energy in British plant species. Plant Cell Env 16:673–679

Juneau P, Harrison PJ (2005) Comparison by PAM fluorometry of photosynthetic activity of nine marine phytoplankton grown under identical conditions. Photochem Photobiol 81:649–653

Kirilovsky D (2007) Photoprotection in cyanobacteria: the orange carotenoid protein (OCP)-related non-photochemical quenching mechanism. Photosynth Res 93:7–16

Kleczkowski LA (1994) Glucose activation and metabolism through UDP-glucose pyrophosphorylase in plants. Phytochemistry 37:1507–1515

Kolber ZS, Prasil O, Falkowski PG (1998) Measurements of variable chlorophyll fluorescence using fast repetition rate techniques: defining methodology and experimental protocols. Biochim Biophys Acta 1367:88–106

Kornyeyev D, Holaday AS, Logan BA (2004) Minimization of the photon energy absorbed by 'closed' reaction centers of photosystem 2 as a photoprotective strategy in higher plants. Photosynthetica 42:377–386

Kroon BM, Thoms S (2006) From electron to biomass: a mechanistic model to describe phytoplankton photosynthesis and steady state growth rates. J Phycol 42:593–609

Kufryk GI, Vermaas WFJ (2006) Sll1717 Affects the redox State of the plastoquinone pool by modulating quinol oxidase activity in thylakoids. J Bact 188:1286–1294

Külheim C, Agren J, Jansson S (2002) Rapid regulation of light harvesting and plant fitness in the field. Science 297:91–93

Lavaud J (2007) Fast regulation of photosynthesis in diatoms: Mechanisms, evolution and ecophysiology. Funct Plant Sci Biotech 1:267–287

Lavaud J, Strzepek RF, Kroth PG (2007) Photoprotection capacity differs among diatoms: Possible consequences on the spatial distribution of diatoms related to fluctuations in the underwater light climate. Limnol Oceanogr 52:1188–1194

Lavaud J, Rousseau B, Etienne A-L (2004) General features of photoprotection by energy dissipation in planktonic diatoms (Bacillariophyceae). J Phycol 40:130–137

Lavaud J, Rousseau B, van Gorkom HJ, Etienne A-L (2002) Influence of the diadinoxanthin pool size on photoprotection in a marine planktonic diatom *Phaeodactylum tricornutum*. Plant Physiol 129:1398–1406

Ledford HK, Niyogi KK (2005) Singlet oxygen and photo-oxidative stress management in plants and algae. Plant Cell Environ 28:1037–1045

Li X-P, Phippard A, Pasari J, Niyogi KK (2002) Structure-function analysis of photosystem II subunit S (PsbS) *in vivo*. Funct Plant Biol 29:1131–1139

Logan BA, Adams WWIII, Demmig-Adams (2007) Avoiding common pitfalls of chlorophyll fluorescence analysis under field conditions. Funct Plant Biol 34:853–859

Mackey K, Paytan A, Grossman A, Bailey S (2008) A photosynthetic strategy for coping in a high-light, low nutrient environment. Limnol Oceanogr 53:900–913

Maxwell K, Johnson GN (2000) Chlorophyll fluorescence-a practical guide. J Exp Bot 51:659–668

McCauley S, Melis A, Tang G, Arnon D (1987) Protonophores Induce Plastoquinol Oxidation and Quench Chloroplast Fluorescence: evidence for a cyclic, proton-conducting pathway in oxygenic photosynthesis. Proc Nat Acad Sci USA 84:8424–8428

Mehler AH (1951) Studies on reactions of illuminated chloroplasts I. Mechanism of the reduction of oxygen and other Hill reagents. Arch Biochem Biophys 33:65–77

Miller AG, Canvin DT (1989) Glycoaldehyde inhibits CO_2 fixation in the cyanobacterium *Synechococcus* UTEX 625 without inhibiting the accumulation of inorganic carbon or the associated quenching of chlorophyll *a* fluorescence. Plant Physiol 91:1044–1049

Miller AG, Espie GS, Canvin DT (1988) Active transport of CO_2 by the cyanobacterium *Synechococcus* UTEX 625. Plant Physiol 86:677–683

Miyake C, Horiguchi S, Makino A, Shinzaki Y, Yamamoto H, Tomizawa K-I (2005) Effects of light intensity on cyclic electron flow around PSI and its relationship to non-photochemical quenching of Chl fluorescence in tobacco leaves. Plant Cell Physiol 46:1819–1830

Moreland DE (1980) Mechanisms of action of herbicides. Annu Rev Plant Physiol 31:597–638

Moss DA, Bendall DS (1984) Cyclic electron transport in chloroplasts. The Q-cycle and the site of action of antimycin. Biochim Biophys Acta 767:389–395

Müller P, Li X-P, Niyogi KK (2001) Non-photochemical quenching. A response to excess light energy. Plant Physiol 125:1558–1566

Mullineaux CW, Emlyn-Jones D (2005) State-transitions: and example of acclimation to low-light stress. J Exp Bot 56:389–393

Munekage Y, Takeda S, Endo T, Jahns P, Hashimoto T, Shikanai T (2001) Cytochrome b_6f mutation specifically affects thermal dissipation of absorbed light energy in *Arabidopsis*. Plant J 28:351–359

Munekage Y, Hashimoto M, Miyake C, Tomizawa K, Endo T, Tasaka M, Shikanai T (2004) Cyclic electron flow around photosystem I is essential for photosynthesis. Nature 429:579–582

Neubauer C, Yamamoto HY (1992) Mehler-peroxidase reaction mediates zeaxanthin formation and zeaxanthin-related fluorescence quenching in intact chloroplasts. Plant Physiol 99:1354–1361

Nixon PJ, Mullineaux CW (2001) Regulation of photosynthetic electron transport. In: Aro E-M, Andersson B (eds) Regulation of photosynthesis. Kluwer, Dordrecht, pp 533–555

Niyogi KK (2000) Safety valves for photosynthesis. Curr Opin Plant Biol 3:455–460

Niyogi KK, Li XP, Rosenberg V, Jung H-S (2005) Is PsbS the site of non-photochemical quenching in photosynthesis? J Exp Bot 56:375–382

Owens TG (1986) Light-harvesting function in the diatom *Phaeodactylum tricornutum* II. A response to excess light energy. xPlant Physiol 80:739–746

Oxborough K, Horton P (1987) Characterization of the effects of Antimycin A upon high energy state quenching of chlorophyll fluorescence (qE) in spinach and pea chloroplasts. Photosynth Res 12:119–128

Pascal AA, Zhenfeng L, Broess K, van Oort B, van Amerongen H, Wang C, Horton P, Robert B, Chang W, Ruban A (2005) Molecular basis of photoprotection and control of photosynthetic light-harvesting. Nature 436:134–137

Peltier G, Cournac L (2002) Chlororespiration. Annu Rev Plant Physiol Plant Mol Biol 53:523–550

Peon J, Tan X, Hoerner D, Xia C, Luk YF, Kohler B (2001) Excited state dynamics of methyl viologen. Ultrafast photoreduction in methanol and fluorescence in acetonitrile. J Phys Chem 105:5768–5777

Perkins R, Mouget J-L, Lefebvre S, Lavaud J (2006) Light response curve methodology and possible implications in the application of chlorophyll fluorescence to benthic diatoms. Mar Biol 149:703–712

Pfannschmidt T (2005) Acclimation to varying light qualities: toward the functional relationship of state transitions and adjustment of photosystem stochiometry. J Phycol 41:723–725

Prasil O, Kolber Z, Berry JA, Falkowski PG (1996) Cyclic electron flow around PSII in vivo. Photosynth Res 48:395–410

Pressman BC, Harris EJ, Jagger WC, Johnson JH (1967) Antibiotic-mediated transport of alkali ions across lipid bilayers. Proc Nat Acad Sci USA 58:1949–1956

Ralph PJ, Polk S, Moore KA, Orth RJ, Smith WA (2002) Operation of the xanthophyll cycle in the seagrass *Zostera marina* in response to variable light. J Exp Mar Biol Ecol 271:189–207

Rich P, Madgwick S, Moss D (1991) The interaction of duroquinol, DBMIB and NQNO with the chloroplast cytochrome *bf* complex. Biochim Biophys Acta 1058:312–328

Rotatore C, Lew RR, Colman B (1992) Active uptake of CO_2 during photosynthesis in the green alga *Eremosphaera viridis* is mediated by a CO_2-ATPase. Planta 188:539–545

Ruban AV, Johnson MP (2009) Dynamics of higher plant photosystem cross-section associated with state-transitions. Photosynth Res 99:173–183

Ruban AV, Berera R, Ilioaia C, van Stokkum IHM, Kennis JTM, Pascal AA, van Amerongen H, Robert B, Horton P, van Grondelle R (2007) Identification of a mechanism of photoprotective energy dissipation in higher plants. Nature 450:575–579

Ruban VA, Lavaud J, Rousseau B, Guglielmi G, Horton P, Etienne A-L (2004) The super-excess energy dissipation in diatom algae: comparative analysis with higher plants. Photosynth Res 82:165–175

Serra T, Borrego C, Quintana X, Calderer R, Lopez R, Colomer J (2009) Quantification of the effect of nonphotochemical quenching on the determination of *in vivo* chl a from phytoplankton along the water column of a freshwater reservoir. Photochem Photobiol 85:321–331

Shelp BJ, Canvin DT (1984) Evidence for bicarbonate accumulation by *Anacystis nidulans*. Can J Bot 62:1398–1403

Shyam R, Raghavendra AS, Sane PV (1993) Role of dark respiration in photoinhibition of photosynthesis and its reactivation in the cyanobacterium *Anacystis nidulans*. Physiol Plant 88:446–452

Sicher RC (1984) Glycolaldehyde inhibition of photosynthetic carbon assimilation by isolated chloroplasts and protoplasts. In: Sybesma C (ed) Advances in photosynthetic research, vol III. Martinus Nijhoff/Dr W Junk, The Hague, pp 413–416

Sigalat C, Haraux F, de Kouchkovsky Y (1993) Force-flow relationship in lettuce thylakoids. 1. Strict control of election flow by internal pH. Biochemistry 32:10201–10208

Simonis W, Urbach W (1973) Photophosphorylation in vivo. Annu Rev Plant Physiol 24:89–114

Singh K, Shyam R, Sane PV (1996) Reactivation of photosynthesis in the photoinhibited green alga *Chlamydomonas reinhardtii:* role of dark respiration and of light. Photosynth Res 49:11–20

Standfuss J, van Scheltinga ACT, Lamborghini M, Kühlbrandt W (2005) Mechanisms of photoprotection and non-photochemical quenching in pea light-harvesting complex at 2.5 Å resolution. EMBO J 24:919–928

Stehfest K, Toepel J, Wilhelm C (2005) The application of micro-FTIR spectroscopy to analyze nutrient stress-related changes in biomass composition of phytoplankton algae. Plant Physiol Biochem 43:717–726

Stokes DM, Walker DA (1972) Photosynthesis by isolated chloroplasts. Inhibition by DL-glyceraldehyde of carbon dioxide fixation. Biochem J 128:1147–1157

Stroch M, Spunda V, Kurasova I (2004) Non-radiative dissipation of absorbed excitation energy within photosynthetic apparatus of higher plants. Photosynthetica 42:323–337

Suggett D, Kraay G, Holligan P, Davey M, Aiken J, Geider R (2001) Assessment of photosynthesis in a spring cyanobacterial bloom by use of a fast repetition rate fluorometer. Limnol Oceanogr 46:802–810

Szabo I, Bergantino E, Giacometti GM (2005) Light and oxygenic photosynthesis: energy dissipation as a protection mechanism against photo-oxidation. EMBO Rep 6:629–634

Tagawa K, Tsujimoto HY, Arnon DI (1963) Role of chloroplast ferredoxin in the energy conversion process of photosynthesis. Proc Nat Acad Sci USA 49:567–572

Toepel J, Domin A, Gilbert M, Wilhelm C (2004) Quantitative Estimates to convert photosynthetic electron transport rates into biomass. Arch Hydrobiol 160:515–526

Trebst A (1980) Inhibitors in electron flow: tools for the functional and structural localization of carriers and energy conservation sites. Meth Enzymol 69:675–715

Trebst A (2007) Inhibitors in the functional dissection of the photosynthetic electron transport system. Photosynth Res 92:217–224

Tyystjarvi E, Koski A, Keranen M, Nevalainen O (1999) The Krautsky curve is a built-in barcode. Biophys J 77:1159–1167

Velthuys B (1981) Electron-dependent competition between plastoquinone and inhibitors for binding to photosystem II. FEBS Lett 126:277–281

Wagner H, Jakob T, Wilhelm C (2005) Balancing the energy flow from captured light to biomass under fluctuating light conditions. New Phytol 169:95–108

Wagner H, Gilbert M, Goss R, Wilhelm C (2006) Light emission originating from photosystem II radical pair recombination is sensitive to zeaxanthin related non-photochemical quenching (NPQ). J Photochem Photobiol 83:172–179

Wilhelm C, Duval JC (1990) Fluorescence induction kinetics as a tool to detect chlororespiratorian activity in the prasinophycean alga *Mantoniella squamata*. Biochim Biophys Acta 1016:197–202

Wilhelm C, Wild A (1984) The variability of the photosynthetic unit in *Chlorella* II. The effect of light intensity and cell development on photosynthesis, P-700 and cytochrome f in homocontinuous and synchronous cultures of *Chlorella*. J Plant Physiol 115:124–135

Zhu S-H, Green BR (2008) Light-harvesting and photoprotection in diatoms: Identification and expression of L818-like proteins. In: Allen J-F, Gantt, E, Golbeck JH, Osmond B (eds) Photosynthesis. Energy from the Sun: 14th International Congress on Photosynthesis. Springer, Dordrecht, pp 261–264

Chapter 5
Microscopic Measurements of the Chlorophyll *a* Fluorescence Kinetics

Ondrej Komárek, Kristina Felcmanová, Eva Šetlíková, Eva Kotabová, Martin Trtílek, and Ondrej Prášil

1 Introduction

Chl*a* fluorescence induction kinetics (FK) is a well established non invasive method for measuring physiological state of oxygen evolving phototrophic organisms (Papageorgiou and Govindjee 2004; Roháček et al. 2008). The methodology can be used theoretically in any space scale from molecular, subcellular (Küpper et al. 2000a) to planetary dimensions as remote sensing probing (Moya and Cernovic 2004). Development of methods with high spatial resolution started with measurements of spatially resolved (two dimensional) fluorescence dynamics imaging using calibrated CCD cameras (reviewed by Nedbal and Whitmarsh 2004; Oxborough 2004). Such measurements allow to directly investigate differences among photosynthetic entities, such as chloroplasts, cells, tissues, organs or individuals. A macroscopic imaging system using the pulse amplitude modulation principle (Nedbal et al. 2000) allowed to study routinely the spatial heterogeneity of photosynthetic processes (Siebke and Weis 1995; Nedbal et al. 2000; Ferimazova et al. 2002). Applying the same principle on the microscopic scale, the Fluorescence Kinetic Microscopy (FKM), was developed by Küpper et al. (2000b) and has led to number of new applications. Nevertheless, the high resolution chl*a* FKM presents several specific engineering and biological problems (e.g. optical limitations, sample stability). Here we review the FKM methodology, recent achievements and propose future novel approaches.

Two dimensional, spatially resolved measurements of the chlorophyll fluorescence signal using the modified fluorescence microscope combined with CCD camera and dedicated software allow to observe and study variable fluorescence dynamics of individual objects, parts of objects or individual pixels and to statistically analyze the measured fluorescence parameters. Number of microscopic chlorophyll FK imaging systems have been developed. In his review, Oxborough (2004) divided them into (i) low resolution systems, which allow to study features down to the level of small group of cells (Omasa et al. 1987; Daley et al. 1989; Fenton and Crofts 1990; Genty and Meyer 1995; Siebke and Weis 1995; Scholes and Rolfe 1996; Nedbal et al. 2000; Zangerl et al. 2002), and (ii) high resolution systems, which can resolve features at the cellular and sub-cellular levels (Oxborough and Baker 1997; Osmond et al. 1999; Küpper et al. 2000a; Rolfe and Scholes 2002). Several of these fluorescence microscopic imaging systems are now available commercially. However, only few can be applied for the fully dynamic fluorescence measurements with proper kinetic resolution and can be used for the true 2D fluorescence kinetic microscopy imaging (FKMI) of algal cells. Here we focus on the system developed in our laboratories. The combination of imaging system with sensitive fluorometer and spectrometer can increase the versatility of the instrument considerably.

In imaging systems, it is possible to employ techniques that are in principle comparable to methods used in standard, non-imaging fluorometers (discussed in Nedbal et al. 2000; Nedbal and Whitmarsh 2004), such as PAM (pulse amplitude modulation; Schreiber et al. 1986; Schreiber 2004) and quenching analysis

O. Komárek (✉), K. Felcmanová, E. Šetlíková, E. Kotabová, and O. Prášil
Laboratory of Photosynthesis, Institute of Microbiology ASCR, Opatovický Mlýn, 379 01, Třeboň, Czech Republic
e-mail: komarek@alga.cz

M. Trtílek
Photon Systems Instruments Ltd, Koláčkova 39, Řečkovice, Brno, 621 00, Czech Republic

D.J. Suggett et al. (eds.), *Chlorophyll a Fluorescence in Aquatic Sciences: Methods and Applications*, Developments in Applied Phycology 4, DOI 10.1007/978-90-481-9268-7_5, © Springer Science+Business Media B.V. 2011

(Bradbury and Baker 1983; Grey et al. 2006; Papageorgiou and Govindjee 2004; Šiffel et al. 2000). In 2D signal harvesting, images are referred to as being 'raw' or 'parameter'. Depending on the measuring protocol, the raw fluorescence image (or images) can represent states corresponding to F_o, F_o' (minimum fluorescence signal in the dark and light adapted states, respectively), F′ (fluorescence signal at any point between F_o' and F_m'), F_m and F_m' (maximal fluorescence in the dark and light adapted states). These are then used in construction of images of number of standard parameters that characterize photochemical and non-photochemical processes. The interpretation of these parameters is then similar like in standard, non-imaging techniques.

In general, the data output from the FKM imaging software is of three types: (i) the time resolved sequence of images containing information about distribution of fluorescence intensity at the particular time of measurement. The time resolution with which these individual pictures are obtained is limited by the used CCD camera system and is usually lower (~tens of ms) than the standard non-imaging kinetic systems (~μs); (ii) the calculated parameter images or histograms computed from the measured images data; (iii) the numerical data in spreadsheets that represent these parameters computed as averages of the selected areas (i.e. individual objects of interest) of the analyzed images.

The individual steps of given measurement can be programmed and stored in the measurement protocol and should be selected according to the phenomenon and object that we want to study. Important parts of the measuring protocol are proper settings of the timing, intensity and spectral quality of the measuring or saturating flashes and actinic irradiation. Sensitivity of the CCD camera needs to be adjusted to match the fluorescence signal intensity.

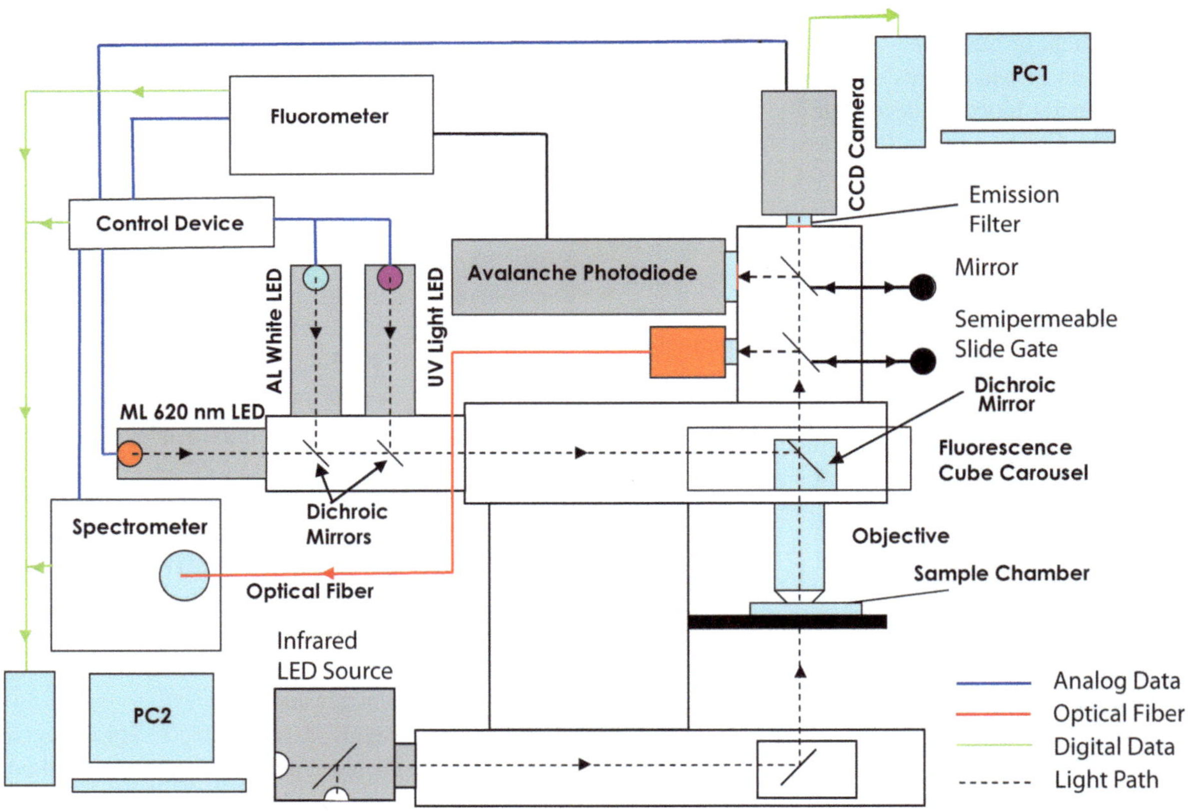

Fig. 1 The schematics of the FKM. The control device selects one of the three independent measuring modes: 2D fluorescence kinetics, non-imaging spectrometer and non-imaging fast fluorometer. Measurements of imaging 2D fluorescence kinetics can be synchronized with the spectrometer

Also the optical path of the microscope-objective parameters, dichroic mirrors, excitation and emission filters have to be optimized for given measurement.

2 Fluorescence Techniques in High Resolution

Here we describe the components of the typical setup of laboratory-based FKM (Fig. 1). Other versions, e.g. portable instruments for field work, can be simpler and do not have to contain all features described.

The FKM uses the light-addition technique to induce the maximal fluorescence yields Fm and Fm′. The light-addition technique requires the use of at least three light sources (Roháček 2002):

1. Low intensity pulses of measuring light (ML). Individual pulses have length of tens to hundreds of μs.
2. Continuous actinic light (AL) of intensity of hundreds of μmol photons m^{-2} s^{-1}.
3. Saturation pulse radiation (SP) of 0.3–1.6 s duration, with intensity 3,000–10,000 μmol photons m^{-2} s^{-1} for measurements of F_m and F_m'.

Additional sources of irradiation can be also employed:

4. Far red illumination (FR) with λ = 730–735 nm and intensity ~10 μmol photons m^{-2} s^{-1}. FR is used to activate preferentially Photosystem I (PS I) and to oxidize the plastoquinone (PQ) pool. If applied as transmitted irradiation, it can be with advantage also used for visual focusing and identification of individual objects, in cases when the transmitted visible illumination light path is missing or not applicable.

Regarding the source of illumination, although the classical xenon and halogen light sources provide the full broadband spectra, they do not allow fast modulation of their intensity. Therefore, in most FKM instruments, highly efficient and powerful light emitting diodes (LEDs) of distinct wavelength are used as controllable sources of illumination. The emission wavelengths of the LEDs used for ML, AL or SP should be optimized according to the spectral properties of the studied organisms. The broadband "white" LEDs are often used as general light source in conjunction with appropriate dichroic filters that select the narrower spectral band for excitation. Some specific applications (e.g. when studying cyanobacteria with generally low emission intensity) require the use of monochromatic LEDs of stronger intensity.

The important limitation of the FKM technique arises from the fact that the measuring light can be quite high and therefore can have actinic effects on microscopic objects. Absolute intensity of the ML increases when using objectives with higher magnification. Neutral filtering of the ML before optical pathway can solve the problem only partly, because the detection sensitivity of the standard CCD cameras limits the fluorescence signal to noise ratio. Certain amount of uneven intensity of illumination of the object is acceptable, provided that the actinic effect of the measuring light is negligible. Since most of the calculated fluorescence parameters are normalized, any inhomogeneities resulting from uneven illumination cancel out.

Important is the proper alignment of the microscope (axis alignment of optical paths for various light sources and for sensors) in order to transfer good quality image and provide high signal. As mentioned above, the signal intensity is on one hand critical for sufficient signal to noise ratio of the camera, but ML intensity has to be kept low to prevent actinic effects. The intensity of incident irradiance in the focus of the objective can be increased by magnification of optical parts up to hundred fold times. The incident light intensities should be measured directly at the sample focus plane or calculated from the measured intensity of the collimated beam and the known area of the focused illuminated spot. The actinic effect of ML can be also tested (if the sample is exposed to a series of ML flashes, the F_0 signal should be constant in time). One should also set and use the intensity of the AL that is comparable to physiological conditions.

In addition to the light source, the fluorescence kinetic microscope is usually equipped with the carousel that houses set of dichroic filters that select the proper spectral wavelength. Another set of emission filters can be placed in front of the detector CCD camera to select different emission spectral bands. The rapid exchange of the filters then allows to measure different fluorophores in the same object (chlorophylls, phycobilins, GFPs etc.). The spectral heterogeneity and variability of photosynthetic antennae among different phytoplankton species can be targeted using multi-spectral illumination sources for ML or AL.

Most chl*a* fluorescence imaging systems use CCD (charge coupled devices) cameras for capture of 2D fluorescence image. The disadvantage of standard CCD cameras with resolution 512 × 512 pixels is their low speed given by time required for integration and processing of the weak fluorescence signal from single cells. This prevents the use of fast (sub-millisecond) measuring methods in 2D (e.g. fluorescence induction curves or FRR – fast repetition rate fluorometry).

The dynamic range of CCD detectors is a function of the well capacity of the sensor array and the design of the readout circuit. It can range from the standard 12 bit resolution (i.e. the image can contain up to 4,096 levels of intensity) up to 18 bit resolution (65,536 levels). High resolution cameras often incorporate active cooling (liquid nitrogen or multi-stage Peltier unit) (Oxborough and Baker 1997), which decreases the dark noise. For example a CCD sensor cooled to 200 K will accumulate only negligible 0.001 counts pixel^{-1} s^{-1} of camera dark noise (Oxborough 2004) (Fig. 2).

Despite the fact that instrument limitations prevent the imaging of fast (sub-millisecond) kinetic changes in chl*a* variable fluorescence, these processes can be still measured on the level of single microscopic objects (individual cells or chloroplasts) if one attaches the non-imaging fluorometer to the microscope (Fig. 1). The object image created by the microscope can be diverted to the sensitive detector (photomultiplier or avalanche photodiode) using optical adapters and light guides. The alignment of the optical pathway allows then either to detect the integral signal from the whole optical field or to select and analyze only relatively small part of the measured field (corresponding to e.g. selected cell). Using this approach, one can determine for example the kinetics of Q_A^- reoxidation following single turnover flashes or to measure the fast kinetics of fluorescence induction (the so-called OJIP curves).

Similarly, the non-imaging highly-sensitive spectrometer can be connected to the microscope and aligned with the optical path by optical fibers to obtain spectrally resolved fluorescence kinetics measurements of the single cells or colony of cells in the wide spectral range (UV–VIS–IR) with ms time resolution (Küpper et al. 2009). This opens possibilities to measure not only emission spectra of single cells, but also the spectral dependence (and heterogeneity) of parameters that are otherwise considered to be spectrally constant (e.g. F_v/F_m, NPQ etc.). In addition, absorption properties of single cells or colonies can be also determined, but the cells need to be on the transparent support (Fig. 3).

Although the standard microscopic slides can be used for observing phytoplankton cells under the FKM, they pose a number of constrains, such as movement of cells, undefined pressure and lack of proper gas exchange that can create semi-anaerobic conditions around the samples or cause possible desiccation which prevent longer measurements. To solve the problem, special measuring chambers were designed that provide optimal conditions for the analyzed phytoplankton (Fig. 4). The chambers allow to maintain cells in environment with well defined and constant temperature, pH, nutrient and gas concentrations and thus allow to study individual cells over prolonged time periods (hours–days).

Movement of the object presents significant problem of the high spatial resolution FKM of biological objects. Mobile flagellate cells need to be immobilized in medium with high density, such as 1–3% agar or agarose (Küpper et al. 2004b). However, the relatively high temperature (>25°C) required to keep the medium fluid at the time of cell immobilization limits its use for low-temperature species (snow- or ice-algae). The contact and exchange of gases and nutrients of the immobilized sample with the flowing medium is provided by the semipermeable foil, e.g. cellophane (Fig. 4). Still, even if cells are properly immobilized, some partial cell movement can occur during longer measuring protocols (tens of s) and high resolution imaging can be impossible. This has been observed for several filamentous and coccal phytoplankton with autonomous moving abilities.

The alternative approach of preventing cell movement is to use optical tweezers (Fig. 5). This method uses lasers for capturing individual cells by the radial and centripetal forces inside the focused laser beam cone. For this, an additional laser beam is introduced in the optical path of the objectives by fiberoptics. For phytoplankton, it is necessary to use laser beam in the near infrared region (>950 nm) which cannot damage photosynthetic apparatus.

Using conventional fluorescence microscopes, it is possible to resolve fluorescence parameters from objects larger than ~1–2 µm. In theory, it is possible to resolve objects as small as 340 nm when imaging chl*a* fluorescence (half the wavelength that is being detected). In practice, chl*a* fluorescence can only be resolved at this level using confocal microscopy (Osmond et al. 1999) where light sources are lasers and laser diodes. However, these light sources can be

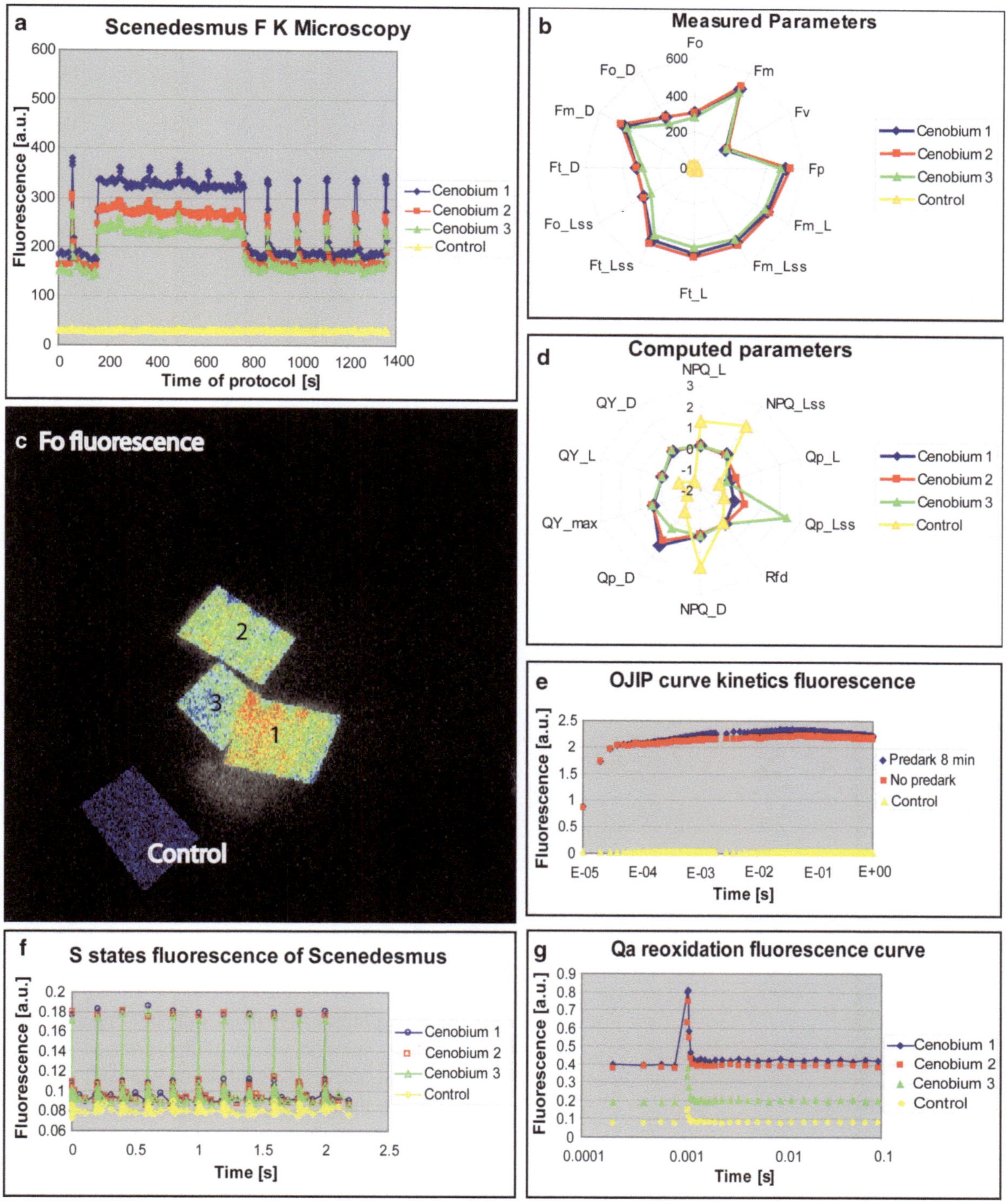

Fig. 2 Potential of the FKM method shown on the example of coenobia of alga *Scendesmus quadricauda*. Individual coenobia were selected manually from the raw fluorescence image (panel **c**) and then treated as single objects (marked 1–3, panel **c**). The control signal represents the background noise. Panel **a** – fluorescence kinetic curves computed for individual coenobia. The measuring protocol first detected F_o, then F_m was measured using the saturating flash, at in the white actinic irradiance (intensity ~100 μmol photons m^{-2} s^{-1}) was switched on for 10 min. The recovery of fluorescence yields was followed in the darkness for another 10 min. Each 150 s a saturating flash was given to measure F_m'. Panel **b** shows intensities of fluorescence measured for individual coenobia, panel **d** shows the computed photochemical and non-photochemical parameters. Panels **e–g** show results of non-imaging fast fluorescence measurements on selected cells (**e** – fast fluorescence induction curve, **f** – oscillations of fluorescence intensity after series of single-turnover flashes, **g** – kinetics of Q_A^- reoxidation following the single turnover flash)

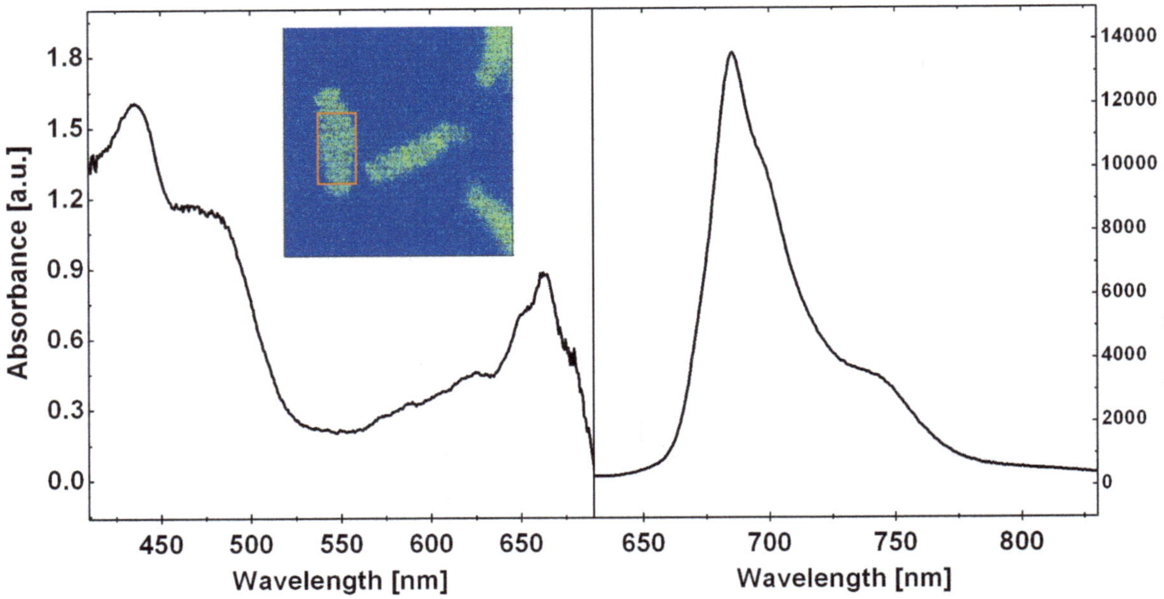

Fig. 3 The absorption (left) and emission fluorescence spectra (right) of the *Scenedesmus quadricauda* single cell measured by spectrometer attached to the FKM. The data were captured from the coenobium cells shown in the inset

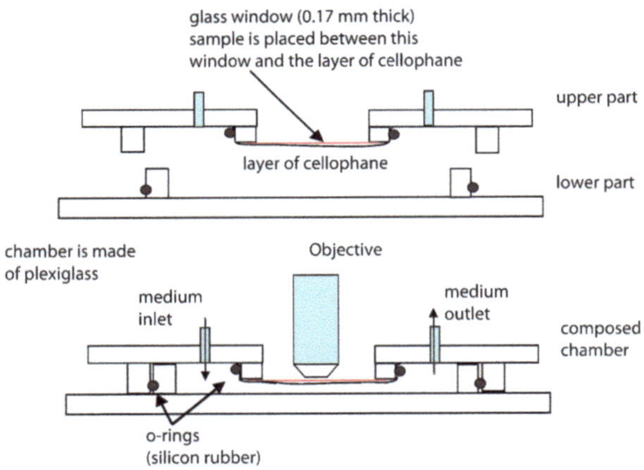

Fig. 4 The membrane microscopic chamber with medium circulation system used for incubations of cells under physiological conditions

damaging to the photosystetic microorganisms due to their high power that is further increased by the optics. Neverthless, 3D confocal laser scanning microscopy has been used for analysis of chl*a* fluorescence parameters by Omasa et al. (2008) using Nipkow disk-type confocal laser scanning microscope for measuring bean mesophyll chloroplasts fluorescence.

The measuring protocols are set of commands that determine the course of the experiment and assure its reproducibility. They allow manipulation with timing, light intensities, sensitivity of the camera and length and frequency of data capture. After the measurement, we obtain the time series of the digital images of the fluorescence as detected by the CCD camera. The result is the kinetic information about the individual pixels of each image. Usually the studied chloroplast, cell, colonies of cells etc., are composed from several pixels that are grouped together and processed as individual objects. The object can be defined and selected manually or the software can perform autoselection by setting the range of the signal intensities that define objects. For each selected object, the kinetic curve of fluorescence intensity is then computed together with the average numeric values of fluorescence parameters. Kinetic curves as well as images and numeric data can be then exported and processed in standard spreadsheets.

For improving the contrast within fluorescence images, the range of data values can be stretched across the entire palette using image equalization and gamma correction (described in detail by Oxborough 2004). Under high resolution in microscopy, the heterogeneity arises from relatively small scale sources. The smaller object or area analyzed means higher noise, but high number of repetitions or equalizing of the phenomenon over many items and averaging data increase the significance of the data.

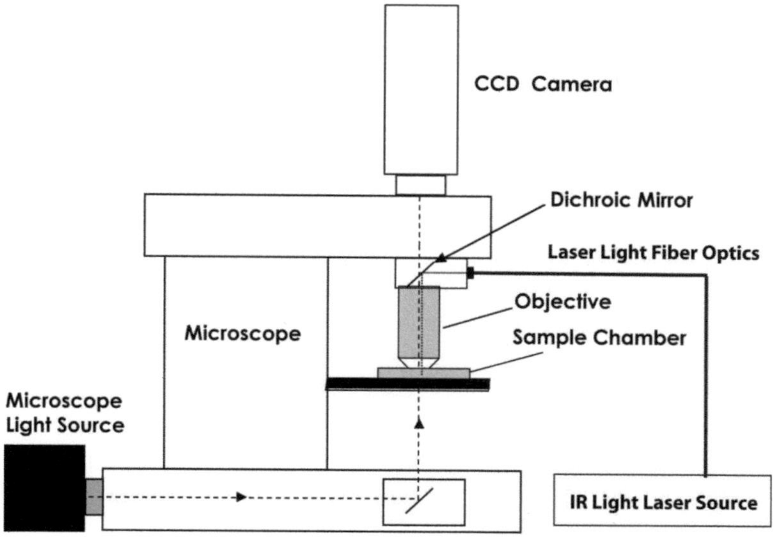

Fig. 5 Optical tweezers in FK microscope. The dichroic mirror reflecting IR is placed between the objective and camera

3 Applications of Fluorescence Kinetic Microscopy

Analysis of the variability and spatial heterogeneity of fluorescence parameters are the major benefits of the fluorescence kinetics imaging. In our laboratory, the primary motivation for the use of FKM was the investigation of the photosynthesis of individual plant and phytoplankton cells. Subsequently, majority of the studies were done on the non-hetrocytous nitrogen fixing cyanobacterium *Trichodesmium* IMS101 where the dynamics of the fluorescence is directly related to the N_2 fixation by individual cells or cell filaments. Only FKM allowed us to observe significant heterogeneity in the variable fluorescence among the individual cells of *Trichodesmium*. The heterogeneity seems to be related with the regulation of photosynthesis in cells that perform nitrogen fixation (Berman-Frank et al. 2001; Küpper et al. 2004a, 2008). The effects of iron limitation on the photosynthesis of *Trichodesmium* were studied by Küpper et al. (2008). Recently, the FKM equipped with spectral detection was used to study the dynamics of individual phycobiliproteins during state-transitions (Küpper et al. 2009).

In parallel, FKM was used to study the inhibition by toxic heavy metals of photosynthesis in aquatic plants (e.g. Küpper et al. 2002). Imaging and spectrally resolved microscopic measurements of chlorophyll FK were used to study cadmium-induced inhibition of photosynthesis and acclimation to cadmium stress in the hyper-accumulator *Thlaspi caerulens* (Küpper et al. 2007).

Long-term time series (longer than hours) of semi-continual measurements of fluorescence parameters were performed using the flow-through chamber and measuring photosynthetic activity of individual cells or chloroplasts. The cell cycle study of the coenobial green algae *Scenedesmus quadricauda* was performed by Šetlíková et al. (2005). The results show that fluorescence kinetics in actinic light are strongly determined by the phase of the cell cycle. High photosynthetic activity of the young cells (coenobia) resulted in fast quenching in the actinic light and quick recovery of the fluorescence yields in the darkness. Although there were quantitative differences between individual cells and coenobia, the character of the kinetics was for all studied objects similar. At the late growth and dividing phase of the cells cycle, the recovery of the fluorescence activity slowed down and finally during cell division only the quenching and strong negative reaction to light was observed (Šetlíková et al. 2005). At the stage of dividing cells the high heterogeneity in NPQ was observed. Similar studies were also performed on cyanobacteria (*Trichodesmium*) and photosynthetic apicomplexa (*Chromera velia*).

Algal cells cultivated for biotechnology application can be pumped into special microscopic chamber and their variable fluorescence signal analyzed by FKM. Based on the on-line software analysis of the 2D chl*a* fluorescence kinetic data, individual cell with set properties (photochemical activity, emission spectrum etc.)

can be classified and sorted. In this way, the most productive or otherwise exceptional cells can be isolated. The manipulation of the cells and particles can be achieved using optical tweezers. The laser beam can keep the object, move it and release it to the microchannel system with the constant flow. Sorting is managed by complex tweezers regulation system based on object dynamic fluorescence parameters.

Some microscopic organisms react to light intensity changes by phototaxis towards or away from the light source. Instant reaction can be followed by fluorescence kinetic microscope. Positive or negative fototaxis can be observed in monadoid algae with flagella, filamentous cyanobacteria or diatoms. Response of algal cells to different light regimes were studied also in coincidence with the synchronization of the cultures in *S. quadricauda* (Šetlíková et al. 2005) (see Fig. 6c, d). Diatom species within the intertidal (litoral) zone migrate vertically due to the complex set of factors including light intensity and quality, UV-B radiation, temperature and salinity. Whole community taken into chamber was studied and contribution of individual participants to photosynthetic parameters was determined (Oxborough 2004).

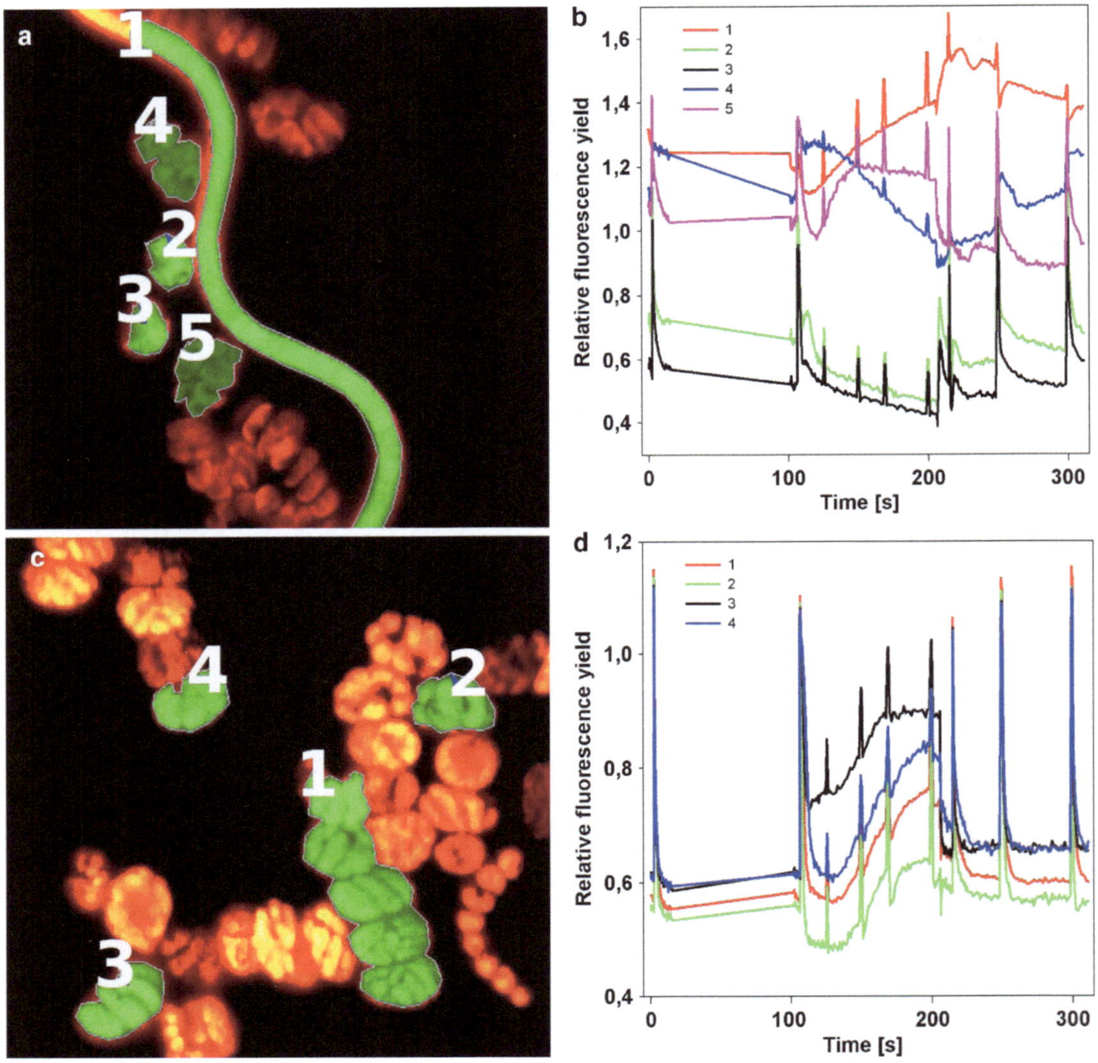

Fig. 6 Heterogeneity in fluorescence response of individual algal species. Panels **a** and **b** show variability of fluorescence kinetic curves among algae *Arthrospira* (1), *Haematococcus* (2, 3) and *Scenedesmus* (4, 5). Selected species are shown in **a** as numbered green areas. Panels **c** and **d** show the differences in fluorescence kinetic curves within a single species (*Scenedesmus*). The individual cells differ in parameters F_v/F_m (NPQ): cell 1 = 0.49 (0.42), cell 2 = 0.50 (0.47), curve 3 = 0.44 (0.24), curve 4 = 0.44 (0.38)

Biotic stress was studied by many authors at the macroscopic level in agricultural plants, the spatial distribution of the infection and fluorescence reaction of the tissue was studied by double photon scanning microscope (Pineda et al. 2008). Interaction among algae and parasites was at the microscopic level performed by Gachon et al. (2006).

Specific groups of microscopic phototrophic organisms differ in taxonomic features which directly correspond to utilization of light and interaction with other environmental conditions. These features were used for creation of basic structure in algal and cyanobacterial taxonomical system. Pigment composition, chloroplast structure and presence of the specific pigment protein complexes determines the photosynthetic properties of the organism. The variable fluorescence can be used in phytoplankton ecophysiology by determining the response of individual taxonomic groups to given conditions. Results of our pilot experiments indicate that each taxonomic group within a mixed community has characteristic response to actinic irradiance that can be observed as differences in photochemical and non-photochemical quenching (Figs. 6a, b, and 7). Some species are quenching variable fluorescence very efficiently, others show longer time of recovery in the darkness. Another approach utilizes multispectral excitation provided by different LEDs and allows to reconstruct the excitation spectrum of individual cells. Using the non-imaging emission spectrometer, one can also determine the absorption and emission spectra of individual cells.

Gorbunov et al. (1999) used the single cell method based on the modified FRR fluorometer. The analyzed sample was in the drop at the tip of the pin at the center of parabolic integration sphere. The emission signal was focused to the detector. This method was sufficiently sensitive, but the single cell measurement was dependent on the proper dilution of the sample. The system was followed by the sorting device with set of micropumps switched according to measured parameters.

The photosynthetic processes of the surface growing phototrophic organisms can be also studied by FKM. Cyanobacterial mats are adapted for life in extreme conditions and provide potential source of biologically active compounds. These strategies can be studied with the portable FKM of natural populations.

In summary, measurement of the chl*a* variable fluorescence kinetics on the level of single algal cells has great

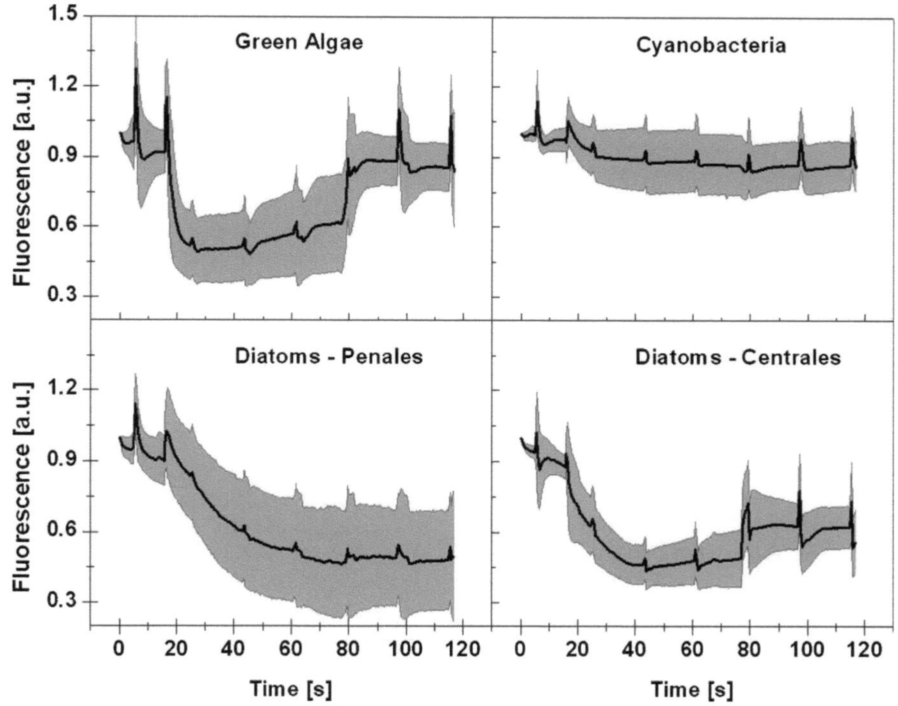

Fig. 7 The statistical evaluation of the kinetics of selected freshwater plankton species – green algae, cyanobacteria, central diatoms and pennate diatoms. The fluorescence kinetic curves of individual species have specific kinetics which can be used for the determination of the species. The standard deviation results from 30 measurement repetitions

potential. FKM is a useful and potent tool for solving questions about phototrophic organisms of several levels of complexity. It provides new dimension in studying photosynthesis and its regulation. The applications can vary from cellular physiology to complex environmental issues, from the molecular to geographical levels.

Acknowledgments Financial support was provided by EEA Fund. No A/CZ0046/1/0021 and by the project GACR GA (206/07/0917).

References

Berman-Frank I, Lundgren P, Chen Y, Küpper H, Kolber Z, Bergman B, Falkowski P (2001) Segregation of nitrogen fixation and oxygenic photosynthesis in the marine cyanobacterium *Trichodesmium*. Science 294:1534–1537

Bradbury M, Baker NR (1983) Analysis of the induction of chlorophyll fluorescence in leaves and isolated thylakoids: contributions of photochemical and non-photo-chemical quenching. Proc Roy Soc Lond B Biol 220:251–264

Daley PF, Rashke K, Ball JT, Berry JA (1989) Topography of photosynthetic activity of leaves obtained from video images of chlorophyll fluorescence. Plant Physiol 90: 1233–1238

Fenton JM, Crofts AR (1990) Computer aided fluorescence imaging of photosynthetic systems: Application of video imaging to the study of fluorescence induction in green plants and photosynthetic bacteria. Photosynth Res 26:59–66

Ferimazova N, Küpper H, Nedbal L, Trtílek M (2002) New insights into photosynthetic oscillations revealed by two-dimensional microscopic measurements of chlorophyll fluorescence kinetics in intact leaves and isolated protoplasts. Photochem Photobiol 76:501–508

Gachon CMM, Küpper H, Küpper FC, Setlik I (2006) Single-cell chlorophyll fluorescence kinetic microscopy of *Pylaiella littoralis* (Phaeophyceae) infected by *Chytridium polysiphoniae* (Chytridiomycota). Eur J Phycol 41:395–403

Genty B, Meyer S (1995) Quantitative mapping of leaf photosynthesis using chlorophyll fluorescence imaging. Aust J Plant Physiol 22:277–284

Gorbunov MY, Kolber Z, Falkowski PG (1999) Measuring photosynthetic parameters in individual algal cells by Fast Repetition Rate fluorometry. Photosynth Res 62:141–153

Grey DW, Zoe G, LA Cardon L (2006) Simultaneous collection of rapid chlorophyll fluorescence induction kinetics, fluorescence quenching parameters, and environmental data using an automated PAM 2000/CR10X data logging system. Photosynth Res 87:295–301

Küpper H, Šetlík I, Trtílek M, Nedbal L (2000a) A microscope for two-dimensional measurements of in vivo chlorophyll fluorescence kinetics using pulsed measuring radiation, continuous actinic radiation, and saturating flashes. Photosynthetica 38:553–570

Küpper H, Spiller M, Küpper F (2000b) Photomethric method for the quantification of chlorophylls and their derivatives in complex mixtures: fitting with gaus-peak-spectra. Anal Biochem 286:247–256

Küpper H, Šetlík I, Spiller M, Küpper FC, Prášil O (2002) Heavy metal-induced inhibition of photosynthesis: Targets of in vivo heavy metal chlorophyll formation. J Phycol 38:429–441

Küpper H, Ferimazova N, Šetlík I, Berman-Frank I (2004a) Traffic lights in Trichodesmium. Regulation of photosynthesis for nitrogen fixation studied by chlorophyll fluorescence kinetic microscopy. Plant Physiol 135:2120–2133

Küpper H, Šetlík I, Hlásek M (2004b) A versatile chamber for simultaneous measurements of oxygen exchange and fluorescence in filamentous and thallous algae as well as higher plants. Photosynthetica 42:579–583

Küpper H, Parameswaran A, Leitenmaier B, Trtílek M, Šetlík I (2007) Cadmium-induced inhibition of photosynthesis and long-term acclimation to cadmium stress in the hyperaccumulator *Thlaspi caerulescens*. New Phytol 175:655–674

Küpper H, Šetlík I, Seiber S, Prášil O, Šetlíková E, Strittmatter M, Levitan O, Lohscheider J, Adamska I, Berman-Frank I (2008) Iron limitation in the marine cyanobacterium *Trichodesmium* reveals new insights into regulation of photosynthesis and nitrogen fixation. New Phytol 179:784–798

Küpper H, Andresen E, Wiegert S, Šimek M, Leitenmaier B, Šetlík I (2009) Revesible coupling of individual phycobiliprotein isoforms during state transitions in the cyanobacterium *Trichodesmium* analysed by single-cell fluorescence kinetic measurements. Biochim Biophys Acta 1787:155–167

Moya I, Cernovic ZG, Cernovic ZG (2004) Remote sensing of chlorophyll fluorescence: Instrumentation and analysis. In: Papageorgiou GC, Govindjee (eds) Chlorophyll *a* fluorescence: A signature of photosynthesis., Springer, Dordrecht. pp 429–445

Nedbal L, Whitmarsh J (2004) Chlorophyll fluorescence imaging of leaves and fruits. In: Papageorgiou GC, Govindjee (eds) Chlorophyll *a* fluorescence: A signature of photosynthesis., Springer, Dordrecht. pp 389–407

Nedbal L, Soukupová J, Kaftan D, Whitmarsh J, Trtílek M (2000) Kinetic imaging of chlorophyll fluorescence using modulated light. Photosynth Res 66:3–12

Omasa K, Shimazaki K-I, Aiga I, Larcher W, Onoe M (1987) Image analysis of chlorophyll fluorescence transients for diagnosting the photosynthetic system of attached leaves. Plant Physiol 84:748–752

Omasa K, Konishi A, Tamura H, Hosoi F (2008) 3D confocal laser scanning microscopy for the analysis of chlorophyll fluorescence parameters of chloroplasts in intact leaf tissues. Plant Cell Physiol 50:90–105

Osmond B, Schwartz O, Gunning B (1999) Photoinhibitory printing on leaves, visualized by chlorophyll fluorescence imaging and confocal microscopy, is due to diminished fluorescence from grana. Aust J Plant Physiol 26:717–724

Oxborough K (2004) Using chlorophyll *a* fluorescence imaging to monitor photosynthetic performance. In: Papageorgiou GC, Govindjee (eds) Chlorophyll *a* fluorescence: A signature of photosynthesis. Springer, Dordrecht, pp 409–428

Oxborough K, Baker NR (1997) An instrument capable of imaging chlorophyll a fluorescence from intact leaves at very low irradiance and at cellular levels. Plant Cell Environ 20:1473–1483

Papageorgiou GC, Govindjee (eds) (2004) Chlorophyll *a* fluorescence: A signature of photosynthesis. Springer, Dordrecht

Pineda M, Soukupova J, Matous K, Nedbal L, Baron M (2008) Conventional and combinatorial chlorophyll fluorescence imaging to tobamovirus-infected plants. Photosynthetica 46:441–451

Roháček K (2002) Chlorophyll fluorescence parameters: The definitions, photosynthetic meaning, and mutual relationship. Photosynthetica 40:13–29

Roháček K, Soukupová J, Barták M (2008) Chlorophyll fluorescence: A wonderful tool to study plant physiology and plant stress. Research Signpost, India, pp 41–104

Rolfe SA, Scholes JD (2002) Extended depth-of-focus imaging of chlorophyll fluorescence from intact leaves. Photosynth Res 72:107–115

Scholes JD, Rolfe SA (1996) Photosynthesis in localized regions of oat leaves infected with crown rust (*Pucinia coronata*): Quantitative imaging of chlorophyll fluorescence. Planta 199:573–582

Schreiber U (2004) Pulse-Amplitude (PAM) fluorometry and saturation pulse method. In: Papageorgiou GC (ed) Chlorophyll*a* fluorescence: A signature of photosynthesis. Springer, Dordrecht, pp 279–319

Schreiber U, Schliwa U, Bilger W (1986) Continuous recording of photochemical and non-photochemical chlorophyll fluorescence quenching with a new type of modulation fluorometer. Photosynth Res 10:51–62

Šetlíková E, Šetlík I, Küpper H, Kasalický V, Prášil O (2005) The photosynthesis of individual algal cells during the cell cycle of *Scenedesmus quadricauda* studied by chlorophyll fluorescence kinetic microscopy. Photosynth Res 84:113–120

Siebke K, Weis E (1995) Imaging of Chlorophyll a fluorescence in leaves: Topography of photosynthetic oscillations in leaves of *Glechoma hederacea*. Photosynth Res 45:225–237

Siffel P, Hunalova I, Rohacek K (2000) Light-induced quenching of chlorophyll fluorescence at 77K in leaves, chloroplasts and Photosystem II particles. Photosynth Res 65:219–229

Zangerl AR, Hamilton JG, Miller TJ, Crofts AR, Oxborough K, Berenbaum MR, de Lucia EH (2002) Impact of folivory on photosynthesis is greater than the sum of its holes. Proc Natl Acad Sci USA 99:1088–1091

Chapter 6
Estimating Aquatic Productivity from Active Fluorescence Measurements

David J. Suggett, C. Mark Moore, and Richard J. Geider

1 Fluorescence as a Probe for Photosynthesis

One of the major uses of active chlorophyll fluorescence in aquatic studies has been to examine primary productivity free from constraints associated with 'conventional' gas exchange-based measurements. Fluorescence can be measured directly *in situ* and without the need to incubate discrete samples that have been removed from natural inherent environmental variability (Kolber and Falkowski 1993). Additionally, the high rate of data collection afforded by fluorometry allows investigations of the coupling of productivity to physical forcing and hence the examination of transient phenomena that can have a disproportionate impact on the functional response of aquatic systems (Kolber et al. 1990; Falkowski et al. 1991; Moore et al. 2003). In spite of these advantages, fluorescence-based productivity applications have, to some extent, taken a considerable time to become firmly established within conventional aquatic sampling disciplines.

Active fluorescence is used to estimate the quantum efficiency of charge separation in the photosystem II reaction centre (designated ϕ_{PSII}), which can be multiplied by the rate of light absorption to obtain the photosynthetic electron transfer rate through PSII (ETR_{PSII}). Work by Genty et al. (1989) is generally well recognized as the major breakthrough study in using fluorescence as a means to examine productivity. Clear empirical evidence was presented for the first time of a linear relationship between fluorescence-based measures of photochemical efficiency (PSII) and independent measures of the quantum yield of CO_2 fixation in maize. Thus, ETR_{PSII} was found to be linearly related to the rate of CO_2 fixation (Genty et al. 1989), with the proportionality coefficient being the electron requirement for CO_2 fixation. Similar observations soon followed (see Baker and Oxborough 2004) and confirmed that PSII photochemical efficiency appeared to be strongly correlated with photosynthetic CO_2 fixation. It was only a matter of time until these same exercises were applied to aquatic organisms (e.g. Kolber and Falkowski 1993; Boyd et al. 1997; Gilbert et al. 2000a; Suggett et al. 2001; see Kromkamp and Forster 2003).

However, there are many reasons why the relationship between ETR_{PSII}, O_2 evolution and CO_2 fixation should not be fixed. For example, cyclic electron flow around PSII (Prášil et al. 1996) or plastid terminal oxidase activity (Bailey et al. 2008; Cardol et al. 2008) can uncouple ETR_{PSII} from the rate of gross O_2 production by PSII. Furthermore, other electron sinks within the photosynthetic electron transfer chain (e.g. O_2 uptake by the plastid terminal oxidase activity and/or the water-water cycle associated with the Mehler reaction (Behrenfeld et al. 2008, Suggett et al. 2009a), within the Calvin Cycle (e.g., oxygenation of RuBP), or associated with nitrate assimilation (Holmes et al. 1989), can lead to uncoupling of net O_2 evolution and CO_2 fixation from ETR_{PSII}. Not surprisingly, a number of studies of vascular plants began to appear demonstrating that linearity between photochemical yields for PSII and carbon fixation did not hold for all environmental conditions (see e.g. Baker and Oxborough 2004). Marine and freshwater phototrophs

D.J. Suggett (✉) and R.J. Geider
Department of Biological Sciences, University of Essex, Colchester, Essex CO4 3SQ, UK
e-mail: dsuggett@essex.ac.uk

C.M. Moore
National Oceanography Centre, Southampton, University of Southampton Waterfront Campus, European Way, Southampton SO14 3ZH, UK

D.J. Suggett et al. (eds.), *Chlorophyll a Fluorescence in Aquatic Sciences: Methods and Applications*, Developments in Applied Phycology 4, DOI 10.1007/978-90-481-9268-7_6, span a much greater taxonomic (evolutionary) range of organisms (Falkowski and Raven 2007). Consequently it would perhaps be surprising if such diversity was not reflected by even greater divergence in the photosynthetic physiology linking ETR_{PSII} and carbon uptake, than had previously been observed for vascular plants.

Whilst several fluorescence approaches can be used to calculate ETR_{PSII}, differences in protocol (e.g. Fast Repetition Rate (FRR) versus Pulse Amplitude Modulation (PAM); single versus multiple turnover pulses) and application (e.g. spectral composition of light sources and measurement timescales) will inevitably lead to different estimates of ETR_{PSII}. Determining the systematic errors associated with any one protocol or application has proven problematic, not least since no independent bench mark exists with which to validate a 'true' ETR_{PSII}. Thus, there is not yet a consensus on the optimal approach for estimating ETR_{PSII} using active fluorescence (e.g. Kolber and Falkowski 1993; Kromkamp and Forster 2003; Suggett et al. 2003, 2009a). None-the-less, substantial progress has been made in documenting and understanding differences amongst approaches, with important implications as researchers move towards a new era of using ETRs to answer complex physiological questions (Behrenfeld et al. 2004; Wagner et al. 2006) rather than simply as a means to provide a proxy for oxygen evolution or carbon uptake.

Technological improvements and increased commercial availability has resulted in a steady rise of the number of active fluorescence based studies estimating linear ETR in aquatic systems. Notable early studies include the introduction of the PAM approach to estimate ETR_{PSII} in aquatic systems by Schreiber et al. (1993) and use of the pump and probe approach to estimate ETR_{PSII} by Kolber and Falkowski (1993). In particular, Kolber and Falkowski's paper represented a breakthrough by not only describing a new approach for determining the ETR_{PSII} and scaling this rate to an ecological level, but also provided a first direct comparison of such ETRs with independent productivity measurements of carbon fixation for phytoplankton from a wide range of oceanic conditions. Whilst a linear relationship between PSII activity and carbon uptake was noted, considerable variability was equally obvious (up to 100% of the mean). Thus, the potential for a variable requirement of electrons to fix carbon was demonstrated on natural samples, although practical considerations associated with amalgamating data sets using a variety of approaches may significantly contribute to such variability (Boyd et al. 1997; Suggett et al. 2001, see below). Similar studies soon followed for the wide range of fluorescence techniques and aquatic systems, in particular for the microphytobenthos (e.g. Morris and Kromkamp 2003; Morris et al. 2008; Perkins et al., Chapter 12 this volume), macroalgae and seagrasses (e.g. Beer et al. 1998, 2000; Franklin and Badger 2001; Longstaff et al. 2002; Enriquez and Borowitzka, Chapter 9 this volume), corals (e.g. Gorbunov et al. 2001; Levy et al. 2004, 2006; Chapter 10 by Warner et al., this volume) and aerobic anoxygenic phototrophs (e.g. Kolber et al. 2001).

2 Overview of the Theory of Calculating ETR_{PSII}

All procedures for estimating ETR_{PSII}, regardless of the aquatic environment in question or fluorometer employed, follow the same theoretical principle. They are essentially bio-optical models in which the ETR is calculated as the product of incident light (E), the proportion of incident light absorbed (a) and the efficiency with which the absorbed light drives photochemistry, which varies with light intensity (ϕ(E)), i.e. in the simplest notation:

$$ETR = a \cdot \phi(E) \cdot E \tag{1a}$$

or on a PSII basis applicable to fluorescence measurements:

$$ETR_{PSII} = a_{PSII} \cdot \phi_{PSII}(E) \cdot E \tag{1b}$$

where a_{PSII} indicates the absolute absorption cross section of PSII and ϕ_{PSII} is the photochemical efficiency of PSII (Kolber and Falkowski 1993; Suggett et al. 2003; Baker and Oxborough 2004; Table 1).

For many ecological applications absolute rates of photosynthetic electron transport are frequently desirable. Thus, for example, with optically thick macrophytes, ETR_{PSII} is typically calculated per unit area of plant surface. This rate can be scaled to the whole plant or ecological level if the leaf area index, designated LAI, is known (LAI = projected foliage area per unit area of ground surface). In contrast, for optically thin phytoplankton suspensions, the ease with which absolute chlorophyll concentrations can be measured *in situ* typically results in ETR_{PSII} being calculated per unit chlorophyll *a* (ETR^{Chl}_{PSII}), with subsequent scaling to the ecological level using the chlorophyll *a* concentration:

$$ETR_{PSII} = ETR^{Chl}_{PSII} \cdot [\mathrm{Chl}a] \quad (2)$$

Unless complemented by other measurements, active fluorescence can only provide information on the properties of PSII (i.e., the effective light absorption cross-section and/or the photochemical efficiency of PSII), hence allowing estimation of ETR_{PSII}. Scaling of ETR_{PSII} to ETR^{Chl}_{PSII} can be further calculated as follows:

$$ETR^{Chl}_{PSII} = a^{Chl}_{PSII} \cdot \phi_{PSII}{}' \cdot \phi_{RCII} \cdot E \quad (3)$$

where a^{Chl}_{PSII} is the chlorophyll a-specific light absorption coefficient of PSII, i.e. it is the total amount of absorbed light delivered to all the PSII reaction centres within a sample normalized to the total chlorophyll present within the sample (Kolber and Falkowski 1993; Suggett et al. 2003, 2004; Baker and Oxborough 2004; Table 1). Note that in going from Eq. 1 to Eq. 3 the prime notation (i.e. $\phi_{PSII}{}'$ rather than ϕ_{PSII} (E)) is now adopted to indicate that measurements are performed under actinic light (E).

Inherent to this generic construction is an additional (assumed) constant ϕ_{RCII}, the quantum yield of photochemistry within PSII (Kolber and Falkowski 1993). This constant represents the ratio of electrons yielded per mol photons delivered to the reaction centre, i.e. has units of mol e^- (mol photon)$^{-1}$, and is taken to be unity since one electron is transferred from P_{680} to Q_A per quanta absorbed and delivered to the reaction center.

Although all active fluorescence techniques can provide direct measurements of $\phi_{PSII}{}'$, additional

Table 1 Summary of equations used to calculate primary productivity (as the electron transport rate, ETR) based on active chlorophyll*a* fluorescence measurements. E is the actinic light intensity (mol photons m^{-2} s^{-1}). Both $\phi_{PSII}{}'$ and $F_q{}'/F_m{}'$ are termed the PSII photochemical efficiency under actinic light and are dimensionless. All other fluorescence factors, $F_v{}'/F_m{}'$ (PSII maximum efficiency under actinic light) and $F_q{}'/F_v{}'$ (PSII operating efficiency) are also dimensionless. The effective absorption cross section under actinic light (dark acclimation), $\sigma_{PSII}{}'$ (σ_{PSII}), is in units of m^2 (mol RCII)$^{-1}$ whilst the concentration of PSII reaction centres (RCIIs) normalised to chlorophyll*a*, n_{PSII}, is in units of mol RCII (mol chl*a*)$^{-1}$. a^* and fAQ_{PSII} are the chlorophyll*a* specific absorption coefficient, (m^2 mol chl*a*)$^{-1}$ and the proportion of a^* that goes to PSII (%), following the terminology from Johnsen and Sakshaug (2007). ϕ_e is the quantum yield for electron transport through PSII, mol O_2 (mol e–)$^{-1}$ (Kolber and Falkowski 1993). ϕ_{RC} is the quantum yield of photochemistry within PSII, mol e^- (mol photon)$^{-1}$, and is assumed to be constant and unity. The constant 3,600 accounts for conversion of seconds to hours

Description	Term (units)	Equation
1. (Relative) Electron Transfer Rate	*r*ETR (a.u.)	$= E \cdot \phi_{PSII}{}' \cdot \phi_{RC}$
2. PSII photochemical efficiency under actinic light	$\phi_{PSII}{}'$ (dimensionless)	$= F_q{}'/F_m{}' (= F_v{}'/F_m{}' \cdot F_q{}'/F_v{}')$ $= \sigma_{PSII}{}'/\sigma_{PSII} \cdot (F_q{}'/F_v{}' \cdot F_v/F_m)$
3. (Absolute) Electron Transfer Rate	ETR_{PSII} (mol e^- (mol RCII)$^{-1}$ s^{-1}) ETR_{PSII} (mol e^- (mol chl*a*)$^{-1}$ h^{-1})	$= E \cdot \phi_{PSII}{}' \cdot \phi_{RC} \cdot a_{PSII}$ $= E \cdot F_q{}'/F_v{}' \cdot \phi_{RC} \cdot \sigma_{PSII}{}'$ $= E \cdot \phi_{PSII}{}' \cdot \phi_{RC} \cdot a^{Chl}_{PSII} \cdot 3600$
4. Chl*a*-specific PSII absorption coefficient	a^{Chl}_{PSII} (m^2 (mol chl*a*)$^{-1}$)	$= \sigma_{PSII}{}' \cdot n_{PSII} \cdot (F_v{}'/F_m{}')^{-1}$ $= a^* \cdot fAQ_{PSII}$
5. Spectral correction to weight absorption relative to light source	$\overline{a}^{Chl}_{PSII}$ (m^2 (mol chl*a*)$^{-1}$)	$\left(\sum_{400}^{700} a^{Chl}_{PSII}(\lambda) \cdot E_o(\lambda)\right) \Delta\lambda \sum_{400}^{700} E_o(\lambda)\,\Delta\lambda \Big/ \sum_{400}^{700} E_o(\lambda)\,\Delta\lambda$
6. Fluorescence based chlorophyll normalized O_2 productivity estimate	P(F) (mol O_2 (mol chl*a*)$^{-1}$ h^{-1})	$= E \cdot \phi_{PSII}{}' \cdot \phi_{RC} \cdot \overline{a}^{Chl}_{PSII} \cdot \phi_e \cdot 3600$ $= E \cdot \phi_{PSII}{}' \cdot \overline{a}^{Chl}_{PSII} \cdot 0.0009$ *

* Indicates reduction of P(F) under the assumptions that ϕ_{RC} and ϕ_e are constant with values of 1.0 mol e^- (mol photon)$^{-1}$ and 0.25 mol O_2 (mol e^-)$^{-1}$, respectively.

Note for *r*ETR: since the absorptance is not included in the calculation, we should not assign units of µmole e^- m^{-2} s^{-1} that are produced by combining the photochemical yield, light and ϕ_{RC}. Even though absorptance is dimensionless, the value of the absorptance scales the value of E to give the ETR. As such we state rETR units as arbitrary (a.u.). Refer to Cosgrove and Borowitzka, Chapter 1 this volume, for more information on the terminology used information may be required to obtain a_{PSII}^{Chl}, as discussed in Sections 3.1–3.2. Often measurements of a_{PSII}^{Chl} are not available, in such cases relative electron transfer (*r*ETR, Table 1) rates are frequently reported, i.e.

$$rETR = \phi_{PSII}' \cdot E \quad (4)$$

This chapter will explore (1) the alternative approaches and applications of active fluorescence for estimating ETR in aquatic environments, and (2) how these approaches correspond with alternative estimates of productivity, i.e. based on 'conventional' measures of gas exchange.

2.1 Measuring ϕ_{PSII}' and Calculating ETR

The quantum efficiency of PSII under actinic light, is designated ϕ_{PSII}' and can be de-composed into two components: the efficiency of excitation energy transfer from the light-harvesting antenna pigments to the reaction centers of PSII (here designated ϕ_{NPQ}', where the NPQ indicates that this efficiency may be reduced by non-photochemical quenching in the antenna pigments), and the efficiency of charge separation in RCII (which we designate ϕ_P', following Kolber and Falkowski 1993; see Suggett et al. 2009a). Once again the primes indicate that these efficiencies depend on actinic light; thus,

$$\phi_{PSII}' = \phi_P' \cdot \phi_{NPQ}' \quad (5)$$

In the PAM approach, values for ϕ_{PSII}' are routinely estimated from measurements of F_q'/F_m' (see Chapter 1 by Cosgrove and Borowitzka, this volume, for analogous terms and measurement practices),

$$\phi_{PSII}' = F_q'/F_m' \quad (6)$$

Alternatively, F_q'/F_m' can be directly derived from fluorescence lifetime measurements free from potential uncertainties in fluorescence induction models (e.g. Lavergne and Trissl 1995). These will not be considered further here, but see Huot and Babin, Chapter 3 this volume). Following Genty et al. (1989) F_q'/F_m' can be expressed as the product of the maximum PSII photochemical efficiency under actinic light (F_v'/F_m') and the PSII operating efficiency (F_q'/F_v'). F_v'/F_m' accounts for non-photochemical dissipation of absorbed excitation energy and F_q'/F_v' accounts for the efficiency of photochemical charge separation, both of which regulate the overall efficiency of PSII photochemistry; thus,

$$\phi_P' = F_q'/F_v', \text{and } \phi_{NPQ}' = F_v'/F_m' \quad (7)$$

and the relative ETR (*r*ETR) may be calculated as,

$$rETR = F_q'/F_m' \cdot E = F_q'/F_v' \cdot F_v'/F_m' \cdot E \quad (8)$$

With absolute ETR_{PSII} as,

$$ETR_{PSII} = a_{PSII} \cdot F_q'/F_m' \cdot E = a_{PSII} \cdot F_q'/F_v' \cdot F_v'/F_m' \cdot E \quad (9)$$

Hence, although it is often informative to dissect F_q'/F_m' into its components to assess the role of non-photochemical quenching in reducing the efficiency of photochemistry, this is not necessary to obtain ϕ_{PSII}' or ETR_{PSII} (Genty et al. 1989; Kromkamp and Forster 2003).

Like the PAM approach, ϕ_{PSII}' can be estimated from F_q'/F_m' using pump and probe (P&P, Kolber and Falkowski 1993) and fast repetition rate (FRR) fluorescence (Kolber et al. 1998) approaches. However, alternative treatments of the product of ϕ_{PSII}' and a_{PSII} are possible using P&P and FRR approaches since the effective absorption cross-section of PSII (designated σ_{PSII}', units of Å^2 quanta^{-1}, or alternatively m^2 (mol RCII)$^{-1}$) can also be directly measured (Kolber and Falkowski 1993; Kromkamp and Forster 2003; Suggett et al. 2006a, b, 2009b). The effective absorption cross section is the product of the optical absorption cross section and the maximum PSII photochemical efficiency, i.e. $\sigma_{PSII}' = a_{PSII} \cdot F_v'/F_m'$ (Mauzerall and Greenbaum 1989; Kolber et al. 1998; Kromkamp and Forster 2003); Consequently, substituting σ_{PSII}' into Eq. 9 it is easy to show that ETR_{PSII} can also be calculated as,

$$ETR_{PSII} = \sigma_{PSII}' \cdot \phi_P' \cdot E = \sigma_{PSII}' \cdot F_q'/F_v' \cdot E \quad (10)$$

Use of this model requires measurement of F_q'/F_v', which can be obtained relatively easily if fluorescence yields are measured simultaneously for the same sample under both light and transient dark conditions (Suggett et al. 2006a). One additional advantage of this F_q'/F_v' derivation is that it can remove the need to use additional independent discrete water samples, this time to quantify the background fluorescence (the "blank", Cullen and Davis 2003; Moore et al. 2005; Suggett et al. 2006a), as,

$$F_q'/F_v' = \frac{(F_q'/F_m' - f)}{(F_v'/F_m' - f)} \tag{11}$$

where f accounts for the proportion of background fluorescence inherent to the absolute fluorescence yield measurements. Here, F_q'/F_m' is measured under actinic light and F_v'/F_m' is measured under transient dark conditions, such as using the dual 'light' and 'dark' chambered FRR fluorometer However, this approach assumes that any rapidly reversible NPQ is not lost over the timescale of the transient dark period used for measurements of F_v'/F_m' (see Suggett et al. 2006a). Alternatively, F_q'/F_v' can be estimated using additional independent measurements of the minimum fluorescence yield under ambient light (F_o') from discrete water samples (Suggett et al. 2003, Baker and Oxborough 2004).

Under field conditions, the approach described by Eq. 11 further removes the need to collect additional independent discrete water samples required to measure light absorption (Kolber and Falkowski 1993; Gorbunov et al. 2001; Kolber et al. 2001; Suggett et al. 2001, 2006b). Measurements of σ_{PSII}' are often subject to noise, especially in high actinic light conditions. Therefore, to overcome the problem of noise in measurements of σ_{PSII}', σ_{PSII} can be measured in darkness, when the magnitude of variable fluorescence is greater due to the pool of RCII being 'open' at the initial point of measurement. ETR_{PSII} can then be calculated from,

$$ETR_{PSII} = \sigma_{PSII} \cdot \frac{F_q'/F_m'}{F_v/F_m} \cdot E \tag{12}$$

However, application of this model may also introduce uncertainties due to the requirement for either the collection of discrete water samples to allow full relaxation of light induced (i.e. reversible) NPQ (typically > 30 min) prior to measurement of σ_{PSII}, or use of measurements of σ_{PSII} made at night (Kolber and Falkowski 1993; Suggett et al. 2006a). The equivalence of Eqs. 10 and 12 relies on NPQ having an identical influence on F_v'/F_m' or σ_{PSII}' (Gorbunov et al. 2001), i.e. under these conditions it is expected that,

$$\sigma_{PSII}'/\sigma_{PSII} = \frac{F_v'/F_m'}{F_v/F_m} \tag{13}$$

an expectation that appears to be in reasonable agreement with data (see Section 3.1).

2.2 Examining Changes to the Quantum Yield Under Actinic Light

As described above, knowledge of the PSII photochemical efficiency under actinic light (ϕ_{PSII}') is required to calculate an ETR and can be conveniently determined by all active fluorescence approaches (chapter by Huot and Babin, this volume). Under dark acclimated conditions, maximum values for the PSII photochemical efficiency (ϕ_{PSII}) are observed. However, the values obtained will depend upon whether a single or multiple turnover (ST or MT) protocol was applied. As discussed in Huot and Babin (this volume), values for ϕ_{PSII} from the same sample are higher from MT than ST protocols. Basically, a MT protocol generates multiple charge separation events, which reduce both primary and secondary quinone (Q_A and Q_B) as well as plastoquinone (PQ) molecules. In contrast, a ST protocol reduces only Q_A. Plastoquinone (PQ) molecules strongly quench fluorescence; therefore, induction of the maximum fluorescence yield using an MT (F_m^{MT}) reduces PQ and so substantially lowers PQ-based quenching (Kramer et al. 1995). This same mechanism operates under actinic light with important implications for interpreting the light response of PSII photochemistry. Higher values of PSII quantum yields under actinic light (ϕ_{PSII}') for MT relative to ST are observed under relatively low light intensities; however, both MT and ST values for ϕ_{PSII} converge as the light intensity (E) increases beyond the saturating light intensity, termed E_K, (Fig. 1, Schreiber et al. 1995; Suggett et al. 2003). E_K signifies the light intensity at which the maximum ETR_{PSII} is reached; consequently, actinic light intensities greater than E_K induce gradual reduction of the PQ pool. Thus, the observed convergence of ϕ_{PSII}' under high light conditions is fully consistent with the mechanistic difference between MT and ST measurements outlined above.

The implications of using these alternative protocols are most apparent when comparing the light response of MT- versus ST-based derivations of the relative electron transfer rate, *r*ETR (Eq. 8, Table 1), (Fig. 1). The estimated initial slope of the ETR vs E curve rises more steeply, and towards a higher maximum rate, for the MT-based compared to the ST-based $rETR_{PSII}$. At the highest actinic light intensities, both *r*ETRs converge. Thus, light dependent trends observed in differences between ST and MT-based ϕ_{PSII}' and $rETR_{PSII}$ would appear to be an artifact associated with assessing the

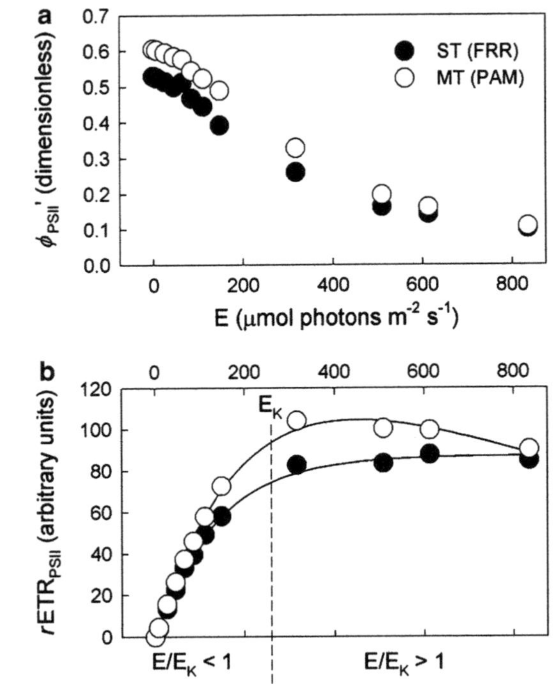

Fig. 1 Fluorescence-based photosynthesis light-response curves obtained from simultaneous PAM (Pulse Amplitude Modulated) and FRR (Fast Repetition Rate) measurements upon an aliquot of *Chaetoceros muelleri* (CCAP/1010/3) grown under ca. 150 μmol photons m^{-2} s^{-1} in f/2 nutrient replete medium (see Suggett et al. 2003). Fluorescence data plotted is (**a**) the PSII photochemical efficiency measured under actinic light (ϕ_{PSII}', $= F_q'/F_m'$, dimensionless) and (**b**) the relative ETR$_{PSII}$ (rETR, Table 1). Each point is the mean of triplicate measurements on different aliquots sampled from subsequent culture dilutions. Solid lines in panel **b** is the light response determined from fitting a model that describes the light dependence of photosynthesis to the rETR versus E model (Suggett et al. 2003). Values of E_K, termed the light saturation parameter, were calculated from the value of the maximum rETR$_{PSII}$ ($r\mathrm{ETR}_{PSII}^{max}$) relative to the value for the initial slope (α), both of which were derived from this model

maximum fluorescence yield using an MT protocol (Suggett et al. 2003, Beer and Axelsson 2004). As such, care must be taken when comparing MT- versus ST-measurements for the same actinic light intensity, or indeed comparing the 'shape' of the ETR$_{PSII}$-light response, i.e., through parameterisation of the initial slope, maximum value and inhibition value.

Ultimately, the choice of approach for quantifying ϕ_{PSII}' will not always be the same because of differences in the choice and application of the fluorometer used. However, it is important to realise inherent assumptions in alternative approaches, particularly in comparison with other measures of photosynthetic activity.

3 Light Absorption by Photosystem II

Measurements of ϕ_{PSII}' in combination with those of incident light intensity (E) enable determination of the relative electron transfer rate, rETR (Table 1; Eq. 8), and so provide a means for examining environmental or taxonomic induced changes to productivity; of course, these rETRs will have limited application for aquatic systems where changes in productivity are being influenced by changes in light absorption (e.g. Flameling and Kromkamp 1998; Hennige et al. 2008; see also Chapters 9,10). To translate a value of rETR$_{PSII}$ into 'absolute' ETR$_{PSII}$ requires that light absorption specific to PSII photochemistry, termed a_{PSII} (Table 1), be quantified either using biophysical measurements, bio-optical-based determinations or some combination of the two.

3.1 *Bio-Physical Measures of PSII Absorption and Calculation of Chlorophyll-Specific ETR*

Single turnover FRR fluorescence induction protocols can directly measure the effective PSII absorption cross section (σ_{PSII}, σ_{PSII}') (chapter by Huot and Babin, this volume), which, as we have already seen, can be used in determining ETR$_{PSII}$ (Eqs. 10 and 12). Use of Eq. 12 requires that Eq. 13 holds. Comparing simultaneous measurements of $\sigma_{PSII}'/\sigma_{PSII}$ with $(F_v'/F_m')/F_v/F_m$) under controlled laboratory conditions indicates good agreement with Eq. 13 for a wide range of microalgae (Fig. 2, Suggett et al. 2009a). In contrast to these observations, many studies that have estimated ETR$_{PSII}$ from Eq. 12 have measured σ_{PSII}' during transient dark conditions to increase the signal to noise, in particular when exposing samples to high light (Kolber and Falkowski 1993; Suggett et al. 2003, 2006a, b, 2007). Here, comparisons of $\sigma_{PSII}'/\sigma_{PSII}$ with $(F_v'/F_m')/F_v/F_m$), using values of σ_{PSII}' obtained from measurements made during short dark intervals between actinic light exposures, reveals that the assumption of Eq. 13 does not appear to hold for some microalgal species (Fig. 2c) when σ_{PSII}' is measured in the dark and F_v'/F_m' is not. However, when both σ_{PSII}' and $F_v'/$

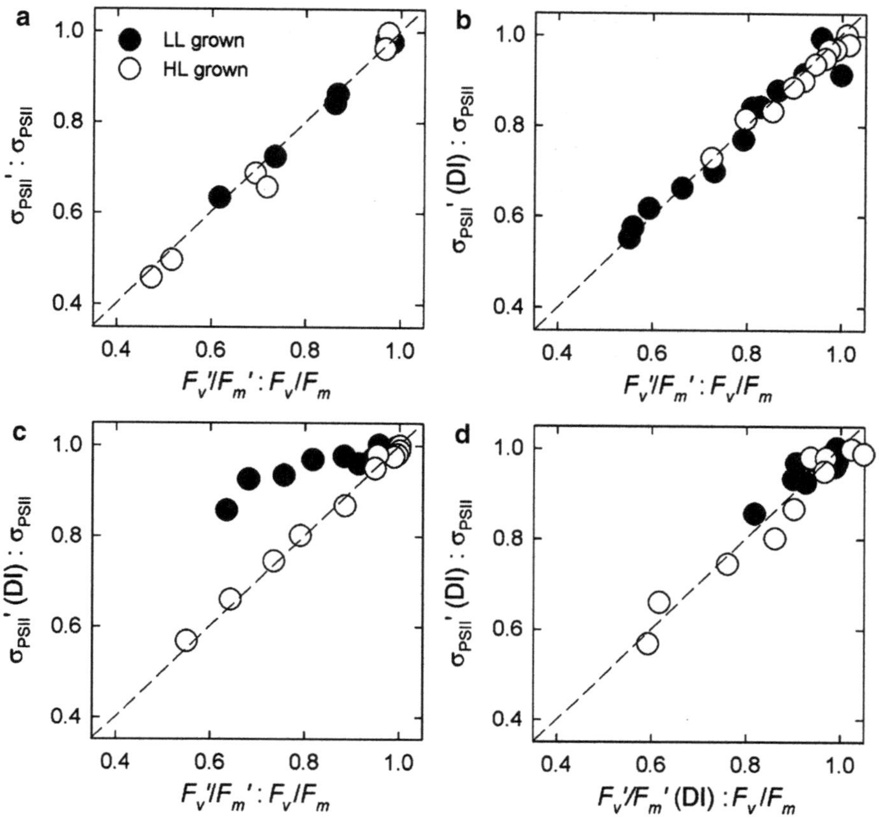

Fig. 2 Comparisons of the extent of non-photochemical quenching measured from alternative (FRR) fluorescence ratios: (**a**) the effective absorption cross section under actinic light (σ_{PSII}') to that under dark acclimation conditions (σ_{PSII}) versus the maximum PSII under actinic light (F_v'/F_m') to that under dark acclimation conditions (F_v/F_m); (**b**, **c**) σ_{PSII}' measured during a 20 s dark interval (DI) following actinic light exposure (σ_{PSII}' (DI)) relative to σ_{PSII} versus F_v'/F_m' : F_v/F_m; (D) σ_{PSII}' (DI) : σ_{PSII} versus the ratio of variable to maximum fluorescence also measured during the same 20 second dark interval (DI) following actinic light exposure, termed F_v'/F_m' (DI), relative to F_v/F_m. Data for panels A-B are from *Prorocentrum minimum* Choptank isolate (Suggett et al. 2004, 2009a) whilst those for panels C-D are from *Emiliania huxleyi* strain B11 (Suggett et al. 2007) both grown under ca. 25 (LL) and 300 (HL) μmol photons m^{-2} s^{-1} in nutrient replete media and exposed to a gradient of actinic light. All data were measured using FRR fluorometry

F_m' are measured following a brief dark interval agreement with Eq. 13 is again found (Fig. 2d). Such observations suggest that, at least for some taxa/growth conditions, some component(s) of rapidly reversible NPQ is(are) not accounted for when measurements are performed in the dark. Presence of such rapidly reversible NPQ presents additional limitations in estimating F_q'/F_v' (Eq. 11). In this specific case, absolute measures of σ_{PSII}', and in turn ETR_{PSII}, may thus be potentially overestimated by up to 30%.

Scaling ETR_{PSII} to the chlorophyll-specific ETR (ETR^{Chl}_{PSII}) or, through independent measurement of chlorophyll *a*, the absolute rate of ETR within a sample (c.f. Eqs. 1 and 2), requires a measure of the ratio of mol RCII to mol Chl (Kolber and Falkowski 1993; Babin et al. 1996; Table 1). Combining Eqs. 3 and 10, the chlorophyll*a*-specific ETR can be obtained from:

$$ETR^{Chl}_{PSII} = \sigma_{PSII}' \cdot F_q'/F_v' \cdot n_{PSII} \cdot E \qquad (14)$$

where n_{PSII} is defined as,

$$n_{PSII} = RCII/Chla \qquad (15)$$

as a measure of the ratio of the (functional) PSII reaction centre (RCII) concentration to total chlorophyll*a* present. Quantifying n_{PSII} remains a challenge to the use of fluorometers for productivity applications. Frequently "typical" values of n_{PSII} have been employed (Kolber and Falkowski 1993). However, laboratory studies have demonstrated variability in n_{PSII} both between taxa and with growth conditions (Dubinsky et al. 1986; Suggett et al. 2004, 2007, 2009a; Table 2) potentially introducing substantial uncertainty to estimates of ETR^{Chl}_{PSII}.

Our knowledge of n_{PSII} variability comes almost exclusively from microalgae as a result of the nature of the biophysical techniques used to determine the photosynthetic unit size. Single turnover excitation kinetics must be used to drive O_2 evolution from P680 to yield the PSII unit size whilst redox changes in the far-red region of peak absorption by the P700+ cation radical yield the PSI unit size (e.g. Falkowski et al. 1981). Within a suspension of single-celled algae, the cells are effectively freely suspended randomly and evenly in the P680/P700 excitation light fields; this approach works so long as the suspension is not too dense relative to the excitation intensity so as to induce self-shading. Therefore, for multicellular/mulit-layered algae and seagrasses, the individual thylakoids must be

Table 2 Summary of reported (and previously unpublished) values for n_{PSII}. Values here are presented as the reciprocal of n_{PSII}, i.e. $1/n_{PSII}$ (mol chl*a* (mol RCII)$^{-1}$), for simplicity; also as highest to lowest values since this is the trend typically observed when the various light intensities (or nutrient concentrations) for growth increase. Only changes in light intensity or nutrient concentration are indicated for growth conditions since all other factors remained constant (see original citations)

Species	Growth Conditions	$1/n_{PSII}$ (range)	Study
Aureococcus anophagefferens	2 × E	951–879	Suggett et al. (2004, 2009a)
Chlorella pyrenoidosa	6 × E	588–390	Myers and Graham (1971)
Chaetoceros muelleri	2 × E	591–520	Suggett et al. (2004)
Dunaliella tertiolecta	2 × E	833–710	Falkowski et al. (1981)
	2 × E	739–528	Sukenik et al. (1990)
	3 × E	742–501	Suggett et al. (2004, 2009a)
Emiliania huxleyi	6 × E	720–488	Suggett et al. (2004, 2007)
Isochrysis galbana	5 × E	637–264	Dubinsky et al. (1986)
	5 × E	756–406	Herzig and Dubinsky (1992)
	9 [NO_3]	478–431	Herzig and Falkowski (1989)
Mychonastes homosphaera	3 × E	243–163	Malinsky-Rushansky et al. (2002)
Nannochloropis sp.	2 × E	830–783	Fisher et al. (1996)
Planktothrix agardhii	constant; fluctuating	175; 125	Fietz and Nicklisch (2002)
Prorocentrum micans	4 × E	725–514	Dubinsky et al. (1986)
Prorocentrum minimum	3 × E	535–488	Suggett et al. (2004, 2009a)
Pycnococcus provasolii	3 × E	938–587	Suggett et al. (2004, 2009a)
Rhodomonas salina	2 × E	510–471	Suggett et al. (2004)
Scenedesmus quadricauda	5 × E	545–406	Herzig and Dubinsky (1992)
Skeletonema costatum	3 × E	605–590	Falkowski et al. (1981)
Stephanodiscus neostraea	constant; fluctuating	553; 282	Fietz and Nicklisch (2002)
Storeatula major	3 × E	518–444	Suggett et al. (2004, 2009a)
Symbiodinium spp.	2 × E	632–432	Iglesias-Prieto and Trench (1994)
	2 × E	635–273	Hennige et al. (2009)
Synechococcus leopoliensis	4 × E	493–405	Herzig and Dubinsky (1992)
Synechococcus sp.	5 × E	133–311*	Barlow and Alberte (1985)
	2 × E	293–236	Suggett et al. (2004)
Tetraedron minimum	2 × E	687–702	Fisher et al. (1989)
	3 × E	470–10	Malinsky-Rushansky et al. (2002)
Thalassiosira oceanica	2 × E	426–313	Suggett, Moore and Lawson (unpubl.)
Thalassiosira pseudonana	2 × E	485–499	Suggett and Moore (unpubl.)
Thalassiosira weisflogii	5 × E	723–553	Dubinsky et al. (1986)
	2 × E	585–420	Suggett et al. (2004, 2009a)
	2 × E	653–522	Suggett, Moore and Lawson (unpubl.)

* Indicates that PSU increased with growth PFD, likely representing the influence of photoinhibition

liberated into a suspension using a complicated and laborious extraction protocol (e.g. Henley and Dunton 1997). Of course, this same issue of sufficiently saturating the reaction centres with excitation energy in multilayered phototrophs, i.e. macroalgae, seagrasses (and potentially corals) raises fundamental questions as to the use of fluorescence techniques to generate ST values of $\sigma_{PSII}^{(\prime)}$ and thus the possibility of bio-physical measures of a_{PSII} from these organisms intact. That said, limited studies have examined photosynthetic unit sizes from the seagrasses and macroalgae (Henley and Dunton 1997; Major and Dunton 2002) and indeed demonstrate similar environmental regulation of P680 and/or P700 as is observed across the microalgae.

The proportion of RCII that are functional, and thus can contribute to linear ETR, may decline due to photoinhibition and/or nutrient stress. Measurements of reaction centre concentration using oxygen flash yields account for such effects as they inherently measure functional complexes. To account for changes in "functionality", without estimating RCII concentration directly, Kolber and Falkowski (1993) proposed that the measurement of F_v/F_m relative to a maximum value observed under "optimal" growth conditions can indicate the proportion of total the RCII concentration that was photochemically active (later Babin et al. (1996) used measurements of F_v/F_o in a similar manner). Initially, the assumption was made that this maximum value for F_v/F_m was a biological constant = 0.65, Kolber and Falkowski (1993), and hence used to normalise changes of F_v/F_m. Thus, assuming a 'typical' n_{PSII} when all the RCII are functional (n_{PSII}^{max}) we have,

$$ETR_{PSII}^{Chl} = \sigma_{PSII}{}' \cdot \phi_p{}' \cdot n_{PSII}^{max} \cdot (F_v/F_m)\,\text{sample}/0.65 \cdot E \quad (16)$$

However, it has subsequently been documented that both n_{PSII}^{max} (Suggett et al. 2004) and maximum values for F_v/F_m (or F_v/F_o) (Kolbížek et al. 2001; Juneau and Harrison 2005; Suggett et al. 2009b) vary considerably between taxa. Consequently the use of Eq. 16 will only strictly hold so long as the taxonomic composition of the community in question does not change (Suggett et al. 2006a) and caution should thus be applied to measurements crossing relatively large hydrographic, and hence ecological boundaries (but see also Kromkamp and Forster 2003). To circumvent this problem, some FRR-based studies have laboriously measured RCIIs in high chlorophyll waters using independent techniques (Moore et al. 2006; Suggett et al. 2006a). However such an approach is not currently practical for many aquatic systems. Alternatively RCII concentrations could be independently derived from combined bio-optical and bio-physical measurements potentially leading to alternative modelling approaches for predicting the variability in n_{PSII} and/or a_{PSII}^{Chl} (Suggett et al. 2004; Moore et al. 2005, 2006).

3.2 Bio-Optical Based Determinations of PSII Absorption

Researchers working with PAM protocols do not have the luxury of convenient $\sigma_{PSII}^{(\prime)}$ measurements; instead, a_{PSII}^{Chl} must be determined from independent (optical) light absorption measurements. These alternative optical light absorption measurements have advantages for calculating ETR_{PSII}^{Chl} since they can often be readily normalised to algal pigment or biomass. However, it is still necessary to assess the proportion of all absorbed light that is directed to PSII.

Absorption itself is conventionally measured upon discrete samples using a spectrophotometer, and can be expressed as the chlorophyll *a*-specific light absorption coefficient, designated a^{Chl} (see Geider and Osborne 1992). Calculation of a_{PSII}^{Chl} from a^{Chl} requires estimation of the proportion of absorbed photons channelled to PSII relative to PSI. Reconstructing the absorption by each individual pigment present, via the product of the pigment concentration and specific absorption coefficient, enables a rough estimate of light absorbed by the photoprotective relative to the photosynthetically active pigments (Bidigare et al. 1990; Moore et al. 2005). However, this approach incorrectly assumes that pigments either transfer light from antennae to reaction centres with 0% or 100% efficiency, respectively (Suggett et al. 2004). Also PSII is routinely assumed to account for 50% of the total light absorbed, as appears to be the case for higher plants, and presumably green algae (Boichenko 1998). However the actual value can deviate from the assumed 50% for other algal taxa and under variable growth conditions (e.g. Lutz et al. 2001; Suggett et al. 2004; Johnson and Sakshaug 2007; Hancke et al. 2008a).

Fluorescence excitation spectra have been applied to more accurately gauge the proportion of total light absorption that is utilised for PSII photochemistry.

Such applications are dependent upon the assumption that fluorescence emission spectra at 730nm (F_{730}) are indicative the PSII action spectrum, consistent with observations of F_{730} spectra and O_2 evolution action spectra (Neori et al. 1988), and yield minimum PSI fluorescence. Consequently, F_{730} spectra provide a relative measure of the proportion of absorbed light transferred to PSII. Scaling F_{730} spectra relative to their corresponding optical absorption spectra can then provide a absolute measure of the pigment transfer efficiency (Johnsen et al. 1997; Johnsen and Sakshaug 1993; Lutz et al. 2001; Suggett et al. 2004, 2007; Hancke et al. 2008a). Thus, in principle, these scaling approaches can provide a relatively simple, quantitative means for correcting the bulk optical absorption (a) to a_{PSII}. In practice this correction is somewhat confounded by the array of alternative scaling approaches that exist (Fig. 3); in particular (1) F_{730} spectra have been normalised to the Chl$a$ concentration and scaled to the value of a^{Chl} at a wavelength that is indicative of light absorption exclusively by only the photochemically active pigments associated with PSII (the "no overshoot" method, Johnsen et al. 1997; Johnsen and Sakshaug 2007); (2) F_{730} spectra can be scaled to a value of σ_{PSII} (m^2 mol RCII^{-1}) for a known wavelength to yield the sprectral dependency of σ_{PSII}, σ_{PSII} (λ); then the entire spectra is multiplied by n_{PSII} to yield the spectrally dependant a^{Chl}_{PSII} (Suggett et al. 2007, 2009a). In both cases, an estimate of the amount of absorbed light partitioned between the two photosystems can thus be obtained (Fig. 3). Alternatively, (3) both F_{730} and a^{Chl} may be normalised to their respective areas to give a measure of the deviation of a_{PSII}: a_{PSI} from unity, i.e., from 50% each (Suggett et al. 2004).

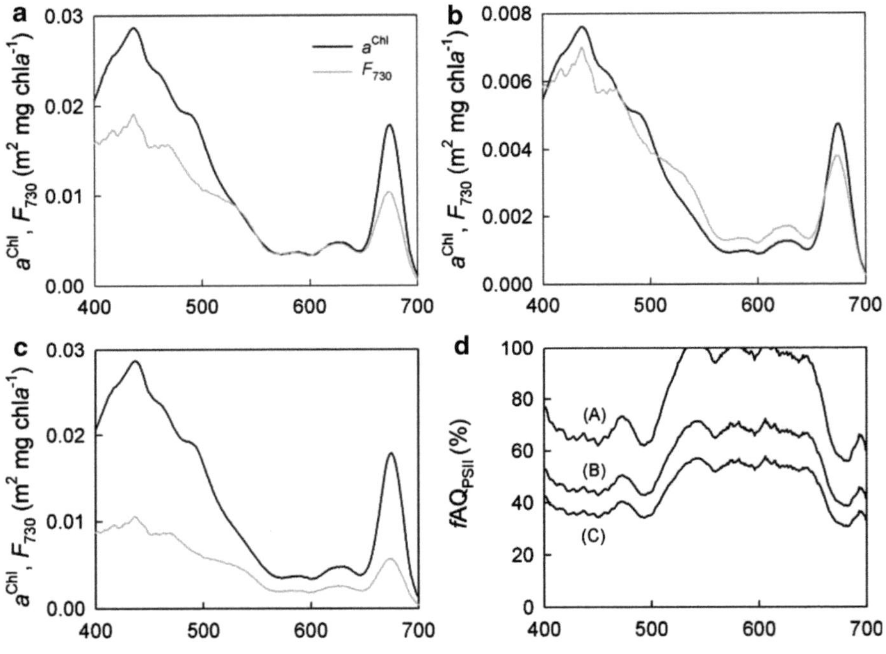

Fig. 3 Fluorescence excitation spectra (emission set to 730 nm, F_{730}) scaled to corresponding chlorophyll$a$ normalised optical absorption spectra (a^{Chl}, m^2 (mg chl$a$)$^{-1}$). Scaling is performed in three different ways: (**a**) F_{730} (instrument units) is scaled to a^{Chl} using the 'no overshoot' approach (Johnsen and Sakshaug 1993, 2007), i.e., values of F_{730} is set to a^{Chl} at a user-defined wavelength so F_{730} is always less than or equal to corresponding values of a^{Chl}. The difference between F_{730} and a^{Chl} is then used to indicate the proportion of all a^{Chl} going to PSII; (**b**) Areas under both F_{730} and a^* are set to an equal value. The ratio of F_{730} to a^{Chl} is then used determine whether more or less than half of all a^{Chl} is directed to PSII, i.e. $(F_{730}/a^{Chl}) \cdot 0.5$ (Suggett et al. 2004); (**c**) F_{730} is scaled according to the absolute measure of the effective absorption cross section (σ_{PSII}, m^2 mol RCII^{-1}) using a specific (single) excitation wavelength to yield the spectrally dependent σ_{PSII} (λ) (see Suggett et al. 2007); σ_{PSII} at each wavelength is then multiplied by n_{PSII} (a constant value for mol RCII (mg chl$a$)$^{-1}$) to yield an absolute estimate of PSII light absorption a^{Chl}_{PSII} (λ) (m^2 mg chla^{-1}) (Suggett et al. 2007). Panel (**d**) shows the various derivations of the proportion of all light absorption (a^{Chl}) directed to PSII, fAQ_{PSII} (%) from each of the three methods in panels A-C. Spectra used are from *Thalassiosira oceanica* CCMP 1003 grown under ca. 300 µmol photons m^{-2} s^{-1} in nutrient replete media (Tracy Lawson and David J. Suggett, unpublished)

Comparing these different scalings demonstrates how important the choice of method is; each one yields a different estimate for the proportion (*f*) of total absorbed quanta (AQ) by only PSII, i.e. termed fAQ_{PSII} (%, Fig. 3), following Johnsen and Sakshaug (2007). Although the various methods result in the same spectrally-dependent variability for fAQ_{PSII}, the "no overshoot" scaling method (Johnsen and Sakshaug 2007) yields values for fAQ_{PSII} that are higher by 40–80% than those of Suggett et al. (2004, 2007). These differences will have important consequences for calculated values of a_{PSII}. It is also important to note that all these methods require measurements of F_{730} posing constraints upon studies of natural communities, as considerable material is required due to the low fluorescence yield (Suggett et al. 2004). Such measurements are again possible on concentrated samples from natural communities of high chlorophyll regions (Moore et al. 2006). However, values of fAQ_{PSII} may vary as cells experience transient changes in environment and partition more or less of their absorbed light energy to PSII. In particular, state transitions would likely affect fAQ_{PSII} and thus lead to errors in estimating ETR_{PSII} if fAQ_{PSII} was assumed always constant.

Studies performed thus far demonstrate that considerable variability for fAQ_{PSII} exists between taxa regardless of the scaling approach employed; for example, values for fAQ_{PSII} appear consistently highest for non phycobilin containing chromophytes and chlorophytes, lowest for cyanophytes and intermediate for cryptophytes and rhosophytes (Johnsen and Sakshaug 2007). Despite differences in absolute value, the same pattern is observed using alterative scaling factors from independent experiments (Lutz et al. 2001; Suggett et al. 2004). These studies demonstrate that growth environment and taxonomy both regulate fAQ_{PSII} and it is clear that fAQ_{PSII} is an important variable that requires considerable attention.

Many of the potential limitations in 'choosing' the best optically-based method reflect limited validation for the absorption-fluorescence scaling approaches for yielding fAQ_{PSII} (see Suggett et al. 2004, 2007). One can assume that light absorption for photochemistry is accounted for from the product of the effective absorption cross section and concentration of reaction centres for both photosystems (Suggett et al. 2007; see also Behrenfeld et al. 2004), i.e., $a^{Chl} = (\sigma_{PSII} \cdot n_{PSII}) + (\sigma_{PSI} \cdot n_{PSI})$ although a further correction for loss of excitation energy due to NPQ in PSII would also strictly be required (see above). Whilst n_{PSII}, n_{PSI} and σ_{PSII} can all be quantified relatively easily, measurements of σ_{PSI} remain challenging and indeed σ_{PSI} is often calculated by assuming $\sigma_{PSI} = [a^{Chl} - (\sigma_{PSII} \cdot n_{PSII})]/n_{PSI}$ (Dubinsky et al. 1986; Suggett et al. 2007). Measurements of σ_{PSI} have been demonstrated using rapid (<10 ms) P700 oxidation kinetics (e.g. MacKenzie et al. 2004) or photochemical thermal efficiencies derived from photoacoustics (Berges et al. 1996). However, both techniques are currently impractical for routine (and high resolution) fluorescence investigations.

Having considered these alternative approaches for estimating PSII absorption (a^{Chl}_{PSII}) it is clear that, as with the PSII photochemical efficiency (ϕ_{PSII}'), potential problems may occur when reconciling ETRs derived using differing techniques. However, independent determinations of a^{Chl}_{PSII} from a bio-optical basis do appear to compare well with those from a bio-physical basis across a wide range of phytoplankton taxa when the influence of fAQ_{PSII} and RCII concentration are fully accounted for (Suggett et al. 2004). Certainly, the accuracy afforded by the approaches appears similar despite the potential for systematic errors in both bio-optical and bio-physical derived variables.

4 Reconciling Active Fluorescence-based Estimates of Productivity with Gas Exchange

A number of studies have compared productivity estimates based on active fluorescence with 'conventional' gas exchange measurements (Hancke et al. 2008b; Suggett et al. 2009a; Table 3). Primarily, such exercises have been undertaken to validate the ETR_{PSII} against independent measurements of photosynthetic gas exchange and so provide a tool for estimating photosynthesis *in situ*. This has largely been performed with the objective of exploiting the high spatial and temporal resolution with which ETR_{PSII} can be measured to provide a more accurate picture of environmental regulation of productivity (e.g. Falkowksi et al. 1991). More recently, researchers have recognized the value of using corresponding ETR_{PSII} and gas exchange measurements to examine how cells utilize photochemical energy for fixing organic carbon (Wagner et al. 2006; Prášil et al. 2008; Mackey et al. 2008; Suggett et al. 2009a). For both applications, appreciation

Table 3 Summary of mean values (and minimum-maximum across the entire data set) of the minimum quantum requirement for O_2 evolution or CO_2 uptake obtained when comparing corresponding measurements of ETR_{PSII} with O_2 evolution or carbon fixation for AAP (aerobic anoxygenic photosynthetic) bacteria, phytoplankton, microphytobenthos, microalgae and seagrasses. Only studies that calculate absolute ETR_{PSII}, i.e., have attempted to account for PSII absorption, are considered here. To date, no such comparative studies have been performed on corals. PP is Pump and Probe, FRR is Fast Repetition Rate and PAM is Pulse Amplitude Modulated fluorometry. This list is not intended to be exhaustive but rather to demonstrate inherent variability that has been observed across the various autotrophic groups. A more complete list of FRR-based studies can be found in Suggett et al. 2009b). In all cases, O_2 evolution and carbon uptake are measured by O_2 electrodes and ^{14}C uptake, unless otherwise indicated. RLC is rapid light curve, SIS is simulated in situ and PE is photosynthesis light-response; also, step refers to an actinic light that is applied in increasing intensities in a step wise fashion (see main text). For many studies, the data has been digitized from source to calculate the mean, minimum and maximum values. A value of 100 mol chl*a* (mol RCII)$^{-1}$ has been assumed for modifying the Kolber et al. (2001) data set here

Method			Mean (range)	Mean (range)	
Techniques	Approach	Organisms	mol e^- : mol O_2	mol e^- :mol CO_2	Study
AAP BACTERIA					
FRR v ^{14}C	RLC v PE	NAP1 isolate		26.2 (12.1–32.2)	Kolber et al. (2001)
PHYTOPLANKTON					
FRR v O_2	*In situ* yield v LD bottle	n/a (natural communities)	4.5 (3.7–5.6)		Sarma et al. (2005)
FRR v $^{16,17,18}O_2$	*In situ* yields		2.6 (2.3–3.0)		
FRR v O_2	*In situ* yield v SIS (24h)	Centric diatoms and pikoeukaryotes	1.2 (0.6–2.5)		Robinson et al. (2009)
FRR v $^{18}O_2$	*In situ* yield v SIS (24h)		0.8 (0.2–1.3)		
FRR v ^{14}C	*In situ* yield v SIS (24h)			3.8 (1.6–6.2)	
FRR v ^{14}C	*In situ* yield v 2h PE			7.2 (2.1–67.2)	
FRR v $^{18}O_2$, $^{14}CO_2$	25–50 min yields on same sample	*Aureococcus anophagefferens*	3.7 (1.6–4.4)	14.3 (7.9–88.4)	Suggett et al. (2009a)
		Dunaliella tertiolecta	3.6 (1.9–6.5)	7.2 (5.2–14.4)	
		Prorocentum minimum	4.3 (3.1–8.1)	10.1 (6.0–22.2)	
		Pycnococcus provasolii	5.7 (3.7–9.6)	14.9 (8.8–25.4)	
		Storeatula major	3.4 (2.6–5.0)	7.4 (4.5–17.0)	
		Thalassiosira weissflogii	4.0 (3.3–6.7)	6.1 (5.1–12.9)	
PP v ^{14}C	*In situ* yield v SIS (4h)	n/a (natural communities)		5.4 (2.5–15.3)	Kolber and Falkowski (1993)
FRR v ^{14}C	*In situ* yield v 1h PE	Cyanobacteria, nanoflagellates		8.7 (1.1–10.1)	Suggett et al. (2001)
	In situ yield v 1–2h PE	Diatoms, dinoflagellates		11.9 (8.2–13.7)	Raateoja et al. (2004)
		Cyanobacteria, flagellates		8.8 (6.2–11.5)	
	In situ yield v SIS (day)	Picoeukaryotes, prochlorophytes		8.8 (3.1–13.6)	Corno et al. (2005)
	In situ yield v SIS (24h)	Coccolithophores, flagellates		10.1 (8.5–14.0)	Pemberton et al. (2006)
	Step PE (5 min) v 20 min PE	*Dunaliella tertiolecta*		6.1 (5–9–7.8)	Fujiki et al. (2007)
	In situ yield v 2–4h PE	Diatoms, flagellates		7.2 (3.6–10.3)	Debes et al. (2008)
	In situ yield v 2h PE	Cyanobacteria, chlorophytes		20.9 (9.1–37.9)	Kromkamp et al. (2008)
FRR v C-uptake	*In situ* yield v *in situ* pH drift (daily)	Diatoms		4.0 (3.2–5.1)	Suggett et al. (2006a)
		Diatoms, flagellates		6.1 (4.9–7.8)	
PAM v O_2	Both light steps (steady state) on same sample	*Chlorella vulgaris*	4.7 (4.0–5.1)		Gilbert et al. (2000b)
		Cryptomonas ovata	3.7 (3.5–4.3)		
		Cyclotella meneghiniana	3.8 (3.5–4.0)		
		Synechococcus leopoliensis	4.4 (3.0–5.3)		
	Both light steps (steady state) on same sample	Cyanobacteria (*Anabaena* and *Microcystis* sp.)	5.3 (3.0–7.3)		Masojídek et al. (2001)
	Both light steps (steady state) on same sample	*Chlorella vulgaris*	8.3 (8.1–8.4)		Wagner et al. (2006)
		Phaeodactylum tricornutum	5.4 (4.5–6.3)		
	Step PE (5 min) v 1h PE	*Prorocentrum minimum*	3.4 (2.6–8.2)		Hancke et al. (2008a, b)
		Prymnesium parvum	9.9 (6.7–11.8)		
		Phaeodactylum tricornutum	6.7 (5.6–8.5)		

(continued)

Table 3 (continued)

Method					
Techniques	Approach	Organisms	Mean (range) mol e⁻ : mol O_2	Mean (range) mol e⁻ :mol CO_2	Study
PAM v ^{14}C	RLC v SIS (3h)	Chlorophytes, Diatoms, Crytpophytes		13.6 (5.3–19.8)	Gilbert et al. (2000a)
	Step PE (5 min) v 1h PE	*Prorocentrum minimum*		9.2 (6.2–23.8)	Hancke et al. (2008b)
		Prymnesium parvum		14.8 (13.3–15.2)	
		Phaeodactylum tricornutum		8.0 (5.8–9.7)	
	RLC v 2h PE	Cyanobacteria, chlorophytes		16.8 (8.2–26.4)	Kromkamp et al. (2008)
MICROPHYTOBENTHOS					
PAM v O_2	Both light steps (steady state) on same sample	*Cylindrotheca closterium*	4.2 (0.73–32.4)		Morris and Kromkamp (2003)
PAM v ^{14}C	RLC v 1h PE	n/a (natural samples)		40.1 (27.0–77.7)	Hartig et al. (1998)
	RLC v 30 min PE	n/a (natural samples)		4.7 (3.6–7.1)	Morris et al. (2008)
MACROALGAE					
PAM v $^{18}O_2$	Both light steps (steady state) on same sample	*Porphyra columbina*	10.7 (6.4–19.1)		Franklin and Badger (2001)
		Ulva australis	6.2 (2.9–10.6)		
		Zonaria crenata	7.8 (4.7–20.8)		
PAM v O_2	Both light steps (steady state) on same sample	*Ulva lactuca*	4.3 (2.6–5.6)		Beer et al. (2000)
		Ulva fasciata	3.8 (3.0–10.1)		
	Both diel in situ (+PAM RLCs)	*Ulva lactuca*	6.6 (5.1–8.2)		Longstaff et al. (2002)
	Both light steps (steady state) on same sample	*Porphyra leucosticta*	2.7 (1.8–3.6)		Figueroa et al. (2003)
		Ulva olivascens	13.7 (12.1–16.1)		
		Ulva rotunda	23.7 (10.5–26.9)		
	Both light steps (steady state) on same sample	*Ulva lactuca*	3.8 (1.7–6.3)		Beer and Axelsson (2004)
		Fucus serratus	4.1 (1.5–7.1)		
		Laminaria saccharina	3.7 (1.9–4.3)		
		Palmaria palmata	3.5 (2.5–4.8)		
		Porphyra umbilicalis	3.9 (6.3–2.1)		
	Both light steps (steady state) on same sample	*Ulva rigida*	3.9 (3.5–4.3)		Cabello-Pasini and Figueroa (2005)
	RLC v step O2	*Desmarestia menziessi*	6.8 (3.0–7.2)		Runcie and Riddle (2006)
		Hinantothallus grandifolius	10.5 (9.5–12.1)		
		Iridaea mawsonii	4.9 (7.0–12.5)		
SEAGRASS					
PAM v O_2	Both light steps (steady state) on same sample (+PAM RLCs)	*Cymodocea nodosa*	2.6 (1.2–5.4)		Beer et al. (1998)
		Halophila stipulacea	8.3 (6.7–11.9)		
		Zostera marina	3.2 (2.0–4.3)		
	Both light steps (steady state) on same sample	*Halodule wrightii*	2.9 (1.8–4.9)		Beer and Björk (2000)
		Halophila ovalis	3.8 (3.4–5.8)		
	Both light steps (steady state) on same sample	*Zostera noltii*	6.7 (4.6–7.9)		Silva and Santos (2004)
	Both light steps (steady state) on same sample	*Thalassia testudinum*	4.3 (2.9–8.7)		Enriquez et al. (2006)

of potential inaccuracies in the chosen ETR_{PSII} algorithm is clearly desirable. However, assessing the systematic and random errors is complicated by no single gold standard existing with which to inter-compare alternative ETR algorithms or alternative measures of productivity. Potentially aquatic research is in danger of becoming inundated with an ever expanding number of productivity comparison studies that are difficult to reconcile (Table 3), with the subsequent potential for increasing confusion and little useful progress. Therefore, whilst no single method is currently universally applicable or accepted, it is important to understand how the alternative methods (and their associated 'currencies') are related.

Gas exchange measurements themselves quantify O_2 evolution or inorganic carbon uptake and thus typically provide a more useful 'currency' than electrons with which to describe ecosystem function (Suggett et al. 2009a). Table 3 summarises the measured electron requirements through PSII for photosynthetic O_2 evolution or CO_2 fixation; expressed as mol e^- (mol $O_2)^{-1}$ and mol e^- (mol $CO_2)^{-1}$. Based on the z-scheme for linear electron flow, a minimum value of 4 e^- linearly transported through PSII is required for gross O_2 production or CO_2 fixation.

Higher values (> 4 mol e^- (mol $O_2)^{-1}$) are expected for net O_2 evolution and CO_2 fixation due to electron sinks operating within the photosynthetic electron transfer chain and Calvin cycle, electron sinks associated with nitrate reduction and perhaps interactions with mitochondrial respiration. All of the reducing equivalents (e.g., NADPH, reduced ferredoxin etc.) required by an oxygenic obligate photoautotrophic cell are ultimately derived from linear electron flow. ATP is produced in parallel to the production of NADPH during linear electron flow, using the energy associated with transfer of protons across the thylakoid membrane (Baker et al. 2007). The generation of NADPH and ATP by the Z-scheme can be summarized by the stoichiometry:

$$2H_2O + 2\ NADP^+ + 2.6\ ADP + 2.6\ Pi + 8\ photons \rightarrow$$
$$O_2 + 2\ NADPH + 2.6\ ATP + 2H^+$$

Carbon dioxide fixation via the Calvin cycle consumes 2 NADPH and 3 ATP for each CO_2 fixed. Thus, linear electron flow does not provide the correct ATP:NADPH ratio for CO_2 fixation (Baker et al. 2007). Thus, photoautotrophs must supplement the ATP generated in parallel to linear electron transfer from H_2O to NADPH with additional light-driven generation of ATP (Fig. 5). Operation of a CO_2 concentrating mechanism, which is necessary to suppress photorespiratory activity in phytoplankton, requires additional ATP exacerbating the mismatch between NADPH and ATP supply by the z-scheme from water to NADPH. If cyclic electron transfer around PSI provides this ATP, then there is no need for additional charge separation in PSII. However, two other processes that can generate additional ATP involve the reduction of O_2 by the photosynthetic electron transfer chain, and thus require charge separation at PSII: these processes (involve either the plastoquinone terminal oxidase (PTOX) or the transfer of e- from reduced ferredoxin to O_2 in the Mehler reaction (e.g. Behrenfeld et al. 2004, 2008). Thus, we should expect systematic differences in the e^- $(O_2)^{-1}$ and e^- $(CO_2)^{-1}$ ratios dependent on whether gross or net gross photosynthesis is measured, and whether nitrate and/or ammonium are being assimilated in parallel with photosynthesis. However practical consideration concerning the derivation of these ratios must also be considered.

4.1 Practical Constraints in Comparing Fluorescence- and Gas Exchange-Based Productivity Measurements

Measurements of ETR and photosynthetic gas exchange are rarely performed simultaneously on the same sample. Such simultaneous measurements are essential if one is to eliminate the possible effects of different sample treatments on the measurements of ETR_{PSII} and photosynthetic gas exchange. More often than not, ETRs have been estimated on a sample in one light regime whilst gas exchange has been measured on a separate sample (or samples) using different light regimes. Light regimes may differ in the actinic light source used, the optical geometries under which ETR and gas exchange are measured, as well as in the duration of exposure to different actinic light levels (MacIntyre and Cullen 2005). These various differences often arise from operational constraints inherent either to the fluorescence measurements themselves or to conventional (gas exchange) productivity measurements, in particular where radioisotope tracers are employed.

In addition to studies comparing alternative productivity measurements under the same controlled conditions in the laboratory 'real time' ETRs *in situ* have been compared with laboratory-based gas exchange incubation experiments. Laboratory-based ETR_{PSII} and gas exchange measurements can be run in parallel and routinely in the form of a photosynthesis-light response experiment. However, such light response experiments can also be performed using different protocols potentially resulting in inconsistencies between investigations (MacIntyre et al. 2000; Suggett et al. 2009a).

ETR-light response curves. Fluorescence (ETR_{PSII})-light response experiments in the laboratory typically monitor changes of ETR in response to stepped increases of light intensity, however, the duration of exposure to different light intensities are delivered can be either 'rapid', i.e. 10–20s in duration (White and Critchley 1999; Ralph and Gademan 2005; Perkins et al. 2006; Hennige et al. 2008), or prolonged, i.e. 3–4 min (Suggett et al. 2003; Serôdio et al. 2006). Both rapid and prolonged ETR_{PSII} approaches offer the user advantages. Rapid ETR_{PSII}-light curves (RLCs) are applied when sampling time is constrained, for example, when examining mobile cells but also when attempting to characterize the fluorescence response to a rapidly changing (micro) environment (Serôdio et al. 2006). This latter point relates to one of the original concepts of RLCs, to enable "in situ" activity to be measured and not the potential activity achieved using longer actinic exposure times (White and Critchley 1999; see also Perkins et al. 2006; this volume; Warner et al., this volume); consequently, RLCs are typically performed without prior dark acclimation. In contrast, prolonged light steps are designed to enable the fluorescence yield and so ETR_{PSII} to reach an approximate 'steady' state: Fluorescence yields, σ_{PSII}' and F_q'/F_m' alter upon immediate exposure to light but may subsequently settle to a new and constant level, the steady state. Most likely this trend reflects the capacity of, and time constants associated with, the donor and acceptor reactions of photochemistry along with NPQ to adjust linear electron flow following a change in light intensity. As such, the steady state approach produces a light response that describes the condition to which cells have acclimated (Suggett et al. 2003). Application of either a rapid or steady state protocol will thus result in a different 'shape' of the ETR-light response (Serôdio et al. 2006). Approaches that apply a stepped increase in irradiance are also limited by the cumulative delivery of the light dose which mayfurther influence the operation of photochemical and non-photochemical dissipation and hence the shape of the ETR light response (e.g. Perkins et al. 2006; Herlory et al. 2007).

Photosynthesis-light response curves. Oxygen light response curves measured using Clark-type electrodes (e.g. Falkowski et al. 1986) or membrane inlet mass spectrometry (MIMS, e.g. Kana 1990) can potentially use exactly the same protocol as for measurements of ETR_{PSII}. In fact, use of the same protocol is a fundamental criterion if electron transport rates and gas exchange rates are to be truly inter-compared but can only be achieved for O_2 methods with relatively dense algal cultures in artificial light sources (Suggett et al. 2009a). Here, optically dense cultures used for such measurements may ultimately compromise attempts at simultaneous measurement of ETR_{PSII} due to difficulties in both defining the light field within the cuvette and saturating PSII photochemistry during the fluorescence measurement. Thus, careful assessment of the response of variable fluorescence to optical density must initially be made. In contrast, $^{14}CO_2$ light response curves are end-point measurements (e.g. Platt and Jassby 1976; MacIntyre et al. 1996), as are O_2 bottle incubations (e.g. Grande et al. 1991), and employ durations of tens of minutes to hours (MacIntyre and Cullen 2005; Table 2).

Comparison of ETR-light and photosynthesis-light responses. Light response protocols that allow fluorescence to reach steady state are thought to be most comparable to 'conventional' gas exchange based photosynthesis light (P-E) response experiments (Suggett et al. 2003; Serôdio et al. 2006). However, even these ETR_{PSII} measurements, whether from a stepped light response experiment or an *in situ* profile, do not fully reproduce the timescales used for gas exchange-based photosynthesis light response experiments.

Processes regulating photosynthesis operate at very different time scales for ETR_{PSII} compared to gas exchange-based light response experiments (Suggett et al. 2009a); for example, the relatively short term ETR experiments may account for rapid maintenance of linear electron transfer by state transitions and xanthophyll activity but not longer term acclimation mechanisms, such as regulation of Rubisco or PSII proteins, that may be inherent to gas exchange experiments (MacIntyre et al. 2000, 2002).

In situ ETR. Not surprisingly, many active fluorescence-based productivity comparison studies have measured ETR_{PSII} *in situ*. *In situ* ETR_{PSII} estimates are based on point measurements in time and space, which often require repeated profiling (e.g. Smyth et al. 2004) or extrapolation (e.g. Suggett et al. 2001) to longer periods to calculate *in situ* productivity on ecologically meaningful time scales (typically 1 day), or for comparison with longer timescale (simulated) *in situ* productivity measurements (typically 6–24 h) (e.g. Kolber and Falkowski 1993). A popular method for examining phytoplankton has been to generate an effective ETR_{PSII} light-response profile *in situ* by profiling a fluorometer throughout the water column (Suggett et al. 2001; Smyth et al. 2004; Suggett et al. 2006b; Fig. 4). Here, the water column structure may affect the 'shape' of the ETR_{PSII} light-response that is generated for several reasons:

Firstly, the rate of change of light intensities as cells circulate with depth will determine whether the fluorescence yield can reach steady state (Moore et al. 2003; 2006). Thus, whilst an ETR_{PSII} light-response may be apparent *in situ* (Fig. 4; Suggett et al. 2006b), those from highly mixed waters may be more representative of a rapid light curve than a steady state induction protocol.

Secondly, well mixed waters will have an ETR_{PSII} light-response profile that is uniquely characteristic of a homogenous community, albeit likely composed of multiple taxa with potentially differing photophysiological properties. In contrast, multiple components of the ETR_{PSII} light-response may be observable in stratified waters reflecting vertical gradients in community structure (Fig. 4). Under these stratified conditions it may be possible to separate the ETR profile according to such differences in water column community structure, or extract 'instantaneous' values of ETR_{PSII} from depths corresponding to discrete water samples, which can be analysed in more detail (e.g., Suggett et al. 2001; Corno et al. 2005, 2008).

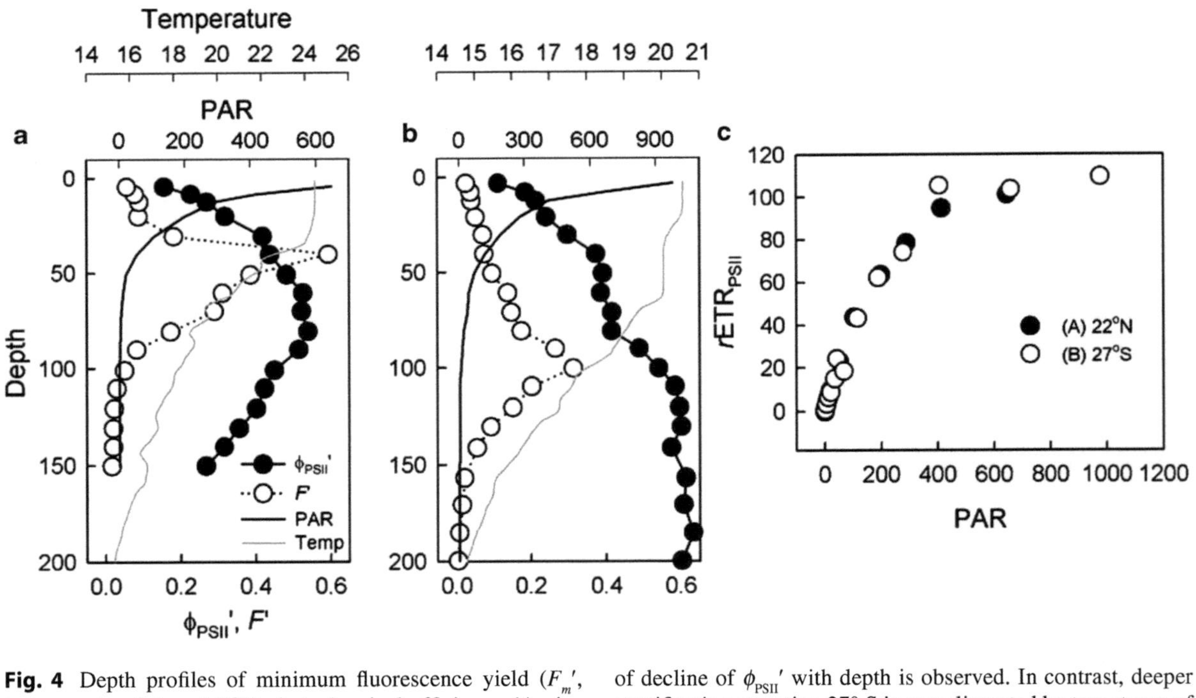

Fig. 4 Depth profiles of minimum fluorescence yield (F_m', instrument units) and PSII photochemical efficiency ($\phi_{PSII}' = F_q'/F_m'$, dimensionless, Table 1) under actinic light, and temperature (°C) and incident light (E, μmol electrons m^{-2} s^{-1}), from two stations during Atlantic Meridional Transect (AMT) cruise 11 (see Suggett et al. 2006b): (**a**) 22.0°N 21.0°W and (**b**) 27.3°S 38.7°W. All data were collected by CTD and FRR fluorescence during the rosette upcast and binned into ca. 4m (temperature, E) and 10m (FRR fluorescence) depths. Whilst stratification was observed at station 22.0° N, a single pattern of decline of ϕ_{PSII}' with depth is observed. In contrast, deeper stratification at station 27° S is complimented by two stages of decline of ϕ_{PSII}' with depth: between ca. 120–90 m and then between ca. 40 m and the surface. This second station had pronounced high light and low light adapted communities; (**c**) relative ETR_{PSII} ($rETR_{PSII}$) calculated for each data point *in situ* for ϕ_{PSII}' and E (following Table 1) and plotted against E for both stations. In all cases, the ETRs are weighted by data collected in the upper water column where light intensities are highest

Finally, ETR estimates require that artifacts associated with instrument operation remain constant throughout the water column. Application of a single "fluorescence blank" for correcting absolute fluorescence yields may or may not be appropriate depending on the water body in question (see above, also chapter by Huot and Babin, this volume; Laney, this volume). Early FRR fluorometers are also known to be subject to the 'red light effect' where the photomultiplier tube is shut down in the near surface where high red light intensities can saturate (and potentially damage) the detector (e.g.. Suggett et al. 2001; Smythe et al. 2004, Huot and Babin, Chapter 3, this volume; Laney, Chapter 2, this volume). Consequently, data from the highest light intensities (where ETR_{PSII} ultimately 'saturates', see Fig. 4, or perhaps even inhibits) may be lost. In this case using the 'shape' of the ETR_{PSII} versus E curve to inform productivity models will of course be limited. This problem will be more problematic for case-2 than for case-1 waters where the light-saturated ETRs are restricted to the near surface.

In situ and simulated in situ primary productivity. Primary productivity within the water column can be calculated from P-E curves and information on the *in situ* light field (e.g. Sathyendranath et al. 1995). However, such calculations are subject to systematic errors associated with matching the light doses within the incubators used to generate the P-E curve to the light environment *in situ*, as well as potential errors associated with sampling the phytoplankton, especially in stratified waters. Often, primary productivity is determined from *in situ* or simulated *in situ* ^{14}C bottle incubations.

Open water estimates of primary productivity. Open water methods employ diel variability of the concentrations of dissolved oxygen and/or TCO_2 (which can potentially be inferred from pH in some cases) to calculate net and gross community productivity (e.g. Suggett et al. 2001, 2006a; Sarma et al. 2005). These methods measure community metabolism, and are free from any potential bottle effects. However, they have limited temporal resolution, are subject to heterogeneity in the water mass under investigation, and need to account for physical exchanges within the water column and between the water column and the atmosphere.

Productivity comparison studies that simultaneously measure ETR_{PSII} and O_2 evolution (e.g. Longstaff et al. 2002; Morris and Kromkamp 2003; Sarma et al. 2005; Wagner et al. 2006; Table 3) and/or CO_2 uptake (Suggett et al. 2006a, 2009a) upon the same sample (or body of water) clearly represent the truest reconciliation of independent photosynthetic 'currencies'. Surprisingly few studies working with phytoplankton have attempted either approach (Table 3) primarily since gas exchange measurements *in situ* require that the same body of water be sampled long enough to detect a significant change. Such a constraint may improve as pH (CO_2) and O_2 sensing technologies continue to improve; however, accurately accounting for the autotrophic component of respiration can still remain problematic (Suggett et al. 2006a). Similarly, quantifying ETR_{PSII} *in situ* over a prolonged time scale for the same body of water as sampled for *in vivo* gas exchange incubations cannot be guaranteed. In this instance researchers working with benthic organisms are potentially at an advantage. For example, important insights into the coupling of electron transfer and net O_2 evolution have been achieved for macroalgae (e.g. Beer et al. 2000; Figueroa et al. 2003), seagrasses (e.g. Beer et al. 1998; Enriquez et al. 2006) and corals (e.g. Levy et al. 2004, 2006) (Table 2). Measurements can be performed quickly since the samples are typically of relatively high biomass, however sample enclosure and hence isolation from natural environmental fluctuations represents a potential weakness. Use of microelectrodes may provide some means to circumvent this limitation (e.g. Glud et al. 2002).

Accounting for spectral quality. Even if studies are able to match the time scales associated with the different productivity measurements, accurately quantifying differences between ETR_{PSII} and gas exchange is further complicated by the need to match the spectra of the actinic irradiances used (Boyd et al. 1997; Geider and Osborne 1992; Suggett et al. 2001; MacIntyre and Cullen 2005). Examining the same sample maintained under a single light source clearly ensures a consistent spectral quality of incident light. However, strict comparisons between *in situ* ETR_{PSII} measurements and incubation measurements and/or incubations performed under different light regimes requires that spectrally resolved optical measures of $a^{Chl}(\lambda)$, or biophysical measures of $\sigma_{PSII}(\lambda)$ (which is initially weighted to the fluorometers light source), be weighted by the incident light spectra (e.g. Gilbert et al. 2000a, b; Suggett et al. 2001, 2007; Moore et al. 2006; see Table 1). Such corrections require detailed quantification of the spectral nature of the corresponding light fields and can prove to be very significant (Suggett et al. 2001; Moore et al. 2006), in particular where 'blue' excitation by fluorometers are applied in environments where natural irradiance is heavily dominated by 'red' wavelengths (see MacIntyre and Cullen 2005). However, in some cases the spectral nature of the underwater light field may be relatively constrained across ocean regions enabling the use of a generic correction factor (e.g. Suggett et al. 2006b).

4.2 Are ETRs Indicative of Gross O_2 Evolution?

ETR_{PSII} provides a direct measure of the photochemical operation of PSII and thus quantifies the capacity for gross O_2 evolution. However, few studies have directly compared corresponding measurements of ETR and gross O_2 evolution by PSII. Conversion of the productivity rate from electron transport to gross O_2 evolution requires knowledge of the quantum yield for electron transport through PSII, termed ϕ_e, mol O_2 (mol e^-)$^{-1}$ (Kolber and Falkowski 1993). Alternatively, the reciprocal of ϕ_e ($1/\phi_e$) describes the minimum quantum (electron) requirement of gross O_2 evolution. The chosen value for $1/\phi_e$ is often set constant at the assumed maximum of 4 mol e^- (mol O_2)$^{-1}$ (Table 1) since a minimum of 4 electron steps are required to evolve one mol of O_2 (Falkowski and Raven 2007). However, higher values have been inferred (5–10 mol e^- (mol O_2)$^{-1}$) based on independent studies that have derived ϕ_e using simultaneous measurements of ETR_{PSII} and O_2 evolution (e.g. Flameling and Kromkamp 1998). Accurate values of $1/\phi_e$ strictly require knowledge of gross O_2 evolution by PSII; however, most to date do not account for possible light-dependent O_2 sinks, as discussed in the next section, and likely result in underestimation of $1/\phi_e$. Few studies have actually quantified gross O_2 evolution, e.g. using mass inlet membrane spectroscopy (Franklin and Badger 2001; see also Suggett et al. 2003, 2009a), but yield interesting insight as to the potential variable nature of $1/\phi_e$.

In seeking to estimate $1/\phi_e$ and thus convert ETR_{PSII} to carbon fixation estimates for natural phytoplankton assemblages, Kromkamp et al. (2008) employed an alternative method by plotting a regression of the quantum yields for carbon fixation (ϕ_c) against comparative FRR-measures of F_q'/F_m'. In this case, only the optical absorption (a) and carbon fixation (or alternatively O_2 evolution) rates are required to yield an electron requirement for carbon fixation, $1/\phi_{e(c)}$ (or O_2 evolution, $1/\phi_e$). Such an approach has the potential to generate a wealth of information in light of the comparative gas exchange and ETR sets that already exist and validation for testing the values of $1/\phi_{e(c)}$ that are ultimately produced will almost certainly be required.

In many cases regressions of $1/\phi_c$ (or indeed $1/\phi_{O2}$) against F_q'/F_m' from data collected across a range of light intensities will be non-linear (Kromkamp et al. 2008; see also Genty et al. 1989; Baker and Oxborough 2004); as such, values of $1/\phi_{e(c)}$ (or $1/\phi_e$) will not be constant. Certainly, the electron requirement for O_2 evolution ($1/\phi_e$) is known to be higher than the theoretical minimum value (Kolber and Falkowski 1993; Franklin and Badger 2001; Kromkamp and Forster 2003) under very low and high actinic irradiances due to electron 'slippage' at- (Quigg et al. 2006) and cycling around- (Prášil et al. 1996; Feikema et al. 2006) PSII, respectively. Electron slippage refers to the decay of unstable intermediate S-states of RCIIs and Q_A/Q_B charge recombination during water splitting whilst electron cycling describes the alternative pathway for electrons to reoxidise $P680^+$ via cytochrome-*b* (Fig. 5). Similarly, the shape of the non-linear regression between $1/\phi_c$ (or indeed $1/\phi_{O2}$) against F_q'/F_m' will not be the same for all taxa or environments (Kromkamp et al. 2008). With these considerations in mind, obtaining algorithms describing the environmental or taxonomic dependency of $1/\phi_{e(c)}$ ($1/\phi_e$) may not be straightforward; however, if achievable such algorithms will transform the confidence with which fluorometers are ultimately used for productivity studies.

Previous discussion of ST- versus MT- based estimates of ETR_{PSII} has important implications in this context. Firstly, derived values for $1/\phi_e$ under relatively low light intensities will inevitably be higher for MT- than for ST- protocols as a result of the higher photochemical efficiencies estimated from MT inductions. Secondly, MT- based ETRs decrease under relatively high light as a result of losses of fluorescence quenching as PQ molecules are increasingly reduced (Fig. 1). Consequently, apparent light-dependent decreases of $1/\phi_e$, which may be inferred under high light may be misinterpreted as a greater capacity for cyclic flow. Beer and Axelsson (2004) recognized this phenomenon

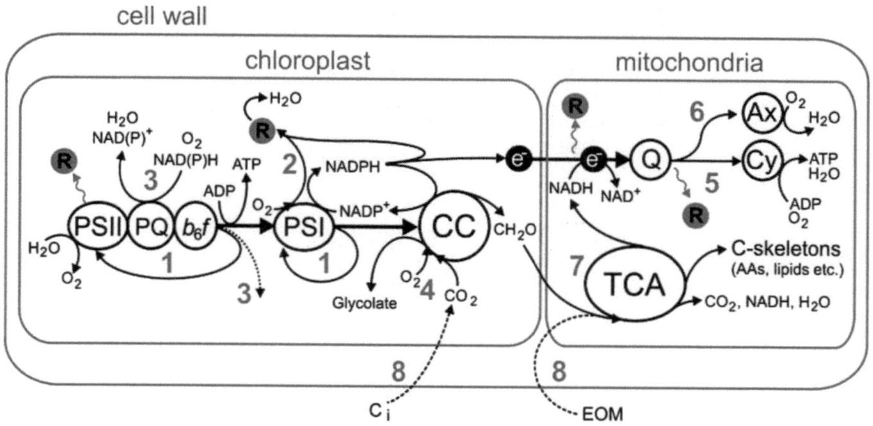

Fig. 5 Simplified photosynthetic and respiratory pathways acting as intracellular electron sinks within the chloroplast and mitochondria (for more detail refer to Falkowski and Raven 2007). Processes highlighted are: (**1**) cyclic electron flow around PSII, via the cytochrome b_6f complex, and PSI (**2**) Mehler reaction, (**3**) terminal oxidase activity (**4**) photorespiration, (**5**, **6**) mitochondrial respiration via cytochrome-c (Cy) or alternative oxidase (Ax), (**7**) incorporation of simple organic molecules (via conversion to sugars) into the TriCarboxylic Acid (TCA) cycle, and (**8**) active uptake of inorganic carbon (C_i, for photosynthesis) or extracellular organic matter (EOM, for respiration). Each of these processes can preferentially alter the pool of energy (ATP) and/or reductant (NADPH, NADP) to meet cellular requirements under temporarily altered growth conditions. Of particular note is: (**a**) terminal oxidases in the chloroplast, such as PTOX (plastid terminal oxidase) operating at the PQ pool (but see also Bailey et al., 2008; Cardol et al., 2008), consume protons and reductant to recycle O_2 to H_2O; in prokaryotes, electron transport from cytochrome b_6f complex is directly coupled to Cy (not indicated), (**b**) Active uptake of C_i will consume ATP or reductant depending upon which form of Ci, i.e. HCO_3 or CO_2, is available, and (**c**) EOM (dissolved/particulate) is broken down and incorporated into TCA for up-regulation of respiration. Several 'fates' can occur for an 'incomplete' TCA cycle: drawing off reductant prior to mitochondrial ETR or drawing off carbon skeletons. If the cycle is 'complete', ATP is generated by mitochondrial ETR. Filled grey circles signify production of reactive oxygen species

when assessing the light-dependency of $1/\phi_e$ for macroalgae; specifically, the quantum yield for electron transport decreased disproportionately relative to that for O_2 once the actinic light intensity was much higher than the growth light intensity. As such, they recommended that ETRs from MT-based approaches were not reliable once values of F_q'/F_m' decreased below a threshold value of ca. 0.1. This value may in fact be more variable depending upon the species and growth conditions in question. Furthermore, on a theoretical basis it is possible to suggest that MT-based approaches are instead not reliable *below* this threshold since it is at the lower actinic light intensities that MT techniques overestimate ϕ_{PSII}' as a result of artificial removal of PQ quenching (Section 2.2).

ST-based studies may also present problems. Although few data exists for $1/\phi_e$ determinations from ST approaches (Suggett et al. 2009a), values up to 25% lower than the theoretical minimum have been estimated, potentially suggesting systematic errors with the ST-based ETR calculation. Whilst current information does suggest that $1/\phi_e$ may not be constant under all conditions (Quigg et al. 2006; Prášil et al. 1996; Feikema et al. 2006), the magnitude and consistency of such potential systematic errors requires further attention.

4.3 Estimating Net O_2 Production and C-Fixation from ETRs

Further obstacles limit the use of ETR_{PSII} to estimate net O_2 production or CO_2 fixation. In addition to any potential variability in $1/\phi_e$, the relationship between gross O_2 evolution by PSII and the ultimate net rate of O_2 production or carbon fixed can also be highly variable (Williams and Robertson 1991; Grande et al. 1991; Wagner et al. 2006; Hancke et al. 2008b). As such, it is not surprising that most studies comparing independent measurements of ETRs with carbon fixation rates do not observe a strict proportionality even if some degree of correlation exists (e.g. Kromkamp et al. 2008; Robinson et al. 2009; Suggett et al. 2009a; Table 3). Scatter and/or non-linearity in the relationship between

ETR and net O_2 production or C-fixation is normally attributed to one or more O_2 consuming pathways that operate in the light and to variability of the 'photosynthetic quotient', describing the ratio of moles O_2 produced per mole CO_2 assimilated (Laws 1991; Williams and Robertson 1991; Falkowski and Raven 2007) (Fig. 5). Unfortunately, operation of these various pathways has rarely been independently quantified and so we have little information on the part they play in accounting for the decoupling observed in studies comparing ETR_{PSII} and O_2 evolution (or) carbon fixation.

Light-dependent oxygen consuming pathways are known to operate in a wide range of primary producers, including prokaryotic microalgae (Kana 1993; Badger et al. 2000; Weng and Shieh 2004; Bailey et al. 2008), eukaryotic microalgae (Kana 1990; Lewitus and Kana 1995; Eriksen and Lewitus 1999; Badger et al. 2000; Cardol et al. 2008; Suggett et al. 2008), macroalgae (Franklin and Badger 2001) and seagrass (Kana 1990). The primary mechanisms responsible are the Mehler reaction at PSI (Badger et al. 2000) and terminal oxidases in thylakoids and miotochondria (McDonald and Vanlerberghet 2006; Bailey et al. 2008; Cardol et al. 2008), which both recycle O_2 back to water, alongside photorespiration (Badger et al. 2000). In this way, accumulation of excess energy and reductant, as well as O_2, with the capacity to form toxic radicals, within the cell is reduced. Operation of these mechanisms can consume as much as 50% of the gross O_2 produced, in particular under 'stress' conditions of high light or temperature, and so provides an effective means for regulating photosynthetic competency. Furthermore, the nutritional status of the growth environment appears to determine which mechanism(s) preferentially operate, e.g. Mehler and terminal oxidase activity within the chloroplast appear to primarily operate under nutrient depleted or stress conditions (Lewitus and Kana 1995; Bailey et al. 2008).

The photosynthetic quotient depends on the reduction level of the primary substrate or product of photosynthesis (Falkowski and Raven 2007). Relatively large changes in photosynthetic quotient, ca. 0.75–2.25, can be observed between growth conditions, largely depending on how inorganic nitrogen is allocated into biochemical constituents (Laws et al. 1991; Wagner et al. 2006). Variability of the photosynthetic quotient is also apparent between species grown under similar conditions (Williams and Robertson 1991; Wagner et al. 2006; Suggett et al. 2009a).

4.4 Reconciliation of $ETR_{PSII}:O_2:CO_2$ Estimates

Taking into account the large array of approaches that have been used in comparing ETRs with conventional gas exchange measurements (Table 3), as well as the apparent variability that exist for both photosynthetic quotient and light-dependent O_2 consumption, leads to difficulty in reconciling many of the studies performed to date. MT (PAM)-based studies have typically compared ETR_{PSII} with O_2 evolution and yield values of mol e^- (mol O_2 evolved)$^{-1}$ that are higher than the mol e^- (mol C-uptake)$^{-1}$ from the ST (FRR)-based studies. This trend is not unexpected given the methodological differences between ETR_{PSII} determinations combined with the potential influence of photosynthetic quotients. Recent analysis of FRR (ST)-based phytoplankton studies to date have demonstrated that some general patterns may exist (Suggett et al. 2009a; Table 3): a higher ratio of mol e^-:(mol C-fixed) for flagellates than diatoms is often observed both in controlled laboratory and field conditions (e.g. Raateoja et al. 2004; Estévez-Blanco et al. 2006; Suggett et al. 2006a); natural populations of cyanobacteria also appear to exhibit values as high as those observed for flagellates (Suggett et al. 2001; Raateoja et al. 2004; Kromkamp et al. 2008). Whilst these patterns cannot yet be considered statistically or methodologically robust due to the substantial range and variability of the data from these various studies alongside the likely systematic uncertainties described above, they certainly warrant more attention. Few obvious patterns seem to hold across MT (PAM)-based studies in microalgae, or indeed across groups of macroalgae or of seagrasses (Table 2). Consequently, as yet, despite an apparent wealth of data amounting from ETR_{PSII}-productivity comparisons, it appears that we are still some way from understanding the underlying nature of any variability between rates of electron transport, net O_2 evolution and C-fixation.

5 Future Application of ETRs to Primary Productivity Studies

Active chlorophyll fluorescence presents exciting opportunities for aquatic photosynthesis studies; how this potential is met will be entirely dependent upon the nature of the study in question. Many applications have aspired to the use of fluorescence as a convenient means to estimate productivity, frequently in the 'currency' of carbon fixation, quickly and with high resolution *in situ*. Unfortunately, it is clear that we have much to learn about how environmental forcing and evolutionary adaptation (taxonomy) have combined to regulate the allocation of electrons into fixed carbon (Behrenfeld et al. 2004). Improved physiological models based on controlled experiments will likely enable active fluorescence to be increasingly used as a tool for examining the flow of photosynthetic energy for cellular processes, of which carbon fixation is just one, rather than as a proxy productivity measurement. Such a consideration is particularly relevant for studies examining the highly variable, often resource limited, growth conditions inherent to natural communities. Laboratory-based studies are certainly moving towards this type of application (e.g. Figueroa et al. 2003; Wagner et al. 2006; Hancke et al. 2008a, b). However, such studies would benefit from improved confidence in the ETRs generated by the different techniques and protocols used across aquatic sciences.

Answering the question as to whether it is appropriate to use ETR_{PSII} to predict C-fixation may ultimately rest with the level of accuracy and precision we are willing to accept and, perhaps importantly, how much effort we are prepared to invest in improving upon these levels. Studies focusing on single biogeochemical provinces or waters dominated by similar taxa often observe some degree of relationship between photosynthetic 'currencies'. Such relationships are likely the product of both methodological consistency within any given study but also constrained systematic variability, i.e. where environmental conditions select for specific communities. Thus, larger scale studies that (spatially and temporally) span different biogeographical conditions will in particular require much better understanding of the environmental control upon factors that must be assumed in calculations of ETR_{PSII} or conversion of ETR_{PSII} to estimates of C-fixation.

A more effective use of ETR_{PSII} for understanding the coupling between light absorption and carbon fixation will likely remain far from straight forward for many working on natural samples. Such studies require accurate independent measurements of productivity, i.e. carbon fixation or oxygen evolution, or at least confidence of what photosynthetic processes are being accounted for when using alternative protocols. Indeed, what we are actually measuring in terms of gross or net C-fixation (O_2 evolution) methods is still far from resolved. Accuracy of the ETRs themselves is also still constrained by inherent limitations, e.g. quantification of fQA_{PSII} or RCII concentration, as discussed previously. Such limitations may be overcome through additional bio-optical information. Future applications of active fluorometry may thus be more powerful if combined with integrated bio-optical instrumentation platforms, such as multispectral fluorometers, hyperspectral radiometers or submersible flow cytometers (Suggett et al. 2009a), which enable better control for taxonomic variability. Using these more applied approaches will certainly enable fluorescence-based ETR_{PSII} measurements to continue to increase our understanding of the physiological-basis for variability in aquatic productivity (Behrenfeld et al. 2004, 2006) and perhaps improve our capacity to examine the regulation of carbon fixation at scales not amenable to alternative techniques. Active fluorescence techniques still potentially provide one of our most promising tools for examining and understanding the environmental control of productivity, in particularly as aquatic systems are subjected to our current intense period of rapid environmental (climate) change. However, we must look to optimising and reconciling the common approaches that exist for using fluorescence to determine ETR_{PSII}. Hopefully we can then begin to move past the various methodological problems and so more successfully capitalise on the rapidly growing fluorescence data sets to address more fundamental physiological and ecological questions.

References

Babin M, Morel A, Claustre H, Bricaud C, Kolber Z, Falkowski PG (1996) Nitrogen- and irradiance-dependent variations of the maximum quantum yield of carbon fixation in eutrophic, mesotrophic and oligotrophic marine systems. Deep Sea Res I 43:1241–1272

Badger MR, von Caemmerer S, Ruuska S, Nakano H (2000) Electron flow to oxygen in higher plants and algae: rates and control of direct photoreduction (Mehler reaction) and rubisco oxygenase. Phil Trans R Soc Lond B 355: 1433–1446

Bailey S, Melis A, Mackey KRM, Cardol P, Finazi G, van Djiken G, Berg GM, Arrigo K, Shrager J, Grossman A (2008) Alternative photosynthetic electron flow to oxygen in marine *Synechococcus*. Biochim Biophys Acta 1777:269–276

Baker NR, Harbinson J, Kramer DM (2007) Determining the limitations and regulation of photosynthetic energy transduction in leaves. Plant Cell Env 30:1107–1125

Baker NR, Oxborough K (2004) Chlorophyll fluorescence as a probe of photosynthetic productivity. In: Papageorgiou GC and Govindjee (eds) Chlorophyll*a* fluorescence: a signature of photosynthesis. Springer, Dordrecht, pp 65–82

Barlow RG, Alberte RS (1985) Photosynthetic characteristics of phycoerythrin-containing marine *Synechococcus* sp. Mar Biol 86:63–74

Beer S, Axelsson L (2004) Limitations in the use of PAM fluorometry for measuring photosynthetic rates of macroalgae at high irradiances. Eur J Phycol 39:1–7

Beer S, Björk M (2000) Measuring rates of photosynthesis of two tropical seagrasses by pulse amplitude modulate (PAM) fluorometry. Aquat Bot 66:69–76

Beer S, Larsson C, Poryan O, Axelsson L (2000) Photosynthetic rates of *Ulva* (Chlorophyta) measured by pulse amplitude modulated fluorometry. Eur J Phycol 35:69–74

Beer S, Weill A, Veste M, Susell L, Eshell A (1998) Measuring photosynthetic rates in seagrasses by pulse amplitude modulated (PAM) fluorometry. Mar Ecol Prog Ser 174:293–300

Behrenfeld MJ, Halsey KH, Milligan AJ (2008) Evolved physiological responses of phytoplankton to their integrated growth environment. Phil Trans R Soc Lond B 363: 2687–2703

Behrenfeld MJ, Prášil O, Babin M, Bruyant F (2004) In search of a physiological basis for covariations in light-limited and light-saturated photosynthesis. J Phycol 40:4–25

Behrenfeld MJ, Worthington K, Sherrell RM, Chavez FP, Strutton P, McPhaden M, Shea DM (2006) Controls on tropical Pacific Ocean productivity revealed through nutrient stress diagnostics. Nature 442:1025–1028

Berges JA, Charlebois DO, Mauzerall DC, Falkowski PG (1996) Differential effects of nitrogen limitation on photosynthetic efficiency of photosystems I and II in microalgae. Plant Physiol 110:689–96

Bidigare RR, Ondrusek ME, Morrow JH, Kiefer DA (1990) In vivo absorption properties of algal pigments. SPIE 1302 (Ocean Optics X):289–302

Boichenko VA (1998) Action spectra and functional antenna sizes of Photosystems I and II in relation to the thylakoid membrane organization and pigment composition. Photosynth Res 58:163–174

Boyd PW, Aiken J, Kolber ZS (1997) Comparison of radiocarbon and fluorescence based (pump and probe) measurements of phytoplankton photosynthetic characteristics in the Northeast Atlantic Ocean. Mar Ecol Prog Ser 149:215–226

Cabello-Pasini A, Figueroa F (2005) Effect of nitrate concentration on the relationship between photosynthetic oxygen evolution and electron transport rate in *Ulva rigida* (Chlorophyta). J Phycol 41:1169–1177

Cardol P, Bailleul B, Rappaport F, Derelle E, Béal D, Breyton C, Bailey S, Wollman FA, Grossman A, Moreau H, Finazzi G (2008) An original adaptation of photosynthesis in the marine green alga *Ostreococcus*. Proc Nat Acad Sci USA 105:7881–7886

Corno G, Letelier RM, Abbott MR, Karl DA (2005) Assessing primary production variability in the north Pacific subtropical gyre: a comparison of fast repletion rate fluorometry and ^{14}C measurements. J Phycol 42:51–60

Corno G, Letelier RM, Abbott MR, Karl DA (2008) Temporal and vertical variability in photosynthesis in the North Pacific Subtropical Gyre. Limnol Oceanogr 53:1252–1265

Cullen JJ, Davis RF (2003) The blank can make a big difference in oceanographic measurements. Limnol Oceanogr Bull 12:29–35

Debes H, Gaard E, Hansen B (2008) Primary production on the Faroe shelf: Temporal variability and environmental influences. J Mar Sys 74:686–697

Dubinsky Z, Falkowski PG, Wyman K (1986) Light harvesting and utilisation by phytoplankton. Plant Cell Physiol 27:1335–1349

Enríquez S, Rodríguez-Román A (2006) Effect of water flow on the photosynthesis of three marine macrophytes from a fringing-reef lagoon. Mar Ecol Prog Ser 323:119–132

Eriksen NT, Lewitus AJ (1999) Cyanide-resistant respiration in diverse marine phytoplankton. Evidence for the widespread occurrence of the alternative oxidase. Aquatic Microb Ecol 17:145–152

Estévez-Blanco P, Cermeño P, Espiñeira M, Fernández E (2006) Phytoplankton photosynthetic efficiency and primary production rates estimated from fast repetition rate fluorometry at coastal embayments affected by upwelling (Rías Baixas, NW of Spain). J Plankton Res 28:1153–1165

Falkowski PG, Owens TG, Ley AC, Mauzerall DC (1981) Effect of growht irradiance levels on the ratio of reaciotn centres in two species of marine phytoplankton. Plant Physiol 68:669–673

Falkowski PG, Raven JA (2007) Aquatic photosynthesis, 2nd ed. Princeton University Press, Princeton, NJ

Falkowski PG, Wyman K, Ley AC, Mauzerall DC (1986) Relationship of steady state photosynthesis to fluorescence in eukaryotic algae. Biochim Biophys Acta 849:183–192

Falkowksi PG, Ziemann D, Kolber ZS, Bienfang PK (1991) Role of eddy pumping in enhancing primary production in the ocean. Nature 352:55–58

Feikema WA, Marosvölgyi MA, Lavaud J, van Gorkom HJ (2006) Cyclic electron transfer in photosystem II in the marine diatom *Phaeodactylum tricornutum*. Biochim Biophys Acta 1757:829–834

Fietz S, Nicklisch A (2002) Acclimaiton of the diatom *Stephanodiscus neostraea* and the cyanobacterium *Planktothrix agardhii* to simulated and natural light fluctuations. Photosynth Res 72:95–106

Figueroa FL, Conde-Álvarez R, Gómez I (2003) Relations between electron transport rates determined by pulse amplitude modulated chlorophyll fluorescence and oxygen evolution in macroalgae under different light conditions. Photosynth Res 75:259–275

Fisher T, Minnaard M, Dubinsky Z (1996) Photoacclimation in the marine alga *Nannochloropsis* sp. (Eustigmatophyta): a kinetic study. J Plankton Res 18:1797–1818

Fisher T, Shurtz-Swirski R, Gepstein S, Dubinsky Z (1989) Changes in the levels of Ribulose-1, 5-biphosphate Carboxylase/Oxygenase (Rubisco) in *Tetraedron minimum* (Chlorophyta) during light and shade adaptation. Plant Cell Phys 30:221–228

Flameling IA, Kromkamp JC (1998) Light dependence of quantum yields for PSII charge separation and oxygen evolution in eucaryotic algae. Limnol Oceanogr 43:284–297

Franklin LA, Badger MR (2001) A comparison of photosynthetic electron transfer rates in macroalgae measured by

pulse amplitude modulated chlorophyll fluorometry and mass spectrometry. J Phycol 37:756–767
Fujiki T, Suzue T, Kimono H, Saino T (2007) Photosynthetic electron transport in *Dunaliella tertiolecta* (Chlorophyceae) measured by fast repetition rate fluorometry: relation to carbon assimilation. J Plankton Res 29:199–208
Genty B, Briantais J-M, Baker NR (1989) The relationship between the quantum yield of photosynthetic electron transport and quenching of chlorophyll fluorescence. Biochim Biophys Acta 990:87–92
Geider RJ, Osborne BA (1992) Algal photosynthesis: the measurements of algal gas exchange. Chapman & HalL, New York
Gilbert M, Domin A, Becker A, Wilhelm C (2000a) Estimation of primary productivity by chlorophyll*a in vivo* fluorescence in freshwater phytoplankton. Photosynthetica 38:111–126
Gilbert M, Wilhelm C, Richter M (2000b) Biooptical modelling of oxygen evolution using in vivo fluorescence: Comparison of measured and calculated photosynthesis-irradiance (P-I) curves in four representative phytoplankton species. J Plant Physiol 157:307–314
Glud RN, Rysgaard S, Kühl M (2002) A laboratory study on O2 dynamics and photosynthesis in ice algal communities: quantification by microsensors, O_2 exchange rates, ^{14}C incubations and a PAM fluorometer. Aquat Microb Ecol 27:301–311
Gorbunov MY, Kolber ZS, Lesser MP, Falkowski PG (2001) Photosynthesis and photoprotection in symbiotic corals. Limnol Oceanogr 46:75–85
Grande KD, Bender ML, Irwin B, Platt T (1991) A comparison of net and gross rates of oxygen production as a function of light intensity in some natural plankton populations and in a *Synechococcus* culture. J Plankton Res 13:1–16
Hancke K, Hancke TB, Johnsen G, Sakshaug E (2008a) Rate of O_2 production derived from pulse amplitude modulated fluorescence: testing three biooptical approaches against measured O_2 production rate. J Phycol 44:803–813
Hancke K, Hancke TB, Olsen LM, Johnsen G, Glud RN (2008b) Temperature effects on microalgal photosynthesis-light responses measured by O_2 production, pulse amplitude modulated fluorescence and ^{14}C assimilation. J Phycol 44:501–514
Hartig P, Wolfstein K, Lippemeier S, Colijn F (1998) Photosynthetic activity of natural microphytobenthos populations measured by fluorescence (PAM) and ^{14}C-tracer methods: a comparison. Mar Ecol Prog Ser 166:53–62
Henley WJ, Dunton KH (1997) Effects of nitrogen supply and continuous darkness on growth and photosynthesis of the arctic kelp *Laminaria solidungula*. Limnol Oceanogr 42:209–216
Hennige SJ, Smith DJ, Perkins R, Consalvey M, Paterson DM, Suggett DJ (2008) Photoacclimation, growth and distribution of massive coral species in clear and turbid waters. Mar Ecol Prog Ser 369:77–88
Hennige SJ, Suggett DJ, Warner ME, McDougall KE, Smith DJ (2009) Photobiology of *Symbiodinium* revisited: bio-physical and bio-optical signatures. Coral Reefs 28:179–195
Herlory O, Richard P, Blanchard EF (2007) Methodology of light response curves: application of chlorophyll fluorescence to microphytobenthic biofilms. Mar Biol 153:91–101
Herzig R, Dubinsky Z (1992) Photoacclimation, photosynthesis and growth in phytoplankton. Isr J Bot 41:199–211
Herzig R, Falkowski PG (1989) Nitrogen limitation in *Isochrysis galbana* (Haptophyceae) 1. Photosynthetic energy conversion and growth efficiencies. J Phycol 25:462–471
Holmes JJ, Weger HG, Turpin DH (1989) Chlorophyll *a* fluorescence predicts total photosynthetic electron flow to CO_2 or NO_3^-/NO_2^- under transient conditions. Plant Physiol 91:331–337
Iglesias-Prieto R, Trench RK (1994) Acclimation and adaptation to irradiance in symbiotic dinoflagellates. I. Responses of the photosynthetic unit to changes in photon flux density. Mar Ecol Prog Ser 113:163–175
Johnsen G, Prezelin BB, Jovine RVM (1997) Fluorescence excitation spectra and light utilization in two red tide dinoflagellates. Limnol Oceanogr 42:1166–77
Johnsen G, Sakshaug E (1993) Bio-optical characteristics and photoadaptive responses in the toxic and bloom forming dinoflagellates *Gyrodinium aureolum, Gymnodinium galatheanum* and two strains of *Prorocentrum minimum*. J Phycol 29:627–642
Johnsen G, Sakshaug E (2007) Bio-optical characteristics of PSII and PSI in 33 species (13 pigment groups) of marine phytoplankton, and the relevance for pulse-amplitude-modulated and fast-repetition-rate fluorometry. J Phycol 43:1236–1251
Juneau P, Harrison PJ (2005) Comparison by PAM fluorometry of photosynthetic activity of nine marine phytoplankton grown under identical conditions. Photochem Photobiol 81:649–653
Kana TM (1990) Light-dependent oxygen cycling measured by an oxygen-18 isotope dilution technique. Mar Ecol Prog Ser 64:293–300
Kana TM (1993) Rapid oxygen cycling in *Trichodesmium thiebautii*. Limnol Oceanogr 38:18–24
Koblížek M, Kaftan D, Nedbal L (2001) On the relationship between the non-photochemical quenching of the chlorophyll fluorescence and the Photosystem II light harvesting efficiency.A repetitive flash fluorescence induction study. Photosynth Res 68:141–152
Kolber ZS, Falkowski PG (1993) Use of active fluorescence to estimate phytoplankton photosynthesis in situ. Limnol Oceanogr 38:1646–1665
Kolber ZS, Plumley FG, Lang AS, Beatty JT, Blakenship RE, VanDover CL, Vetriani C, Koblížek M, Rathgeber C, Falkowski PG (2001) Contribution of aerobic photoheterotrophic bacteria to the carbon cycle in the ocean. Science 292:2492–2495
Kolber ZS, Prášil O, Falkowski PG (1998) Measurements of variable chlorophyll fluorescence using fast repetition rate techniques: defining methodology and experimental protocols. Biochim Biophys Acta 1367:88–106
Kramer DM, DiMarco G, Loreto F (1995) Contribution of plastoquinone quenching to saturation pulse-induced rise of chlorophyll fluorescence in leaves. In: Mathis P (ed) Photosynthesis from light to the biosphere, vol 1. Kluwer, Dordrecht, pp 147–150
Kromkamp JC, Forster RM (2003) The use of variable fluorescence measurements in aquatic ecosystems: differences between multiple and single turnover measuring protocols and suggested terminology. Eur J Phycol 38:103–112
Kromkamp JC, Dijkman NA, Peene J, Simis SGH, Gons HJ (2008) Estimating phytoplankton primary production in Lake IJsselmeer (The Netherlands) using variable fluorescence (PAM-FRRF) and C-uptake techniques. Eur J Phycol 43:327–344
Lavergne J, Trissl HW (1995) Theory of fluorescence induction in Photosystem II: derivation of analytical expressions in a model including Exciton-Radical-pair equilibrium and

restricted energy transfer between photosynthetic units. Biophys J 68:2474–2492

Laws EA (1991) Photosynthetic quotients, new production and net community production in the open ocean. Deep Sea Res 38:143–167

Levy O, Achituvi Y, Yacobi YZ, Dubinsky Z, Stambler N (2006) Diel 'tuning' of coral metabolism: physiological responses to light cues. J Exp Biol 209:273–283

Levy O, Dubinsky Z, Schneider K, Achituvi Y, Zakai D, Gorbunov MY (2004) Diurnal hysteresis in coral photosynthesis. Mar Ecol Prog Ser 268:105–117

Lewitus AJ, Kana TM (1995) Light respiration in six estuarine phytoplankton species: contrasts under photoautotrophic and mixotrophic growth conditions. J Phycol 31:754–791

Longstaff B, Kildea T, Runcie JW, Cheshire A, Dennison WC, Hurd C, Kana T, Raven JA, Larkum AWD (2002) An in situ study of photosynthetic oxygen exchange and electron transport rate in the marine macroalga *Ulva lactuca* (Chlorophyta). Photosynth Res 74:281–291

Lutz VA, Sathyendranath S, Head EJH, Li WKW (2001) Changes in the in vivo absorption and fluorescence excitation spectra with growth irradiance in three species of phytoplankton. J Plankton Res 23:555–569

MacIntyre HL, Cullen JJ (2005) Using cultures to investigate the physiological ecology of microalgae. In: Andersen RA (ed) Algal culturing techniques. Elsevier Academic Press, New York, pp 287–326

MacIntyre HL, Geider RJ, McKay RM (1996) Photosynthesis and regulation of Rubisco activity in net phytoplankton from Delaware Bay. J Phycol 32:718–731

MacIntyre HL, Kana TM, Anning T, Geider RJ (2002) Photoacclimation of photosynthesis irradiance response curves and photosynthetic pigments in microalgae and cyanobacteria. J Phycol 38:17–38

MacIntyre HL, Kana TM, Geider RJ (2000) The effect of water motion on short-term rates of photosynthesis by marine phytoplankton. Trends Plant Sci 5:12–17

MacKenzie TDB, Burns RA, Campbell DA (2004) Carbon status constrains light acclimation in the cyanobacterium *Synechococcus elongates*. Plant Physiol 136:3301–3312

Mackey KRM, Paytan A, Grossman AR, Bailey S (2008) A photosynthetic strategy for coping in a high-light, low-nutrient environment. Limnol Oceanogr 53:900–913

Major KM, Dunton KH (2002) Variation in light-harvesting characteristics of the seagrass *Thalassia testudinum*: evidence for photoacclimation. J Exp Mar Biol Ecol 275:173–189

Malinsky-Rushansky N, Berman T, Berner T, Yacobi YZ, Dubinsky Z (2002) Physiological characteristics of phytoplankton isolated from Lake Kinneret: responses to temperature and light. J Plankton Res 24:1173–1183

Masojídek J, Grobbelaar JU, Pechar L, Koblížek M (2001) Photosystem II electron transport rates and oxygen production in natural waterblooms of freshwater cyanobacteria during a diel cycle. J Plankton Res 23:57–66

Mauzerall D, Greenbaum NL (1989) The absolute size of a photosynthetic unit. Biochim Biophys Acta 974:119–140

McDonald AE, Vanlerberghet GC (2006) Origins, evolutionary history, and taxonomic distribution of alternative oxidase and plastoquinol terminal oxidase. Comp Biochem Physiol D 1:357–364

Morris EP, Kromkamp JC (2003) Influence of temperature on the relationship between oxygen- and fluorescence-based estimates of photosynthetic parameters in a marine benthic diatom (*Cylindrotheca closterium*). Eur J Phycol 38:133–142

Morris EP, Forster RM, Peene J, Kromkamp JC (2008) Coupling between Photosystem II electron transport and carbon fixation in microphytobenthos. Aquat Microb Ecol 50:301–311

Moore CM, Lucas MI, Sanders R, Davidson R (2005) Basin-scale variability of phytoplankton bio-optical characteristics in relation to bloom state and community structure in the Northeast Atlantic. Deep Sea Res I 53:104–409

Moore CM, Suggett DJ, Hickman AE, Kim Y-N, Tweddle JF, Sharples J, Geider RJ, Holligan PM (2006) Phytoplankton photoacclimation and photoadaptation in response to environmental gradients in a shelf sea. Limnol Oceanogr 51:936–949

Moore CM, Suggett DJ, Holligan PM, Sharples J, Abraham ER, Lucas MI, Rippeth TP, Fisher NR, Simpson JH, Hydes DJ (2003) Physical controls on phytoplankton physiology and production at a shelf sea front: An FRRF-based field study. Mar Ecol Prog Ser 259:29–45

Myers J, Graham J-R (1971) The photosynthetic unit in *Chlorella* measured by short repetitive flahses. Plant Physiol 48:282–286

Neori A, Vernet M, Holm-Hansen O, Haxo FT (1988) Comparison of chlorophyll far-red and red fluorescence excitation spectra with photosynthetic oxygen action spectra for photosystem II in algae. Mar Ecol Prog Ser 44:297–302

Pemberton KL, Clarke R, Joint I (2006) Quantifying uncertainties associated with the measurement of primary production. Mar Ecol Prog Ser 322:51–59

Perkins RG, Mouget J-L, Lefebvre S, Lavaud J (2006) Light response curve methodology and possible implications in the application of chlorophyll fluorescence to benthic diatoms. Mar Biol 149:703–712

Platt T, Jassby AD (1976) Relationship between photosynthesis and light for natural assemblages of costal marine phytoplankton. J Phycol 12:421–430

Prášil O, Kolber Z, Berry JA, Falkowski PG (1996) Cyclic electron flow around photosystem II in vivo. Photosynth Res 48:395–410

Prášil O, Suggett DJ, Cullen JJ, Babin M, Govindjee (2008) Aquafluo 2007: chlorophyll fluorescence in aquatic sciences, an international conference held in Nove Hrady. Photosynth Res 95:111–115

Quigg A, Kevekordes K, Raven JA, Beardall J (2006) Limitations on microalgal growth at very low photon fluence rates: the role of energy slippage. Photosynth Res 88:299–310

Raateoja M, Seppälä J, Kuosa H (2004) Bio-optical modelling of primary production in the SW Finnish coastal zone, Baltic Sea: fast repetition rate fluorometry in Case 2 waters. Mar Ecol Prog Ser 267:9–26

Ralph PJ, Gademan R (2005) Rapid light curves: a powerful tool to assess photosynthetic activity. Aquat Bot 82:222–237

Robinson C, Tilstone GH, Rees AP, Smythe TJ, Fishwick JR, Tarran GA, Luz B, Barkan E, David E (2009) Comparison of *in vitro* and *in situ* plankton production determinations. Aquat Microb Ecol 54:13–34

Runcie JW, Riddle MJ (2006) Photosynthesis of marine macroalgae in ice covered and ice free environments in East Antarctica. Eur J Phycol 41:223–234

Sarma VVSS, Abe O, Hashimoto S, Inhuma A, Saino T (2005) Seasonal variations in triple oxygen isotopes and gross oxygen production in Sagami Bay, central Japan. Limnol Oceanogr 50:544–552

Sathyendranath S, Longhurst A, Caverhill CM, Platt T (1995) Regionally and seasonally differentiated primary production in the North Atlantic. Deep Sea Res 42:1773–1802

Schreiber U, Hormann H, Neubauer C, Klughammer C (1995) Assessment of Photosystem II photochemical quantum yield by chlorophyll fluorescence quenching analysis. Aust J Plant Physiol 22:209–220

Schreiber U, Neubauer C, Schuwa U (1993) PAM fluorometer based on medium-frequency pulsed Xe-flash measuring light: A highly sensitive new tool in basic and applied photosynthesis research. Photosynth Res 36:65–72

Serôdio J, Vieira S, Cruz S, Coelho H (2006) Microphytobenthos vertical migratory photoresponse as characterised by light-response curves. Photosynth Res 90:29–43

Silva J, Santos R (2004) Can chlorophyll fluorescence be used to estimate photosynthetic production in the seagrass *Zostera noltii*? J Exp Mar Biol Ecol 307:207–216

Smythe TJ, Pemberton KL, Aiken J, Geider RJ (2004) A methodology to determine primary production and phytoplankton photosynthetic parameters from Fast Repetition Rate Fluorometry. Plankton Res 24:1337–1350

Suggett DJ, Kraay G, Holligan P, Davey M, Aiken J, Geider RJ (2001) Assessment of photosynthesis in a spring cyanobacterial bloom by use of a fast repetition rate fluorometer. Limnol Oceanogr 46:802–810

Suggett DJ, Le Floc'H E, Harris GN, Leonardos N, Geider RJ (2007) Different strategies of photoacclimation by two strains of *Emiliania huxleyi* (Haptophyta). J Phycol 43:1209–1222

Suggett DJ, Maberly SC, Geider RJ (2006a) Gross photosynthesis and lake community metabolism during the spring phytoplankton bloom. Limnol Oceanogr 51:2064–2076

Suggett DJ, MacIntyre HL, Geider RJ (2004) Evaluation of biophysical and optical determinations of light absorption by photosystem II in phytoplankton. Limnol Oceanogr Meth 2:316–332

Suggett DJ, MacIntyre HL, Kana TM, Geider RJ (2009a) Comparing electron transport with gas exchange: parameterising exchange rates between alternative photosynthetic currencies for eukaryotic phytoplankton. Aquat Microb Ecol 56:147–162

Suggett DJ, Moore CM, Hickman AE, Geider RJ (2009b) Interpretation of Fast Repetition Rate (FRR) fluorescence: signatures of community structure versus physiological state. Mar Ecol Prog Ser 376:1–19

Suggett DJ, Moore CM, Marañón E, Omachi C, Varela RA, Aiken J, Holligan PM (2006b) Photosynthetic electron turnover in the tropical and subtropical Atlantic Ocean. Deep Sea Res II 53:1573–1592

Suggett DJ, Oxborough K, Baker NR, MacIntyre HL, Kana TM, Geider RJ (2003) Fast Repetition Rate and Pulse Amplitude Modulation chlorophyll a fluorescence measurements for assessment of photosynthetic electron transport in marine phytoplankton. Eur J Phycol 38:371–384

Suggett DJ, Warner ME, Smith DJ, Davey P, Hennige SJ, Baker NR (2008) Photosynthesis and production of hydrogen peroxide by *Symbiodinium* (Pyrrhophyta) phylotypes with different thermal tolerances. J Phycol 44:948–956

Sukenik A, Bennet J, Mortain-Bertrand A, Falkowski PG (1990) Adaptation of the photosynthetic apparatus to irradiance in *Dunaliella tertiolecta*. Plant Physiol 92:891–898

Wagner H, Jakob T, Wilhelm C (2006) Balancing the energy flow from captured light to biomass under fluctuating light conditions. New Phytol 169:95–108

Williams PJ, Robertson JE (1991) Overall plankton oxygen and carbon dioxide metabolism: the problem of reconciling observations and calculations of photosynthetic quotients. J Plankon Res 13:153–169

Weng J-H, Shieh Y-J (2004) Salicylhydroxamic acid (SHAM) inhibits O_2 photoreduction which protects nitrogenase activity in the cyanobacterium *Synechococcus* sp. RF-1. Photosynth Res 82:151–164

White AJ, Critchley C (1999) Rapid light curves: a new fluorescence method to assess the state of the photosynthetic apparatus. Photosynth Res 53:63–72

Chapter 7
Taxonomic Discrimination of Phytoplankton by Spectral Fluorescence

Hugh L. MacIntyre, Evelyn Lawrenz, and Tammi L. Richardson

1 Introduction

Chlorophyll fluorescence techniques are used widely in both laboratory and field studies to assess the abundance and physiological responses of cyanobacteria, microalgae, macroalgae and vascular plants, as described in other chapters in this volume. Most of the instruments used in these studies excite fluorescence in the blue region of the spectrum and measure chlorophyll fluorescence (peak ca. 685 nm) at ambient temperature. Fluorescence is generally detected using a photomultiplier tube (PMT), which is very sensitive to intensity but insensitive to spectral quality. Cross-talk between the light source used to excite fluorescence and the detector is prevented by the use of cut-off filters on both the emitter and the PMT, or by the use of emitters with narrow wavebands, such as light-emitting diodes (LEDs) or lasers, and a long-pass filter on the detector. With the advent of LEDs, which have a very high efficiency (intensity of light output per unit power input) compared to the xenon flash-lamps used in many older instruments, commercially-available fluorometers can have very low power demands and be both small and sensitive (detection limits are typically <1 mg m^{-3} of Chl*a*[1]). This makes them ideal for unattended monitoring such as on platforms, moorings or gliders.

Most such instruments use a single waveband to excite fluorescence and a single waveband to detect it. Those that use LEDs with peak emission at 460 or 470 nm excite Chl*a* fluorescence both directly and indirectly. Because the LED's emission spectrum has only a partial overlap with Chl*a*'s absorption spectrum (Fig. 1), direct excitation is less efficient than it would be for a light source with maximum emission at Chl*a*'s absorption peak at 440 nm. However, the 470 nm LED's emission is absorbed efficiently by the accessory *b* and *c* chlorophylls (Chl*b* and Chl*c*) and photosynthetic carotenoids (PSC) and is absorbed with much lower efficiency by phycoerythrins (PE). These can transfer the excitation energy to Chl*a*, which gives rise to its fluorescence. Although phycocyanins (PC) can also transfer energy to Chl*a*, their very low absorption of emission from the 470 nm LED (Fig. 1) would result in a negligible contribution to Chl*a*'s fluorescence with this excitation spectrum. Conversely, were the excitation delivered by

H.L. MacIntyre (✉)
Department of Oceanography, Dalhousie University, Halifax, NS B3H 4J1, Canada
e-mail: hugh.macintyre@dal.ca

E. Lawrenz and T.L. Richardson
Marine Science Program, University of South Carolina, Columbia, SC 29208, USA

T.L. Richardson
Department of Biological Sciences, University of South Carolina, Columbia, SC 29208, USA

[1] *Abbreviations*: AOA, Algae Online Analyzer (*bbe* Moldaenke); CCMP, Guillard-Provasoli Center for the Culture of Marine Phytoplankton (Boothbay Harbor, ME, USA); CFC, cellular fluorescence capacity, an analog of the maximum quantum yield of photosynthesis; Chl*a*, *b*, *c*; chlorophylls *a*, *b* and *c*; DPS, de-epoxidation state; FRR fast repetition rate (fluorometry); LHC, light-harvesting complex; K_E, saturating parameter of the growth-irradiance curve; MgDVP, Mg-3, 8-divinyl phaeoporphyrin a_5 monomethyl ester; NPQ, non-photochemical quenching; PAM, pulse amplitude modulated (fluorometry); PC, phycocyanin; PE, phycoerythrin; PPC, photoprotective carotenoids; PSC, photosynthetic carotenoids; PSI, photosystem I; PSII, photosystem II; SFS, spectral fluorescence signature(s)

D.J. Suggett et al. (eds.), *Chlorophyll* a *Fluorescence in Aquatic Sciences: Methods and Applications*, Developments in Applied Phycology 4, DOI 10.1007/978-90-481-9268-7_7, © Springer Science+Business Media B.V. 2011

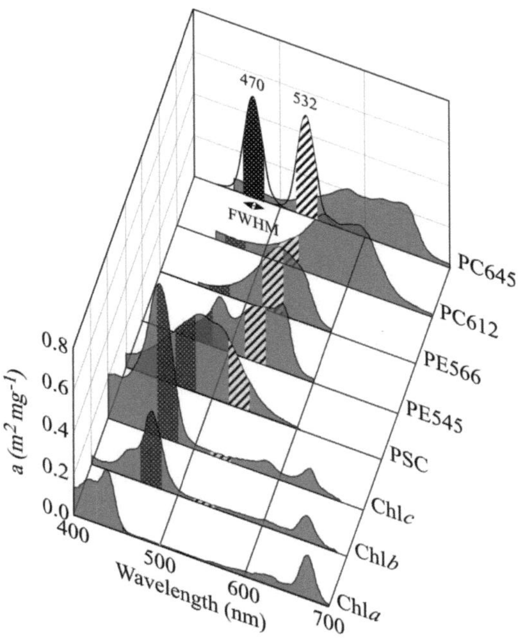

Fig. 1 Absorption spectra for representative photosynthetic pigments. The emission spectra of two light sources, centered at 470 and 532 nm and each with 20 nm half-power bandwidth, are also shown. The bands corresponding to the full width at half maximum emission (FWHM) are superimposed on the absorption spectra. Spectra for chlorophylls and photosynthetic carotenoids (PSC) are approximations of absorption *in vivo* (Bidigare et al. 1990). Spectra for phycoerythrins (Hoef-Emden 2008) and phycocyanins (MacColl and Guard-Friar 1983) (PE and PC) are for absorption *in vitro*

an LED with maximum emission at 532 nm (Fig. 1), photosynthetic carotenoids and phycobilins would be able to absorb the excitation energy and transfer it to Chl*a*, resulting in Chl*a* fluorescence, but there would be negligible excitation of chlorophylls themselves.

Because the complement of accessory pigments varies between phytoplankton taxa, differences in the wavelengths at which Chl*a* fluorescence is stimulated and/or differences in the emission spectra have been used to infer taxonomic structure. Taxonomic differences in the absorption and fluorescence excitation spectra were described more than 50 years ago (Haxo and Blinks 1950; McLeod 1958) and have been documented in depth since (e.g. Neori et al. 1988; Johnsen and Sakshaug 2007). Spectral fluorescence as a discriminatory tool was first applied to phytoplankton in the late 1970s, notably by Yentsch and co-workers (Yentsch and Yentsch 1979; Yentsch and Phinney 1985), and the technique has since been developed extensively, largely by research groups in Estonia (Poryvkina et al. 1994; 2000; Babichenko et al. 1999; 2000); Germany (Kolbowski and Schreiber 1995; Gerhardt and Balode 1998; Beutler et al. 2002; 2003; 2004; Bodemer 2004; Jakob et al. 2005) and Finland (Seppälä and Balode 1998; Seppälä et al. 2007; Seppälä and Olli 2008). The technique has also been proposed to assay community structure in macroalgae (Topinka et al. 1990; Kieleck et al. 2001) and benthic microalgae (Gerhardt and Balode 1998; Aberle et al. 2006). Integrated multi-channel fluorometers are available commercially from several manufacturers, including Walz (Effeltrich, Germany), *bbe* Moldaenke (Kiel, Germany), Photon Systems Instruments (Brno, Czech Republic), LDI^3 (Tallinn, Estonia) and JFE ALEC (Kobe, Japan). Most approaches use excitation of Chl*a* fluorescence at multiple wavelengths but others have used excitation-emission matrices (Oldham et al. 1985; Desiderio et al. 1997; Zhang et al. 2006). In what follows, we review the principles of taxonomic fingerprinting by spectral fluorescence in microalgae and cyanobacteria. We do not address the application of spectral fluorescence in other contexts, such as flow cytometry (see Sosik et al., Chapter 8, this volume) or differentiation of corals (Mazel 1995), nor do we consider the applications of specialized techniques, such as picosecond time-resolved fluorescence spectra, 88 K fluorescence spectra or fluorescence recovery after bleaching, that are used to investigate energy transfer between pigments, state transitions and protein mobility (reviewed by Allen and Mullineaux 2004; Mimuro 2005).

2 The Principles of Taxonomy by Spectral Fluorescence

The intensity of fluorescence excited by a given light source can be expressed in general terms as:

$$Fl^{Chl}(\lambda_{Em}) = \sum \left[E(\lambda_{Ex}) \cdot a_{PS}^{Chl}(\lambda_{Ex}) \right] \phi(\lambda_{Ex,Em}) \quad (1)$$

where $Fl^{Chl}(\lambda_{Em})$ is the Chl*a*-specific fluorescence intensity at emission wavelength λ_{Em}; $E(\lambda_{Ex})$ is the intensity of the excitation at wavelength λ_{Ex}; $a_{PS}^{Chl}(\lambda_{Ex})$ is the Chl*a*-specific absorption coefficient by photosynthetic pigments at λ_{Ex}; and $\phi(\lambda_{Ex,Em})$ is the effective quantum yield of fluorescence for the given excitation and emission wavelengths. Of these terms, differences in the excitation source and the optical geometry of the detector largely determine $E(\lambda_{Ex})$. Secondary variations arise from the presence of other optically-active materials (i.e., those that absorb and scatter light) within the interrogation volume. The effective Chl*a*-specific absorption coefficient integrates both direct absorption by Chl*a* and absorption by accessory pigments that transfer energy to it. Its magnitude varies with wavelength, as specified, and with both genotypic and phenotypic variability in pigment complement and the degree of pigment packaging. Last, the effective quantum yield of fluorescence at the given excitation/emission wavelengths also depends on the proportion of energy that is directed to photosynthesis vs fluorescence vs other energy-dissipating pathways, which is under both genotypic and phenotypic control. It also depends on the degree to which the fluorescence emission is re-absorbed by the pigments and absorbed or scattered by any other optically-active materials in the interrogation volume. We consider below how variability in these parameters determines the degree to which spectral fluorescence measurements can be used to estimate the abundance and taxonomic affiliations of a phytoplankton assemblage.

2.1 Energy Transfer Between Pigments

As described in Eq. 1, fluorescence emission depends in part on the wavelength-dependence of the photosynthetic absorption coefficient, a_{PS}^{Chl}. Because room-temperature fluorescence at 680–690 nm, which is the standard window for commercial instruments, is dominated by fluorescence from PSII, (see Itoh and Sugiura 2004), it is more accurate to say that it depends on the photosynthetic absorption cross-section of PSII. This in turn is related to the effective photosynthetic cross-section per reaction center (σ in darkness and σ' in the presence of actinic illumination) via the size of the pigment antenna, which is the mean number of pigment molecules associated with each reaction center. Both σ and σ' can be estimated by curve-fitting the dependence of fluorescence intensity on excitation intensity at a given waveband, using an FRR fluorometer (Laney 2003; Laney and Letelier 2008); σ can also be estimated by spectral correction of the absorption cross-section for the contribution of non-photosynthetic pigments, given a specified distribution of pigments between PSI and PSII (Suggett et al. 2004).

The mechanisms of energy transfer between pigments have been reviewed in detail by Falkowski and Raven (2007), and what follows reiterates their account. Absorption of light energy by Chl*a* raises an electron

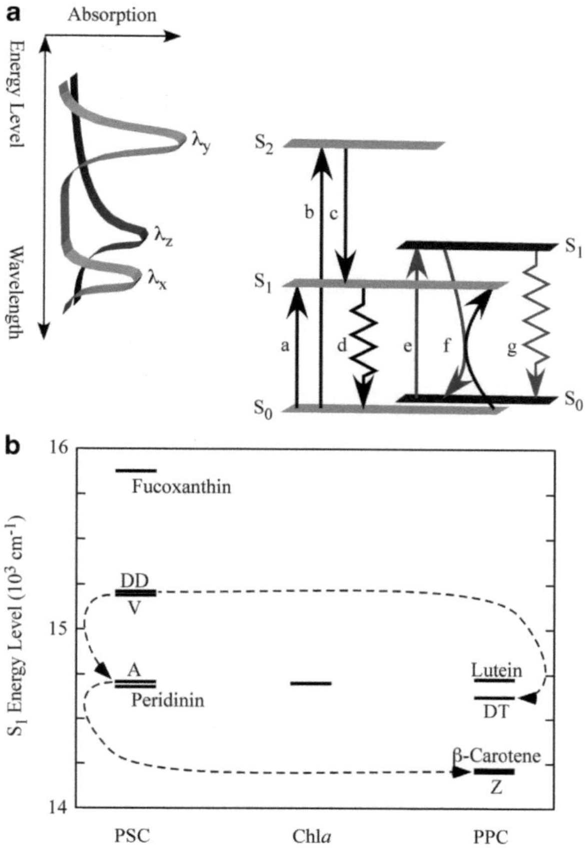

Fig. 2 (**a**) Schematic representation of electron energy levels and corresponding absorption bands for Chl*a* and an accessory pigment, shown as light and dark grey, respectively (modified from Falkowski and Raven 2007). *a, b, e*: raise in electron energy to the S_1 or S_2 level after absorption of a photon. *c*: spontaneous decay from the S_2 to S_1 level. *d, g*: decay from the S_1 to S_0 level, with emission of fluorescence. *f*: exciton migration. (**b**) Modeled S_1 energy levels of select photosynthetic and photoprotective carotenoids (PSC and PPC), relative to Chl*a*. Data are from Frank et al. (1997). Dotted arrows show the inter-conversion by de-epoxidation of the xanthophylls violaxanthin, antheraxanthin and zeaxanthin (V, A and Z) and diadinoxanthin and diatoxanthin (DD and DT)

from its ground state, S_0, to either the S_1 or S_2 level, indicated by *a* and *b* in Fig. 2a. Note that the absorption maxima for each pigment in Fig. 1 correspond to the energy required to achieve these changes, as a photon's energy declines with wavelength. An electron at the S_2 level drops rapidly to the S_1 level (*c* in Fig. 2a). From the S_1 level, the electron may either return to the ground state (*d* in Fig. 2a), emitting radiation with a lower energy and longer wavelength (i.e., fluorescence) or by transferring its energy to another Chl*a* molecule in raising its electron to the S_1 level. The latter is termed "exciton migration" and refers to the transfer of the energy associated with the electron, not migration of the electron itself (i.e., ionization). Exciton migration can also occur between accessory pigments and Chl*a*. In the example that is shown, a pigment with an S_1 energy level higher than Chl*a*'s (e.g., PC) can absorb energy at wavelengths intermediate between the blue and red peaks of Chl*a*'s absorption spectrum (*e* in Fig. 2). If the S_1 energy band overlaps Chl*a*'s, exciton migration can occur between the accessory pigment and Chl*a* (*f* in Fig. 2), extending Chl*a*'s effective absorption spectrum and so contributing to the photosynthetic absorption cross-section (a_{PS}^{Chl} in Eq. 1). In the absence of exciton migration, the accessory pigment's excited electron may also return to the ground state with a loss of energy as fluorescence (*g* in Fig. 2a).

The phycobilins have easily-observed fluorescence peaks. Because these reflect S_1 levels with higher energies than Chl*a*'s, the fluorescence has a correspondingly shorter wavelength. Although carotenoids do fluoresce, their quantum yields of fluorescence are very low and can not be detected with most instruments (Mimuro 2005). Their dominant absorption peak, centered at about 500 nm (Fig. 1), corresponds to the S_2 energy level. The S_1 energy levels are optically forbidden and so lack corresponding absorption peaks. The energy level of the S_1 state determines whether the carotenoid is photosynthetic or photoprotective. Because entropy does not permit exciton migration from a lower to a higher S_1 level, the energy transfer is unidirectional. Carotenoids with an S_1 level that is higher than but still overlapping with Chl*a*'s can transfer energy to it and, like the phycobilins, contribute directly to the spectral dependence of the photosynthetic absorption cross-section (a_{PS}^{Chl}). In contrast, those with an S_1 level that is lower than, but still overlapping with, Chl*a*'s can only accept electrons from it. These carotenoids absorb light but do not contribute to the photosynthetic cross-section and account for much of the difference between absorption and PSII fluorescence excitation spectra (see Lutz et al. 2001; Suggett et al. 2004). The cooperativity between photosynthetic carotenoids and Chl*a* has been visualized by comparing the excitation spectra of isolated Chl*a*-Chl*c*-peridinin binding proteins, functioning thylakoid micelles and whole cells (Jovine et al. 1995; Johnsen et al. 1997).

The modeled S_1 energy levels for several common carotenoids (Frank et al. 1997) are shown in Fig. 2b. Note that the S_1 energy level modeled for peridinin is lower than Chl*a*'s while lutein's is higher, although peridinin does transfer excitons to Chl*a* and lutein does not. The classification into photosynthetic and photoprotective groups is based on more direct means of estimating energy flow between Chl*a* and carotenoids such as picosecond time-resolved fluorescence spectroscopy (Mimuro 2005) and resonance Raman spectroscopy (Ruban et al. 2007), rather than the modeled energy levels alone.

2.2 Taxonomic Differences in Fluorescence Spectra

Taxonomic differences in the complement of PSII accessory pigments determine the differences in both the fluorescence excitation and emission spectra that are illustrated in Fig. 3. The fluorescence excitation spectra of a representative chlorophyte (*Dunaliella tertiolecta*), chromophyte (the diatoms *Chaetoceros muelleri* and *C. affinis*) and cryptophyte (*Storeatula major*) differ most obviously in the region between 500 and 600 nm. This is a consequence of differences in the complement of photosynthetic accessory pigments: the non-photosynthetic (photoprotective) carotenoids are distributed across taxa but do not contribute to the excitation spectra (Johnsen and Sakshaug 2007). In the chlorophyte, the carotenoids are dominated by lutein and β-carotene, both of which are photoprotective (see Fig. 2b) and therefore do not contribute to the excitation spectrum. Dominance of the carotenoid pool by photoprotective forms is also found across the chlorophyte lineages (the Prasinophyceae, Chlorophyceae, Euglenophyceae and Eustigmatophyceae, Jeffrey et al. 1997). The exception is for those taxa that contain siphonaxanthin, which is a major carotenoid in some Siphonales (e.g. Benson and Cobb 1981) and which may occur in some Micromonadophyceae (Fawley and Lee 1990; Wright et al. 1991). Variation in the shape of the excitation spectra within the chlorophytes is driven primarily by the relative amounts of Chl*a* and the accessory chlorophylls, Chl*b* and, in some prasinophytes, MgDVP (Jeffrey et al. 1997). The latter is an analog of Chl*c* and can give rise to pronounced secondary peaks in the excitation and emission spectra at c. 470 and 650 nm, respectively (Haxo and Blinks 1950), particularly when present in high quantities (e.g. in some prasinophytes, see Suggett et al. 2004; Johnsen and Sakshaug 2007).

The diatoms in Fig. 3 have relatively high quotas of the photosynthetic carotenoid fucoxanthin, which

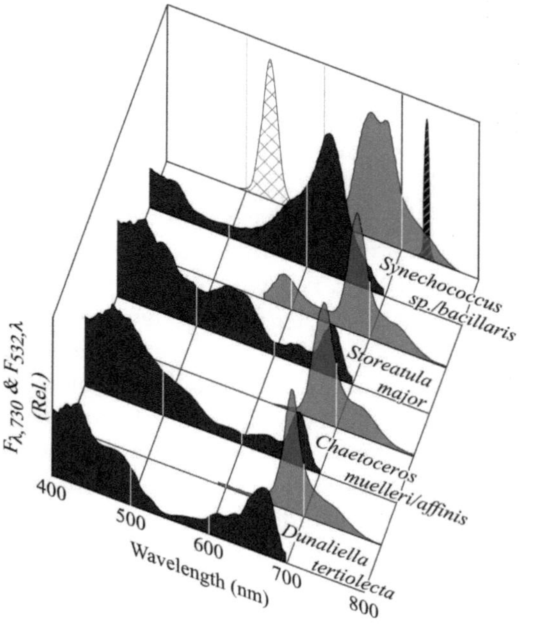

Fig. 3 Fluorescence excitation and emission spectra (dark and light grey, respectively) for a representative chlorophyte (*Dunaliella tertiolecta*), chromophytes (the diatoms *Chaetoceros muelleri*, excitation, and *C. affinis*, emission), cryptophyte (*Storeatula major*) and phycocyanin-rich cyanobacteria (*Synechococcus sp.*, excitation, and *S. bacillaris*, emission). Excitation spectra (dark grey) were driven with a variable excitation and detected at 730 nm (hatched spectrum on the back panel). Emission spectra (light grey) were driven with a 532 nm laser (cross-hatched spectrum on the back panel) and detected with a charge-coupled device. Excitation spectra are from Suggett et al. (2004). Emission spectra are unpublished data (MacIntyre and R. Cox)

contributes to a_{PS}^{Chl} and accounts for the higher fluorescence excitation at 500–530 nm relative to that of the chlorophyte. Fucoxanthin and its derivatives, 19′-hexanoyloxyfucoxanthin and 19′-butanoyloxyfucoxanthin, or peridinin are found in all the chromophytic lineages (the Coscinodiscophyceae, Bacillariophyceae, Prymnesiophyceae, Chrysophyceae, Raphidophyceae, Pelagophyceae, Dinophyceae and Phaeophyceae. Hooks et al. 1988; Andersen et al. 1996; Jeffrey and Vesk 1997). In both the chlorophyte and diatom, the fluorescence emission spectrum is similar, with a single dominant peak at c. 680 nm attributable to Chl*a*. The peak wavelength can vary by 10 nm or more (Millie et al. 2002; Johnsen and Sakshaug 2007), which has been ascribed to differences in pigment-protein binding between taxa.

The excitation and emission spectra for the cryptophyte differ from those of the chlorophyte and chromophyte in having a marked peak in the excitation at c 550 nm and a second distinct peak in the emission spectrum at c. 590 nm (Fig. 3). These are due to the presence of phycobilins, in this case a PE (see Fig. 1), which give members of the class their characteristic colors. In most cryptophyte lineages, which are pink to brown in appearance, the phycobilins are PEs but some contain PCs (Hoef-Emden 2008) and may be blue-green.

Although the rhodophytes (not shown) and cyanobacteria also have Chl*a* and phycobilins as the photosynthetic pigments, they are unlike cryptophytes in lacking *c* chlorophylls. The Prochlorococcales are an exception among the cyanobacteria with regard to accessory chlorophylls, as they also contain divinyl Chl*b* and divinyl Chl*a* (Burger-Wiersma and Post 1989; Chisholm et al. 1992). The differences in excitation spectra between the cyanobacteria and cryptophytes arise both from differences in the pigments (Apt et al. 1995) and their arrangement in phycobilisomes in the cyanobacteria (and rhodophytes, which are not shown) vs. the LHC in the cryptophytes. The cyanobacterial PSII reaction center contains relatively small quantities of Chl*a* and light harvesting by PSII is dominated by the bilins in the phycobilisome. Chlorophyll *a* is primarily associated with PSI in cyanobacteria, it makes little contribution to the PSII excitation spectrum (Lutz et al. 2001; Suggett et al. 2004; Johnsen and Sakshaug 2007). In contrast to cyanobacteria, the phycobilins in cryptophytes are packed inside the thylakoid lumen (Spear-Bernstein and Miller 1989) and transfer energy efficiently to both PSI and PSII (Wit et al. 2008). There can also be significant differences between the emission spectra for cryptophytes and cyanobacteria. The spectra shown for the representative species in Fig. 3 are for a PE-containing cryptophyte and a PC-rich cyanobacterium. The latter lacks the 590 nm emission peak of PE and has instead a dominant peak at c. 655 nm due to PC. While both cryptophytes and cyanobacteria have PE-rich (pink) and PC-rich (blue-green) members, there are persistent differences in the excitation spectra because of the differences in the distribution of pigments in the phycobilisome vs the antenna/lumen and because of the presence of *c* chlorophylls as accessory pigments in the cryptophytes. There are also differences in the emission spectra of PE-rich representatives of the two classes because of taxonomic differences in the type of phycoerythrins (Apt et al. 1995). The PE emission peak for the pink cyanobacteria *Synechococcus sp.* CCMP833 and *Synechocystis sp.* CCMP843 (not shown) was centered at c. 570 nm vs 590 nm in the cryptophytes *Storeatula major* (Fig. 3) and *Rhodomonas lens* CCMP739 (not shown).

Differences in both the excitation and emission spectra have been used to characterize the taxonomic structure of both microalgae and macroalgae *in vivo*. In many, if not most, cases, the approach has also been used to estimate the abundance of each group. We review the approach below, illustrating some potential pitfalls with experimental data.

2.3 Taxonomic Discrimination by Spectral Fluorescence

The basis of fluorometric taxonomic discrimination lies in classifying differences in fluorescence signatures that arise from consistent differences in pigment complement. Multiple approaches have been used that vary with the capabilities of the instruments at hand, but all rely on comparing the responses of unknown samples to the responses of representative taxa. At the simplest level, this can be achieved by examining the difference in emission, using excitation at two different wavelengths. This is illustrated in Fig. 4a, which shows discrimination between a diatom and a chlorophyte using the ratio of fluorescence excited at 530 and 450 nm (Yentsch and Yentsch 1979). The two wavelengths preferentially stimulate photosynthetic carotenoids (which the chlorophytes largely lack but which the diatoms have) and Chl*a* (Fig 1) and so the ratio exploits the dip in the chlorophyte's excitation spectrum relative to a diatom's at 530 nm (Fig. 3). The ratio approach works well in 2-species mixtures, but it has limited ability to discriminate between more complex mixtures, is sensitive to changes in background fluorescence, and is easily biased by acclimative changes in the representative organisms' excitation spectra (Oldham et al. 1985; Beeler SooHoo et al. 1986; Hilton et al. 1989).

By careful targeting of the regions of the spectra that have the highest between-taxon variability (Hilton et al. 1988), a fairly high degree of discrimination can be achieved with as few as 3–5 excitation wavelengths and a single emission wavelength. This is illustrated in Fig. 4b, which shows a cluster analysis of the spectral signatures of 44 strains of microalgae from seven classes, which were generated by excitation at 410, 510 and 540 nm (Gaevsky et al. 2005). The spectral fluorescence signatures (SFS) were reported as the ratios of Fl_{540}:Fl_{410}, Fl_{510}:Fl_{410} and Fl_{510}:Fl_{540}, which were standardized to the mean value before constructing a matrix of pair-wise Bray-Curtis similarity coefficients in PRIMER-E (Clarke and Warwicke 2001). With three exceptions, the cultures clustered into three taxonomically-appropriate groups; chromophytes (almost all the diatoms and the dinoflagellates), chlorophytes (the chlorophytes *sensu stricto*, euglenophytes and xanthophytes) and the cyanobacteria. The lone cryptophyte and two of the diatoms clustered with the cyanobacteria. The dispersion within and between these groups is shown in a multidimensional scaling (MDS) plot (Fig. 4c) that is based, like the cluster analysis, on the matrix of pair-wise similarity coefficients. The ordination is analogous to a 2-D plot of principal components based on fully spectral or discrete data (Zhang et al. 2006; Seppälä and Olli 2008); the differences being that MDS plots can be based on non-parametric variables, are not scaled and are explicitly multivariate (in this case a 2-D representation of a 43-dimensional distribution rather than of just the first two dimensions of a multi-dimensional distribution). The spatial distribution is a representation of difference, in which distance between samples is inversely proportional to their similarity. Note that while the cryptophyte clusters with the cyanobacteria at 80% similarity, it forms an out-group with one of the two anomalous diatoms, *Synedra sp.*, found within the cyanobacterial cluster (Fig. 4c).

The discriminatory power of the approach can be assessed by subjecting subsets of the similarity matrix to the permutative ANOSIM analysis, which is analogous to *a posteriori* pair-wise testing of differences after univariate 1-way Analysis of Variance, to test for significant differences. (The subsets are defined *a priori* as the taxonomic groups, rather than the groupings from the cluster analysis, to avoid a circular argument.) There are highly significant ($p < 0.001$) differences between the three groups that contain enough samples to give unbiased results, the chlorophytes, diatoms, and cyanobacteria (Table 1). The small sample sizes of the remaining groups precludes high confidence but their similarities to their taxonomic affiliates (dinoflagellates to diatoms, euglenophytes and xanthophytes to chlorophytes) are

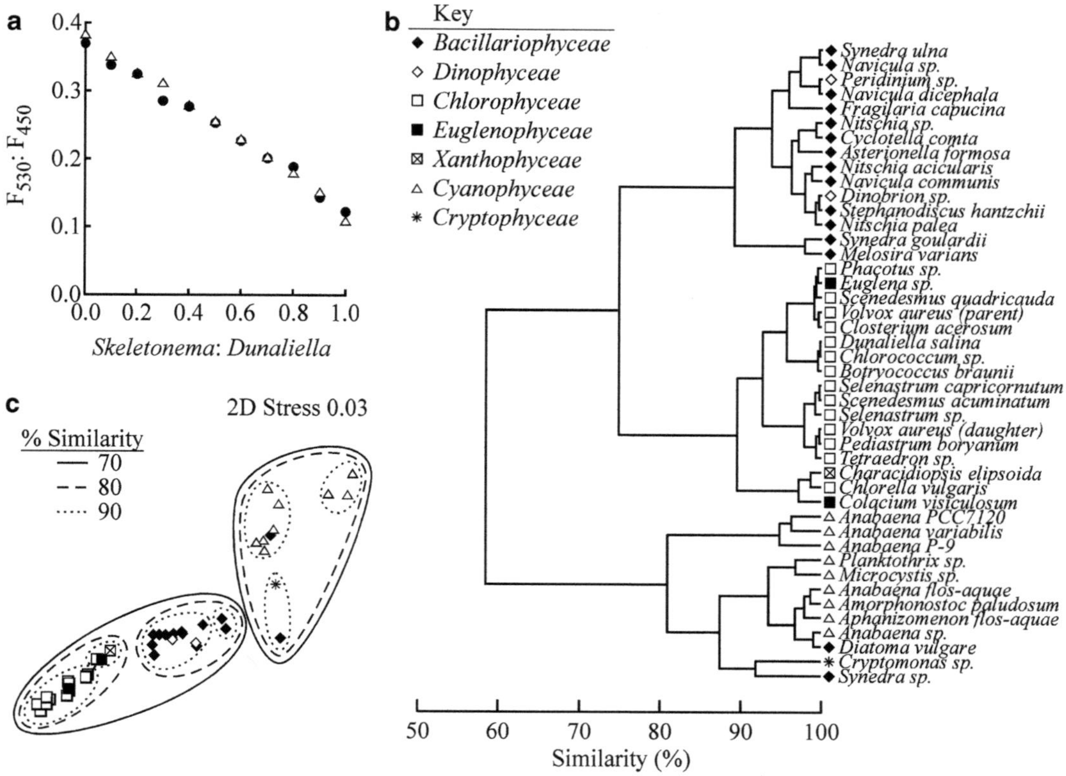

Fig. 4 (**a**) Changes in the ratio of $Fl_{530}:F_{450}$ in binary mixtures of the diatom *Skeletonema costatum* and the chlorophyte *Dunaliella tertiolecta* (redrawn from Yentsch and Yentsch 1979). Circles and triangles represent replicated experiments. (**b**) Cluster analysis of average species- or strain-specific fluorescence signatures of 44 cyanobacteria and microalgae. Each signature is the average of 3 – 107 replicates. Data are from Gaevsky et al. (2005). See text for details of the analysis. (**c**) Dimensionless ordination of the similarity matrix from (b) in a multidimensional scaling plot (MDS). The distance between any pair of points is inversely proportional to their similarity. Contours are the similarity levels derived from the cluster analysis. The very low value of stress refers to the (minor) distortion imposed by reducing a 43-dimensional matrix to a 2-dimensional representation

comparable to the within-group similarities and their differences from taxonomic outgroups (dinoflagellates vs chlorophytes and cyanobacteria etc.) are also comparable to the differences between their affiliates and the out-groups (Table 1). Three conclusions can be drawn from the analysis of this data-set:

1. The spectral signatures reflect statistically-significant ($p < 0.01$) differences between organisms (Table 1)
2. There is variability within the statistically-different groups such that within-group similarity is less than 100% (i.e., signatures are not identical, Table 1) and
3. Outliers within the taxonomic groups can be classified as belonging to the wrong taxonomic group (Fig 4b, c)

The within-group variability means that classification between groups would probably best be handled by fuzzy logic, although to our knowledge, the approach has yet to be used. Instead, a variety of statistical techniques has been employed to analyze the spectral signatures of an unknown sample in terms of the contributions of known taxa. The spectral signature may be based on a continuous excitation spectrum (Babichenko et al. 1994; Gerhardt and Balode 1998; Seppälä and Olli 2008), a continuous emission spectrum (Cowles et al. 1993), and either a full (Oldham et al. 1985) or reduced excitation/emission matrix (Desiderio et al. 1997; Zhang et al. 2006). Because of the high level of redundancy in the spectra and matrices, comparable discriminatory power can be derived from discrete measurements that target areas of maximum variability in the excitation spectra (Kolbowski and Schreiber 1995; Beutler et al. 2002, 2004; Parésys et al. 2005) or in both the excitation and emission spectra (Murphy and Cowles 1997). In the case of spectral data, the signal derived from an unknown sample can be analyzed to estimate the contributions of known

Table 1 Similarity analysis of taxonomic clusters based on standardized SFS (Gaevsky et al. 2005). Groupings are based on taxonomy, not on the cluster analysis in Fig. 4 (i.e. *Diatoma vulgare* and *Synedra sp.* are grouped with the diatoms rather than the cyanobacteria). N refers to the number of taxa within the group, not to the number of samples analyzed within a species, which ranged from 3 to 107 (mean 27). Similarity is the within-group similarity, defined by the SIMPER test of PRIMER-E. Dissimilarity is between-group dissimilarity (= 100% – Similarity). Testing for between-group differences vs the groups with the highest replication (i.e., the Bacillariophyceae, Chlorophyceae and Cyanophyceae) was by ANOSIM. The ANOSIM R statistic and p level are reported for each pair-wise comparison. The ANOSIM test was only run for those groups in which 55 or more permutative tests were possible (i.e., not for the Xanthophyceae and Cryptophyceae, which had only one representative each)

	N	% Similarity	% Dissimilarity with Bacillariophyceae	% Dissimilarity with Chlorophyceae	% Dissimilarity with Cyanophyceae
Bacillariophyceae	15	89.3		29.4 R = 0.91; p < 0.001	28.6 R = 0.79; p < 0.001
Chlorophyceae	14	94.4	29.4 R = 0.91; p < 0.001		56.0 R = 1.00; p < 0.001
Cyanophyceae	9	88.8	28.6 R = 0.79; p < 0.001	56.0 R = 1.00; p < 0.001	
Dinophyceae	2	95.3	7.3 R = –0.25; p = 0.86	27.0 R = 1.00; p < 0.01	30.6 R = 0.97; p < 0.02
Euglenophyceae	2	89.3	23.2 R = 0.65; p < 0.03	7.8 R = 0.20; p = 0.18	49.8 R = 1.00; p < 0.02
Xanthophyceae	1	N/A	19.6	9.7	46.2
Cryptophyceae	1	N/A	19.7	47.9	12.7

contributors via Fourier-transformed pattern recognition (Oldham et al. 1985), discriminant analysis (Johnsen et al. 1994), similarity indices (Millie et al. 1997, 2002; Staehr and Cullen 2003) that are based on 4th-derivative transformations (Butler and Hopkins 1970a, b), principal components analysis (Moberg et al. 2002; Seppälä and Olli 2008), Gaussian decomposition of the spectra (Bodemer 2004; Chekalyuk and Hafez 2008) or linear unmixing (Dickinson et al. 2001; Beutler et al. 2002; Beutler et al. 2003; Beutler et al. 2004; Gaevsky et al. 2005; Parésys et al. 2005; Zhang et al. 2006; Seppälä and Olli 2008). The last is the most widely-used approach and is the method implemented in commercial instruments manufactured by Walz and *bbe* Moldaenke. The basis is that the fluorescence of an unknown sample at a wavelength *x* is the sum of the contributions of *n* known groups. The contribution of each group is equal to its concentration, as Chl*a*, and its Chl*a*-specific fluorescence (Eq. 1):

$$Fl(\lambda_x) = \sum_{n=1}^{n=i} Fl^{Chl}(\lambda_x)_n \bullet Chl_n \qquad (2)$$

The spectral fluorescence signature, SFS, which comprises the signals at multiple wavelengths ($x_1, x_{2...} x_m$), is then defined in terms of three matrices (Seppälä and Olli 2008):

$$SFS = CF + E \qquad (3)$$

where *SFS* is a matrix of y sample spectra (*y* by *x*), *C* is a matrix of the Chl*a* concentrations of the known taxa (*y* by *n*), *F* is a matrix of the Chl*a*-specific fluorescence intensities of the known taxa (*x* by *n*), and *E* is an error matrix. In the case of ambiguity or mutual exclusion between similar taxa in the solution, the approach can be qualified with a decision-tree (Parésys et al. 2005). Here, one or more criteria are used to select alternative sets of *C* and *F* matrices (e.g. for distinguishing between PE-rich and PC-rich cyanobacteria, in the case where one or the other dominates).

Figure 5 illustrates linear unmixing, using Chl*a* fluorescence excited by LEDs in five wave-bands. The SFS of the four known taxa are shown normalized to the mean value, as Norm SFS, in Fig. 5a. Differences in the Norm SFS can be attributed to differences in pigment complements. The representative cyanobacterium, *Synechococcus bacillaris*, is PC- rather than PE-rich, so its SFS is dominated by the excitation spectrum of PC

(Fig. 1). The Norm SFS of the representative diatom, chlorophyte and cryptophyte are more similar to each other than to the cyanobacterium, reflecting differences in the distribution of pigments between eukaryotes vs procaryotes (see above) and differ from each other primarily in their relative magnitudes at 470–570 nm. This is the region of the spectrum that is most sensitive to the presence and abundance of photosynthetic carotenoids (the diatom), phycoerythrin (the cryptophyte) or their absence (the chlorophyte). The Norm SFS of two "unknowns", the diatom *Entomoneis* cf. *alata* and the unicellular rhodophyte *Porphyridium cruentum*, are shown in Fig. 5b. Note the similarity between *Entomoneis'* Norm SFS and those of the known diatom, the cryptophyte, and the chlorophyte. The Norm SFS for the second unknown, *Porphyridium*, is not obviously similar to any of four known taxa.

The similarity matrix is represented in an MDS plot (Fig. 5c), where the four known Norm SFS define each apex of a tetrahedron. The lines joining any two points represent binary mixtures of those two taxa. The planes defined between any three points represent ternary mixtures of the three. The internal volume of the tetrahedron represents all quarternary mixtures for positive solutions of Eq. 3. The solution to Eq. 3 for *Entomoneis* is a binary mixture of the representative diatom, *Thalassiosira pseudonana*, (65%) and the cryptophyte, *Rhodomonas lens*, (35%) as shown by its position on the line connecting their SFS (Fig. 5c). In contrast, ordination of *Porphyridium* Norm SFS on the MDS plot places it outside the tetrahedron of positive solutions to Eq. 3, and closest to the cryptophyte and the representative cyanobacterium, *Synechococcus bacillaris* (Fig. 5c). Intuitively, this makes sense: the rhodophyte, like the cyanobacterium, has a phycobilisome and lacks Chl*c* and PSC, so would have a relatively low contribution from them at 470–525 nm (Figs. 1 and 5b). However, unlike this cyanobacterium, but like the cryptophyte, it is PE-rich, resulting in high excitation at 525 nm (Fig. 5b). Because the fit lies outside the region of positive fits to the four Norm SFS, it must be constrained to

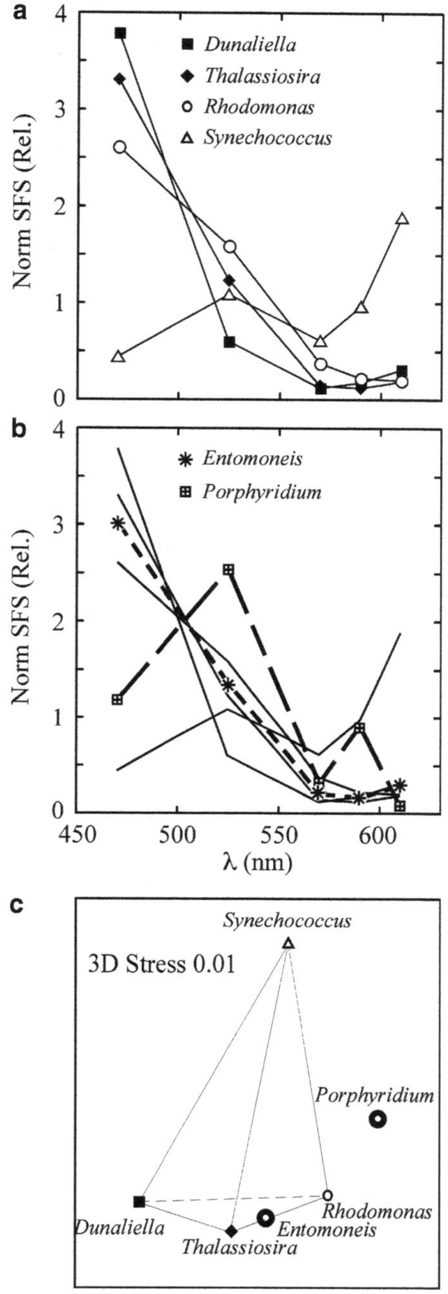

Fig. 5 (**a**) Normalized spectral fluorescence-excitation signatures (Norm SFS) for nutrient-replete cultures of a representative chlorophyte (*Dunaliella tertiolecta* CCMP1320), chromophyte (the diatom *Thalassiosira pseudonana* CCMP1335), cryptophyte (*Rhodomonas lens* CCMP739) and cyanobacteriaum (*Synechococcus bacillaris* CCMP1333). The spectra were collected with *bbe* Moldaenke's Algae Online Analyzer (AOA), corrected according to Kopf and Heinz (1984) and normalized to the average of the weighting coefficients to emphasize differences in shape. (**b**) Normalized SFS for a second diatom (*Entomoneis cf alata* CCMP1522) and a unicellular rhodophyte (*Porphyridium cruentum* CCMP1328). These are superimposed on the calibration

Fig. 5 (continued) signatures from (a), which are shown as fine lines. (**c**) Multidimensional scaling plot (MDS) of the calibration signatures from (**a**) and the unknown signatures from (**b**). The lines connecting each pair of calibration signatures represents the mixing line of two-component mixtures (see text for details) positive contributions, which translates to a mixture of the cyanobacterium (24%) and the cryptophyte (76%).

Three assumptions must be met for accurate identification with linear unmixing, (Beutler et al. 2002):

1. The signatures of the known taxa must be linearly independent so that it is not possible to reproduce the signature of one taxon by the weighted sum of the others.
2. The signatures of the known groups must be invariant (i.e. the terms on the right-hand side of Eq. 1 must be constant or must have compensatory changes, so that $Fl^{Chl}(\lambda)$ is constant).
3. The spectral fluorescence signatures must be determined with sufficient accuracy.

The first assumption has been validated by Beutler and co-workers (2002), for an instrument using five excitation wavelengths and a single detection wavelength in what are now commercially-available configurations, *bbe* Moldaenke's Fluoroprobe and Algae Online Analyzer (Fig. 5). Note that assumption (1) must be tested for any different configuration when using other wavelengths for excitation and/or emission. We present data to evaluate assumptions (2) and (3) below, using the Algae Online Analyzer (AOA). The instrument can be trained on up to five SFS without being under-constrained but we focused on the four characteristic of our local estuarine waters, PC-rich cyanobacteria, chlorophytes, cryptophytes and chromophytes.

3 Variation in Chlorophyll-specific Fluorescence, F^{Chl}

It became clear very soon after the introduction of the *in vivo* fluorescence as an estimate of Chl*a*, (Lorenzen 1966), that the ratio of fluorescence to Chl*a* (Fl^{Chl}) is not constant. The ratio was shown to vary between species (Strickland 1968; Loftus and Seliger 1975), with light exposure and nutrient availability (Kiefer 1973a, b; Loftus and Seliger 1975; Vincent 1979), and with cell size (Alpine and Cloern 1985). For recent reviews, see Babin (2008) and Chapter 3. The observation of variable Fl^{Chl} has been reiterated frequently in the decades since (e.g. Kruskopf and Flynn 2006) and has been explicitly recognized by almost all researchers working on the use of SFS-based taxonomy since Yentsch and Yentsch (1979). Strictly, this invalidates Assumption (2), above. In addition to the effect that variability in Fl^{Chl} has on the estimate of Chl*a* concentration, changes in the spectral dependence of Fl^{Chl} will result in mis-identification through linear unmixing where a single representative signature is used for each taxon.

3.1 Inter-Specific Variability

We illustrate the effect of inter-specific variability (cf. the data of Gaevsky et al., Fig. 4) on estimates of taxonomic composition and abundance in Fig. 6. The AOA was calibrated with one representative of each of the chlorophytes, chromophytes (diatom), cryptophytes and cyanobacteria (Fig. 5) and then challenged with 11 chromophytes from four classes. All cultures were nutrient-replete and grown at the same irradiance in semi-continuous culture (MacIntyre and Cullen 2005). Comparison of the Norm SFS of the chromophytes with those of the calibration species showed that they clustered with the representative chromophyte but did show significant dispersion (Fig. 6a). The similarity comparisons between the unknown chromophytes and calibration representative, *Thalassiosira pseudonana*, (not shown) were higher than between the unknowns and the other calibration taxa and higher than between *T. pseudonana* and the other calibration taxa. In short, as shown with the data of Gaevsky and coworkers (Fig. 4, Table 1), the classification of SFS followed the unknown species' taxonomic affiliation. Because this comparison is based on normalized signatures (i.e., the shape of the SFS alone), it reflects differences in the spectral dependence of excitation arising from differences in $a_{PS}^{Chl}(\lambda_{Ex})$ and $\phi(\lambda_{Ex,Em})$ in Eq. 1. These in turn reflect differences in the ratio of pigments, the internal concentration of pigments and the degree of pigment packaging, the quantum yield of photosynthesis, and the likelihood of excess energy being re-emitted as fluorescence rather than directed into other dissipative pathways.

When linear unmixing is used to estimate the taxonomic composition, the monocultures were identified as dominated by chromophytes, with lesser contributions from the 3 other groups. In all but one case, the component of the population that was mis-identified was attributed to chlorophytes or cryptophyes (Fig. 6b), reflecting the proximity of the unidentified species'

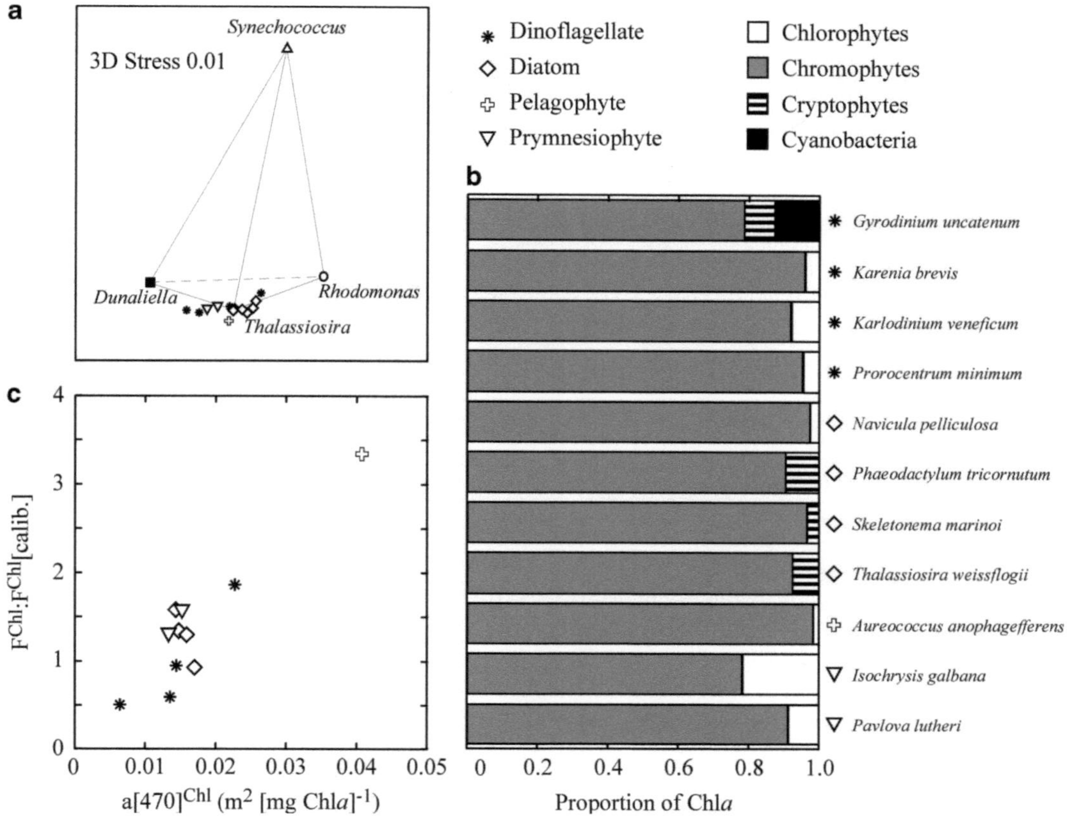

Fig. 6 (**a**) Multidimensional scaling plot (MDS) of the AOA's calibration signatures from Fig. 5c and the signatures of 11 'unknown' chromophytes (dinoflagellates, diatoms, prymnesiophytes and a pelagophyte). The unknown chromophytes clustered most closely with the calibration chromophyte (the diatom *Thalassiosira pseudonana*) but, as indicated by the dispersion in the MDS, were not identical. (**b**) The AOA's classification of the 11 unknown chromophytes, based on linear unmixing fits to the calibration SFS (Fig. 5a). Because of the variance in their SFS (cf. Fig. 5b), they were identified as being 79–98% chromophytes, with 2–21% contributions of other taxa. (**c**) F^{Chl} at 470 nm for the 11 unknown chromophytes, expressed as the ratio to F^{Chl} of the calibration chromophyte, as a function of Chl*a*-specific absorption at 470 nm

Norm SFS to the eukaryote mixing plane (the chlorophyte-chromophyte-cryptophyte surface, Fig. 6a). Differences between the SFS of the three eukaryotes used to calibrate the AOA are driven primarily by changes in the ratio of PSC or PE to accessory chlorophylls (Figs. 1 and 5a). The estimates of relative abundance (Fig. 6), which ranged from 79–98% chromophytes and 2–21% other taxa, depend on both the shape of the signature (the norm SFS) and each of the 4 calibration species' spectrally-weighted values of Fl^{Chl} (Eq. 2). The low Fl^{Chl} of the calibration cyanobacterium relative to the eukaryotes (0.05–0.13x of the eukaryotes' values at 470 nm, and 0.33–1.67x at 525 nm), is the reason that a relatively minor deviation from the eukaryote mixing plane towards the cyanobacterial apex can result in a disproportionate apparent contribution by the cyanobacteria. This is illustrated by the 12% contribution of cyanobacteria to the estimated composition of the dinoflagellate *Gyrodinium uncatenum* (Fig. 6b), although its SFS, which is the right-most of the unknown chromophytes' (Fig. 6a), is little further from the eukaryotes' mixing plane than those of the adjacent diatoms.

The degree to which the linear unmixing over- or under-estimates the chlorophyll concentration of the unknown samples depends on the relative magnitude of Fl^{Chl} for each species (Eq. 2). This in turn reflects the Chl*a*-specific photosynthetic cross-section, a_{PS}^{Chl}, which depends on both the relative abundance of photosynthetic vs photoprotective pigments and the degree of pigment packaging. The importance of the latter is illustrated in Fig. 6c, which shows the variation in Fl^{Chl}

as a function of Chl*a*-specific absorption at 470 nm. The values of $a(470)^{Chl}$ decrease by 6.4*x* between the smallest cell, *Aureococcus anophagefferens* (c. 1.5 μm) and the largest, *Gyrodinium uncatenum* (c. 45 x 60 μm) as pigment packaging (internal self-shading) increases. The residual variation in Fl^{Chl} reflects differences in the relative abundance of photosynthetic vs photoprotective pigments and the effective quantum yield of fluorescence (a_{PS}^{Chl} vs a^{Chl} and $\phi(\lambda_{Ex,Em})$, Eq. 1). The same pattern holds for the other 4 wavelengths used by the AOA.

3.2 Intra-specific Variability

Intra-specific variability in photosynthetic architecture (the constituents of the photosynthetic apparatus) occurs in response to variability in nutrient availability (Laws and Bannister 1980; Kolber et al. 1988; Sosik and Mitchell 1991), temperature (Verity 1981; Sosik and Mitchell 1994; Moore et al. 1995; Anning et al. 2001) and irradiance (reviewed by MacIntyre et al. 2002). The effect on Fl^{Chl} of acclimation to changing light intensity and nutrient availability in the diatom *Thalassiosira pseudonana* are illustrated in Fig. 7. There is relatively little variation in either the quantum yield of PSII activity or Fl^{Chl} for nutrient-replete cultures of *Thalassiosira pseudonana* (Fig. 7a, b): coefficients of variation are 2% and 15%, respectively. The quantum yield of PSII activity is determined here with DCMU (3-[3,4-dichlorophenyl]-1,1-dimethylurea) and is termed cellular fluorescence capacity, CFC (Vincent 1980), to distinguish it from the analog that is measured using PAM or FRR fluorescence, F_v/F_m. Under nutrient-limited growth (i.e., when cultures are in balanced growth and fully acclimated to reduced nutrient availability in chemostats, MacIntyre and Cullen 2005),

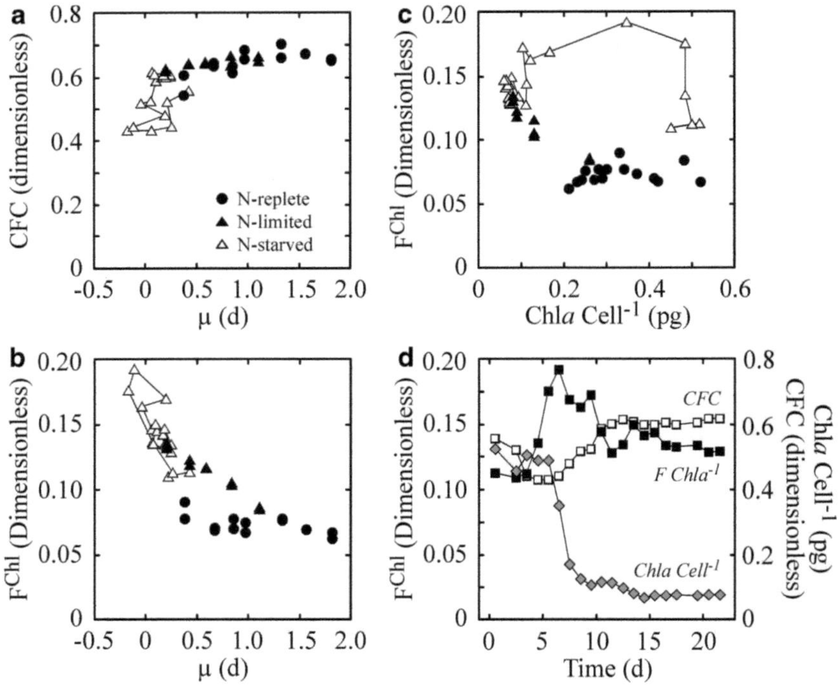

Fig. 7 Variation in (**a**) quantum yield of PSII activity (CFC) and (**b**) F^{Chl} in the diatom *Thalassiosira pseudonana* as a function of growth rate in balanced (N-replete and N-limited) and unbalanced (N-starved) growth. Data for the N-starved culture represent the trajectory through acclimation to N-limitation (see **d**). F^{Chl} is parameterized as the dimensionless ratio of fluorescence *in vivo* to fluorescence *in vitro* (Kiefer 1973a). (**c**) Variation in F^{Chl} with pigment quota in the same samples. (**d**) Time-course of CFC, F^{Chl} and pigment quota during acclimation to N-limitation after an exponential-phase batch culture was switched to chemostat mode (see text for details). Data of Yang, MacIntyre and Cullen, reported in part by Cullen and co-workers (Cullen et al. 1992; Zhu et al. 1992; Parkhill et al. 2001; MacIntyre et al. 2002; MacIntyre and Cullen 2005). Panel (**d**) is based on figures from MacIntyre and Cullen (2005)

Fl^{Chl} is correlated with the degree of nutrient limitation, expressed as the reduction in growth rate in the chemostats. There is a significant ($p < 0.01$) but relatively minor reduction in CFC: a 5% decline over an 82% reduction in growth rate, as compared to a 53% increase in Fl^{Chl}. These trends may be over-estimates of the true variability, due to an actinic effect of the excitation beam in the fluorometer causing some photosynthetic induction (Parkhill et al. 2001).

Although down-regulation of pigments occurs under both high irradiance and low nutrient availability (Geider et al. 1997, 1998a, b), the reduction in pigment, on a per-cell or per-carbon basis, is more pronounced over the range of nutrient limitation than over the range of nutrient-replete growth. The increase in Fl^{Chl} is significantly correlated with the reduction in pigmentation ($p < 0.001$; $R = -0.93$: Fig. 7c), suggesting that an increased Chl*a*-specific efficiency of light harvesting through reduction of the package effect is largely responsible for the increase (cf. Fig. 6c). (We note that the inverse relationship between these variables could be a spurious correlation (Jackson and Somers 1991), driven by the presence of chlorophyll concentration in both variables. However the quotient $F\ Cell^{-1}$ for N-limited cultures has a higher range than does chlorophyll concentration, 5.0*x* vs 2.7*x*, indicating that the relationship is robust.)

While variation in Fl^{Chl} and CFC under N-limitation (balanced growth) is modest, the range under nutrient starvation (unbalanced growth) is less so. Fig. 7d shows the response of a batch culture in late exponential phase having severe limiting nutrient supply imposed on it by being switched to chemostat mode (see MacIntyre and Cullen 2005, for details). The nutrient supply was at 0.15 of the nutrient-replete growth rate. In the initial phase of nutrient-starvation, there is a rapid reduction in CFC and a corresponding rise in Fl^{Chl}. Given that CFC is correlated with the maximum quantum yield of photosynthesis, a reduction in CFC reflects a reduced efficiency of energy transfer into photosynthesis, which will lead to a compensatory increase in energy dissipation as fluorescence or heat. The initial phase of nutrient starvation is succeeded by an acclimative down-regulation of Chl*a* (Fig. 7d) during resource re-allocation to balance light harvesting and growth (Geider et al. 1998a). After the initial very rapid phase of down-regulation, both CFC and Fl^{Chl} recover and all three eventually reach the new equilibrium condition of balanced, nutrient-limited growth. The trajectories of both CFC and Fl^{Chl} through the transition from nutrient starvation to nutrient limitation are superimposed on the data for acclimated cultures (Fig. 7a–c). Two observations are obvious. First, the range of values falls well outside the envelope of responses for cultures in balanced growth. The range in Fl^{Chl} is 42% higher than the maximum observed under extreme nutrient limitation (growth at 15% of the nutrient-replete growth rate at an irradiance 1.75*x* K_E, the irradiance at which nutrient-replete growth becomes light-saturated). The second observation is that Fl^{Chl} does not covary with pigment quota and pigment packaging during nutrient starvation (Fig. 7c), in contrast to both the inter- and intra-specific trends for acclimated cultures (Fig. 6c, 7c). We can infer that variability in Fl^{Chl} (Eq. 1) is driven primarily by variability in $a_{PS}^{Chl}(\lambda_{Ex})$ in balanced growth but must also be driven by variability in $\phi(\lambda_{Ex,Em})$ in unbalanced growth.

The intra-specific variability in Fl^{Chl} shown in Fig. 7 is for data collected at a single wavelength. Given the relationship between Fl^{Chl} and the estimate of Chl*a* (Eq. 2), we would predict that fluorescence-based estimates of biomass will be biased by variation in Fl^{Chl} resulting from changes in photosynthetic architecture and efficiency (Fig. 6c, 7c). The prediction is borne out by repeated observation of variability in the relationship between fluorescence and Chl*a* over the past 30 years (see above). It is difficult to predict if the taxonomic classification based on SFS will be as sensitive. Accurate identification depends on invariant spectral signatures, which in turn depend on constant ratios of functional pigment arrays. What constitutes a functional array depends on the interrogation wavelengths. If accessory pigments and Chl*a* are both excited, then the ratios of accessory pigments to Chl*a* will determine the constancy of the SFS (or the lack of constancy). However, for instruments that target only the accessory pigments (e.g. the AOA: compare the excitation wavelengths in Fig. 5a with the absorption spectra in Fig. 1), then the critical ratios will be the abundances of accessory chlorophylls, photosynthetic carotenoids and phycobilins relative to each other.

Studies of chromophytes growing under nutrient-replete conditions show relatively little variation in the ratios of Chl*c* and PSC to Chl*a* when growing below K_E (reviewed by MacIntyre et al. 2002). At irradiances above K_E, there is a selective loss of both Chl*c* and PSC relative to Chl*a*. However, ratios of PSC to Chl*c* are more constrained than the ratios of either to Chl*a* (Table 2). In all cases, the ratios of PPC to photosynthetic pigments have a wide range, increasing with irradiance. There are few reports of pigments in

Table 2 Variations in pigment ratios (g g^{-1}) for cells in nutrient-replete growth over a range of irradiances. The irradiance range is expressed relative to the saturating irradiance for growth (E/K_E – see MacIntyre et al. 2002). Data are for the chromophytes *Thalassiosira weissflogii* (diatom), *Amphidinium carteri* (dinoflagellate), *Emiliania huxleyi, Hymenomonas carterae* and *Phaeocystis antartica* (prymnesiophytes) and the cyanobacterium *Synechococcus* sp. (clones WH7803). PB is the sum of PE and PC

Species & Source	E/K_E	Chl*c*: Chl*a*	PSC: Chl*a*	PE:Chla	PC:Chla	PPC: Chl*a*	PSC: Chl*c*	PPC: Chl*c*	PSC: PPC	PB:PPC
Thalassiosira weissflogii (Sosik 1988)	0.1–10.3 (79x)	0.6–1.0 (1.7x)	0.34–0.37 (1.1x)	N/A	N/A	0.05–0.29 (5.8x)	0.37–0.57 (1.5x)	0.05–0.46 (9x)	1.2– .0 (5.7x)	N/A
Thalassiosira weissflogii (Goericke and Welschmeyer, 1992a, b)	0.2–10.1 (51x)	0.4–1.0 (2.8x)	0.25–0.37 (1.5x)	N/A	N/A	0.06–0.40 (6.7x)	0.29–1.0 (3.6x)	0.06–1.12 (19x)	0.9–5.3 (5.8x)	N/A
Amphidinium carteri (Sosik 1988)	0.7–9.5 (14x)	0.8–1.0 (1.2x)	0.60–0.68 (1.1x)	N/A	N/A	0.15–0.39 (2.6x)	0.61–0.80 (1.3x)	0.15–0.46 (3.1x)	1.7–4.2 (2.4x)	N/A
Emiliania huxleyi (Stolte et al. 2000)	0.2–7.8 (34x)	0.5–1.0 (2.0x)	0.6–1.2 (1.9x)	N/A	N/A	0.1–0.8 (5.6x)	1.1–1.3 (1.2x)	0.2–1.6 (10x)	0.8–8.2 (10x)	N/A
Hymenomonas carterae (Sosik 1988)	0.14–8.5 (61x)	0.8–1.0 (1.3x)	0.28–0.36 (1.3x)	N/A	N/A	0.08–0.26 (3.3x)	0.34–0.42 (1.2x)	0.10–0.32 (3.2x)	1.1–4.3 (4.0x)	N/A
Phaeocystis antarctica (Moisan and Mitchell 1999)	0.2–6.2 (39x)	0.4–1.0 (2.3x)	0.3–0.7 (2.4x)	N/A	N/A	0.02–0.4 (19x)	0.4–0.7 (1.9x)	0.02–0.9 (43x)	0.8–8.2 (57x)	N/A
Synechococcus sp. WH7803 (Kana and Glibert 1987; Kana et al. 1988)	0.3–13.8 (44x)	N/A	N/A	4.4–38.6 (8.8x)	1.5–2.7 (1.8x)	0.55 – 1.6 (2.8x)	N/A	N/A	N/A	0.01–0.24 (17x)

chlorophytes grown over comparable ranges of irradiance but the Chl*b*:Chl*a* ratio in *Dunaliella tertiolecta* and *D. salina* varies about 2*x* (Sukenik et al. 1990; Neidhardt et al. 1998) over a 24*x* range in growth irradiance. It varies 3.3*x* in the Chl*b*-containing cyanobacterium *Prochlorococcus marinus* (Moore et al. 1995) grown at 0.5–15.0 E/K_E. There is a preferential loss of phycobilins relative to Chl*a* with increasing irradiance in nutrient-replete cryptophytes (Lewitus and Caron 1990; Sciandra et al. 2000; Hammer et al. 2002) and cyanobacteria (Table 2) and a preferential loss of PE over PC in the latter (Kana and Glibert 1987; Six et al. 2004). The losses in cyanobacteria reflect the progressive shortening of the phycobilin rods in the phycobilisome vs the stable Chl*a*-containing reaction center (Grossman et al. 1993). There are comparable or greater changes in pigment ratios under both nutrient-limitation and nutrient-starvation in representatives of the chromophytes, crytophytes and cyanobacteria (Collier et al. 1993; Latasa and Berdalet 1994; Latasa 1995; Goericke and Montoya 1998; Sciandra et al. 2000; Henriksen et al. 2002; Staehr et al. 2002). In contrast, there was only 1.1*x* variability in Chl*b*:Chl*a* under a 9*x* range of nutrient-limited growth rates in the chlorophyte *D. tertiolecta* (Sosik and Mitchell 1991), as compared to the c. 2*x* range when grown over a 24*x* range of irradiance (Sukenik et al. 1987).

Assessing the effect of changing pigment quotas on taxonomic classification requires knowledge of whether the change makes them more or less like members of other taxonomic groups. This depends on both the magnitude/type of change in the pigment ratios and on the wavelengths at which the SFS are determined. We tested the effect of simultaneous nutrient starvation and increased irradiance (both of which drive down-regulation of pigments) on the four calibration species (Fig. 5), using the AOA. Nutrient-replete cultures in mid- to late-exponential phase were subject to a 10% daily dilution with N-free medium and monitored for biochemical composition and their taxonomic signatures. Nitrate was exhausted within 6 days in all cultures and Chl*a* concentrations declined as the growth rate declined below the dilution rate and the cells underwent washout (Fig. 8a–d). This raised the effective irradiance in the 10-L (0.22 m diameter) culture vessels so that the cultures were subject to the synergistic effects of reduced nutrients and increased irradiance on regulation of light harvesting. In all cases, the increase in C:N and decrease in Chl*a*:C ratios (Fig. 8e–h) indicated progressive down-regulation of light harvesting, although the time-courses differed between species. There were only minor changes in accessory chlorophyll:Chl*a* ratios in the eukaryotes (Fig. 8i–k) and a minor increase in photosynthetic carotenoid:Chl*a* ratios in the diatom *Thalassiosira pseudonana* (Fig. 8j). Estimates of phycobilins from whole-cell absorption indicated an initial increase as cells continued to grow, followed by a decrease during the combined nutrient-stress and increased irradiance, in both the cryptophyte *Rhodomonas lens* and the cyanobacterium *Syncechococcus bacillaris* (Fig. 8k–l). In all cases, the PPC:Chl*a* ratio increased (Fig. 8i–l). These changes were easily visible as the cultures turned from deep green (the chlorophyte *Dunaliella tertiolecta*), pink (*Rhodomonas*) or blue-green (*Synechococcus*) to yellow-green or from brown to orange-brown (*Thalassiosira*).

The changes in pigment ratios were reflected in the Norm SFS (Fig. 9a-d). In *Dunaliella*, there was a minor steepening of the decrease between 470 and 525 nm, presumably as a consequence of increased masking of Chl*b*'s absorption by lutein (Fig. 8i) and/or β-carotene (which was not quantified). In *Thalassiosira*, the Norm SFS were effectively invariant (Fig. 9b). As with *Dunaliella, Rhodomonas* showed a progressive steepening of the Norm SFS between 470 and 570 nm (Fig. 9c), reflecting both the progressive loss of PE and the increase in alloxanthin (Fig. 8k) and/or β-carotene relative to Chl*a* and Chl*c*. In *Synechococcus*, the Norm SFS were driven by the changes in PC:Chl*a* (Fig. 8l), with the peaks at 525 and 610 nm becoming more pronounced as the ratio increased and then progressively flatter as it fell again. The high variation at 470 nm in the Norm SFS (Fig. 9d) is largely an artifact of normalization: the change in F^{Chl} at 610 nm was 27*x* that at 470 nm, so normalizing to the mean F^{Chl} across wavelengths resulted in the values at all wavelengths being driven by variability at 610 and, to a lesser extent, 570 and 590 nm.

The matrix of the pair-wise similarity coefficients is displayed in an MDS plot (Fig. 9e), in which the varying Norm SFS are compared with the apices of the tetrahedron formed by the Norm SFS of the cultures used for the calibration. Because the chlorophyte already has the steepest decline in Norm SFS of the eukaryotes (Fig. 5a), the progressive steeping of the Norm SFS makes the nutrient-starved cultures less, rather than more, like the other eukaryotes. In the calibration, the

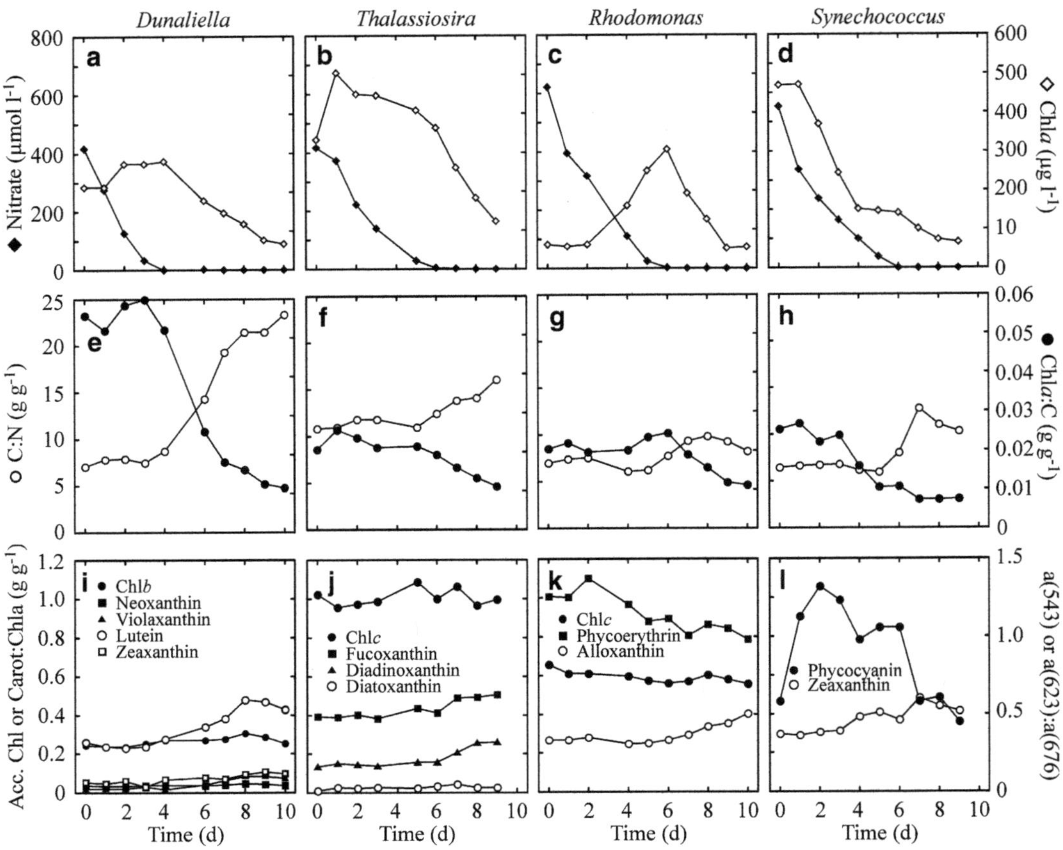

Fig. 8 Time-courses of (**a-d**) nitrate drawdown and biomass (as Chl*a*); (**e–h**) C:N and Chl*a*:C; and (**i–l**) pigment:Chl*a* ratios in the chlorophyte *Dunaliella tertiolecta* (A, E and I), the chromophyte (diatom) *Thalassiosira pseudonana* (B, F and J), the cryptophyte *Rhodomonas lens* (C, G and K), and the cyanobacterium *Synechococcus bacillaris* (D, H and L). Mid- to late-exponential phase cultures were diluted 10% daily with N-free medium. Chlorophylls and carotenoids were determined by HPLC. Concentrations of β-carotene were not determined. Phycobilins are approximated as the ratio of peak absorption (543 and 623 nm) to absorption at the Chl*a* red peak (676 nm) *in vivo*

cryptophyte has the most gradual decline in Norm SFS between 470 and 570 nm (Fig. 5a). Consequently, the progressive steepening caused by selective loss of PE, in the absence of photosynthetic carotenoids, makes it more like the chlorophyte: note that the change lies very close to the cryptophyte-chlorophyte mixing line (Fig. 9e). The largely invariant Norm SFS of the chromophyte is evident from the minimal dispersion between samples. Last, the initial increase in PC:Chl*a* in the cyanobacterium (Fig. 8l) was manifested as more pronounced peaks at 525 and 570–610 nm. These are less, not more, similar to the Norm SFS of the eukaryotes (Fig. 5a), causing the initial move away from their mixing plane (Fig. 9e). The subsequent loss of PE, again in the absence of photosynthetic carotenoids, also resulted in the cyanobacterium's Norm SFS becoming more like the chlorophyte's, as evidenced from their proximity to the cyanobacterium-chlorophyte mixing line (Fig. 9e).

The changes in Norm SFS caused negligible inaccuracies in taxonomic classification in the chlorophyte and chromophyte (Fig. 10a, b). Because their Norm SFS became increasingly similar to that of the calibration chlorophyte (Fig. 9e), both the cyanobacterium and, particularly, the cryptophyte had progressive inaccuracies in taxonomic classification. In both, linear unmixing of the SFS converged on a mixture of the correct taxon and chlorophytes (Fig. 10c, d). As predicted, Fl^{Chl} increased

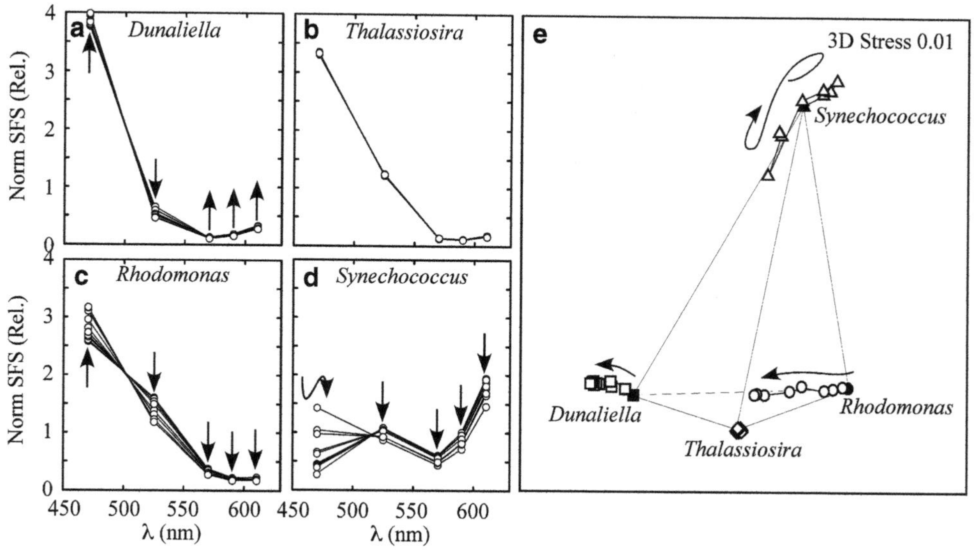

Fig. 9 Changes in the Norm SFS for (**a**) the chlorophyte *Dunaliella tertiolecta,* (**b**) the chromophyte (diatom) *Thalassiosira pseudonana,* (**c**) the cryptophyte *Rhodomonas lens* and (**d**) the cyanobacterium *Synechococcus bacillaris* under progressive N- and light-stress (see text for details). Arrows denote the net direction of change over the time-course at each wavelength. (**e**) Multidimensional scaling plot of the Norm SFS from A-D (open symbols) superimposed on the calibration signatures (closed symbols, cf. Fig. 5c). Arrows denote the time-course

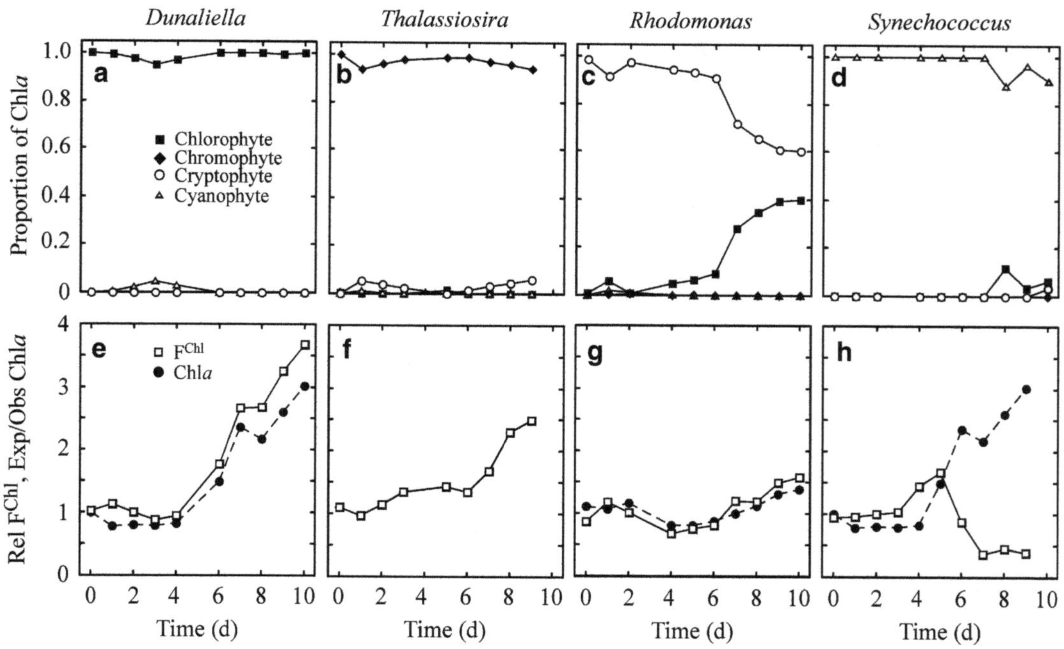

Fig. 10 Time-courses of (**a-d**) taxonomic classification (as the proportion of total Chl*a*); and (**e–h**) relative F^{Chl} and estimated Chl*a* in the chlorophyte *Dunaliella tertiolecta* (**a** and **e**), the chromophyte (diatom) *Thalassiosira pseudonana* (**b** and **f**), the cryptophyte *Rhodomonas lens* (**c** and **g**), and the cyanobacterium *Synechococcus bacillaris* (**d** and **h**) under progressive N- and light-stress (see text for details). Taxonomic classification is the output of the linear unmixing solution to the fluorescence signature. Values of F^{Chl} are at the peak wavelength (470 nm for the eukaryotes and 610 nm for the cyanobacterium) and are expressed as ratios to the calibration values. Chl*a* is expressed as the ratio of the value predicted from the linear unmixing to the value determined by HPLC

at all wavelengths, which was reflected in over-estimates of Chl*a* by up to 3*x* (Fig. 10e–h). The over-estimate of Chl*a* in the cyanobacterium (*Synechococcus*, Fig. 10h) was caused by the mis-assigned classification and the taxonomic differences in F^{Chl}.

3.3 Short-Term Quenching

So far, we have considered only the effects of relatively slow changes in photosynthetic architecture, involving regulation of pigment quota, on SFS. Other changes that might also affect SFS are much more rapid (MacIntyre et al. 2000). Short-term variation in Fl^{Chl} reflects the balance between light harvesting and downstream utilization of reductant, which is the basis of efforts to estimate photosynthetic rates from variable fluorescence (Falkowski et al. 1988; Genty et al. 1989; 1990; Krause and Weis 1991), discussed by Suggett, Moore and Geider in Chapter 5 (see also Suggett et al. 2009). The magnitude of Fl^{Chl} depends both on the balance between the activities of photosynthetic and dissipative pathways within the photosynthetic apparatus and on the intensity and duration of the excitation (reviewed by Kromkamp and Forster 2003). This is illustrated in Fig. 11a, which shows fluorescence measured before (F_0, F) and after (F_m, F_m') the application

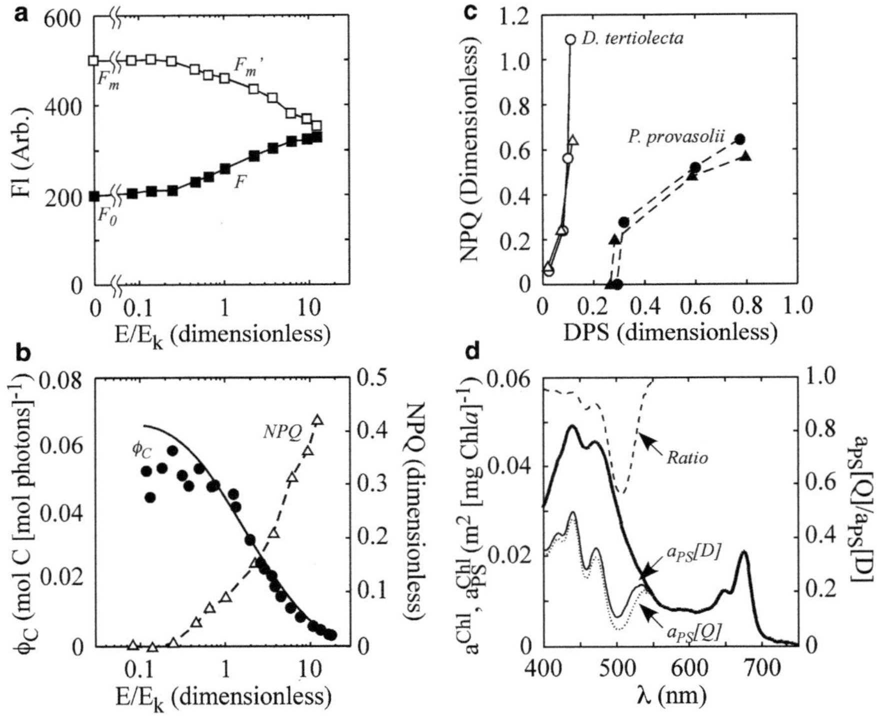

Fig. 11 (**a**) Minimum (F_0 and F) and maximum (F_m and F_m') fluorescence in darkness (F_0, F_m) and with actinic illumination (F, F_m') in the prasinophyte *Pycnococcus provasolii*. Fluorescence was measured with a Walz PAM101. Background irradiance is expressed in terms of the saturating parameter of a parallel ^{14}C photosynthesis-irradiance (PE) curve, E_k, after reconciliation for differences in light quality (MacIntyre and Cullen 2005). Note log scale. (**b**) The quantum yield of carbon fixation (ϕ_C, based on ^{14}C uptake) and non photochemical quenching (NPQ) as a function of relative irradiance in the same culture. The deviation of the fit from the distribution of φ_C at low irradiance (cf. Johnson and Barber 2003) is a consequence of forcing the fit of the PE curve through the origin. (**c**) Relationship between NPQ, measured with a Chelsea Fastracka FRR, and de-epoxidation state (DPS) in cultures of *P. provasolii* (closed symbols) and the chlorophyte *Dunaliella tertiolecta* (open symbols) incubated in darkness and at c. 60, 300 and 1000 µmol photons m^{-2} s^{-1} (fluorescence data from Suggett et al. 2009). Circles and triangles denote replicate cultures. (**d**) Chl*a*-specific absorption spectra for *P. provasolii* (a^{Chl}, heavy line) and reconstructed absorption spectra for photosynthetic pigments (a_{PS}^{Chl}, light solid and dotted lines). Reconstruction was based on partitioning a^{Chl} using HPLC pigment data (Bidigare et al. 1990), on the assumption that the spectrum for MgDVP was the same as Chlc's. Photosynthetic absorption spectra are from samples with minimal and maximal DPS (darkness, $a_{PS}[D]$, solid line; high light, $a_{PS}[Q]$, dotted line). The ratio of photosynthetic absorption with maximum DPS to photosynthetic absorption in the dark-acclimated state is also shown (dashed line)

of a saturating flash, in the dark (F_0, F_m) or in the presence of actinic illumination (F, F_m'). F rises with irradiance as the quantum yield of photosynthesis, and hence the probability of energy transfer into photosynthetic pathways, declines (Fig. 11b). In contrast, the maximum fluorescence (F_m, F_m') declines with irradiance, a change that can be expressed as non-photochemical quenching ($NPQ = [F_m - F_m']/F_m'$), as shown in Fig. 11b. The effect of varying irradiance on Fl^{Chl} can be minimized by blocking electron transfer with DCMU (e.g., Gaevsky et al. 2005) or by dark-adapting the sample being interrogated. Measuring F_m rather than F_0 has the advantage that it maximizes signal strength. In practice, any instrument that is used in flow-through mode will allow at least partial relaxation of photosynthetic induction (F tends to F_0) or NPQ (F_m' tends to F_m), provided that the plumbing is opaque. Instruments that interrogate natural samples *in situ* without shielding can and will be confounded by the effect, as noted shortly after fluorometry was proposed as a tool for studying phytoplankton (Kiefer, 1973a, b).

Although variability in Fl^{Chl} can and does bias the estimate of Chl*a* from fluorescence (Eq. 2), the degree to which it might bias the taxonomic classification will depend on the magnitude of NPQ and its underlying cause. One mechanism that correlated with NPQ is the xanthophyll cycle (Demmig-Adams 1990), in which rapid de-epoxidation converts a photosynthetic carotenoid into a photoprotective one. In the chlorophytes, sequential de-epoxidations convert violaxanthin to antheraxanthin to zeaxanthin; and in chromophytes, a single de-epoxidation converts diadinoxanthin to diatoxanthin. As can be seen (Fig. 2b), there is a reduction in the S_1 energy level that results in a change in the ratio of these photoprotective to photosynthetic carotenoids. The degree of interconversion can be expressed as the de-epoxidation state, DPS (Casper-Lindley and Björkman 1998):

$$DPS = \frac{0.5[A]+[Z]}{[V]+[A]+[Z]} \qquad (4)$$

or

$$DPS = \frac{0.5[DT]}{[DD]+[DT]} \qquad (5)$$

in which $[V]$, $[A]$, $[Z]$, $[DD]$ and $[DT]$ are the concentrations of, respectively, violaxanthin, antheraxanthin, zeaxanthin, diadinoxanthin and diatoxanthin. The magnitude of NPQ may be correlated with DPS, as is shown for *Pycnococcus provasolii* (Fig. 11c) or may be independent of it (Arsalane et al. 1994; Casper-Lindley and Björkman 1998; Koblizek et al. 2001), as is shown for *Dunaliella tertiolecta* (Fig. 11c) and as is seen in the cyanobacteria, rhodophytes, cryptophytes and euglenophytes, which lack a xanthophyll cycle but still generate NPQ. In one sense, the de-epoxidation is spectrally neutral, as the dissipatory step is invoked at the S_1 energy level of Chl*a*. This is lower than the S_2 or higher absorption maxima of the accessory pigments whose relative abundance defines the SFS (Fig. 2a), so photons absorbed by any pigment that contributes to the SFS should have an equal opportunity of being dissipated as heat rather than being re-emitted as fluorescence. In this sense, fluorescence quenching by the xanthophyll cycle will operate through the effective quantum yield, $\phi(\lambda_{Ex,Em})$, rather than the absorption term, $a_{PS}^{Chl}(\lambda_{Ex})$ in Eq. 1. The contribution of the xanthophylls may also act through the absorption term where there is a significant change in the xanthophyll:Chl*a* ratio, as can occur in response to short-term exposure to bright light (Olaizola and Yamamoto 1994), under acclimation over a light gradient (reviewed by MacIntyre et al. 2002), or in response to nutrient limitation (Sosik and Mitchell 1991).

However, if the epoxidated forms of the xanthophylls are significant contributors to photosynthetic absorption, then de-epoxidation could alter the SFS. This is illustrated by modeling the photosynthetic absorption spectrum by partitioning total absorption between the contributions of photosynthetic and non-photosynthetic pigments (Bidigare et al. 1990), and assuming that the effect of pigment packaging is constant at a given wavelength (cf., Geider et al. 1998). The predicted change in $a_{PS}^{Chl}(\lambda_{Ex})$ for *Pycnococcus* (Fig. 11c) is shown in Fig. 11d. The observed xanthophyll interconversion makes no difference to $a_{PS}^{Chl}(\lambda_{Ex})$ at wavelengths above 550 nm but could cause a reduction of up to 40% at lower wavelengths. In the case of SFS generated with the AOA (see previous sections), this would cause a steepening of the SFS between the 470 and 525 nm excitation bands: the modeled reduction in $a_{PS}^{Chl}(\lambda_{Ex})$ is 10% at 470 nm but 26% at 525 nm. Because *Pycnococcus* is a prasinophyte, its SFS is closest to the chlorophyte's, so this would be analogous to the steepening of *Dunaliella's* SFS under nutrient starvation (Fig. 9a). The xanthophyll cycle would make *Pycnococcus'* SFS less, not more, similar to the other eukaryotes' (Fig. 9e), and there should be no consequent error in taxonomic classification due to xanthophyll cycle activity (cf. Fig. 10a).

Cryptophytes and cyanobacteria lack the xanthophyll cycle, but NPQ is correlated with DPS in chromophytes (Casper-Lindley and Björkman 1998; Koblizek et al. 2001). As chromophytes' SFS are typically differentiated on the basis of the presence of photosynthetic carotenoids, xanthophyll cycle activity might bias their signatures towards the chlorophytes', resulting in a bias in taxonomic classification. The degree to which this might occur is difficult to predict: it would depend on the irradiance to which the cells are acclimated (hence the quota of xanthophylls relative to other pigments, see MacIntyre et al. 2002), the intensity and duration of the irradiance to which they are exposed (Fig. 11b, c, cf. Arsalane et al. 1994; Olaizola and Yamamoto 1994), and to adaptive differences between species (Sakshaug et al. 1987; Strzepek and Harrison 2004; Lavaud et al. 2007).

In addition to the xanthophyll cycle, short-term quenching may involve state transitions. This is likely the factor that causes the large change in NPQ in the absence of significant de-epoxidation in *Dunaliella* (Fig. 11c). State transitions involve the migration of a proportion of light-harvesting complexes or phycobilisomes between PSII and PSI. The response is driven by redox control (reviewed by Kruse 2001; Allen 2002), and induction of State 2 (migration of antennae to PSI) occurs when PSII activity exceeds that of PSI at high irradiance (Schubert et al. 1995; Campbell and Oquist 1996) or in response to the demands of nitrite reduction (Sahay et al. 2006). State transitions are well documented in chlorophytes and cyanobacteria and may play a dominant role in energy dissipation in the former, even though they have a functional xanthophyll cycle (Garcia-Mendoza et al. 2002; Ross et al. 2008). State transitions also occur in cyanobacteria during light/dark transitions (Schubert et al. 1995; Campbell and Oquist 1996). They are not thought to occur in chromophytes (Owens 1986). State transitions reduce a_{PS}^{Chl} at PSII, so will reduce Fl^{Chl} under State 2 conditions relative to State 1. They have the potential to alter the SFS because of the unequal distribution of pigments between the antenna and the reaction center. Chlorophyll *b* and the (trace quantities of) photosynthetic carotenoids are associated with the antenna in chlorophytes (Yamamoto and Bassi 1996) and the phycobilins are associated with the phycobilisome's rods in cyanobacteria (Gantt et al. 1979; Gantt 1996). Consequently the relative abundance of accessory pigments to that of Chl*a* at PSII will decline in State 2. Because of the co-location of accessory pigments, their relative abundance at PSII should not be affected. The sensitivity of the SFS to state transitions will depend on the degree to which it is weighted by accessory pigment to Chl*a* ratios.

We performed empirical tests of the effect of inducing NPQ on both the taxonomic classification and estimates of biomass. The tests were performed on the calibration species; the chlorophyte *Dunaliella tertiolecta*, the cryptophyte *Rhodomonas lens*, the cyanobacterium *Synechococcus bacillaris*, and the diatom *Thalassiosira pseudonana*; as well as on two other chromophytes (the dinoflagellate *Amphidinium carterae*, and the prymnesiophyte *Emiliania huxleyi*). The cultures were grown at optically-thin densities under nutrient-replete conditions at c. 80 μmol photons m^{-2} s^{-1}. Sample cultures were then dark-acclimated for 20 min to allow NPQ to relax, illuminated at 800 μmol photons m^{-2} s^{-1} for 40–50 min, then returned to darkness. Continuous assessments of taxonomic status and estimates of Chl*a* were made using the AOA. Pigment concentrations were determined at the beginning, midpoint and end of the light period. In all cases, there was no significant (>1%) change in the taxonomic classification, although *Amphidinium* and *Emiliania* were consistently identified as 15% chlorophytes/85% chromophytes and 60% chlorophytes/40% chromophytes, respectively. We emphasize that this result is not general: the sensitivity of the SFS to the short-term quenching mechanism will depend on both the physiological status of the organisms being investigated and on the wavelengths used to construct the SFS. Although the experimental irradiance was 10*x* the growth irradiance, the magnitude of DPS was relatively low (<0.4 in those species with a xanthophyll cycle). Had the cultures been acclimated to higher irradiance, a comparable shift in irradiance might have caused higher quenching and biased the SFS. Also, had the SFS included significant direct excitation of Chl*a*, taxonomic bias would have been more likely. Although there was no significant change in taxonomic assignment, estimates of biomass fell as fluorescence was quenched and rose as quenching was relaxed. The response of the diatom *Thalassiosira pseudonana* is shown in Fig. 12a.

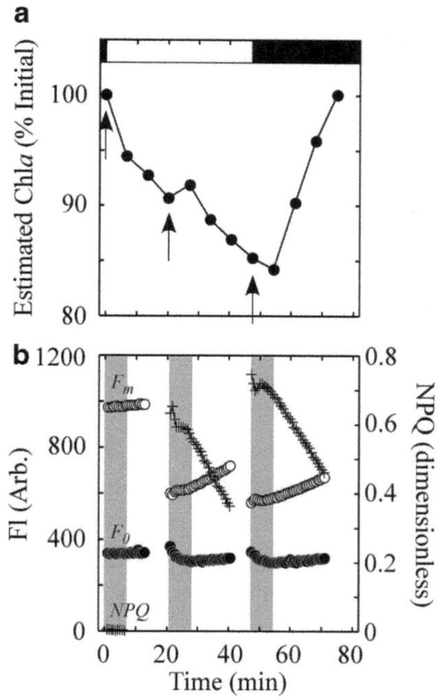

Fig. 12 (**a**) Time-course of Chl*a* estimates on a dark-acclimated culture of the diatom *Thalassiosira pseudonana* exposed to c. 50 min of high irradiance (10*x* growth irradiance) and then returned to darkness (light and dark bars at top of panel). Estimates were made with *bbe* Moldaenke's AOA. Chl*a* is expressed relative to the T_0 value. Arrows indicate sub-samples removed to darkness. (**b**) Time-courses of minimum and maximum fluorescence (F_0 and F_m) and non photochemical quenching (NPQ) in subsamples from (a). Samples were held in darkness and monitored continuously with a Walz WaterPAM. Grey bars show the duration of the repeated measurements made on a single sample by the AOA (a)

A factor that affects the sensitivity to short-term quenching is the degree to which relaxation is possible before or while the measurement is being made. The data shown in Fig. 12a are averages of repeated interrogations over an interval of ca. 7 min, and made on a sample drawn into a darkened interrogation chamber rather than left under the ambient illumination. The first-order time constants for induction/relaxation of the xanthophyll cycle are relatively short, *ca.* 0.1 min^{-1} in the chlorophyte *Chlorella vulgaris* (Goss et al. 2006), which would allow a significant relaxation of the fluorescence quenching during measurement. Time constants for state transitions are comparably short (Schubert et al. 1995). This is illustrated in Fig. 12b, which shows short-term changes in samples removed from the time-course experiment. The first was taken pre-illumination and shows the dark-adapted levels of F_0, F_m, and NPQ. The second, taken after 20 min of illumination shows the relaxation of quenching. The kinetics of relaxation of F to F_0 and F_m' to F_m are both biphasic. The former had a rapid exponential decline (τ = 0.57 ± 0.04 min^{-1}), followed by a slower and approximately linear rise (0.33 ± 0.04% of F_0 min^{-1}). Relaxation of F_m' was slower. There was an initial rise over 2 min that was too short to resolve by curve-fitting, followed by a linear rate (0.76 ± 0.01% of F_m min^{-1}). Consequently, while the initial values of F and F_m' were 108 and 61% of F_0 and F_m, respectively, the values integrated over 7 min were 95 and 64%. Similar patterns, but with slower kinetics, were observed after 47 min of illumination (Fig. 12b). Because the magnitude of the quenching depends on both the intensity and duration of illumination (Fig. 11a, 12a), and because there are taxon-specific differences in the irradiance-dependence of both F and F_m' (see next section), we can not be more specific about the effect of allowing some relaxation to occur on the estimates of fluorescence. In general, though, measurements taken in darkness will have less quenching than those taken under actinic illumination and will be less likely to bias the SFS.

4 Optical Indices and Application of the SFS Approach in the Field

Phytoplankton are not the only colored materials in their environment. Other optically-active materials, which include water itself, chromophoric dissolved organic material (CDOM) and suspended sediments and particulate organic material (detritus), all attenuate light, so have the ability to modulate both excitation and emission. Alteration of the excitation beam affects Fl^{Chl} through changes in the excitation irradiance ($E(\lambda_{Ex})$ in Eq. 1). Alteration of the emission beam affects Fl^{Chl} through variation in the effective quantum yield ($\phi(\lambda_{Ex,Em})$ in Eq. 1). Because the excitation beam in a fluorometer is typically collimated, both absorption and scattering (the inherent optical properties that determine attenuation) will tend to attenuate $E(\lambda_{Ex})$. However, scattering can also augment $E(\lambda_{Ex})$ through the phenomenon of path-length amplification: the probability of a photon impinging on a particle and being absorbed is increased when photons that have not been absorbed are scattered back into the optical path. In contrast to the excitation beam, fluorescence emission is isotropic (i.e. has an equal probability of occurring in all directions). If the photon is emitted within the cone of detection and towards the detector, it will be observed. In the case of a closed-cavity instrument, the signal can be amplified by use of reflectors that redirect fluorescence emitted in other directions towards the detector. Particles that scatter the fluorescence emission back towards the detector operate in similar fashion.

Because both absorption and scattering are spectrally-dependant, they have the potential to bias the SFS. The spectral variation for a representative estuarine water-mass is shown in Fig. 13a. While the shape of the phytoplankton absorption spectrum depends on the relative abundance of pigments with differing absorption characteristics, the shapes of the CDOM and detrital absorption curves are highly conservative (Roesler and Perry 1995), following an approximately exponential decline through the visible wavelengths:

$$a(\lambda) = a(400)e^{-\sigma(\lambda-400)} \quad (6)$$

where $a(\lambda)$ and $a(400)$ are absorption at λ and 400 nm (m^{-1}); and σ is the spectral slope (nm^{-1}). Absorption by phytoplankton, CDOM and detritus depends on their relative concentrations: absorption by water is constant, as is its scattering. The latter is very small, less than 0.006 m^{-1} between 400 and 700 nm (Buiteveld et al. 1994), which is less than 0.2% of the total scattering shown in Fig. 13a, and makes no significant contribution in most estuarine and coastal waters. The spectral dependence of scattering has the potential to be more variable than the dependencies of the components of

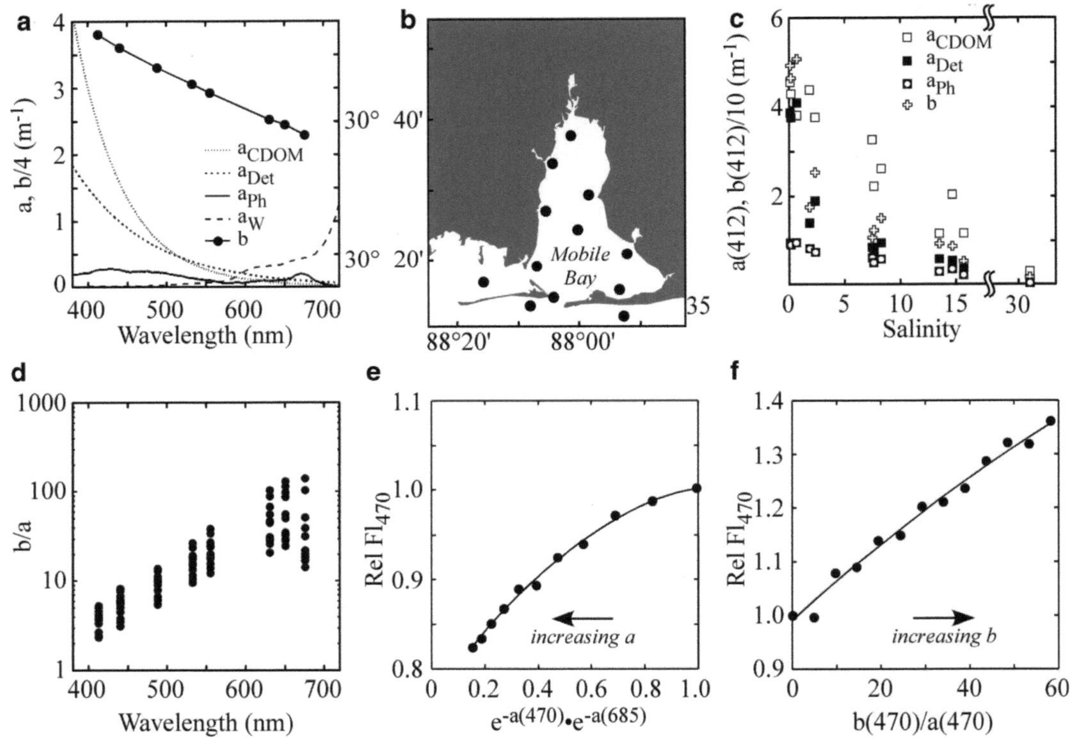

Fig. 13 (**a**) Absorption (a) and scattering (b) spectra for a representative sample from Mobile Bay, USA. Note different scales for a and b. Absorption by CDOM (a_{CDOM}), detritus (a_{Det}) and phytoplankton (a_{Ph}) were measured on material retained on (a_{Det} and a_{Ph}) and passing (a_{CDOM}) a GF/F filter. Absorption enhancement on the filter was corrected using coefficients from Lohrenz (2000). Absorption by water (a_w) is from Pope and Fry (1997) and van Kee (cited by Mueller 2003). Scattering coefficients were estimated using an ac-9 (WETLabs, Philomath, OR), corrected according to Gallegos and Neale (2002). (**b**) Map of Mobile Bay, USA, showing locations of sample sites. Latitude and longitude are °N and °W. (**c**) Variation of absorption and scattering at 412 nm ($a(412)$ and $b(412)$) as a function of salinity at the sample sites. Note different scales for a and b. (**d**) Variation in the ratio of scattering to total absorption (b/a) with wavelength at the sample sites. Note log scale. (**e**) Variation in relative fluorescence of Basic Blue 3 excited at 470 nm with increasing concentrations of CDOM, as a function of the absorption parameter (see text for details). The line is a 2nd-order polynomial fit. (**f**) Variation in relative fluorescence of Basic Blue 3 excited at 470 nm with increasing concentrations of Maalox as a function of b/a (see text for details). The line is a 2nd-order polynomial fit

absorption, as the efficiency of scattering depends on the ratio of particle diameter to wavelength (van de Hulst 1957). The scattering coefficient may therefore increase, decrease or be approximately constant through the visible wavelengths, depending on the size of the particles (Morel 1991; Ahn et al. 1992).

The relative contributions of the materials other than water to absorption and scattering vary with water type. Absorption is dominated by CDOM and detritus in most coastal and estuarine (Case II) waters and dominated by water and phytoplankton in most pelagic (Case I) waters. In the former, there is usually a strong variation with salinity, reflecting terrestrial inputs, as is illustrated with data collected on a single day in Mobile Bay, a semi-tropical estuary on the northern Gulf of Mexico (Fig. 13b, c). Note that absorption at 412 nm by phytoplankton is relatively minor compared to absorption by CDOM and detritus (11–18% of the sum of a_{CDOM} and a_{Det}). The high load of suspended sediments is also manifested in the ratio of scattering to total absorption (the sum of a_{CDOM}, a_{Det}, a_{Ph} and a_{w}), b/a (Fig. 13d). Because the magnitudes of absorption and scattering are not constant, between sites or between wavelengths, they have the potential to bias the SFS in two ways. The first is through direct optical interference in the spectral fluorescence measurements that are compared with the calibration SFS for taxonomic classification. The second is indirect and is related to differences in light absorption and quenching between representatives of different taxa. The spectral characteristics of the underwater light regime are driven by variability in the ratio of chromophores that absorb strongly at short wavelengths (CDOM, detritus and phytoplankton) vs water, which absorbs strongly at long wavelengths. Consequently, the light absorbed by different taxa, which reflects their pigment complement and determines F^{Chl}, will not be the same, either between taxa or between water bodies. We model the magnitude of these in turn, using data collected in Mobile Bay (Fig. 13).

4.1 Bias in SFS by Background Absorption and Scattering

We can investigate the potential interference of background absorption and scattering by modeling optical interference under the conditions at each site, using the SFS generated with the AOA (Fig. 5a). Although the effects of absorption and scattering are interactive, we can treat them independently as a first approximation. The effect of variable absorption in the absence of changes in scattering was tested by measuring fluorescence from solutions of the fluorescing aniline dye Basic Blue 3 (Kopf and Heinze 1984) with additions of purified CDOM (Suwannee River humic acid, International Humic Substances Society). The additions corresponded to the range observed at the Mobile Bay stations, 0–6 m^{-1} at 400 nm (0–4.91 m^{-1} at 412 nm, cf. Fig. 13c). From first principles, we would expect the signal to be attenuated for both the excitation and emission in proportion with an absorption parameter defined as $exp(-a(\lambda_{Ex}))\bullet exp(-a(\lambda_{Em}))$. A full description of the attenuation would require definition of the effective optical path-length(s). In practice these are very difficult to define as the signal is the integral of multiple effective path-lengths and weighted by the response of fluorophores that are closest to the detector (Serôdio 2004). The relationship between fluorescence, expressed relative to fluorescence in the absence of a CDOM addition, and the absorption parameter is shown in Fig. 13e, calculated for 470 nm excitation and 685 emission. Note that the value of the parameter falls as absorption by CDOM increases. The relationship could be fit well with a second-order polynomial ($R^2 = 0.995$). The effect of scattering in the absence of absorption was tested using dilutions of Maalox, a suspension of aluminum hydroxide and magnesium hydroxide particles. The dilutions were made to match the range of scattering observed at 412 nm in the Mobile Bay samples (0–36 m^{-1}, b/a 0 – 58.2; Fig. 13c, d). As anticipated, the path-length enhancement caused relative fluorescence to rise. A second-order polynomial provided a marginally better fit than a first-order (linear) equation (RMS = 0.014 vs 0.015; $R^2 = 0.993$ vs 0.991, Fig. 13f).

The effect of optical interference by background absorption and scattering on the SFS was then modeled by calculating the cumulative effects of absorption (Fig. 13e) and scattering (Fig. 13f) for each excitation wavelength, parameterized with the spectral data collected at each site (Fig. 13a, c and d). The effect of total non-water absorption (the sum of a_{CDOM}, a_{Det} and a_{Ph}) was assumed to follow the same relationship as was observed for the Suwannee River CDOM in the test. The contributions of water were not included as the instrument was calibrated on aqueous suspensions. The effect of scattering was assumed to follow the same relationship as the Maalox, although the particles present in the natural samples will likely include a wider size-range. As both the efficiency and angular distribution of scattering vary with size, the Maalox is likely to be an imperfect substitute for the natural assemblage of particles. If the latter have a mean volume scattering function that is more or less isotropic (i.e., with an angular distribution that is weighted less or more towards the forward angle) than the Maalox, the relationship in Fig. 13f will tend to under- or over-estimate the enhancement of fluorescence, respectively.

Because the contributions of CDOM and detritus to absorption decline more rapidly with wavelength than does scattering (Fig. 13a), the relative reduction in fluorescence due to absorption declines with wavelength more rapidly than does the enhancement due to scattering. In short, the bias is more negative at short than long wavelengths. When the calibration SFS (Fig. 5a) were weighted for the reduction by absorption and enhancement by scattering at each wavelength and re-normalized, the spectral bias was manifested as an anti-clockwise rotation in the Norm SFS between the 470 and 535 nm channels in the eukaryotes and between 525 and 570 nm in the cyanobacterium (Fig. 14a–d). The biased SFS that would be detected from the chlorophyte (*Dunaliella*) have a higher similarity to the cyanobacterium (*Synechococcus*) because the chlorophyte's SFS has the steepest decline between 470 and 570 nm and the steepest rise between 570 and 610 nm (Fig. 5a). This can be seen in the ordination of the similarity matrix (Fig. 14e), and would result in a degree of mis-identification as a mixture of chlorophytes and cyanobacteria. The bias in signal from the chromophyte (*Thalassiosira*) makes it more similar to the cryptophyte (*Rhodomonas*), which has the flattest SFS of the eukaryotes between 470 and 570 (Fig. 5a). In consequence, the biased signals would result in a degree of mis-identification as a mixture of chromophytes and cryptophytes. Because the cryptophyte has the flattest decline between 470 and 570 nm of the three eukaryotes, a further flattening makes it less, not more, like the other eukaryotes. Because its SFS declines between 570 and 610 nm, the rotational bias will make it slightly more like the cyanobacterium (Fig. 14e). The biased signals would result in a degree of mis-identification as a mixture of cryptophytes and cyanobacteria. Only the cyanobacterium would be immune from mis-classification. As it has the SFS with the steepest rise between 570 and 610, the steepening of the rise caused by the rotational bias makes the signal less, not more, like the

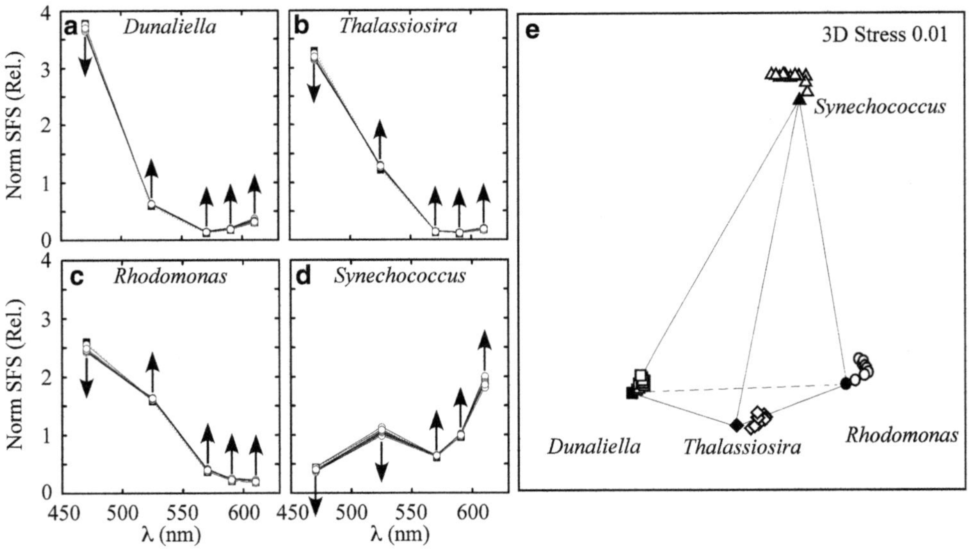

Fig. 14 Changes in the Norm SFS for (**a**) the chlorophyte *Dunaliella tertiolecta,* (**b**) the chromophyte (diatom) *Thalassiosira pseudonana,* (**c**) the cryptophyte *Rhodomonas lens* and (**d**) the cyanobacterium *Synechococcus bacillaris* modeled with varying degrees of optical interference (see text for details). Arrows denote the net direction of change relative to the calibration values at each wavelength. (**e**) Multidimensional scaling plot of the Norm SFS from **a–d** (open symbols) superimposed on the calibration signatures (closed symbols, cf. Fig. 5c)

eukaryotes (Fig. 14e) and the biased signals should cause no change in the taxonomic classification.

4.2 Quenching In Situ and Taxonomic Assessment

The spectral dependence of light penetration through a water column is a function of the relative abundance of the optically-active materials in it. Because representatives of the different taxa differ in the distribution of photosynthetic pigments, the light-absorption by each ($a^{Chl}_{PS}(\lambda_{Ex})$) would differ, depending on whether attenuation were weighted towards long wavelengths (domination by water; Case I waters) or towards the short wavelengths (domination by CDOM and detritus; Case II). Because fluorescence is in turn determined by the rate of light absorption (Fig. 11a), this could be manifested as differences in apparent abundance of each group. We consider this by modeling light attenuation and absorption for the four calibration species used so far. Their specific absorption spectra are shown in Fig. 15a, as is a mean curve for a population containing equal biomass (as Chl*a*) of each. This mean spectrum was used with measurements from the clearest and most turbid samples in Fig. 13 to model downwelling irradiance profiles (Fig. 15b, c). Briefly, incident noon-time spectral irradiance was modeled according Greg and Carder (1990). The model was parameterized with records of air temperature, wind speed, atmospheric pressure and relative humidity from the center of Mobile Bay (http://www.mymobilebay.com/) and with UV and visibility data that were obtained from NOAA's National Weather Service. Spectral attenuation of the incident irradiance was modeled from sun angle (determined from latitude and date/time), and the spectral dependencies of total absorption and scattering (Kirk 1994), as described in more detail by MacIntyre et al. (2004). Total absorption was defined as the sum of a_{CDOM}, a_{Det}, and a_{Ph}, measured for the field samples, and a_W (cf. Fig. 13a). Variation in absorption was driven by differences in the magnitude rather than shape of the spectra: $a_{CDOM}(400)$ and $a_{Det}(400)$ varied by 12*x* and 60*x* (max/min), respectively, while the spectral slopes, σ (Eq. 6), differed by only 1.2*x* and 1.3*x*. Absorption by phytoplankton, $a_{Ph}(\lambda)$, was defined by the product of the mean Chl*a*-specific absorption spectrum (Fig. 15a) and Chl*a* concentration measured at the site. Absorption by water was given by Pope and Fry (1997). For comparison, at the clearest site (Fig. 15b) $a_{CDOM}(400) = 0.41\ m^{-1}$; $a_{Det}(400) = 0.16\ m^{-1}$; Chl*a* = 1.99 mg m^{-3}; at the most turbid site (Fig. 15c) $a_{CDOM}(400) = 3.66\ m^{-1}$; $a_{Det}(400) = 9.72\ m^{-1}$; Chl*a* = 1.05 mg m^{-3}. The water columns were assumed to be vertically homogenous.

The spectral dependency of irradiance is shown at 0, 1, 2, 3, 4 and 5 optical depths (surface and the depths at which irradiance in the visible, 400–700 nm, is attenuated by five successive *e*-foldings, to 36.8, 13.5, 4.98, 1.83 and 0.67% of surface irradiance) in Fig. 15b and c. Note the spectral shift between the subsurface dominance of irradiance at c. 550 nm (green light) at the clear site and the dominance of irradiance at c. 660 nm (red light) at the turbid site. At any depth, the light absorbed by each of the four taxa that constitutes the assemblage can be calculated as:

$$E_{abs} = \sum_{\lambda=400}^{\lambda=700} E_z(\lambda) a^{Chl}(\lambda) \qquad (7)$$

where E_{abs} is Chl*a*-specific light absorption (μmol photons [mg Chl*a*]$^{-1}$ s^{-1}); $E_z(\lambda)$ is spectral irradiance at depth *z* and wavelength λ (μmol photons m^{-2} s^{-1} nm^{-1}); and $a^{Chl}(\lambda)$ is Chl*a*-specific light absorption at wavelength λ (m^2 [mg Chl*a*]$^{-1}$). However, this description includes absorption by non-photosynthetic pigments, which over-estimates the light that can be re-emitted as fluorescence.

Light absorption by photosynthetic pigments was modeled by decomposing $a^{Chl}(\lambda)$ into the contributions of photosynthetic and non-photosynthetic pigments using published estimates of *in vivo* absorption by different pigment classes (Bidigare et al. 1990) and the spectral dependence of the pigment packaging coefficient (Geider et al. 1998), illustrated in Fig. 11d. The approach is an approximation because of variation in the packaging coefficient that is not accounted for in the spectral reconstruction (Johnsen et al. 1997; Bricaud et al. 2004). Absorption by PC in the cyanobacterium (*Synechococcus*) was estimated from the absorption of Phycocyanin 645 in acidic urea (MacColl and Guard-Friar 1983). The photosynthetic absorption spectra, $a^{Chl}_{PS}(\lambda)$, are shown in Fig. 15d. Note the peak at c. 550 nm due to PE in the cryptophyte (*Rhodomonas*) and the peak at c. 645 due to PC in *Synechococcus*. Note also the difference between the chlorophyte (*Dunaliella*) and diatom (*Thalassiosira*) at c. 450–550 nm

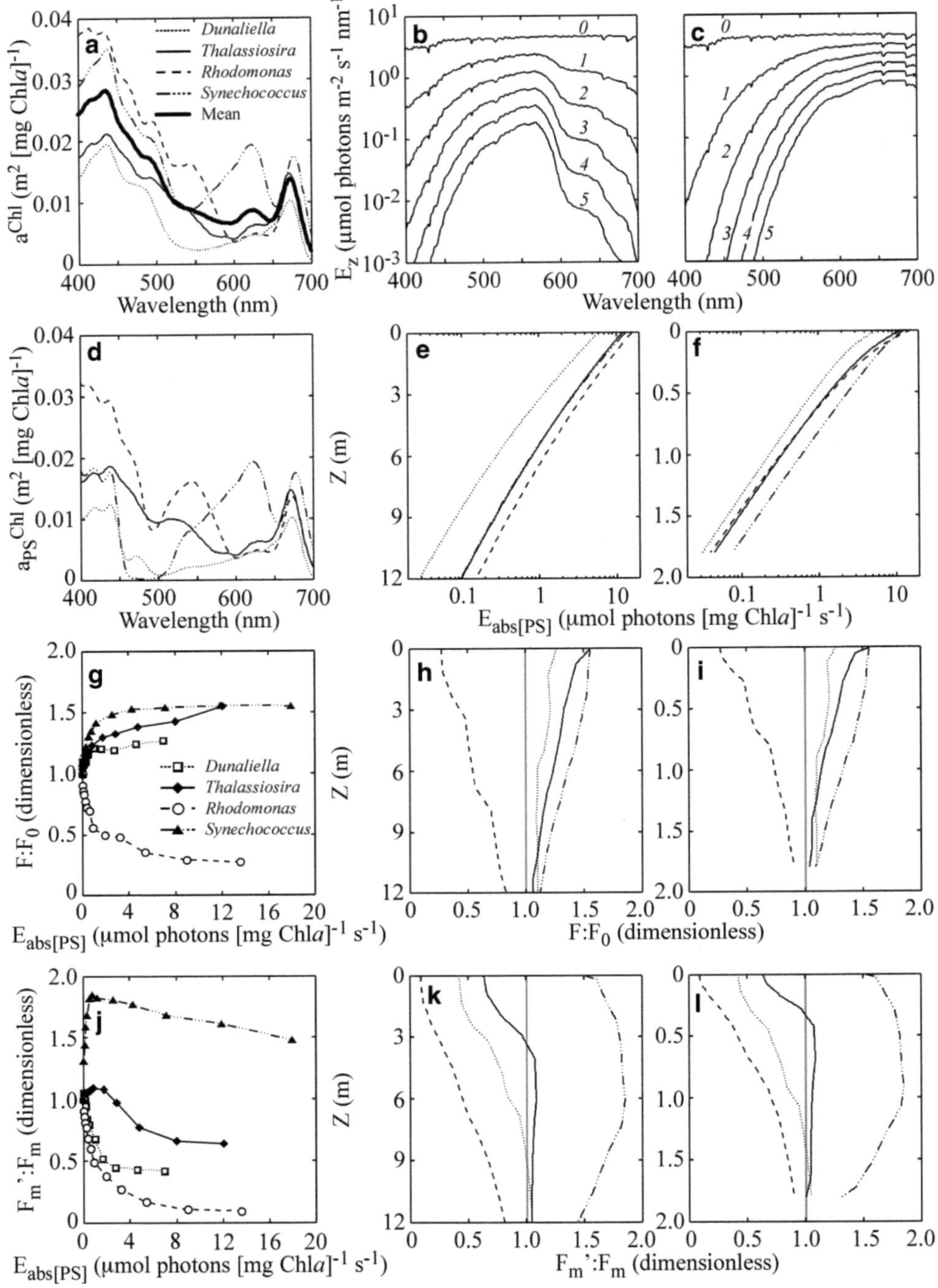

Fig. 15 Effect of submarine irradiance on fluorescence, modeled for the least (**b, e, h** and **k**) and most (**c, f, i** and **l**) turbid samples from Mobile Bay (Fig. 13) See text for details. (**a**) Chl*a*-specific absorption spectra for the chlorophyte *Dunaliella tertiolecta,* the chromophyte (diatom) *Thalassiosira pseudonana,* the cryptophyte *Rhodomonas lens* and the cyanobacterium *Synechococcus bacillaris* (light lines). The heavy line is the mean spectrum for an equal-Chl*a* mixture. The same line patterns are used in **d–l**. (**b, c**) Spectral dependence of irradiance at the surface (0) and 1, 2, 3, 4 and 5 optical depths. (**d**) Modeled Chl*a*-specific photosynthetic absorption spectra for *Dunaliella, Thalassiosira, Rhodomonas* and *Synechococcus*. (**e, f**) Depth-dependence of photosynthetically-useable light, $E_{abs[PS]}$ absorbed by the four taxa. (**g**) Variation in minimum fluorescence, expressed as a ratio to the dark-adapted state ($F:F_0$), as a function of irradiance in the four taxa. Irradiance is expressed as photosynthetically-useable light, $E_{abs[PS]}$. (**h, i**) Depth-dependence of modeled minimum fluoresence, expressed as $F:F_0$ for the four taxa. Vertical grey lines show the true (i.e. dark-adapted) value. (**j**) Variation in maximum fluorescence, expressed as a ratio to the dark-adapted state ($F_m':F_m$), as a function of irradiance in the four taxa. Irradiance is expressed as photosynthetically-useable light, $E_{abs[PS]}$. Symbols are as in (**g**). (**k, l**) Depth-dependence of modeled maximum fluoresence, expressed as $F_m':F_m$ for the four taxa. Vertical grey lines show the true (i.e. dark-adapted) value

due to the respective absence and presence of photosynthetic carotenoids. Substitution of the absorption by photosynthetic pigments, a^{Chl}_{PS}, for a^{Chl} in Eq. 7 allows the depth-distribution of photosynthetic light absorption ($\sum[E(\lambda)\bullet a^{Chl}_{PS}(\lambda)]$ in Eq. 1) to be calculated for each taxon. These are shown in Fig. 15e and f. Light absorption at the clear site, where the subsurface irradiance peaks at wavelengths absorbed efficiently by PE, is highest for the cryptophyte (Fig. 15e). Light absorption at the turbid site, where the subsurface irradiance peaks at wavelengths absorbed efficiently by PC, is highest for the cyanobacterium (Fig. 15f). Absorption is lowest for the chlorophyte at both sites.

We modeled the fluorescence that would be emitted by each taxon under two conditions, where the excitation intensity is low enough not to cause an actinic effect and where it is high enough to saturate fluorescence (F and F_m', respectively). The water column was assumed to be well-mixed and the kinetics of mixing were assumed to be such that the rate of change of irradiance would permit photosynthetic induction, state transitions and xanthophyll cycling but not acclimative pigment regulation (cf. MacIntyre et al. 2000; Raateoja et al. 2009). Critically, we assume that there is a perfect correction for the confounding effects of background attenuation and scattering (see previous section). In both cases, there is no dark-acclimation, so the signal reflects the variability in F^{Chl} in response to ambient irradiance, a reflection of the relative activities of photosynthetic and photoprotective mechanisms (cf. Fig. 11a, b). For the first case (no actinic effect, F), the fluorescence can be estimated in dimensionless terms as the ratio to F_0. The variation in $F{:}F_0$ for the four taxa is shown in Fig. 15g, expressed as a function of irradiance absorbed by photosynthetic pigments (E_{abs}, Eq. 7). Note the difference between the response of *Rhodomonas* and the other taxa. The vertical variation in fluorescence for each species is modeled for the clear and turbid water columns (Fig. 15h and i) from the depth-distributions of E_{abs} and the relationships between $F{:}F_0$ and E_{abs}. The signals are shown in relation to the fully-relaxed condition used for calibration ($F{:}F_0 = 1$). Differences between the water-masses, driven by differences in the light regimes, are minor compared to differences between the taxa. The quenching of F at all irradiances for *Rhodomonas* is the most obvious feature but the differences between the other three taxa are site-dependent. The difference between *Dunalliela* and *Thalassiosira* are more pronounced at the clear site (Fig. 15h) than at the turbid site (Fig. 15i) because the spectral shift towards long wavelengths obviates the differences in light absorption by photosynthetic carotenoids. The differences between *Synechococcus* and *Dunalliela* or *Thalassiosira* are more pronounced at the turbid site because of its enhanced relative absorption by PC (Fig. 15f vs e).

The same exercise was performed with the signals that would be excited with a saturating flash. Variation in $F_m'{:}F_m$ for the four taxa is shown in Fig. 15j and the vertical profiles, calculated as for $F{:}F_0$, are shown for the clear and turbid sites in Fig. 15k and l. The dispersion of responses among the three eukaryotes is greater at the clear site than at the turbid one because of the wider dispersion in light absorption (Fig. 15e *vs* f). The increased difference between *Synechococcus* and the eukaryotes is due to the much higher range in $F_m'{:}F_m$ in the former (Fig. 15j), which reflects in part the state transition that cyanobacteria undergo in darkness (Campbell and Oquist 1996).

The effect of the quenching on taxonomic assessment is shown in Fig. 16. The contribution of each taxon is calculated from the ratio of its fluorescence to total fluorescence. When modeled as above for equal contributions of the four taxa, there is a progressive error due to taxonomic differences in the magnitude of quenching. The error is greatest at high irradiance (i.e. near the surface). Although each taxon contributed 25% of biomass (Chl*a*), the estimated percentages of chlorophytes: chromophytes: cryptophytes: cyanobacteria were 27:33:6:33% when estimated from $F{:}F_0$ and 16:24:3:57% when estimated from $F_m'{:}F_m$ (Fig. 16a, b). The estimates were more accurate with decreasing irradiance, reaching 27:26:20:27 when estimated from $F{:}F_0$ and 24:24:19:33% when estimated from $F_m'{:}F_m$ at 5 optical depths. The estimate of Chl*a* ranged from 116% to 103% when measured from $F{:}F_0$ and from 109 to 67% when estimated from $F_m'{:}F_m$ (Fig. 16c). Distributions modeled for the turbid water site varied by less than a percentage point from these ratios in almost all cases.

The bias in taxonomic weighting and estimates of biomass depends on how representative the fluorescence-irradiance relationships are and on what combination of taxa are used in the model. The quenching observed for *Rhodomonas lens* (Fig. 15g, j), although very different from the responses of the other taxa, is not unique: we have observed similar patterns in the cryptophytes *Storeatula major* and *Proteonomas sulcata*.

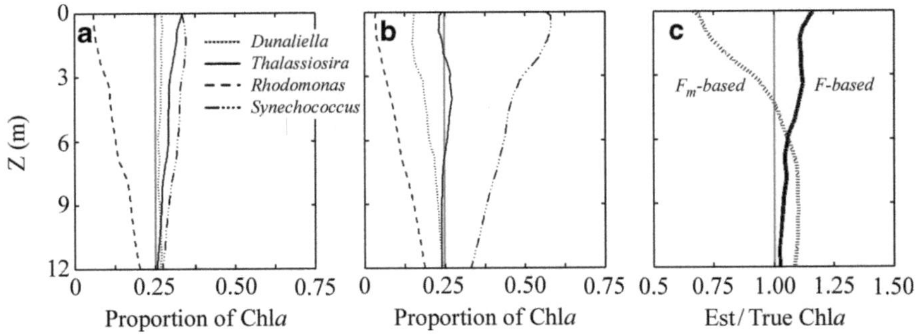

Fig. 16 Modeled error in taxonomic assignment for an equal mixture of the chlorophyte *Dunaliella tertiolecta,* the chromophyte (diatom) *Thalassiosira pseudonana,* the cryptophyte *Rhodomonas lens* and the cyanobacterium *Synechococcus bacillaris*. (See text for details.) (**a, b**) Depth-dependence in the estimated proportion of total Chl*a* assigned to each taxonomic group. The true proportions (vertical grey lines) are all 0.25. Data are based on minimum (**a**) and maximum (**b**) fluorescence. (**c**) Depth-dependence of the error in total Chl*a*, estimated from minimum (F) and maximum (F_m) fluorescence. Chl*a* is expressed as relative to the true value (vertical grey line)

The biases shown in Fig. 16 would be weighted differently, and the bias less severe, if the model were run for a binary mixture of *Thalassiosira* and *Dunaliella*, because of the overall similarity of their fluorescence responses (Fig. 15g, j). The biases would be more severe if the model were run for a binary mixture of *Rhodomonas* and *Synechococcus* because they have the most pronounced differences in quenching responses (Fig. 15g, j).

The significance of these modeling exercises is that:

1. Errors of misidentification can be introduced into taxonomic partitioning if the algorithms are not corrected for biases introduced by other optically-active compounds;
2. Taxon-specific patterns of quenching can confound the taxonomic partitioning, causing in this case a severe underestimate of the contribution of cryptophytes; and
3. The biases in taxonomic classification depend on flash intensity.

Although using a saturating flash has the advantage of increasing the signal:noise ratio, increasing the sensitivity of the method, the errors in taxonomic assignment are more pronounced for the responses measured here. A simple solution is to make measurements at night, when quenching is relaxed (Chekalyuk et al. 2000), or to use a darkened, flow-through system that permits some relaxation of NPQ before the measurement is made (cf. Fig. 12b). While this is easily accommodated for mapping along a cruise-track or on a mooring, it is more difficult for depth-profiling instruments, where the residence time of cells in the interrogation volume can be short enough to bias measurements of the fluorescence ratio F_q/F_v (Schimanski et al. 2006). In the absence of an interrogation that allows F and F_m' to relax towards F_0 and F_m, apparent depth-dependent variations in taxonomic composition and biomass should be treated with caution (Fig. 16). Under the modeled conditions, inter-specific variation in F^{Chl} with irradiance had more of an effect on the taxonomic assessment than did differences in the spectral dependence of irradiance. Note that the model assumed a perfect correction for the biases that would be due to background absorption and scattering, which would otherwise further bias the predicted output.

5 A Field Test of the SFS Approach

The SFS approach has been field-tested in several locales and found to give qualitatively similar results to taxonomic classification based on microscopy or HPLC analysis of pigments (Cowles et al. 1993; Poryvkina et al. 1994; Seppälä and Balode 1998; Gaevsky et al. 2005; Gregor et al. 2005; Jakob et al. 2005; Parésys et al. 2005; Richardson et al. 2010). Robust relationships require both constrained variability in F^{Chl}, hence covariance between fluorescence intensity and Chl*a*, and consistent SFS. However, the past 30 years of research have shown that there is variability in F^{Chl} associated with cell size, taxonomic affiliation and physiological status (re-iterated in Fig. 6c, 7, 10–12) and we have demonstrated that physiologically-driven variability in SFS can confound taxonomic assessment (Fig. 9–10). Further interference can arise from optical interference and irradiance-driven variability in quenching (Fig. 13–16). On the face of it, the observations of qualitative similarities between fluorescence-based and microscopic- or pigment-based taxonomy and variability in F^{Chl} and SFS seem difficult to reconcile.

We suggest that there are three factors that underlie the paradox. The first is that variability in F^{Chl} in natural systems may be lower than the extremes observed in some lab studies. Our observations of F^{Chl} suggest variation of c. 5*x* between species (Fig. 6c) and c. 2–4*x* within species subjected to simultaneous light- and nutrient-stress (Fig. 9). However, there may be less variation in the same species over wide ranges of light- or nutrient-availability when in balanced growth. While the range in F^{Chl} under both nutrient-starved and nutrient-limited growth was similar, c. 2*x*, in the diatom *Thalassiosira pseudonana* (Fig. 7, 9), variability in the chlorophyte *Dunaliella tertiolecta* was 3.7*x* under nutrient starvation but only 1.7*x* under a 9*x* range of nutrient-limitated growth rates (Sosik and Mitchell 1991). We would expect to see the wide ranges characteristic of nutrient starvation under the boom-and-bust dynamics typical of blooms but these are the conditions under which top-down control is reduced. Otherwise, the reduction in growth rate associated with nutrient starvation will result in a loss of biomass, given that biomass accumulation or loss depends on the balance between productivity and grazing. Rapid removal of slowly-dividing, stressed cells by grazing will tend to dampen variation in F^{Chl}, as stressed cells are rapidly removed when the grazing rate exceeds the growth rate.

A second consideration is the synergistic effects of taxonomic affiliation and biomass. In many water-masses, the phytoplankton are dominated by cyanobacteria and/or pico-eukaryotes under low-biomass (oligotrophic) conditions and by eukaryotic nano- or micro-plankton under high-biomass (meso- or eutrophic) conditions (e.g. Bouman et al. 2005; Ras et al. 2008; Uitz et al. 2008). This will enhance the slope of the fluorescence-chlorophyll relationship in bulk measurements of Chl*a* fluorescence excited via chlorophylls or photosynthetic carotenoids but dampen it in bulk measurements of fluorescence excited via phycobilins.

An allied point is the importance of dynamic range. Even if F^{Chl} varies by 2–5*x*, then fluorescence and Chl*a* will almost certainly covary if the variability in Chl*a* approaches an order of magnitude, as can occur down salinity gradients within an estuary or with depth in a stratified water-column.

The third consideration concerns covariance of interfering chromophores. Although the potential for optical inference is highest in estuaries where both CDOM and suspended material reach high levels, the effects of absorption and scattering on signal detection may be largely compensatory. This is illustrated in Fig. 17a and b, which show variation in $a_{CDOM}(400)$ and $a_{Det}(400)$ down the salinity gradient in Mobile Bay. Samples were collected at the sites shown in Fig. 13b at approximately monthly intervals over a year. Although there are multiple sources of CDOM from the different tributaries, there is a significant correlation between CDOM absorption and salinity in each month. Regression slopes for untransformed data vary by 5.6*x* between months but the overall correlation ($R = -0.76$) is still statistically-significant ($p < 0.01$). Similarly, although detrital absorption can be driven by benthic resuspension (which is largely independent of the salinity structure), there are statistically-significant relationships between $a_{Det}(400)$ and salinity in each month and a significant overall correlation ($R = -0.63$, $p < 0.01$). Because the suspended particulate material both scatters and absorbs light, there is a significant relationship between detrital absorption and scattering (not shown). Because both CDOM and detrital absorption covary, there are also significant relationships between total absorption and scattering (Fig. 17c). The overall correlation is 0.89 ($p < 0.01$). Their effects are therefore compensatory (cf Fig. 13e, f), reducing the potential bias on detection of fluorescence signals. Although the assemblage in Mobile Bay was variously dominated by diatoms, chlorophytes, cryptophytes and cyanobacteria in different months, the regression slopes between *F* and Chl*a* (i.e. F^{Chl}) only varied by 2.5*x* between months and fluorescence was significantly correlated with Chl*a* concentration (Fig. 17d; overall $R = 0.86$, $p < 0.01$). This was likely because of the compensatory effects of absorption and scattering and the high dynamic range (Chl*a* concentrations varied 11*x* to 41*x* by month): changes in fluorescence are due primarily to changes in Chl*a* and secondarily to changes in Fl^{Chl}. We would not expect to see similar agreement in locales where the optical interference by absorption is not compensated

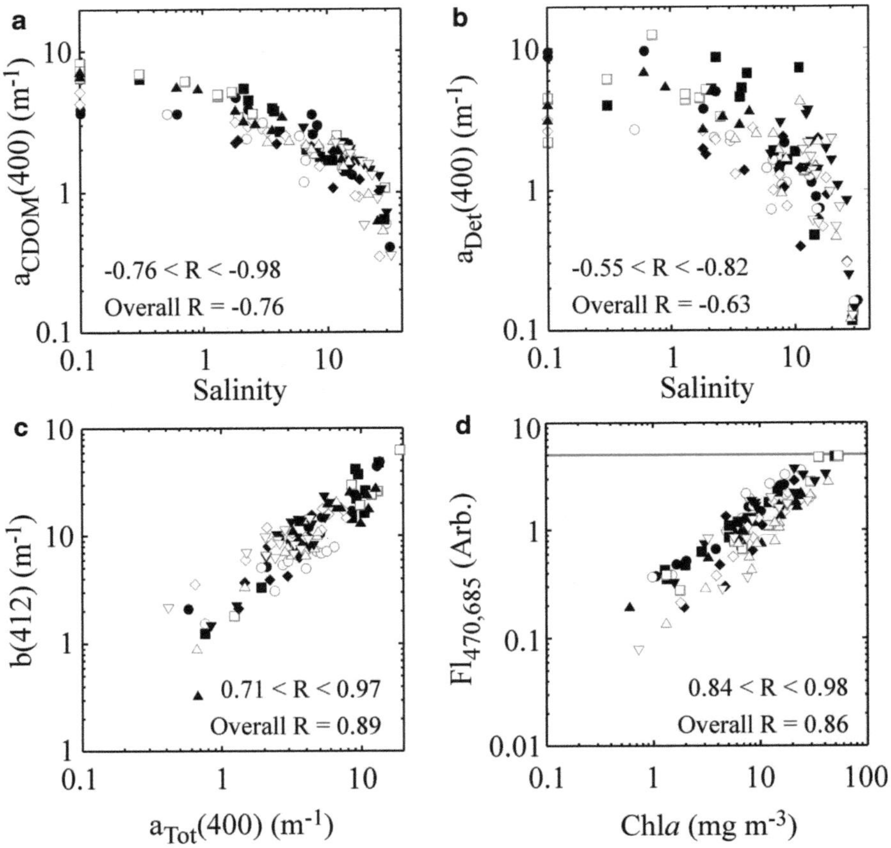

Fig. 17 Variation in optical parameters and Chl*a* in Mobile Bay. Sites (Fig. 13B) were sampled approximately monthly between December, 2004, and October, 2005. Different symbols are for each sample trip. Correlation coefficients are shown for each trip and for the combined dataset (*Overall*). All correlations were statistically significant ($p < 0.01$). (**a**) Variation in absorption by CDOM with salinity. Regression slopes varied by 5.6*x* between trips. (**b**) Variation in absorption by detritus with salinity. Regression slopes varied by 3.9*x* between trips. (**c**) Variation in scattering with total absorption (the sum of $a_{CDOM}(400)$, $a_{Det}(400)$, $a_{Ph}(400)$ and $a_W(400)$). Regression slopes varied by 3.8*x* between trips. (**d**) Variation in fluorescence with Chl*a* concentration. Fluorescence was measured with a SeaPoint Sensors Chl*a* Fluorometer. The grey line indicates sensor saturation. Regression slopes varied by 2.5*x* between trips

for by scattering (e.g. in chromophoric lakes with low seston concentrations), nor where the dynamic range in Chl*a* was small.

Given that there were robust relationships between fluorescence and phytoplankton abundance at the study site, we conducted a field test of the AOA's ability to discriminate between taxa on a transect from Mobile Bay into the Gulf of Mexico. Discrete samples were taken at 11 stations (Fig. 18a) to ground-truth the AOA. Pigment concentrations of material retained on Whatmann GF/F filters were measured by HPLC (Van Heukelem and Thomas 2001) and CDOM absorption was measured on the filtrate. The water was sampled at c. 0.5 m below the surface with a diaphragm pump and flowed through a YSI CT sonde and the AOA. Variations in salinity and temperature are shown in Fig. 18b. The AOA was used with the calibrations derived from laboratory testing (see above), based on nutrient-replete cultures of a representative chlorophyte (*Dunaliella tertiolecta*), chromophyte (the diatom *Thalassiosira pseudonana*), cryptophyte (*Rhodomonas lens*) and the PC-rich cyanobacterium (*Synechoccus bacillaris*). The AOA estimates CDOM concentration from CDOM fluorescence, excited with a UV LED, and uses the data to correct the classification of phytoplankton type and abundance. A further correction is based on transparency in the interrogation cavity, which provides a first-order correction for absorption and scattering (cf. Fig. 13). The AOA's estimates of Chl*a* and CDOM are shown in Fig. 18c. Both rose with proximity to the

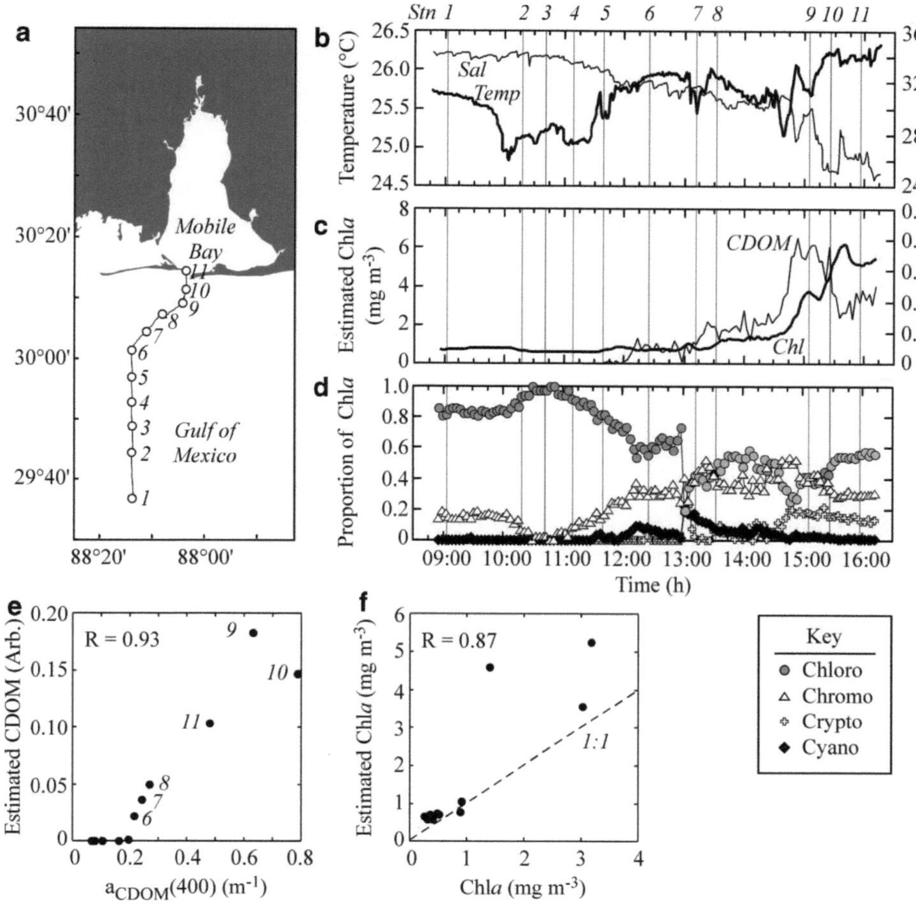

Fig. 18 Variation in physical and optical parameters along a coastal transect. (**a**) Cruise-track offshore from Mobile Bay. Latitude and longitude are in °N and °W. Numbers correspond to stations where discrete samples were taken for ground-truthing. (**b**) Along-track variation in salinity and temperature measured in flow-through mode with a YSI MIDAS CT 6700 sensor. Vertical lines refer to the sampling stations. (**c**) Along-track variation in CDOM and Chl*a* concentrations measured in flow-through mode with *bbe* Moldaenke's AOA. (**d**) Along-track variation in the abundance of four taxonomic groups measured in flow-through mode with *bbe* Moldaenke's AOA. Data are given as the proportion of total estimated Chl*a*. The instrument was calibrated with the chlorophyte *Dunaliella tertiolecta,* the chromophyte (diatom) *Thalassiosira pseudonana,* the cryptophyte *Rhodomonas lens* and the cyanobacterium *Synechococcus bacillaris* (Fig. 5). The key to symbols is below the panel. (**e**) Estimates of CDOM made with the AOA as a function of measured CDOM absorption. Station numbers are indicated beside the data. (**f**) Estimates of Chl*a* made with the AOA as a function of measured concentrations

mouth of Mobile Bay, reflecting the typical variation along the salinity gradient within the bay (cf. Fig 13, 17). Features in the CDOM estimate were coherent with the salinity record, although CDOM was below the AOA's limit of detection at Stations 1–5 (Fig. 18c). Estimates of Chl*a* varied by more than an order of magnitude along the cruise-track and also showed coherence with features in the physical hydrography, although these were less pronounced than the CDOM record.

The AOA's taxonomic classification is shown in Fig. 18d. The data have been normalized to Chl*a* to emphasize the differences in the assigned classification rather than differences in apparent biomass. The classification was dominated by chlorophytes between Stations 1 and 6, with a minor contribution by chromophytes and negligible contributions of cryptophytes and cyanobacteria. Enhanced dominance of the classification by chlorophytes is evident at Stations 2–4, which are hydrographically distinct, based on their temperature regime (Fig. 18a). The classification shows a major discontinuity between Stations 7 and 8, coherent with a disjunction in the salinity record. Stations 7–11 were classified with higher ratios of chromophytes to chlorophytes, and with an increasing contribution of cryptophytes at Stations 9–11. There was a peak in phytoplankton classified as cyanobacteria at Stations 7 and 8, in and landward of the discontinuity in the salinity record.

The AOA's estimates of CDOM and Chl*a* relative to analytical measures on the discrete samples are shown in Fig. 18e and f. The AOA's estimates of CDOM were highly correlated with $a_{CDOM}(400)$ for estimates above the apparent limit of detection of c. 0.2 m^{-1}. Although Pearson's R for these six samples was 0.93, there are obvious differences in the ratio of observed to expected values in the three stations closest to Mobile Bay. These were drawn from hydrographically-distinct samples (Fig. 18b, c) and presumably reflect differences in the quantum yield of CDOM in samples from distinct sources (Del Castillo et al. 1999; Chen et al. 2007). The AOA's estimates of Chl*a* were significantly correlated ($p < 0.05$, $R = 0.87$) with HPLC-based measurements. For the more appropriate correlation between log-transformed data, which conform to the requirement of a normal distribution, $R = 0.91$. There was a 3.7*x* range in the ratio of estimated to observed Chl*a*, suggesting that the correlation was due as much of the 12.5*x* dynamic range in Chl*a* as to the implicit assumption of constant F^{Chl}.

The HPLC data were used to classify the taxonomic structure by MDS ordination and cluster analysis. The pigments that present were chlorophylls *b*, c_2 and c_3, prasinoxanthin, fucoxanthin and its derivatives 19′-butanoyloxyfucoxanthin and 19′-hexanoyloxyfucoxanthin, peridinin, gyroxanthin di-ester, diadinoxanthin and diatoxanthin, alloxanthin and zeaxanthin and the ubiquitous pigments Chl*a* and α- plus β-carotene. Diadinoxanthin and diatoxanthin were grouped, as samples were not filtered rapidly enough to preserve the *in situ* level of de-epoxidation and as the duration of the AOA's interrogation allows some relaxation of NPQ (Fig. 12b). Although Chl*b* was detected at all sites and prasinoxanthin (a marker for a subset of the prasinophytes) was found at all but Stations 2–4, lutein was above the limit of detection in only one sample (Station 9).

Because the objective was to compare the pigment-based community structure with the AOA's classification (i.e., removing the effect of varying biomass, cf. Fig. 18d), concentrations of pigments that could be used to classify phytoplankton at the class level or better (i.e. all but α- plus β-carotene) were expressed as ratios to Chl*a*. Pigments were standardized and square-root transformed to reduce the influence of abundant forms on the analysis. The matrix of pairwise Bray-Curtis similarity coefficients was grouped by cluster analysis. Ordination of the matrix in a MDS plot shows the existence of 3 well-supported groups (Fig. 19a). These were the three stations closest to Mobile Bay (Stations 9–11), 3 stations in a hydrographically-distinct offshore region (Stations 2–4, see Fig. 18b), and the stations in the mid-section of the transect plus that furthest from land (Stations 1 and 5–8). The AOA's classifications were treated in a similar fashion, standardized to Chl*a* and square-root transformed, prior to cluster analysis and ordination. The analysis shows the existence of two dispersed clusters, containing 5 well-supported sub-groups (Fig. 19b). The two main clusters encompass the offshore and nearshore sites (Stations 1–5 vs 6–11). Two of the clusters, A and E (Fig. 19b), correspond to Clusters I (Stations 2–4) and III (Stations 9–11) in the pigment data (Fig. 18b). The remaining cluster in the analysis of pigments, Cluster II, corresponded to Clusters B–D in the AOA data, which were distributed between the offshore and onshore groupings (Fig. 18b).

The mean pigment ratios of Clusters I–III, based on the pigment data, are compared with Clusters A–E, based on the AOA classification in Fig. 19c. Although there are strong trends in zeaxanthin, cyanobacteria were only a minor component of the AOA's classification. This may reflect the relatively high zeaxanthin:Chl*a* ratios in cyanobacteria (e.g., 0.4–1.8 g g^{-1} in *Synechococcus* WH7803, Kana and Glibert 1987; Kana et al. 1988) or might reflect mis-classification of PE-rich forms based on the calibration with a PC-rich isolate. The latter seems unlikely, as PE-rich forms would likely be classified as a mixture of cryptophytes and cyanobacteria (cf. the rhodophyte *Porphyridium cruentum*, Fig. 5) and the estimates of cryptophytes were low. The latter were broadly consistent with the distributions of alloxanthin (Fig. 19c). Alloxanthin is not a particularly useful index for comparison with the AOA's cryptophyte signature, which is weighted heavily by PE, because the PE:alloxanthin ratio is highly variable, as shown earlier for *Rhodomonas lens* (Fig. 8k).

Although the AOA's classifications were dominated by chlorophytes (Fig. 18d, 19c), the accessory pigments contained high levels of chlorophylls and carotenoids associated with multiple classes of chromophytes. The absolute concentrations should be interpreted with

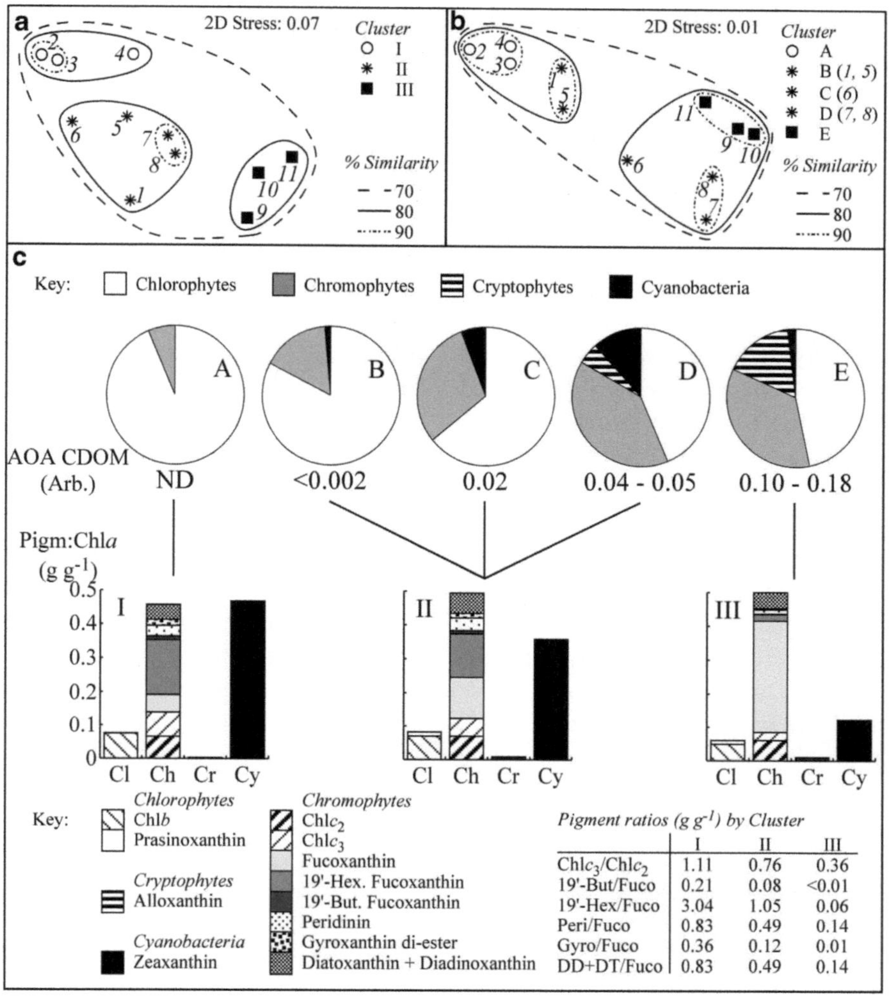

	I	II	III
Chlc3/Chlc2	1.11	0.76	0.36
19'-But/Fuco	0.21	0.08	<0.01
19'-Hex/Fuco	3.04	1.05	0.06
Peri/Fuco	0.83	0.49	0.14
Gyro/Fuco	0.36	0.12	0.01
DD+DT/Fuco	0.83	0.49	0.14

Fig. 19 (**a**) Multidimensional scaling plot of the pigment composition from Stations 1–11 (see Fig. 18a). Data were normalized to Chl*a*, standardized and square-root transformed. Similarities based on cluster analysis are overlaid on the symbols. (**b**) Multidimensional scaling plot of the AOA classification from the same stations. Data were normalized to Chl*a*, standardized and square-root transformed. Similarities based on cluster analysis are overlaid on the symbols. The symbols are those from (**a**). (**c**) Comparison of the mean proportions of taxa in Clusters **a–e** from (**b**), upper row, and the pigment ratios in the corresponding Clusters I – III from (**a**), lower row. The keys are above and below the clusters. The connecting lines indicate correspondence. Columns within each histogram correspond to diagnostic pigments for chlorophytes (Cl), chromophytes (Ch), cryptophytes (Cr) and cyanobacteria (Cy). Ratios of pigments within the chromophytes in Clusters I – III are tabulated at the lower right

caution as the characterisitic ratios to Chl*a* vary widely, but the types of chromophytes which can be inferred from the pigment data could lead to a bias in their classification by the AOA. The presence of Chlc_3, 19′-butanoyloxyfucoxanthin, 19′-hexanoyloxyfucoxanthin and gyroxanthin di-ester are indicative of fucoxanthin-containing dinoflagellates in the family Kareniaceae (Millie et al. 1997), although the pigments are found individually in both the pelagophytes and prymnesiophytes (Bidigare 1989; Jeffrey and Vesk 1997). *Karenia brevis* and *Karlodinium veneficum*, two bloom-forming members of the Kareniaceae, are frequently observed at the study sites. Their presence was not confirmed by microscopy, but the concentrations of gyroxanthin di-ester (<5–17 pg L^{-1}) suggest densities of ca. 60 cells L^{-1} of *Karenia* (Richardson and Pinckney 2004). The presence of peridinin is diagnostic of dinoflagellates (Jeffrey and Vesk 1997). The presence of high fucoxanthin levels relative to the other chromophyte pigments in Cluster III (Fig 19c) is consistent with the dominance by diatoms. As we have shown (Fig. 6b), some chromophytes are classified by the AOA as a mixture of chlorophytes and chromophytes, notably the two prymnesiophytes and two members of the Kareniaceae (*Karenia brevis* and *Karlodinium veneficum*). The apparent dominance by chlorophytes may therefore be due in part to shifts within the chromophytes as well as changes in the abundance of chlorophytes.

We can test this by comparing the structure of the matrices based on the pigment and AOA data, using the BEST test in PRIMER-E (Clarke et al. 2008). The test ranks the similarity coefficients within each matrix and then tests for correlation between the two ranked arrays. (Note that it works from the individual samples, not from aggregated data, so is not biased by groupings defined by the cluster analysis in Fig. 19.) After ranking, it is possible to use permutative testing, creating new similarity matrices of all combinations of single through multiple input parameters. In this case, matrices based on single pigments to combinations of up to 5 pigments were compared with the matrix of AOA classifications. The test was significant at $p = 0.001$, indicating a vanishingly-low probability that the AOA's classification resembled the pigment classification (Fig. 19a, b) through random chance. Examination of the correlations between subsets of the pigment data and the AOA classification (Table 3) shows that the best agreement ($R = 0.802$) was found using five pigments that are distributed between and within the chlorophytes and chromophytes (Chlc_2, Chlc_3, fucoxanthin, 19′-hexanoyloxyfucoxanthin and prasinoxanthin – note that Chl*b* was not one of these and that prasinoxanthin is present only in a subset of the Chl*b*-containing chlorophytes). The high correlation shows that the AOA is sensitive to changes in the community structure but also suggests that the classification is driven in part by shifts within the chromophyte community. The predominance of Chlc_3 and 19′-hexanoyloxyfucoxanthin within the best fits with permutative testing, and the 59x change in the ratio of 19′-hexanoyloxyfucoxanthin to fucoxanthin between Clusters I–III (Fig. 19c) are consistent with varying levels of prymnesiophyes (which may be classified as chlorophyte/chromophyte mixtures, Fig. 6b) vs diatoms driving much of the variability in classification. The higher proportion of prymnesiophytes offshore might therefore be responsible for the higher apparent abundance of chlorophytes in Cluster A (Stations 2–4). Although the ratios of 19′-butanoyloxyfucoxanthin and gyroxanthin

Table 3 Results of BEST test of PRIMER-E (Clarke et al. 2008) comparing pigment ratios, and pigment ratios plus the AOA's estimate of CDOM to the pair-wise similarity matrix of the AOA's taxonomic classification. Both sets of comparisons were significant ($p \leq 0.001$). Permutation testing of successive subsets of the parameters was conducted to isolate those with highest explanatory power, using Spearman's R for the rank correlation method. The parameters with the highest R value for N factors (single and combinations of 2–5) are shown for each test. The combination with the highest R value in each test is in bold. Abbreviations: Prasino, prasinoxanthin, 19′-Hex, 19′-hexanoyloxyfucoxanthin; Fuco, fucoxanthin; DD+DT, sum of diadinoxanthin and diatoxanthin

Inputs	N	R	Parameters
Pigments	1	0.606	Chlc_3
		0.595	19′-Hex
	2	0.749	Prasino,
		0.682	19′-Hex19′-Hex, Fuco
	3	0.785	Chlc_3, Prasino, 19′-Hex
		0.779	Prasino, 19′-Hex, Fuco
	4	0.801	Chlc_2, Chlc_3, Prasino, 19′-Hex
		0.796	Chlc_3, Prasino, 19′-Hex, Fuco
	5	**0.802**	**Chlc_2, Chlc_3, Fuco, Prasino, 19′-Hex**
		0.778	Chlc_3, Prasino, 19′-Hex, Fuco, DDT+DT
Pigments, AOA CDOM	1	0.712	CDOM
		0.606	Chlc_3
	2	0.874	Chlc_2, CDOM
		0.869	Fuco, CDOM
	3	0.855	Prasino, 19′-Hex, CDOM
		0.854	Chlc_2, Fuco, CDOM
	4	**0.885**	**Chlc_2, Prasino, 19′-Hex, CDOM**
		0.876	Chlc_3, Prasino, 19′-Hex, CDOM
	5	0.879	Chlc_2, Chlc_3, Prasino, 19′-Hex, CDOM
		0.872	Chlc_3, Fuco, Prasino, 19′-Hex, CDOM

di-ester to fucoxanthin varied by 51*x* and 28*x* between Clusters I and III (Fig. 19c), neither had any significant correlation with the changes in the AOA's classification (Table 3). It is unlikely therefore that the pattern was due to shifts in the relative abundance of *Karenia* or *Karlodinium*.

Although the comparison of the pigment and classification matrices shows a highly significant relationship and high correlation, it offers little insight into the division of pigment Cluster II between classification Clusters B–D (Fig. 19c). Cursory inspection of the classification clusters shows that the most significant division, between Clusters A, B *vs* C–E, follows the gradient in CDOM, which was below the AOA's limit of detection in the former but detectable in the latter. The division of pigment Cluster II between classification Clusters B–D also follows the gradient in CDOM, with progressively lower proportions of chromophytes occurring as CDOM increases. The trend might reflect covariance, as a trend in decreasing nutrient availability and increasing water clarity as CDOM declines with distance from land. However, the lab-based tests predict that this would result in identification of higher proportions of chlorophytes and lower proportions of cryptophytes offshore as cryptophytes become progressively more nutrient- and light-stressed (Fig. 10c). This is the opposite of what we observe in Clusters B–D. Instead, the data are consistent with the effects of optical interference. The AOA classifications are corrected for interference by CDOM but we did not calibrate the instrument with local CDOM. Under-correction for CDOM interference would cause an anti-clockwise rotation in the Norm SFS (Fig. 14a–d) and cause an under-estimate of chlorophytes and an over-estimate of cyanobacteria and cryptophytes. This is consistent with the trends in these taxa with CDOM between Clusters B and D (Fig. 19c). Under-correction could cause an increase in chromophytes through mis-classification of chlorophytes (be they true chlorophytes or inaccurately assigned chromophytes) or an underestimate through mis-classification of chromophytes (Fig. 14e). The increasing trend in chromophytes with CDOM suggests the former is predominant. The trends evident within Clusters B–D are consistent with lower and higher interference by CDOM in Clusters A and E (Fig. 19c), although the pigment ratios are different.

We can test for the effect of the CDOM correction by including the AOA's estimate of CDOM in the pigment matrix and repeating the permutative tests. The permutative tests show that the estimate of CDOM has higher explanatory power as a single factor than any pigment (Table 3) and is consistently present in the highest-ranked combinations of 2–5 input parameters. When it is included, the correlation coefficients between the [pigment + CDOM] matrix and the AOA classification matrix are consistently higher than the coefficients for the pigment matrix alone (Table 3). The optimum occurs with 4 parameters: the ratios of Chlc_2, prasinoxanthin, and 19′-hexanoyloxyfucoxanthin to Chl*a*, and CDOM. Although the most highly-ranked pigments are consistent between both sets of comparisons, the correlation is higher when CDOM is included: R = 0.885 *vs* 0.802. The best correlation occurred with 4 inputs, rather than 5, indicating that forcing inclusion of the variance structure associated with the fifth variable reduced the goodness of fit. The correlation with 5 inputs was still higher than when CDOM was not included. The results indicate that the bias reflects inadequate correction for the CDOM interference. We did not calibrate the instrument with locally-occurring CDOM and infer that the classification was under-corrected, and that shifts in the classification within Cluster II, as well as between Clusters I–III, reflect incomplete correction for the effects of background optical interference recognition of the SFS.

These patterns are broadly consistent with differences in classification by SFS vs pigment matrices reported by See et al. (2005), who also conducted a study in the Gulf of Mexico, using a Fluoroprobe (*bbe* Moldaenke). This instrument differs from the AOA in having an open interrogation volume, and would therefore be more susceptible to variation in F^{Chl} due to actinic irradiance when used *in situ* (Fig. 15–16). See et al. (2005) used the Fluoroprobe on shipboard, using a ca. 1-min interrogation of surface samples that were shrouded to prevent actinic interference, so it is not clear to what degree quenching would have relaxed. It was calibrated on a slightly different suite of phytoplankton than we used (the diatom *Thalassiosira weissflogii* rather than *T. pseudonana*, and the cyanobacterium *Synechococcus sp.* rather than *S. bacillaris*). Pigments were used to construct taxonomic classifications using CHEMTAX (Mackey et al. 1996). The results were similar to those we report here: the SFS-based classifications showed dominance by chlorophytes (57–81%), with substantial contributions from cryptophytes (13–20%), while the pigment-based classification showed dominance by diatoms and dinoflagellates (61–95%), with negligible contributions by cryptophytes (0–2%) except at one station (25%). Cyanobacteria were not detected with the Fluoroprobe and were estimated at only 0–6% of the population by CHEMTAX. It is not clear if the *Synechococcus sp.* used to calibrate the Fluoroprobe was a PE- or PC-rich strain but the complete absence of cyanobacteria from the SFS classification suggests that they were not responsible for the overestimate of cryptophytes. We would predict classification as mixed cyanobacteria/cryptophytes if the instrument were calibrated with PC-rich strain of *Synechococcus* and challenged with PE-rich strains (cf. *Porphyridium*, Fig. 5b and c) and mixed cyanobacteria/chlorophytes if it were calibrated with a PE-rich strain and challenged with PC-rich ones. (This assumption is based on re-alignment of the calibration tetrahedron in Fig. 5b so that the cyanobacterial apex is defined by PE-rich *Porphyridium* and the PC-rich *Synechoccocus* becomes the unknown sample.) The results reported by See and co-workers (See et al. 2005) are consistent with chromophytes that were mis-classified as either chlorophyte/chromophyte or chromophyte/cryptophyte mixtures (cf. Fig. 6a, b). As two of the five stations occupied were adjacent to estuaries, there may also have been interference by CDOM causing some fraction of chromophytes to be classified as cryptophytes.

6 Conclusion

Taxonomic classification by SFS is subject to bias in the signatures from inter- and intra-specific variation in the shape of signatures and in F^{Chl}, and in the relationship between fluorescence intensity and Chl*a*. These observations are well-recognized, as noted by Poryvkina et al. (2000), "all attempts to provide universal calibration of phytoplankton fluorescence have been doomed to failure". Beyond variability in the shape and magnitude of the SFS, the effect of interfering chromophores (CDOM and other materials that absorb or scatter light) must also be considered. Even so, like other researchers using the method, we find that relationships between fluorescence and Chl*a* are robust, provided that the dynamic range of the signal is large enough, and that there is qualitative agreement between classification by SFS and other analytical methods. We have demonstrated in field tests that, although the accuracy of identification from SFS is subject to (somewhat predictable) bias, it is sensitive to changes in community structure: the inaccuracy is inherent in the use of unique and invariant classification algorithms, not in the fluorescence method *per se*. Because it is capable of monitoring communities continuously and at relatively minor cost, the SFS approach is well-suited as a trigger for adaptive sampling, as a low-level form of screening that can optimize the use of more accurate but more expensive or laborious analyses. In closing, we reiterate the observation made by John Cullen, an early advocate for the use of fluorescence in studying phytoplankton (and a graduate advisor to both HM and TR), that, "we may not know exactly what we are measuring, but the patterns observed are too strong to ignore" (Cullen and Renger 1979).

Acknowledgements We thank Geir Johnsen and an anonymous reviewer for comments that improved this work. We thank Adrienne Stutes, Andy Canion, Alison Rellinger, Preston Kendrick and Emily Goldman for assistance in field sampling and lab work. HLM was supported by the US National Atmospheric and Oceanographic Administration's Cooperative Institute for Coastal and Estuarine Environmental Technology (Grant number NA06NOS4190167). HLM acknowledges support for supplementary fieldwork from the US EPA (Grant numbers X-93190401 and R-83065101-1-6) and US Department of Commerce (Grant number NA17FZ2602-A3-08), administered through the Alabama Center for Estuarine Studies and the Alabama Oyster Reef Restoration Program. TLR acknowledges support from the US National Science Foundation (Grant numbers OCE06234001 and CBET0606940) and the South Carolina Sea Grant Consortium (Grant number P/M-2J-V410).

References

Aberle N, Beutler M, Moldaenke C, Wiltshire KH (2006) 'Spectral fingerprinting' for specific algal groups on sediments *in situ*: a new sensor. Arch Hydrobiol 167:575–592

Ahn Y-H, Bricaud A, Morel A (1992) Light backscattering efficiency and related properties of some phytoplankters. Deep Sea Res 39:1835–1855

Allen JF (2002) Plastoquinone redox control of chloroplast thylakoid protein phosphorylation and distribution of excitation energy between photosystems: discovery, background, implications. Photosynth Res 73:139–148

Allen JF, Mullineaux CW (2004) Probing the mechanism of state transitions in oxygenic photosynthesis by chlorophyll fluorescence spectroscopy, kinetics and imaging. In: Papageorgiou GC, Govindjee (eds) Chlorophyll *a* fluorescence: a signature of photosynthesis. Springer, Dordrecht, pp 447–461

Alpine AE, Cloern JE (1985) Differences in *in vivo* fluorescence yield between three phytoplankton size classes. J Plankt Res 7:381–390

Andersen RA, Bidigare RR, Keller MD, Latasa M (1996) A comparison of HPLC pigment signatures and electron microscopic observations for oligotrophic waters of the North Atlantic and Pacific Oceans. Deep Sea Res 43:517–537

Anning T, Harris G, Geider RJ (2001) Thermal acclimation in the marine diatom *Chaetoceros calcitrans* (Bacillariophyceae). Eur J Phycol 36:233–241

Apt KE, Collier JL, Grossman AR (1995) Evolution of the phycobiliproteins. J Mol Biol 248:79–96

Arsalane W, Rousseau B, Duval J-C (1994) Influence of the pool size of the xanthophyll cycle on the effects of light stress in a diatom: competition between photoprotection and photoinhibition. Biochem Photobiol 60:237–243

Babichenko S, Leeben A, Poryvkina L, van der Wagt R, de Vos F (1994) Fluorescent screening of phytoplankton and organic compounds in sea water. J Environ Monit 2:378–383

Babichenko S, Kaitala S, Leeben A, Poryvkina L, Seppala J (1999) Phytoplankton pigments and dissolved organic matter distribution in the Gulf of Riga. J Mar Syst 23:69–82

Babichenko S, Leeben A, Poryvkina L, van der Wagt R, de Voss F (2000) Fluorescent screening of phytoplankton and organic compounds in sea water. J Environ Monit 2:378–383

Babin M (2008) Phytoplankton fluorescence: theory, current literature and *in situ* measurement. In: Babin M, Roesler CS, Cullen JJ (eds) Real-time coastal observing systems for ecosystem dynamics and harmful algal blooms. UNESCO Publishing, Paris, pp 237–280

Beeler SooHoo J, Kiefer DA, Collins DJ, McDermid IS (1986) *In vivo* fluorescence excitation and absorption spectra of marine phytoplankton: I. Taxonomic characteristics and responses to photoadaptation. Plankton Res 8:197–214

Benson EE, Cobb AH (1981) Pigment/protein complexes of the intertidal alga *Codium fragile* (Suringar) Hariot. New Phytol 88:627–632

Beutler M, Wiltshire KH, Meyer B, Moldaenke C, Lüring C, Meyerhöfer M, Hansen U-P, Dau H (2002) A fluorometric method for the differentiation of algal populations *in vivo* and *in situ*. Photosynth Res 72:39–53

Beutler M, Wiltshire KH, Arp M, Kruse J, Reineke C, Moldaenke C, Hansen U-P (2003) Reduced model of the fluorescence from the cyanobacterial photosynthetic apparatus designed for the *in situ* detection of cyanobacteria. Biochim Biophys Acta 1604:33–46

Beutler M, Wiltshire KH, Reineke C, Hansen U-P (2004) Algorithms and practical fluorescence models of the photosynthetic apparatus of red cyanobacteria and Cryptophyta designed for the fluorescence detection of red cyanobacteria and cryptophytes. Aquat Microb Ecol 35:115–129

Bidigare RR (1989) Photosynthetic pigment composition of the brown tide alga: unique chlorophyll and carotenoid derivatives. In: Cosper EM, Bricelj VM, Carpenter EJ (eds) Novel phytoplankton blooms: causes and impacts of recurrent brown tides and other unusual blooms. Springer, Berlin, pp 57–76

Bidigare RR, Ondrusek ME, Morrow JH, Kiefer DA (1990) *In vivo* absorption properties of algal pigments. SPIE 1302:290–302

Bodemer U (2004) Variability of phycobiliproteins in cyanobacteria detected by delayed fluorescence excitation spectroscopy and its relevance for determination of phytoplankton composition of natural water samples. J Plankton Res 26:1147–1162

Bouman H, Platt T, Sathyendranath S, Stuart V (2005) Dependence of light-saturated photosynthesis on temperature and community structure. Deep Sea Res 52:1284–1299

Bricaud A, Claustre H, Ras J, Oubelkheir K (2004) Natural variability of phytoplanktonic absorption in oceanic waters: Influence of the size structure of algal populations. J Geophys Res 109:C11010

Buiteveld H, Hakvoort JHH, Donze M (1994) The optical properties of pure water. SPIE, Ocean Optics XIII

Burger-Wiersma T, Post AF (1989) Functional analysis of the photosynthetic appratus of *Prochlorothrix hollandica* (Prochlorales), a Chlorophyll *b* containing procaryote. Plant Physiol 91:770–774

Butler WL, Hopkins DW (1970a) An analysis of fourth derivative spectra. Photochem Photobiol 12:451–456

Butler WL, Hopkins DW (1970b) Higher derivative analysis of complex absorption spectra. Photochem Photobiol 12:439–450

Campbell D, Oquist G (1996) Predicting light acclimation in cyanobacteria from nonphotochemical quenching of photosystem II fluorescence, which reflects state transitions in these organisms. Plant Physiol 111:1293–1298

Casper-Lindley C, Björkman O (1998) Fluorescence quenching in four unicellular algae with different light-harvesting and xanthophyll-cycle pigments. Photosynth Res 56:277–289

Wit CDv-D, Doust AB, van Stokkurn LHM, Dekker JP, Wilk KE, Curmi PMG, van Grondelle R (2008) Phycocyanin sensitizes both photosystem I and photosystem II in cryptophyte *Chroomonas* CCMP270 cells. Biophys J 94:2423–2433

Chekalyuk AM, Hafez M (2008) Advanced laser fluorometry of natural aquatic environments. Limnol Oceanogr Methods 6: 591–609

Chekalyuk AM, Hoge FE, Wright CW, Swift RN, Yungel JK (2000) Airborne test of laser pump-and-probe technique for assessment of phytoplankton photochemical characteristics. Photosynth Res 66:45–56

Chen ZQ, Hu CM, Comny RN, Muller-Karger F, Swarzenski P (2007) Colored dissolved organic matter in Tampa Bay, Florida. Mar Chem 104:98–109

Chisholm SW, Frankel SL, Goericke R, Olson RJ, Palenik B, Waterbury JB, West-Johnsrud L, Zettler ER (1992) *Prochlorococcus marinus* nov. gen. nov. sp.: An oxytrophic procaryote containing divinyl-chlorophyll b. Arch Microbiol 157:297–300

Clarke KR, Warwicke RM (2001) Change in marine communities: an approach to statistical analysis and interpretation. Plymouth, PRIMER-E

Clarke KR, Somerfield PJ, Gorley RN (2008) Testing of null hypotheses in exploratory community analyses: similarity profiles and biota-environment linkage. J Exp Mar Biol Ecol 366:56–69

Collier JL, Herbert SK, Fork DC, Grossman AR (1993) Changes in the cynaobacterial photosynthetic apparatus during acclimation to macronutrient deprivation. Photosynth Res 42:173–183

Cowles TJ, Desiderio RA, Neuer S (1993) *In situ* characterization of phytoplankton from vertical profiles of fluorescence emission spectra. Mar Biol 115:217–222

Cullen JJ, Renger EH (1979) Continuous measurement of the DCMU-induced fluorescence response of natural phytoplankton populations. Mar Biol 53:13–20

Cullen JJ, Yang X, MacIntyre HL (1992) Nutrient limitation and marine photosynthesis. In: Falkowski PG, Woodhead AD (eds) Primary productivity and biogeochemical cycles in the sea. Plenum Press, New York, pp 69–88

Del Castillo CE, Coble PG, Morell JM, López JM, Corredor JE (1999) Analysis of the optical properties of the Orinoco River plume by absorption and fluorescence spectroscopy. Mar Chem 66:35–51

Demmig-Adams B (1990) Carotenoids and photoprotection in plants: a role for xanthophyll zeaxanthin. Biochim Biophys Acta 1020:1–24

Desiderio RA, Moore CC, Lantz C, Cowles TJ (1997) Multiple excitation fluorometer for *in situ* oceanographic applications. Appl Opt 36:1289–1296

Dickinson DE, Bearman G, Tille S, Lansford R, Fraser SE (2001) Multi-spectral imaging and linear unmixing add a whole new dimension to laser scanning fluorescence microscopy. BioTechniques 31:1272–1276

Falkowski PG, Raven JA (2007) Aquatic photosynthesis. Blackwell Science, Malden

Falkowski PG, Kolber Z, Fujita Y (1988) Effects of redox state on the dynamics of Photosystem II during steady-state photosynthesis in eucaryotic algae. Biochim Biophys Acta 933:432–443

Fawley MW, Lee CM (1990) Pigment composition of the scaly green flagellate *Mesostigma viridae* (Micromonadophyceae) is similar to that of the siphonous green alga *Bryopsis plumosa* (Ulvophyceae). J Phycol 26:666–670

Frank HA, Chynwat V, Desamero RZB, Farhoosh R, Erikson J, Bautista J (1997) On the photophysics and photochemical propoerties of carotenoids and their role as light-haresting pigments in photosynthesis. PureAppl Chem 69:2117–2124

Gaevsky NA, Kolmakov VI, Anishchenko OV, Gorbaneva TB (2005) Using DCMU-fluorescence method for the identification of dominant phytoplankton groups. J Appl Phycol 17:483–494

Gallegos CL, Neale PJ (2002) Partitioning spectral absorption in case 2 waters: discrimination of dissolved and particulate components. Appl Opt 41:4220–4233

Gantt E (1996) Pigment protein complexes and the concept of the photosynthetic unit: chlorophyll complexes and phycobilisomes. Photosynth Res 48:47–53

Gantt E, Lipschultz CA, Grabowski J, Zimmerman BK (1979) Phycobilisomes from blue-green and red algae. Plant Physiol 63:615–620

Garcia-Mendoza E, Matthijs HCP, Schubert H, Mur LR (2002) Non-photochemical quenching of chlorophyll fluorescence in *Chlorella fusca* acclimated to constant and dynamic light conditions. Photosynth Res 74:303–315

Geider RJ, MacIntyre HL, Kana TM (1997) Dynamic model of phytoplankton growth and acclimation: responses of the balanced growth rate and the chlorophyll *a*: carbon ratio to light, nutrient-limitation and temperature. Mar Ecol Prog Ser 148:187–200

Geider RJ, MacIntyre HL, Graziano LM, McKay RML (1998a) Responses of the photosynthetic apparatus of *Dunaliella teriolecta* (Chlorophyceae) to nitrogen and phosphorus limitation. Eur J Phycol 33:315–332

Geider RJ, MacIntyre HL, Kana TM (1998b) A dynamic regulatory model of phytoplanktonic acclimation to light, nutrients and temperature. Limnol Oceanogr 43:679–694

Genty B, Briantais J-M, Baker NR (1989) The relationship between the quantum yield of photosynthetic electron transport and quenching of chlorophyll fluorescence. Biochim Biophys Acta 990:87–92

Genty B, Harbinson J, Briatais J-M, Baker NR (1990) The relationship between non-photochemical quenching of chlorophyll fluorescence and the rate of photosystem 2 photochemistry in leaves. Photosynth Res 25:249–257

Gerhardt V, Balode M (1998) Delayed fluorescence excitation spectroscopy: a method for automatic determination of phytoplankton composition of freshwaters and sediments (interstitial) and of algal composition of benthos. Limnologica 28:313–322

Goericke R, Montoya JP (1998) Estimating the contribution of microalgal taxa to chlorophyll *a* in the field – variations of pigment ratios under nutrient- and light-limited growth. Mar Ecol Prog Ser 169:97–112

Goericke R, Welschmeyer NA (1992a) Pigment turnover in the marine diatom *Thalassiosira weissflogii*. I. The $^{14}CO_2$-labeling kinetics of chlorophyll *a*. J Phycol 28:498–507

Goericke R, Welschmeyer NA (1992b) Pigment turnover in the marine diatom *Thalassiosira weissflogii*. II. The $^{14}CO_2$-lableling kinetics of carotenoids. J Phycol 28:507–517

Goss R, Lepetit B, Wilhelm C (2006) Evidence for a rebinding of antheraxanthin to the light-harvesting complex during the epoxidation reaction of the violaxanthin cycle. J Plant Physiol 163:585–590

Gregg WW, Carder KL (1990) A simple spectral solar irradiance model for cloudless maritime atmospheres. Limnol Oceanogr 35:1657–1675

Gregor J, Geris R, Marsálek B, Hetesa J, Marvan P (2005) *In situ* quantification of phytoplankton in reservoirs using a submersible spectrofluorometer. Hydrobiologia 548:141–151

Grossman AR, Schefer MR, Chiang GG, Collier JL (1993) The responses of cyanobacteria to environmental conditions: light and nutrients. In: Bryant DA (ed) The molecular biology of cyanobacteria. Kluwer, The Netherlands, pp 641–675

Hammer A, Schumann R, Schubert H (2002) Light and temperature acclimation of *Rhodomonas salina* (Cryptophyceae): photosynthetic performance. Aquat Microb Ecol 29:287–296

Haxo FT, Blinks LR (1950) Photosynthetic action spectra of marine algae. J Gen Physiol 33:389–422

Henriksen P, Riemann B, Kaas H, Sorensen H, Sorensen H (2002) Effects of nutrient-limitation and irradiance on marine phytoplankton pigments. J Plankt Res 24:835–858

Hilton J, Rigg E, Jaworski G (1988) *In vivo* algal fluorescence, spectral change due to light intensity changes and the automatic characterization of algae. Freshw Biol 21:375–382

Hilton J, Rigg E, Jaworski G (1989) Algal identification using *in vivo* fluorescence spectra. J Plankt Res 11:65–74

Hoef-Emden K (2008) Molecular phylogeny of phycocyanin-containing cryptophytes: Evolution of biliproteins and geographical distribution. J Phycol 44:985–993

Hooks CE, Bidigare RR, Keller MD, Guillard RRL (1988) Coccoid eukaryotic ultraplankters with four different HPLC pigment signatures. J Phycol 24:571–580

Itoh S, Sugiura K (2004) Fluorescence of photosystem I. In: Papageorgiou GC, Govindjee (eds) Chlorophyll *a* fluorescence: a signature of photosynthesis. Springer, Dordrecht, pp 231–250

Jackson DA, Somers KM (1991) The spectre of 'spurious' correlations. Oecologia 86:147–151

Jakob T, Schreiber U, Kirschesch V, Langner U, Wilhelm C (2005) Estimation of chlorophyll content and daily primary production of the major algal groups by means of multiwavelength-excitation PAM chlorophyll fluorometry: performance and methodological limits. Photosynth Res 83:343–361

Jeffrey SW, Vesk M (1997) Introduction to marine phytoplankton and their pigment signatures. In: Jeffrey SW, Mantoura RFC, Wright SW (eds) Phytoplankton pigments in oceanography: guidelines to modern methods. UNESCO Publishing, Paris, pp 37–84

Jeffrey SW, Mantoura RFC, Wright SW (1997) Phytoplankton pigments in oceanography: guidelines to modern methods. UNESCO Publishing, Rome

Johnsen G, Sakshaug E (2007) Biooptical characteristics of PSII and PSI in 33 species (13 pigment groups) of marine phytoplankton, and the relevance for pulse-amplitude-modulated and fast-repetition-rate fluorometry. J Phycol 43:1236–1251

Johnsen G, Samset O, Granskog L, Sakshaug E (1994) *In-vivo* absorption characteristics in 10 classes of bloom-forming phytoplankton – taxonomic characteristics and responses to photoadaptation by means of discriminant and HPLC analysis. Mar Ecol Prog.Ser 105:149–157

Johnsen G, Prézelin BB, Jovine RVM (1997) Fluorescence excitation spectra and light utilization in two red tide dinoflagellates. Limnol Oceanogr 42:1166–1177

Johnson Z, Barber RT (2003) The low-light reduction in the quantum yield of photosynthesis: potential errors and biases when calculating the maximum quantum yield. Photosynth Res 75:85–95

Jovine RVM, Johnsen G, Prézelin BB (1995) Isolation of membrane bound light-haryesting-complexes from the dinoflagellates *Heterocapsa pygmaea* and *Prorocentrum minimum*. Photosynth Res 44:127–138

Kana TM, Glibert PM (1987) Effect of irradiances up to 2000 μE m^{-2} s^{-1} on marine *Synechococcus* WH7803 – I. Growth, pigmentation, and cell composition. Deep Sea Res 34:479–495

Kana TM, Glibert PM, Goericke R, Welschmeyer NA (1988) Zeaxanthin and β-carotene in *Synechococcus* WH7803 respond differently to irradiance. Limnol Oceanogr 33:1623–1627

Kiefer DA (1973a) Chlorophyll *a* fluorescence on marine centric diatoms: responses of chloroplasts to light and nutrient stress. Mar Biol 23:39–46

Kiefer DA (1973b) Fluorescence properties of natural phytoplankton populations. Mar Biol 22:263–269

Kieleck C, Bousquet B, Le Brun G, Cariou J, Lotrian J (2001) Laser induced fluorescence imaging: application to groups of macroalgae identification. J Phys D: Appl Phys 34:2561–2571

Kirk JTO (1994) Light and photosynthesis in aquatic ecosystems. Cambridge University Press, Cambridge

Koblizek M, Kaftan D, Nedbal L (2001) On the relationship between the non-photochemical quenching of the chlorophyll fluorescence and the Photosystem II light harvesting efficiency. A repetitive flash fluorescence induction study. Photosynth Res 68:141–152

Kolber Z, Zehr J, Falkowski P (1988) Effects of growth irradiance and nitrogen limitation on photosynthetic energy conversion in photosystem II. Plant Physiol 88:923–929

Kolbowski J, Schreiber U (1995) Computer-controlled phytoplankton analyzer based on 4-wavelengths PAM chlorophyll fluorometer. In: Mathis P (ed) Photosynthesis: from light to biosphere, vol V. Kluwer, The Netherlands, pp 825–828

Kopf U, Heinze J (1984) 2, 7-bis(diehylamino)phenazoxonium chloride as a quantum counter for emission measurements between 240 and 700 nm. Anal Chem 56:1931–1935

Krause GH, Weis E (1991) Chlorophyll fluorescence and photosynthesis: the basics. Ann Rev Plant Physiol Plant Mol Biol 42:313–349

Kromkamp JC, Forster RM (2003) The use of variable fluorescence measurements in aquatic ecosystems: differences between multiple and single turnover measuring protocols and suggested terminology. Eur J Phycol 38:103–112

Kruse O (2001) Light-induced short-term adaptation mechanisms under redox control in the PS II-LHCII supercomplex: LHC II state transitions and PSII repair cycle. Naturwissenschaften 88:284–292

Kruskopf M, Flynn KJ (2006) Chlorophyll content and fluorescence responses cannot be used to gauge reliably phytoplankton biomass, nutrient status or growth rate. New Phytol 169:525–536

Laney SR (2003) Assessing the error in photosynthetic properties determined with Fast Repetition Rate fluorometry. Limnol Oceanogr 48:2234–2242

Laney SR, Letelier R (2008) Artifacts in measurements of chlorophyll fluorescence transients, with specific application to fast repetition rate fluorometry. Limnol Oceanogr Methods 6:40–50

Latasa M (1995) Pigment composition of *Heterocapsa* sp. and *Thalassiosira weissflogii* growing in batch cultures under different irradiances. Sci Mar 59:25–37

Latasa M, Berdalet E (1994) Effect of nitrogen or phosphorus starvation on pigment composition of cultured *Heterocapsa* sp. J Plankt Res 16:83–94

Lavaud J, Strzepek RF, Kroth PG (2007) Photoproprotection capacity differs among diatoms: Possible consequences on the spatial distribution of diatoms related to fluctuations in the underwater light climate. Limnol Oceanogr 52:1188–1194

Laws EA, Bannister TT (1980) Nutrient- and light-limited growth of *Thalassiosira fluviatilis* in continuous culture, with implications for phytoplankton growth in the sea. Limnol Oceanogr 25:457–473

Lewitus AJ, Caron DA (1990) Relative effects of nitrogen or phosphorus depletion and light intensity on the pigmentation, chemical composition, and volume of *Pyrenomonas salina* (Cryptophyceae). Mar Ecol Prog Ser 61:171–181

Loftus ME, Seliger HH (1975) Some limitations of the *in vivo* fluorescence technique. Chesapeake Science 16:79–92

Lohrenz SE (2000) A novel theoretical approach to correct for pathlength amplification and variable sampling loading in measurements of particulate spectral absorption by the quantitative filter technique. J Plankt Res 22:639–657

Lorenzen CJ (1966) A method for the continuous measurement of *in vivo* chlorophyll concentration. Deep Sea Res 13:223–227

Lutz VA, Sathyendranath S, Head EJH, Li WKW (2001) Changes in the *in vivo* absorption and fluorescence excitation spectra with growth irradiance in three species of phytoplankton. J Plankt Res23:555–569

MacColl R, Guard-Friar D (1983) Phycocyanin 645. The chromophore assay of Phycocyanin 645 from the cryptomonad protozoa *Chroomonas* species. J Biol Chem 258:14327–14329

MacIntyre HL, Cullen JJ (2005) Using cultures to investigate the physiological ecology of microalgae. In: Andersen RA (ed) Algal Cultureing Techniques. Amsterdam, Elsevier, pp 287–326

MacIntyre HL, Kana TM, Geider RJ (2000) The effect of water motion on short-term rates of photosynthesis by marine phytoplankton. Trends Plant Sci 5:12–17

MacIntyre HL, Kana TM, Anning T, Geider RJ (2002) Photoacclimation of photosynthesis irradiance response curves and photosynthetic pigments in microalgae and cyanobacteria. J Phycol 38:17–38

MacIntyre HL, Lomas MW, Cornwell JC, Suggett DJ, Koch EW, Gobler CJ, Kana TM (2004) Mediation of benthic-pelagic coupling by microphytobenthos: An energy- and material-based model for initiation of blooms of *Aureococcus anophagefferens*. Harmful Algae 3:403–438

Mackey MD, Mackey DJ, Higgins HW, Wright SW (1996) CHEMTAX – a program for estimating class abundances from chemical markers: application to HPLC measurements of phytoplankton. Mar Ecol Prog Ser 144:265–283

Mazel CH (1995) Spectral measurements of fluorescence emission in Caribbean cnidarians. Mar Ecol Prog Ser 120:185–191

McLeod GC (1958) Delayed light action spectra of several algae in visible and ultraviolet light. J Gen Physiol 42:243–250

Millie DF, Om S, Kirkpatrick GJ, Johnsen G, Tester PA, Vinyard BT (1997) Detection of harmful algal blooms using photopigments and absorption signatures: a case study of the Florida red tide dinoflagellate, *Gymnodinium breve*. Limnol Oceanogr 42:1240–1251

Millie DF, Schofield OME, Kirkpatrick GJ, Johnsen G, Evens TJ (2002) Using absorbance and fluorescence spectra to discriminate microalgae. Eur J Phycol 37:313–322

Mimuro M (2005) Visualization of excitation energy transfer processes in plants and algae. In: Govindjee, Beatty JT, Gest H, Allen JF (eds) Discoveries in photosynthesis. Springer, Dordrecht, pp 171–176

Moberg L, Karlber B, Sørensen K, Källqvist T (2002) Assessment of phytoplankton class abundance using absorption spectra and chemometrics. Talanta 56:153–160

Moisan TA, Mitchell BG (1999) Photophysiological acclimation of *Phaeocystis antarctica* Karsten under light limitation. Limnol Oceanogr 44:247–258

Moore LR, Goericke R, Chisholm SW (1995) "Comparative physiology of *Synechococcus* and *Prochlorococcus*: influence of light and temperature on growth, pigments, fluorescence and absorptive properties. Mar Ecol Prog Ser 116:259–275

Morel A (1991) Optics of marine particles and marine optics. In: Demers S (ed.) Particle Analysis in Oceanography. NATO ANSI Series G 27:141–188

Mueller JL (2003) Ocean Optics Protocols for Satellite Ocean Color Sensor Validation, Revision 4, Volume IV, Err. 1. Ocean Optics Protocols for Satellite Ocean Color Sensor Validation, Revision 4. Fargion GS, Mueller JL . Goddard Space Flight Space Center, Greenbelt, Maryland, National Aeronautical and Space Administration. NASA/TM—2003-211621/Rev4-Vol.IV (ERRATUM 1)

Murphy A, Cowles T (1997) Effects of darkness on multi-excitation *in vivo* fluorescence and survival in a marine diatom. Limnol Oceanogr 42:1444–1453

Neidhardt J, Benemann JR, Zhang L, Melis A (1998) Photosystem-II repair and chloroplast recovery from irradiance stress: relationship between chronic photoinhibition, light-harvesting chlorophyll antenna size and photosynthetic productivity in *Dunaliella salina* (green algae). Photosynth Res 56:175–184

Neori A, Vernet M, Holm-Hansen O, Haxo FT (1988) Comparison of chlorophyll far-red fluorescence excitation spectra with photosynthetic oxygen action spectra for photosystem II in algae. Mar Ecol Prog Ser 44:297–302

Olaizola M, Yamamoto HY (1994) Short-term response of the diadinoxanthin cycle and fluorescence yield to high irradiance in *Chaetoceros muelleri* (Bacillariophyceae). J Phycol 30:606–612

Oldham PB, Zillioux EJ, Warner IM (1985) Spectral "fingerprinting" of phytoplankton populations by two-dimensional fluorescence and Fourier-transform-based pattern recognition. J Plankton Res 43:893–906

Owens T (1986) Light-harvesting function in the diatom *Phaeodactylum tricornutum*. II. Distribution of excitation energy between the photosystems. Plant Physiol 80:739–746

Parésys G, Rigart C, Rousseau B, Wong AWM, Fan F, Barbier J-P, Lavaud J (2005) Quantitative and qualitative evaluation of phytoplankton communities by trichromatic chlorophyll fluorescence excitation with special focus on cyanobacteria. Wat Res 39:911–921

Parkhill JP, Maillet G, Cullen JJ (2001) Fluorescence-based maximal quantum yield for PSII as a diagnostic of nutrient stress. J Phycol 37:517–529

Pope RM, Fry ES (1997) Absorption spectrum (380–700 nm) of pure water. II. Integrating cavity measurements. Appl Opt 36:8710–8723

Poryvkina L, Babichenko S, Kaitala S, Kurosa H, Shalapyonok A (1994) Spectral fluorescent signatures in the characterization of phytoplankton community composition. J Plankton Res 16:1315–1327

Poryvkina L, Babichenko S, Leeben A (2000) Analysis of phytoplankton pigments by excitation spectra of fluorescence. Proceedings of EARSeL-SIG-Workshop LIDAR, Dresden/FRG

Raateoja M, Mitchell BG, Wang H, Olivo E (2009) Effect of water column light gradient on phytoplankton fluorescence transients. Mar Ecol Prog Ser 376:85–101

Ras J, Claustre H, Uitz J (2008) Spatial variability of phytoplankton pigment distributions in the Subtropical South Pacific Ocean: comparison between *in situ* and predicted data. Biogeosciences 5:353–369

Richardson TL, Lawrenz E, Pinckney JL, Guajardo RC, Walker EA, Paerl HW, MacIntyre HL (2010) Spectral fluorometric characterization of phytoplankton community composition using the Algae Online Analyser. Water Research (in press)

Richardson TL, Pinckney JL (2004) Monitoring of the toxic dinoflagellate *Karenia brevis* using gyroxanthin-based detection methods. J Appl Phycol 16:315–328

Roesler CS, Perry MJ (1995) *In situ* phytoplankton absorption, fluorescence emission, and particulate backscattering spectra determined from reflectance. J Geophys Res 100(C7):13279–13294

Ross ON, Moore CM, Suggett DJ, MacIntyre HL, Geider RJ (2008) A model of photosynthesis and photo-protection based on reaction center damage and repair. Limnol Oceanogr 53:1835–1852

Ruban AV, Berera R, Ilioaia C, van Stokkum IHM, Kennis JTM, Pascal AA, van Amerongen H, Robert B, Horton H, van Grondelle R (2007) Identification of a mechanism of photoprotective energy dissipation in higher plants. Nature 450:475–478

Sahay A, Jajoo A, Singh P, Bharti S (2006) Nitrite regulates distribution of excitation energy between the two photosystems by causing state transition. Plant Physiol Biochem 44:7–12

Sakshaug E, Demers S, Yentsch CS (1987) *Thalassiosira oceanica* and *Thalassiosira pseudonana*: two different photoadaptational responses. Mar Ecol Prog Ser 41:275–282

Schimanski J, Beutler M, Moldaenke C, Hansen U-P (2006) A model for correcting the fluorescence signal from a free-falling depth profiler. Water Res 40:1616–1626

Schubert H, Forster RM, Sagert S (1995) *In situ* measurement of state transition in cyanobacterial blooms: kinetics and extent of the state change in relation to underwater light and vertical mixing. Mar Ecol Prog Ser 128:99–108

Sciandra A, Lazzara L, Claustre H, Babin M (2000) Responses of growth rate, pigment composition and optical properties of *Cryptomonas* sp. to light and nitrogen stresses. Mar Ecol Prog Ser 201:107–120

See JH, Richardson TL, Pinckney JL, Shen RJ, Guinasso NL (2005) Combining new technologies for determination of phytoplankton community structure in the northern Gulf of Mexico. J Phycol 41:305–310

Seppälä J, Balode M (1998) The use of spectral fluorescence methods to detect changes in the phytoplankton community. Hydrobiologia 363:207–217

Seppälä J, Olli K (2008) Multivariate analysis of phytoplankton spectral *in vivo* fluorescence: estimation of phytoplankton biomass during a mesocosm study in the Baltic Sea. Mar Ecol Prog Ser 370:69–85

Seppälä J, Ylöstalo P, Kaitala S, Hällfors S, Raateoja M, Maunula P (2007) Ship-of-opportunity based phycocyanin fluorescence monitoring of the filamentous cyanobacteria bloom dynamics in the Baltic Sea. Est Coast Shelf Sci 73:489–500

Serôdio J (2004) Analysis of variable chlorophyll fluorescence in microphytobenthos assemblages: implications of the use of depth-integrated measurements. Mar Ecol Prog Ser 36:137–152

Six C, Thomas JC, Brahamsha B, Lemoine Y, Partensky F (2004) Photophysiology of the marine cyanobacterium *Synechococcus* sp. WH8102, a new model organism. Aquat Microb Ecol 35:17–29

Sosik HM (1988) Analysis of chlorophyll fluorescence in marine phytoplankton: interpretation of flow cytometric signals. Cambridge, MA, Massachussetts Institute of Technology: 88

Sosik HM, Mitchell BG (1991) Absorption, fluorescence, and quantum yield for growth in nitrogen-limited *Dunaliella tertiolecta*. Limnol Oceanogr 36:910–921

Sosik HM, Mitchell BG (1994) Effects of temperature on growth, light absorption, and quantum yield in *Dunaliella tertiolecta* (Chlorophyceae). J Phycol 30:833–840

Spear-Bernstein L, Miller KR (1989) Unique location of the phycobiliprotein light-harvesting pigment in the Cryptophyceae. J Phycol 25:412–419

Staehr PA, Cullen JJ (2003) Detection of *Karenia mikimotoi* by spectral absorption signatures. J Plankton Res 25:1237–1249

Staehr PA, Henriksen P, Markager S (2002) Photoacclimation of four marine phytoplankton species to irradiance and nutrient availability. Mar Ecol Prog Ser 238:47–59

Stolte W, Kraay GW, Noordeloos AAM, Rieman R (2000) Genetic and physiological variation in pigment composition of *Emiliania huxleyi* (Prymnesiophyceae) and the potential use of its pigment ratios as a quantitative physiological marker. J Phycol 36:529–539

Strickland JDH (1968) Continuous measurement of *in vivo* chlorophyll: A precautionary note. Deep Sea Res 15:225–227

Strzepek RF, Harrison PJ (2004) Photosynthetic architecture differs in coastal and oceanic diatoms. Nature 431:689–692

Suggett DJ, MacIntyre HL, Geider RJ (2004) Biophysical and optical determinations of light absorption by photosystem II in phytoplankton. Limnol Oceanogr Methods 2:316–332

Suggett DJ, MacIntyre HL, Kana TM, Geider RJ (2009) Comparing electron transport with gas exchange: parameterising exchange rates between alternative photosynthetic currencies for eukaryotic phytoplankton. Aquat Microb Ecol 56:147–162

Sukenik A, Bennett J, Falkowski PG (1987) Light-saturated photosynthesis – limitation by electron transport or carbon fixation? Biochim Biophys Acta 891:205–215

Sukenik A, Bennett J, Mortain-Bertrand A, Falkowski PG (1990) Adaptation of the photosynthetic apparatus to irradiance in *Dunaliella tertiolecta*. Plant Physiol 92:891–898

Topinka JA, Korjeff Bellows W, Yentsch CS (1990) Characterization of marine macroalgae by fluorescence signatures. Int J Rem Sens 11:2329–2335

Uitz J, HuotY BF, Babin M (2008) Relating phytoplankton photophysiological properties to community structure on large scales. Limnol Oceanogr 53:614–630

van de Hulst HC (1957) Scattering by small particles. Wiley, New York, 470 p

Van Heukelem L, Thomas CS (2001) Computer-assisted high-performance liquid chromatography method development with applications to the isolation and analysis of phytoplankton pigments. J Chromatogr 910:1–49

Verity PG (1981) Effects of temperature, irradiance, and day-length on the marine diatom *Leptocylindrus danicus* Cleve. I. Photosynthesis and cellular composition. J Exp Mar Biol Ecol 55:79–91

Vincent WF (1979) Mechanisms of rapid photosynthetic adaptation in natural phytoplankton communities. I. Redistribution of excitation energy between photosystems I and II. J Phycol 15:429–434

Vincent WF (1980) Mechanisms of rapid photosynthetic adaptation in natural phytoplankton communities. II. Changes in photochemical capacity as measured by DCMU-induced chlorophyll fluorescence. J Phycol 16:568–577

Wright SW, Jeffrey SW, Mantoura RFC, Llewellyn CA, Bjørnland T, Repeta DJ, Welschmeyer NA (1991) Improved HPLC method for the analysis of chlorophylls and carotenoids from marine phytoplankton. Mar Ecol Prog Ser 77:183–196

Yamamoto HY, Bassi R (1996) Carotenoids: localization and function. In: Ort DR, Yocum CF (eds) Oxygenic photosynthesis: the light reactions. Kluwer, The Netherlands, pp 539–563

Yentsch CS, Phinney DA (1985) Spectral fluorescence – an ataxonomic tool for studying the structure of phytoplankton populations. J Plankt Res 7:617–632

Yentsch CS, Yentsch CM (1979) Fluorescence spectral signatures – characterization of phytoplankton populations by the use of excitation and emission-spectra. J Mar Res 37:471–483

Zhang QQ, Lei SH, Wang XL, Zhua CJ (2006) Discrimination of phytoplankton classes using characteristic spectra of 3D fluorescence spectra. Spectrochim Acta Part A 63:361–369

Zhu M, Yang X, Cullen JJ (1992) The study of fluorescence characteristics and biochemical composition of a marine diatom *Thalassiosira pseudonana* 3H in light and dark cycles. Acta Oceanol Sin 12:457–464

Chapter 8
Flow Cytometry in Phytoplankton Research

Heidi M. Sosik, Robert J. Olson, and E. Virginia Armbrust

1 Introduction

The capability of flow cytometry for rapid quantitative observations of individual cells has made it an important tool in plankton research since the mid-1980s. One indication of the impact of flow cytometry has been the appearance of special journal issues focused on applications in aquatic sciences (Cytometry, Yentsch and Horan 1989; Scientia Marina, Reckermann and Colijn 2000; Cytometry, Courties and Troussellier 2001). Since these publications, major advances have been made in in situ applications and in new approaches enabled by molecular and genomic techniques. In this chapter we provide a brief overview of the history of flow cytometry in plankton research, summarize basic measurement principles and review specific applications, and conclude with an assessment of some emerging areas and future prospects. We stress uses that depend on inherent fluorescence characteristics, with some related reference to uses involving fluorescent markers, stains or probes. Because a number of previous reviews and methodology guides are available (e.g., Chisholm et al. 1986; Legendre and Yentsch 1989; Yentsch and Horan 1989; Burkhill and Mantoura 1990; Olson et al. 1991, 1993; Troussellier et al. 1993; Collier and Campbell 1999; Collier 2000; Veldhuis and Kraay 2000; Vives-Rego et al. 2000; Campbell 2001; Legendre et al. 2001; Marie et al. 2005), our primary focus will be on recent and forward-looking approaches.

H.M. Sosik(✉) and R.J. Olson
Biology Department, Woods Hole Oceanographic Institution, Woods Hole, MA, 02543, USA
e-mail: hsosik@whoi.edu

E.V. Armbrust
School of Oceanography, University of Washington, Seattle, WA, 98105, USA

2 Background and Historical Perspective

In essence flow cytometers measure light scattered and fluoresced from individual particles as they pass through a region of focused excitation light. Typical configurations involve particles carried in a fluid stream (e.g., seawater for marine samples) in the center of a particle free sheath which is flowing at high speed (~1–10 m s^{-1}) perpendicular to the illumination beam. Hydrodynamic focusing leads to particles flowing in single file as they intersect the beam. Typical sample analysis rates of commercially available instruments are on the order of 0.1 mL min^{-1}, well suited for natural concentrations of picoplankton and small nanoplankton (~10^4–10^6 mL^{-1}). Light scattering and fluorescence originating from each particle are measured with detectors (typically photomultiplier tubes) positioned around the sensing region, such that lenses and/or detector geometry define the collection angles. For those interested in more detail, Shapiro (2003) provides a thorough consideration of flow cytometry measurement and analysis principles.

The advantages of flow cytometry for analysis of individual cells and similar-sized particles in environmental samples are multi-fold and this potential was

D.J. Suggett et al. (eds.), *Chlorophyll a Fluorescence in Aquatic Sciences: Methods and Applications*, Developments in Applied Phycology 4, DOI 10.1007/978-90-481-9268-7_8, © Springer Science+Business Media B.V. 2011

recognized in the late 1970s. The first concrete steps to apply flow cytometry for plankton studies occurred in the early 1980s when it was adapted from successful applications in biomedical research (where it also remains an established tool). Olson et al. (1983) developed an inexpensive system for analyzing phytoplankton around the same time that a group of marine researchers met to evaluate the utility of an existing commercial system (Yentsch et al. 1983). Within a few years, flow cytometers were being operated on board oceanographic research vessels (Olson et al. 1985; Li 1989), leading to important new insights especially regarding spatial and temporal variability in picophytoplankton communities. Without doubt the most spectacular contribution from this early period was the discovery and characterization of *Prochlorococcus* (Chisholm et al. 1988; Olson et al. 1990a; Chisholm et al. 1992), previously unrecognized and now regarded as both the smallest photosynthetic organism and among the most abundant organisms on the planet (Partensky et al. 1999; Coleman and Chisholm 2007).

Biomedical uses of flow cytometry usually involve treatment of cells with fluorescent dyes or probes prior to analysis, but the most common oceanographic applications entail measurement of untreated phytoplankton cells that naturally exhibit fluorescence associated with their photosynthetic pigments (principally chlorophyll a and some phycobiliproteins; most accessory pigments do not produce measurable fluorescence in vivo). For typical phytoplankton studies, flow cytometers are configured with blue or blue-green excitation wavelengths and measurements include red fluorescence (associated with chlorophyll a) and orange fluorescence (associated with phycoerythrin), plus light scattering from a range of near forward angles and a range of side angles (i.e., near right angle from incident). Variations include addition of green fluorescence, detection of polarized scattered light, and use of dual excitation beams (Olson et al. 1988, 1989). More details can be found in a number of methods reviews focused on plankton applications (e.g., Olson et al. 1993; Campbell 2001; Marie et al. 2005).

Presence of chlorophyll fluorescence makes it possible to discriminate phytoplankton from other particles and flow cytometric analysis permits enumeration, quantification of cell properties such as size and pigmentation, and some level of taxonomic, optical and/or sized-based discrimination (e.g., *Synechococcus*, *Prochlorococcus*, picoeukaryotes, coccolithorids) (e.g., Olson et al. 1989, 1990b). As with any optical measurement, flow cytometric signals reflect complex interactions of light with the target cells and quantitative interpretation in terms of cellular properties has been an on-going subject of study (e.g., Neale et al. 1989; Sosik et al. 1989; DuRand et al. 2002; Green et al. 2003b). Nonetheless, relative signals can be readily exploited to explore abundance patterns, and capabilities for examining spatial and temporal variability in natural phytoplankton communities were greatly advanced once methods were developed to store samples with adequate preservation of flow cytometric signals (Vaulot et al. 1989; Lepesteur et al. 1993; Marie et al. 2005).

Commercial flow cytometers are generally optimized for biomedical or clinical requirements with relatively narrow ranges of signal sizes and high particle concentrations. Consequently, many oceanographic uses require modifications (e.g., custom sample pumps, dual sheath, dual photomultipliers) to increase sample volume throughput and dynamic range of signals. These challenges have been addressed both in custom systems (e.g., Dubelaar et al. 1989; Olson and Sosik 2007) and through modifications to commercial flow cytometers (e.g., Cavender-Bares et al. 1998; Green et al. 2003a; Zubkov and Burkhill 2006). In addition, because commercial instruments often lack the sensitivity to quantify signals from particles such as picophytoplankton (especially those growing in high light environments that lead to low cellular pigment levels) and viruses, customized optics and high laser power configurations have been adapted (e.g., Olson et al. 1990a; Dusenberry and Frankel 1994).

With analyses that include use of stains or probes, flow cytometry has been used in aquatic studies for assessments of nucleic acid content, cell cycle status and growth rate (e.g., Chisholm et al. 1986; Vaulot et al. 1995; Marie et al. 1997), cell viability (e.g., Jochem 2000; Grégori et al. 2001; Veldhuis et al. 2001), and enumeration of other microbes such as heterotrophic bacteria, viruses, and protozoa (e.g., Marie et al. 1997, 1999; Li and Dickie 2001; Rose et al. 2004; Zubkov et al. 2007). Light scattering properties for particles with low or unmeasurable fluorescence has also been used to characterize non-living detrital and mineral particles (Green et al. 2003a, b).

In specially designed flow cytometers, it is possible to physically sort particles out of the sample stream on the basis of their light scattering and fluorescence characteristics. This capability has been critical for the success of flow cytometry in aquatic sciences for several reasons. First, it has permitted verification of how optical signatures (combinations of light scattering and fluorescence characteristics) are interpreted; particles associated with specific optical characteristics can be examined with other techniques such as microscopy and pigment analysis to confirm identity (e.g., Olson et al. 1990b; Goericke and Repeta 1992). Second, because sorted cells can remain viable, this technique has enabled isolation of new cell types into culture (e.g., Reckermann 2000), enabling further details of their physiology and genetics to be examined. Finally, sorting enables a variety of biochemical, physiological and genetic analyses on specific populations extracted from natural assemblages. To date, these analyses have included such varied characteristics as carbon fixation (Chisholm et al. 1988; Li 1994), uptake of various nitrogen forms (Lipschultz 1995; Zubkov et al. 2003), and genetic diversity (Wallner et al. 1997; Urbach and Chisholm 1998).

3 Select Research Applications

Here we highlight a few research areas where flow cytometry has made fundamental contributions to understanding sources and consequences of variability in natural plankton communities. These examples were selected to encompass the breadth of topics from microbial ecology to ocean optics and approaches from comparatively routine shipboard analysis to cell sorting to specialized signal analysis.

3.1 Picophytoplankton Community Structure and Dynamics

The most productive use of flow cytometry in aquatic science has been the study of marine picophytoplankton communities. For these tiny (< 2 μm) fluorescent organisms, flow cytometry is equal to or better than conventional light microscopy in its specificity for identification of cell types, and at the same time it is much more rapid and quantitative not only with respect to cell abundance but also individual cell properties. Several types of picophytoplankton can be enumerated with epifluorescence microscopy, although this technique is not reliable for *Prochlorococcus* in natural samples due to their dim and unstable fluorescence (Chisholm et al. 1988). Once flow cytometers are modified for enhanced sensitivity to study *Prochlorochoccus*, they are extremely capable tools and have greatly expanded studies of picoplankton community structure and dynamics. Flow cytometry's advantages are less for larger phytoplankton, especially microplankton, for which morphological characteristics visible with light microscopy tend to be much better discriminators than integrated cell optical properties (but see below on emerging approaches).

Use of flow cytometry to characterize cells according to size and pigmentation has led to new understanding of horizontal, vertical and temporal variability in a range of picophytoplankton taxa, as well as in their physiological adaptations and ecological roles. It is now well established that *Synechococcus* and *Prochlorococcus* are the numerically dominant phototrophs in open ocean (and even some coastal) environments, with *Synechococcus* exhibiting a somewhat wider (extending poleward) latitudinal range (Olson et al. 1988; Olson et al. 1990a; Campbell et al. 1994, 1997; Partensky et al. 1999; DuRand et al. 2001; Agusti 2004). Flow cytometric measurements have provided insights into vertical and temporal differentiation between open ocean *Synechococcus* and *Prochlorococcus*, with the former tending to be relatively more important in the upper euphotic zone and seasonally during winter or early spring (e.g., Campbell et al. 1997; DuRand et al. 2001).

Beyond abundance patterns, the quantitative single-cell measurements provided by flow cytometry have provided new insights into relationships between particular species and their environments. For example, measurements of individual cell fluorescence often reveal photoacclimation to depth-related changes in light intensity (e.g., Dusenberry et al. 2000), with deep cells more highly pigmented than surface cells. This phenomenon (combined with knowledge of its kinetics) allowed *Prochlorococcus* cells to be used as dynamic tracers for water column mixing, by examining distributions of fluorescence (normalized to light scattering to remove effects of growth) in a population over time (Dusenberry et al. 2001). Flow cytometry has proved integral in documenting high growth rates, impact of grazers, and contributions to primary production (Li 1994; Vaulot et al. 1995; Liu et al. 1997, 1998;

Jacquet et al. 1998; Shalapyonok et al. 1998, 2001; Sosik et al. 2003; Worden and Binder 2003a; Pel et al. 2004). It has also made it possible to understand picophytoplankton diversity and variability in a larger macroecological context (Li 2002).

In early work with flow cytometric analysis of *Synechococcus*, Wood et al. (1985) and Olson et al. (1988, 1990b) highlighted the diversity of pigment types within the genus and began to explore distributions of pigment types in natural environments. The unique quantitative capabilities of flow cytometry made this approach tractable and productive. The findings motivated a variety of investigations involving culture experiments, molecular characterization, and most recently genomic studies that support a maturing view about the importance of sub-genus level diversity in enabling the success of this group over a range of niches (Rocap et al. 2002; Fuller et al. 2003; Zwirglmaier et al. 2008). A similar picture has emerged for *Prochlorococcus*, within which a range of genetically prescribed "ecotypes" have now been described and associated with particular environmental conditions (Palenik and Haselkorn 1992; Moore et al. 1998; Urbach and Chisholm 1998; West and Scanlan 1999; Rocap et al. 2003; Ahlgren et al. 2006; Bouman et al. 2006; Johnson et al. 2006). Most recently this line of inquiry is being applied to study of the more poorly characterized picoeukaryotic phytoplankton (e.g., Rodríguez et al. 2005; Zhu et al. 2005; Foulon et al. 2008), whose abundances are revealed by conventional flow cytometry but without any capability for taxonomic resolution (e.g., Campbell et al. 1997; DuRand et al. 2001; Worden et al. 2004). Some of the newest studies involve complementary use of flow cytometry and molecular methods to increase this resolution (see below on emerging approaches).

3.2 Time Resolved Pulses for Physiological and Ecological Studies

While conventional flow cytometric analysis provides resolution of maximum (peak) or integrated optical signals from each particle, the potential to resolve more information has been exploited in some specialized applications. One example is the use of fluorescence and light scattering pulse shapes to reveal aspects of cell (or chain) size and shape (Cunningham 1990; Dubelaar and Gerritzen 2000). Takabayashi et al. (2006) have shown how this can be exploited to examine links between nutrient availability and chain morphology in diatoms. With a more elaborate custom flow cytometer, Olson et al. (1996, 1999) showed that is it possible to capture fluorescence induction associated with saturation of photosynthetic reaction centers in individual cells, thus permitting characterization of photosynthetic efficiency (or variable fluorescence yield) for flow cytometrically defined populations within natural assemblages. With this "pump-during-probe" flow cytometry approach, we (Olson et al. 2000; Sosik and Olson 2002) were able to document a general lack of group-specific differences in physiological responses to iron stress in Southern Ocean phytoplankton communities (Fig. 1). This kind of

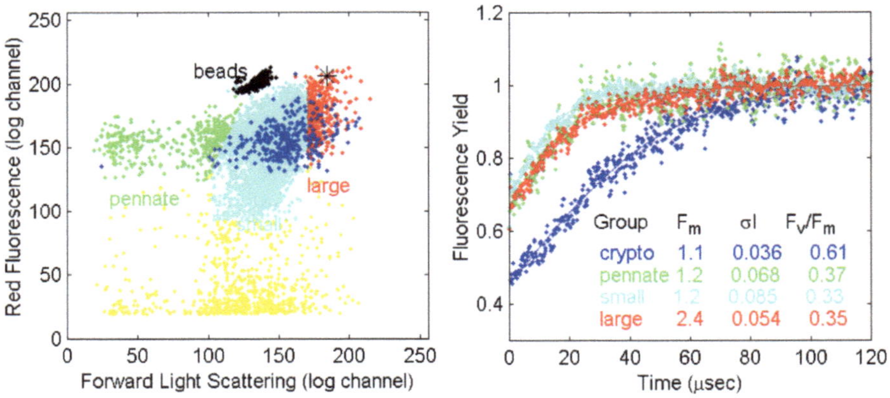

Fig. 1 Pump-during-probe flow cytometry results for a water sample from the Polar Front region of the Southern Ocean. Different groups of phytoplankton cells were separated according to multi-parameter (red fluorescence, orange fluorescence, forward light scattering, and side light scattering) conventional flow cytometry measurements (*left panel*). Associated group-specific variable fluorescence yields compiled from individual cell measurements show that only cryptophytes, which were rare, exhibited higher than average photosynthetic efficiency (Modified from Olson et al. 1999)

information is important for testing hypotheses about the relative importance of bottom up and top down factors in regulating community structure.

3.3 Cell Sorting for Physiology and Diversity

Flow cytometric cell sorting offers a nearly unique means to study purified phytoplankton from natural systems, though the small size of phytoplankton cells and the relatively high sample requirements of most analytical techniques have limited its use. Early work with cell sorting at sea included spectrophotometric and genetic studies of picocyanobacterial ecotypes as discussed above, and measurements of carbon fixation by different phytoplankton groups (Li 1994).

Other early work involved measurements of elemental ratios in sorted fractions of eukaryotic phytoplankton from oligotrophic waters to investigate nutrient limitation (Fig. 2). The carbon to nitrogen (C:N) ratio of unsorted particles was found to be close to Redfield proportions (Redfield 1958) at the surface as well as at the chlorophyll maximum (similar to most previous measurements of total particulate matter). In contrast, most of the phytoplankton samples obtained by sorting from the nitrate-depleted

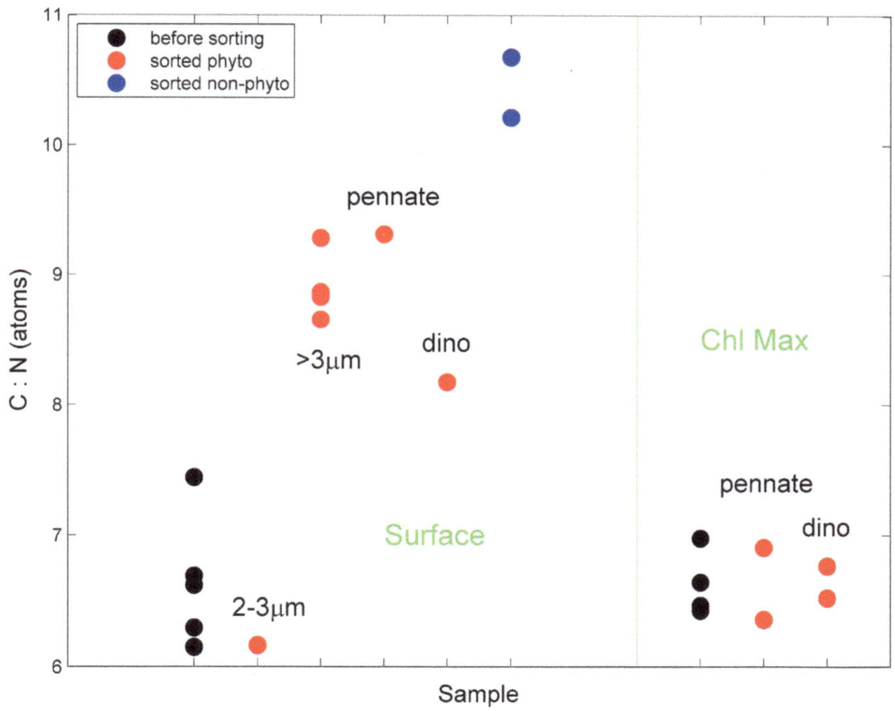

Fig. 2 Carbon to Nitrogen (C:N) ratios of sorted particles from the western Sargasso Sea (unpublished data of Olson). Particles from the surface or chlorophyll maximum layer were concentrated by gravity filtration onto 2- or 3-μm pore-size filters and then purified by flow cytometric cell sorting (Coulter EPICS) based on chlorophyll fluorescence and forward light scattering (as a proxy for cell size). The samples (typically 10^5–10^6 cells, containing 0.5 μmole C and 0.05 μmol N, were obtained by combining a series of freshly-concentrated and -sorted aliquots, which were kept on ice after sorting. The particles acquired during several hours of sorting were collected on quartz-fiber filters and quick-frozen in liquid N_2; C and N were determined by barometry after Dumas combustion. "Before sorting" samples were obtained by filtering aliquots of some of the concentrated samples (> 3-μm) on the same kind of quartz-fiber filters; the amounts of C and N in these samples were always greater than those in sorted samples. In most cases all phytoplankton (i.e., particles with chlorophyll fluorescence) were sorted, but in some samples, discrete subpopulations dominated by particular cell types were sorted: pennate diatoms were recognized by their low forward light scattering signals (Olson et al. 1989) and dinoflagellates dominated the larger phytoplankton in some samples. These identifications were confirmed by light microscopy of sorted cells. Non-phytoplankton particles (presumably mostly detritus) were N-depleted relative to phytoplankton, and surface populations of at least some larger cells (including pennate diatoms and dinoflagellates) were N-depleted relative to similar populations from the chlorophyll maximum layer. C:N of the smallest eukaryotes (which accounted for most of the biomass) were near the Redfield ratio even at the surface, suggesting they were not N-starved

surface layer had C:N ratios of 8.2–9.4, well above the Redfield ratio of 6.6 expected for nutrient-replete phytoplankton. The C:N ratios of similar cells from the chlorophyll maximum layer (where nitrate was present at about 0.5 μM) were between 6 and 7, suggesting that the surface phytoplankton were indeed N-stressed. It is worth noting that a single sample of sorted smaller phytoplankton had low C:N; the high concentration of such small cells may account for the relatively low C:N ratio in total particulates at the surface. In addition, sorted non-fluorescent particles had the highest C:N observed, suggesting this material consists largely of detritus.

More recently, in Sargasso Sea samples sequencing of rRNA genes from sorted fractions (based on both cell size and chlorophyll content) has been carried out to link diversity in the plankton to cell morphologies (Figs. 3 and 4), which is often difficult to do from bulk sequencing. The smallest eukaryotic phytoplankton fractions were shown to be dominated by chlorophytes, while alveolates (including dinoflagellates and ciliates) dominated the largest fraction. Stramenopiles (which include diatoms), haptophytes (such as coccolithophores), and cryptophytes dominated the mid-size fractions, which were also the most taxonomically diverse. The distributions of taxa in sorted samples of non-chlorophyll-containing cells (i.e., heterotrophs) appeared to be more diverse than those of the autotrophs. This finding corroborates results from bulk analyses of picoplankton by Vaulot et al. (2002) where heterotroph/autotroph assignments were inferred on the basis of sequence similarity to known organisms. The strength of the sorting approach lies in use of direct measurements of cell properties to establish trophic designation.

In both of these sorting examples, obtaining sufficient sample material entailed many hours of sorting, but the advent of new higher-speed yet compact sorters (Ibrahim and van den Engh 2003) should greatly facilitate these kinds of studies. At the same time, the development of new and more sensitive analytical approaches continues to expand the prospects for cell sorting applications. For instance, specialized protocols for isotope ratio mass spectrometry of sorted samples allowed Pel et al. (2004) to characterize how group-specific differences in growth rate influence community structure.

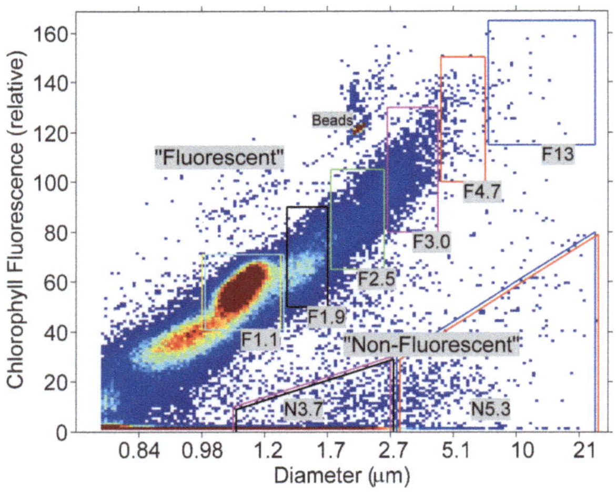

Fig. 3 Flow cytometric signature of particle size (from forward light scattering) and chlorophyll fluorescence for a seawater sample from the Sargasso Sea (with 1-μm fluorescent beads added for reference; unpublished data of Armbrust and Olson). The smallest population visible here contained *Synechococcus* cells which could be distinguished from eukaryotic cells by their phycoerythrin fluorescence (not shown); these prokaryotic cells were not sorted. Sort windows are indicated by colored boxes, and named according to fluorescence and mode particle size (μm) from post-sort purity checks. DNA was extracted from samples from each sort window; rRNA genes were amplified and sequenced, and examples from the resulting libraries presented in Fig. 4

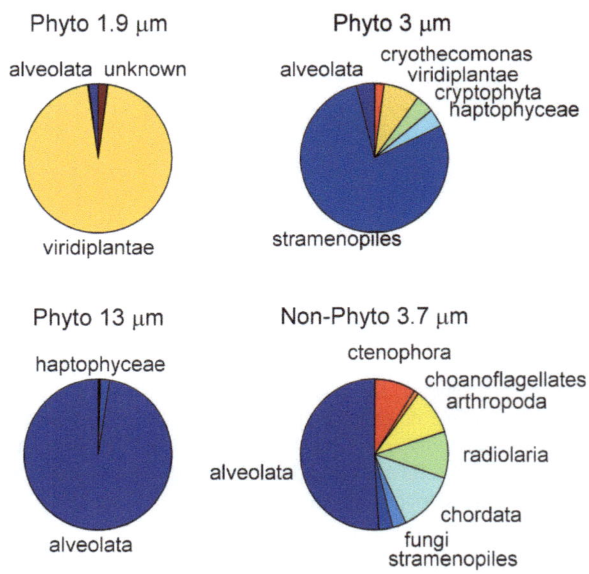

Fig. 4 Distributions of taxa in a subset of the samples sorted as described in Fig. 3: three size fractions of cells containing chlorophyll (with the mode cell size indicated) and one size fraction of non-chlorophyll-containing cells. Note that the specificity of PCR amplification is subject to bias from a variety of sources, so these results should be considered semi-quantitative

3.4 Interpretation of Optical Variability in the Ocean

As anticipated by Platt (1989) a productive line of inquiry with flow cytometry has focused on understanding variability in bulk optical and ecological properties on the basis of statistical properties of populations of phytoplankton and other particles. As discussed above, some studies have focused on this kind of approach for investigations of community structure and contributions to biomass, especially for pico- and small nanophytoplankton in open ocean systems. Other important efforts have explored the links between particles and upper ocean optical properties that determine characteristics such as light penetration and ocean color remotely sensed from satellites or aircraft. For example, Dandonneau et al. (2004) combined satellite-derived maps of chlorophyll concentration with flow cytometry and other approaches to characterize phytoplankton composition for exploration of regional to global-scale patterns likely to affect carbon cycling and export.

A strategy for mechanistically understanding optical variability has built up from investigations aimed at developing quantitative links between flow cytometric measurements and individual particle cross sections for absorption and scattering (Perry and Porter 1989; Sosik et al. 1989; DuRand and Olson 1998). Particle size is a particularly important determinant of light scattering and this has motivated efforts to characterize particle size distributions from flow cytometric measurements (Ackleson and Spinrad 1988; Yentsch and Phinney 1989; DuRand et al. 2002; Green et al. 2003b). With a combination of individual particle size and refractive index distributions estimated from flow cytometric analysis of natural samples (Green et al. 2003b), it is possible to quantify how different types of particles contribute to variability in bulk absorption and scattering coefficients (Fig. 5), as well as other optical properties such as diffuse attenuation and surface reflectance (DuRand and Olson 1998; Green et al. 2003a; Green and Sosik 2004). This insight can contribute to better models for light penetration in the ocean and better retrievals of particle properties from remotely sensed ocean color, possibly including retrieval of phytoplankton community composition or functional types (Kamykowski et al. 2002; Sathyendranath et al. 2004; Alvain et al. 2005; Uitz et al. 2006; Aiken et al. 2007).

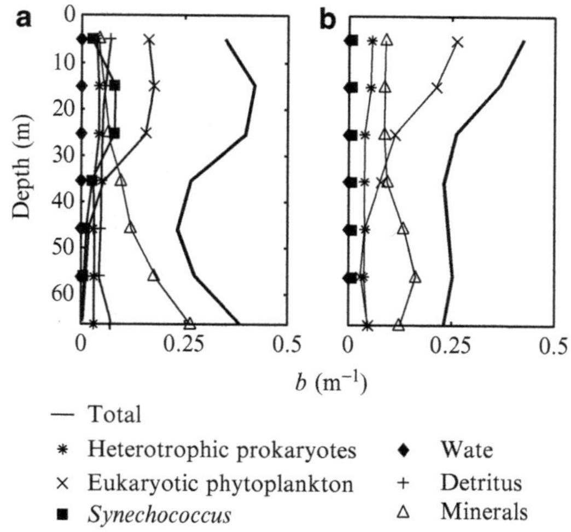

Fig. 5 Flow cytometry derived depth profiles of constituent contributions to total scattering coefficients (b) at 488 nm for New England Shelf waters during (**a**) summer 1996 and (**b**) spring 1997. Results reflect average conditions at 40° 30′ N 70 ° 30′ W from multiple vertical profiles each day for approximately 3 week periods during the Coastal Mixing and Optics Experiment (Dickey and Williams III 2001; Sosik et al. 2001). In these waters, eukaryotic phytoplankton (~1–50 μm) and mineral particles (e.g., from sediment) generally dominate scattering signals, but other types of particles must be considered at particular depths and times (e.g., subsurface maximum *Synechococcus* in summer). Scattering contribution of pure seawater water is from Morel (1974) (Modified from Green et al. 2003a)

Evidence of distinct diel patterns of variability in the abundance and optical properties of phytoplankton has also emerged from flow cytometric analyses. These have been exploited both for understanding diel variations in light transmission (or beam attenuation coefficients) in open ocean waters (Chung et al. 1996; DuRand and Olson 1996; Binder and DuRand 2002) and for characterizing growth rates in particular phytoplankton groups (André et al. 1999; Vaulot and Marie 1999; Binder and DuRand 2002; Sosik et al. 2003). This kind of approach requires high temporal resolution sampling and can be impacted by advection and patchiness, but it has the advantage of being independent of artifacts associated with incubation experiments typically used to quantify rates of photosynthesis, growth, and grazing.

4 Emerging Approaches and Applications

The vitality of flow cytometry as a technology for plankton studies is demonstrated by the steady exploration of expanded capabilities, with applications driven by new analytical and analysis approaches in areas such as imaging and genomics as well as advances in technical capabilities for flow cytometers themselves.

Despite the rapid and quantitative analysis of particles permitted by shipboard flow cytometry, sampling constraints continue to limit conventional applications. Analysis remains relatively labor-intensive and is confined to samples collected from discrete points. Because many problems in plankton ecology and marine particle optics involve processes that are highly variable in space and time, this limitation is serious and has motivated a long-standing interest in development of automated instruments with capabilities to perform outside the laboratory. Progress in this area has been made by several groups (Dubelaar and Gerritzen 2000; Olson et al. 2003; Wang et al. 2005; Dubelaar et al. 2007) and new measurement capabilities are emerging. These include spatial mapping with instruments deployed on underwater vehicles (Cunningham et al. 2003) and extended high resolution time series observations (Sosik et al. 2003; Thyssen et al. 2008). Essential adaptations for these applications include automated sample and data handling as well as onboard performance monitoring (e.g., standard bead analysis) and anti-fouling measures; these capabilities are emerging and providing unprecedented observations (Fig. 6).

Recently, imaging has emerged as a viable enhancement to submersible flow cytometry (Olson and Sosik 2007). Previous work in the last decade demonstrated feasibility of integrating video imaging with flow cytometry in laboratory-based systems for study of plankton and other microbes (Sieracki et al. 1998; Kachel and Wietzorrek 2000; Rutten et al. 2005; Brehm-Stecher 2007). When combined with the

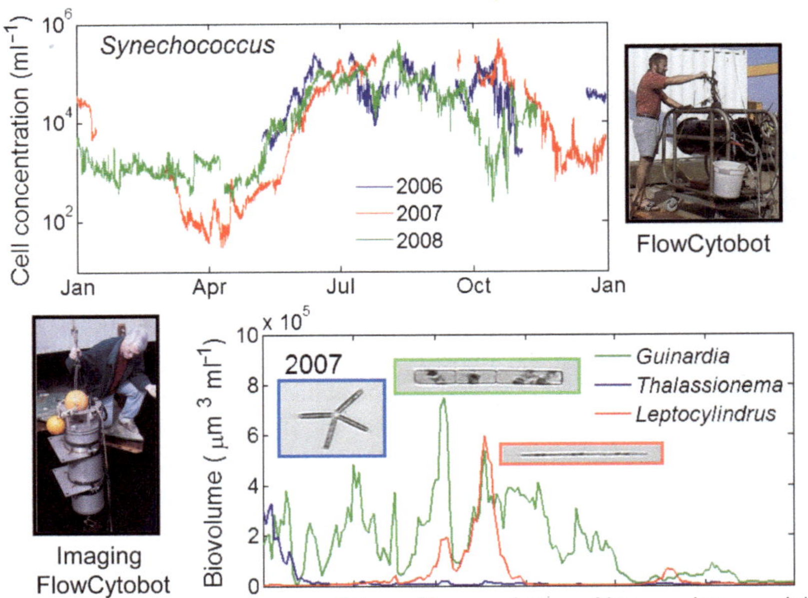

Fig. 6 Automated submersible flow cytometers are being used to generate high resolution time series of the phytoplankton community at the Martha's Vineyard Coastal Observatory.[1] Two instruments, FlowCytobot (Olson et al. 2003) and Imaging FlowCytobot (Olson and Sosik 2007), have been deployed side-by-side for periods as long as 6 months (before servicing and redeployment) to measure single cells from pico- to microplankton. Upper panel emphasizes seasonal and interannual variations in *Synechococcus* abundance from hourly resolved FlowCytobot measurements. Lower panel shows daily resolved fluctuations in selected diatom genera that can be quantified with automated image processing and classification of Imaging FlowCytobot observations (Sosik and Olson 2007); insets show example cell images from the data set for these taxa. Overall in this temperate shelf environment picoplankton dominate in summer and diatoms are important during winter and early spring bloom events, but submersible flow cytometry emphasizes that taxon-specific bloom patterns are extremely variable

[1]http://www.whoi.edu/mvco

growing expertise in extended underwater flow cytometry and with new automated image analysis and classification approaches (e.g., Sosik and Olson 2007), this capability can provide unique observations of plankton communities at ecologically relevant scales (Fig. 6). The emphasis on analysis of picoplankton with flow cytometry in the interval since the mid-1980s was driven largely by the advantages that conventional flow cytometers offer for that class of organisms; in the coming decades it is likely more attention will be brought to larger cells with the proliferation of these new instrument systems with higher sample volume throughput and expanded measurement capabilities such as cell imaging.

An intriguing advance in flow cytometric imaging, that of fluorescence imaging, has been developed in a commercial laboratory-based instrument (ImageStream, Amnis Corp.), in which precise measurements of particle velocity allow synchronized delayed acquisition of an image as it flows past the camera chip, effectively providing ~1,000-fold increase in exposure time for the weak fluorescence emission (Ortyn et al. 2006). High-resolution fluorescence images, combined with brightfield images and conventional flow cytometric measurements, can provide a wealth of new information on individual cells. To date, tests with phytoplankton cells (Fig. 7) have been carried out only with highly-concentrated samples because a very narrow core stream (i.e., slow sample introduction) is required to keep cells stable and in tight focus during the long exposure. Recent developments to extend depth of field (Ortyn et al. 2007) may allow increased sample throughput, making analysis of natural plankton samples more feasible in the future. If this proves fruitful in laboratory and ship-based settings, the approach may ultimately be incorporated into submersible systems.

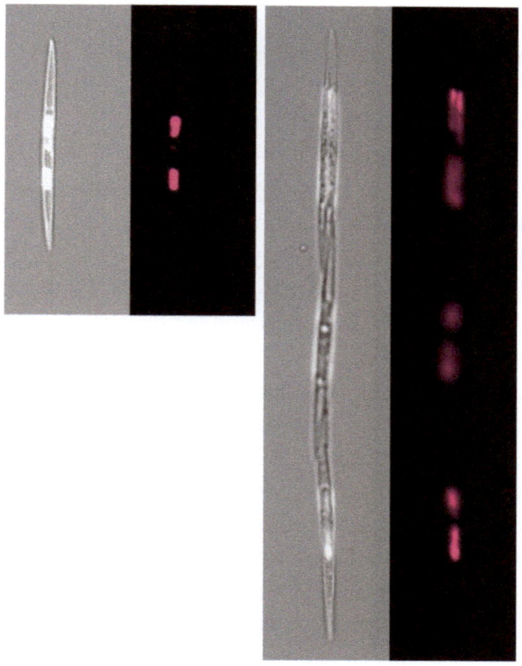

Fig. 7 Brightfield (*left image* in each pair) and red-fluorescence (*right image* in each pair) images of diatoms of the genus *Pseudonitzschia* (single cell and chain of three cells) obtained with ImageStream flow cytometer. In the fluorescence images, a distinct pair of chloroplasts is visible in each cell. Images supplied courtesy of Amnis Corporation and investigators at the Institute of Marine Sciences, University of California, Santa Cruz (B. Carter, R. Kudela, and P. Miller)

There are other developing areas that may particularly advance underwater applications of flow cytometry. Designs that eliminate use of sheath fluid and flow cells are one promising aspect (Wang et al. 2005; van den Engh, personal communication 2008). These systems involve use of multiple light beams or position sensitive detectors to target particles in optical focus and have potential advantages for simple access to particles in natural suspension (e.g., eliminating the need to pump seawater into a pressure housing). Expanded measurement capabilities may also prove increasingly important, such as resolution of pulse shape (Dubelaar and Gerritzen 2000) and laser Doppler velocimetry (Wang et al. 2005) for characterizing aspects of particle size and shape. In addition, continued investment in instrument designs that are smaller and require less power promise to expand the deployment capabilities of submersible flow cytometry. Beyond advances in research capabilities, this kind of progress is likely to accelerate use of flow cytometry for monitoring aquatic systems with focus on societally relevant issues such as nuisance or harmful algal blooms (Rutten et al. 2005; Campbell et al. 2008).

The recent and rapid expansion of molecular and genomic approaches for study of marine plankton in combination with flow cytometry now make it possible to gain species-specific information about organisms and their activity without the need for culturing. The earliest examples of these combined approaches is the use of flow cytometry with Fluorescence In Situ Hybridization (FISH) to examine bacterial diversity and activity in environmental samples (e.g., Wallner et al. 1995; Zubkov et al. 2001; Sekar et al. 2004;

Kalyuzhnaya et al. 2006; Amann and Fuchs 2008); some progress is also being made in this area with eukaryotes (Not et al. 2002; Biegala et al. 2003). Peptide nucleic acid (PNA) probes, which can provide more fluorescence per molecule than FISH DNA probes, have also been explored for bacteria (Brehm-Stecher et al. 2005; Lenaerts et al. 2007; Søgaard et al. 2007), prokaryotic picophytoplankton (Worden et al. 2000; Worden and Binder 2003b) and for selected eukaryotes such as toxic dinoflagellates (Huang et al. 2008). Current protocols for FISH and PNA probes commonly require multiple wash steps making them less robust for use with the relatively low abundance natural populations of eukaryotic communities. However, continued development of adapted or related methods (see below) promise to expand observational capabilities with flow cytometry.

Some of the most promising new approaches rely on recent advances in whole genome amplification techniques that now make it feasible to generate genome sequence data from a few or even single sorted cells (Podar et al. 2007; Stepanauskas and Sieracki 2007), reducing or perhaps even eliminating the need for cell cultivation. The initial studies were of heterotrophic bacteria, but phytoplankton cells are not far behind. For instance, the first attempts at genome analyses from single sorted *Prochlorococcus* cells were unsuccessful due to contamination introduced during sorting (Zhang et al. 2006), but these problems have recently been overcome (Chisholm, personal communication 2008). With eukaryotes, progress is being made rapidly. Zhang et al. (2008) used PCR to amplify phylogenetically informative genes from specific dinoflagellates sorted from a bloom. Duff et al. (2008) have adapted single cell extraction techniques for several groups of protists captured from natural samples; in future work it might be possible to use these approaches in conjunction with cell sorting.

Advances in combining flow cell imaging with flow cytometric detection of fluorescence now make it possible to analyze eukaryotic phytoplankton communities with genus-level and in some instances species-level resolution (Olson and Sosik 2007; Sosik and Olson 2007). Over the past decade, advances have been made in understanding the population genetics of select groups of eukaryotic phytoplankton including diatoms (Rynearson and Armbrust 2005; Rynearson et al. 2006), coccolithophorids (Iglesias-Rodríguez et al. 2006), and dinoflagellates (Shankle et al. 2004); but, in each of these studies, single cells were isolated by hand and commonly cultured for tens of generations making these studies extremely labor intensive. Incorporation of image-based sorting into next generation flow cytometers will greatly accelerate progress in population genetics and genomic studies of the uncultivated eukaryotic phytoplankton. Over the coming years, it will be possible to further explore links between genetic diversity, physiology, and niche differentiation in microplankton in a manner similar to what has been possible for prokaryotic phytoplankton.

In summary, flow cytometry continues to be a standard tool for analysis of aquatic phytoplankton as well as an evolving technique with broad potential for innovation. In the coming decades, plankton ecologists can look forward to new discovery and insights coming from applications of flow cytometry, especially in combination with emerging analytical techniques and developments in multidimensional, multidisciplinary ocean observing capabilities.

Acknowledgments We would like to acknowledge support from the US National Science Foundation and Gordon and Betty Moore Foundation. Brandon Carter, Raphael Kudela, and Peter Miller and Amnis Corp. graciously provided fluorescence images and Rhonda Marohl generated the sequence data.

References

Ackleson SG, Spinrad RW (1988) Size and refractive index of individual marine particulates: a flow cytometric approach. Appl Opt 27:1270–1277

Agusti S (2004) Viability and niche segregation of *Prochlorococcus* and *Synechococcus* cells across the Central Atlantic Ocean. Aquat Microb Ecol 36:53–59

Ahlgren NA, Rocap G, Chisholm SW (2006) Measurement of *Prochlorococcus* ecotypes using real-time polymerase chain reaction reveals different abundances of genotypes with similar light physiologies. Environ Microbiol 8:441–454

Aiken J et al (2007) Validation of MERIS reflectance and chlorophyll during the BENCAL cruise October 2002: preliminary validation of new demonstration products for phytoplankton functional types and photosynthetic parameters. Int J Remote Sensing 28:497–516

Alvain S, Moulin C, Dandonneau Y, Bréon FM (2005) Remote sensing of phytoplankton groups in case 1 waters from global SeaWiFS imagery. Deep Sea Res I 52:1989–2004

Amann R, Fuchs BM (2008) Single-cell identification in microbial communities by improved fluorescence in situ hybridization techniques. Nat Rev Micro 6:339–348

André J-M, Navarette C, Blanchot J, Radenac M-H (1999) Picophytoplankton dynamics in the equatorial Pacific:

Growth and grazing rates from cytometric counts. J Geophys Res 104:3369–3380
Biegala IC, Not F, Vaulot D, Simon M (2003) Quantitative assessment of picoeukaryotes in the natural environment by using taxon-specific oligonucleotide probes in association with tyramide signal amplification-fluorescence in situ hybridization and flow cytometry. Appl Env Microbiol 69:5519–5529
Binder BJ, DuRand MD (2002) Diel cycles in surface waters of the equatorial Pacific. Deep Sea Res II 49:2601–2617
Bouman HA et al (2006) Oceanographic basis of the global surface distribution of *Prochlorococcus* ecotypes. Science 312:918–921
Brehm-Stecher B (2007) New technologies for imaging individual microbial cells. In: Frischknecht F, Shorte S (eds) Imaging cellular and molecular biological function. Springer-Verlag, Berlin, pp 307–337
Brehm-Stecher B, Hyldig JJ, Johnson EA (2005) Design and evaluation of 16S rRNA-targeted peptide nucleic acid probes for whole-cell detection of members of the genus *Listeria*. *Appl Env Microbiol* 71:5451–5457
Burkhill PH, Mantoura RFC (1990) The rapid analysis of single marine cells by flow cytometry. Phil Trans R Soc Lond A 333:99–112
Campbell L (2001) Flow cytometric analysis of autotrophic picoplankton. In: Paul JH (ed) *Methods in microbiology: marine microbial ecology*, vol 30. Academic Press, New York, pp 317–343
Campbell L, Hongbin L, Vaulot D (1997) Annual variability of phytoplankton and bacteria in the subtropical North Pacific Ocean at Station ALOHA during the 1991–1994 ENSO event. Deep Sea Res I 44:167
Campbell L, Nolla H, Vaulot D (1994) The importance of *Prochlorococcus* to community structure in the central North Pacific Ocean. Limnol Oceanogr 39:954–961
Campbell L, Olson RJ, Sosik HM (2008) First toxic *Dinophysis* bloom observed in the Gulf of Mexico, USA. Harmful Algae News 36:10–11
Cavender-Bares KK, Frankel SL, Chisholm SW (1998) A dual sheath flow cytometer for shipboard analyses of phytoplankton communities from the oligotrophic ocean. Limnol Oceanogr 43:1383–1388
Chisholm SW, Armbrust EV, Olson RJ (1986) The individual cell in phytoplankton ecology: cell cycles and application of flow cytometry. In: Platt T, Li WKW (eds) Photosynthetic picoplankton. Can Bull Fish Aquat Sci 214:343–369
Chisholm SW, Frankel SL, Goerike R, Olson RJ, Palenik B, Waterbury JB, Westjohnsrud L, Zettler ER (1992) *Prochlorococcus marinus* nov. gen. sp.: an oxyphototrophic marine prokaryote containing divinyl chlorophyll a and b. Arch Microbiol 157:297–300
Chisholm SW, Olson RJ, Zettler ER, Goericke R, Waterbury J, Welschmeyer N (1988) A novel free-living prochlorophyte abundant in the oceanic euphotic zone. Nature 334:340–343
Chung SP, Gardner WD, Richardson MJ, Walsh ID, Landry MR (1996) Beam attenuation and micro-organisms: spatial and temporal variations in small particles along 140° W during the 1992 JGOFS EqPac transects. Deep Sea Res II 43:1205–1226
Coleman ML, Chisholm SW (2007) Code and context: *Prochlorococcus* as a model for cross-scale biology. Trends Microbiol 15:398–407
Collier JL (2000) Flow cytometry and the single cell in phycology. J Phycol 36:628–644
Collier JL, Campbell L (1999) Flow cytometry in molecular aquatic ecology. Hydrobiologia 401:33–53
Courties C, Troussellier M (2001) Flow cytometry in the marine environment. Cytometry 44:163
Cunningham A (1990) Fluorescence pulse shape as a morphological indicator in the analysis of colonial microalgae by flow cytometry. J Microbiol Meth 11:27–36
Cunningham A, McKee D, Craig S, Tarran G, Widdicombe C (2003) Fine-scale variability in phytoplankton community structure and inherent optical properties measured from an autonomous underwater vehicle. J Mar Sys 43:51–59
Dandonneau Y, Deschamps P, Nicholas J, Loisel H, Blanchot J, Montel Y, Thieuleux F, Becu G (2004) Seasonal and interannual variability of ocean color and composition of phytoplankton communities in the North Atlantic, equatorial Pacific and South Pacific. Deep Sea Res II 51:303–318
Dickey TD, Williams AJ III (2001) Interdisciplinary ocean process studies on the New England shelf. J Geophys Res 106:9427–9434
Dubelaar GBJ, Casotti R, Tarran GA, Biegala IC (2007) Phytoplankton and their analysis by flow cytometry. In: Dolezel J, Greilhuber J, Suda J (eds) Flow cytometry with plant cells: Analysis of genes, chromosomes and genomes. Wiley, New York, pp 287–322
Dubelaar GBJ, Gerritzen PL (2000) CytoBuoy: a step forward towards using flow cytometry in operational oceanography. Sci Mar 64:255–265
Dubelaar GBJ, Groenewegen AC, Stokdijk W, van den Engh GJ, Visser JWM (1989) Optical plankton analyser: a flow cytometer for plankton analysis, II: Specifications. Cytometry 10:529–539
Duff RJ, Ball H, Lavrentyev PJ (2008) Application of combined morphological-molecular approaches to the identification of planktonic protists from environmental samples. J Euk Microbiol 55:306–312
DuRand MD, Green RE, Sosik HM, Olson RJ (2002) Diel variations in optical properties of *Micromonas pusilla* (Prasinophyceae). J Phycol 38:1132–1142
DuRand MD, Olson RJ (1996) Contributions of phytoplankton light scattering and cell concentration changes to diel variations in beam attenuation in the equatorial Pacific from flow cytometric measurements of pico-, ultra- and nanoplankton. Deep Sea Res II 43:891–906
DuRand MD, Olson RJ (1998) Diel patterns in optical properties of the chlorophyte *Nannochloris* sp.: relating individual-cell to bulk measurements. Limnol Oceanogr 43:1107–1118
DuRand MD, Olson RJ, Chisholm SW (2001) Phytoplankton population dynamics at the Bermuda Atlantic time-series station in the Sargasso Sea. Deep-Sea Res II 48:1983–2003
Dusenberry JA, Frankel SL (1994) Increasing the sensitivity of a FACScan flow cytometer to study oceanic picoplankton. Limnol Oceanogr 39:206–209
Dusenberry JA, Olson RJ, Chisholm SW (2000) Field observations of oceanic mixed layer dynamics and picophytoplankton photoacclimation. J Mar Sys 24:221–232
Dusenberry JA, Olson RJ, Chisholm SW (2001) Photoacclimation kinetics of single-cell fluorescence in laboratory and field populations of *Prochlorococcus*. Deep Sea Res I 48:1443–1458
Foulon E, Not F, Jalabert F, Cariou T, Massana R, Simon N (2008) Ecological niche partitioning in the picoplanktonic green alga *Micromonas pusilla*: evidence from environmental surveys using phylogenetic probes. Environ Microbiol 10:2433–2443

Fuller NJ, Marie D, Partensky F, Vaulot D, Post AF, Scanlan DJ (2003) Clade-specific 16S ribosomal DNA oligonucleotides reveal the predominance of a single marine *Synechococcus* clade throughout a stratified water column in the Red Sea. Appl Env Microbiol 69:2430–2443

Goericke R, Repeta DJ (1992) The pigments of *Prochlorococcus marinus*: the presence of divinyl chlorophyll a and b in a marine procaryote. Limnol Oceanogr 37:425–433

Green RE, Sosik HM (2004) Analysis of apparent optical properties and ocean color models using measurements of seawater constitutents in New England continental shelf surface waters. J Geophys Res 109:C03026. doi:03010.01029/02003JC001977

Green RE, Sosik HM, Olson RJ (2003a) Contributions of phytoplankton and other particles to inherent optical properties in New England continental shelf waters. Limnol Oceanogr 48:2377–2391

Green RE, Sosik HM, Olson RJ, DuRand MD (2003b) Flow cytometric determination of size and complex refractive index for marine particles: Comparison with independent and bulk estimates. Appl Opt 42:526–541

Grégori G et al (2001) Resolution of viable and membrane-compromised bacteria in freshwater and marine waters based on analytical flow cytometry and nucleic acid double staining. Appl Env Microbiol 67:4662–4670

Huang B, Hou J, Lin S, Chen J, Hong H (2008) Development of a PNA probe for the detection of the toxic dinoflagellate *Takayama pulchella*. Harmful Algae 7:495–503

Ibrahim SF, van den Engh G (2003) High-speed cell sorting: fundamentals and recent advances. Curr Opin Biotech 14:5–12

Iglesias-Rodríguez MD, Schofield O, Batley J, Medlin L, Hayes P (2006) Intraspecific genetic diversity in the marine coccolithophore *Emiliania huxleyi* (Prymnesiophyceae): the use of microsatellite analysis in marine phytoplankton population studies. J Phycol 42:526–536

Jacquet S, Lennon J, Marie D, Vaulot D (1998) Picoplankton population dynamics in coastal waters of the northwestern Mediterranean Sea. Limnol Oceanogr 43:1916–1931

Jochem F (2000) Probing the physiological state of phytoplankton at the single-cell level. Sci Mar 64:183–195

Johnson ZI, Zinser ER, Coe A, McNulty NP, Woodward EMS, Chisholm SW (2006) Niche partitioning among *Prochlorococcus* ecotypes along ocean-scale environmental gradients. Science 311:1737–1740

Kachel V, Wietzorrek J (2000) Flow cytometry and integrated imaging. Sci Mar 64:247–254

Kalyuzhnaya MG, Zabinsky R, Bowerman S, Baker DR, Lidstrom ME, Chistoserdova L (2006) Fluorescence in situ hybridization-flow cytometry-cell sorting-based method for separation and enrichment of type I and type II methanotroph populations. Appl Env Microbiol 72:4293–4301

Kamykowski D, Zentara SJ, Morrison JM, Switzer AC (2002) Dynamic global patterns of nitrate, phosphate, silicate, and iron availability and phytoplankton community composition from remote sensing data. Glob Biogeochem Cyc 16:1077. doi:10.1029/2001GB001640

Legendre L, Courties C, Troussellier M (2001) Flow cytometry in oceanography 1989–1999: environmental challenges and research trends. Cytometry 44:164–172

Legendre L, Yentsch CM (1989) Overview of flow cytometry and image analysis in biological oceanography and limnology. Cytometry 10:501–510

Lenaerts J, Lappin-Scott HM, Porter J (2007) Improved fluorescent in situ hybridization method for detection of bacteria from activated sludge and river water by using DNA molecular beacons and flow cytometry. Appl Env Microbiol 73:2020–2023

Lepesteur M, Martin JM, Fleury A (1993) A comparative study of different preservation methods for phytoplankton cell analysis by flow cytometry. Mar Ecol Prog Ser 93:55–63

Li WKW (1989) Shipboard analytical flow cytometry of oceanic ultraplankton. Cytometry 10:564–579

Li WKW (1994) Primary production of prochlorophytes, cyanobacteria, and eucaryotic ultraphytoplankton: Measurements from flow cytometric sorting. Limnol Oceanogr 39:169–175

Li WKW (2002) Macroecological patterns of phytoplankton in the northwestern North Atlantic Ocean. Nature 419: 154–157

Li WKW, Dickie PM (2001) Monitoring phytoplankton, bacterioplankton, and virioplankton in a coastal inlet (Bedford Basin) by flow cytometry. Cytometry 44:236–246

Lipschultz F (1995) Nitrogen-specific uptake rates of marine phytoplankton isolated from natural populations of particles by flow cytometry. Mar Ecol Prog Ser 123:245–258

Liu H, Campbell L, Landry MR, Nolla H, Brow SL, Constantinou J (1998) *Prochlorococcus* and *Synechococcus* growth rates and contributions to production in the Arabian Sea during the 1995 southeast monsoons. Deep Sea Res II 45:2327–2352

Liu H, Nolla H, Campbell L (1997) *Prochlorococcus* growth rate and contribution to primary production in the equatorial and subtropical North Pacific Ocean. Aquat Microb Ecol 12:39–47

Marie D, Brussaard CPD, Thyrhaug R, Bratbak G, Vaulot D (1999) Enumeration of marine viruses in culture and natural samples by flow cytometry. Appl Env Microbiol 65:45–52

Marie D, Partensky F, Jacquet S, Vaulot D (1997) Enumeration and cell cycle analysis of natural populations of marine picoplankton by flow cytometry using the nucleic acid stain SYBR Green I. Appl Env Microbiol 63:186–193

Marie D, Simon N, Vaulot D (2005) Phytoplankton cell counting by flow cytometry. In: Andersen RA (ed) Algal culturing techniques. Elsevier, Amsterdam, pp 253–267

Moore LR, Rocap G, Chisholm SW (1998) Physiology and molecular phylogeny of coexisting *Prochlorococcus* ecotypes. Nature 393:464–467

Morel A (1974) Optical properties of pure water and pure seawater. In: Jerlov NG, Nielsen ES (eds) *Optical aspects of oceanography*. Academic Press, New York, pp 1–24

Neale PJ, Cullen JJ, Yentsch CM (1989) Bio-optical inferences from chlorophyll a fluorescence: What kind of fluorescence is measured in flow cytometry? Limnol Oceanogr 34:1739–1748

Not F, Simon N, Biegala IC, Vaulot D (2002) Application of fluorescent in situ hybrization coupled with tyramide signal amplification (FISH-TSA) to assess eukaryotic picoplankton composition. Aquat Microb Ecol 28:157–166

Olson RJ, Chekalyuk A, Sosik HM (1996) Phytoplankton photosynthetic characteristics from fluorescence induction assays of individual cells. Limnol Oceanogr 41:1253–1263

Olson RJ, Chisholm SW, Zettler ER, Altabet MA, Dusenberry JA (1990a) Spatial and temporal distributions of prochlorophyte picoplankton in the North Atlantic Ocean. Deep Sea Res 37:1033–1051

Olson RJ, Chisholm SW, Zettler ER, Armbrust EV (1988) Analysis of *Synechococcus* pigment types in the sea using single and dual beam flow cytometry. Deep Sea Res 35:425–440

Olson RJ, Chisholm SW, Zettler ER, Armbrust EV (1990b) Pigments, size, and distribution of *Synechococcus* in the North Atlantic and Pacific Oceans. Limnol Oceanogr 35:45–58

Olson RJ, Frankel SL, Chisholm SW, Shapiro HM (1983) An inexpensive flow cytometer for analysis of fluorescence signals in phytoplankton: chlorophyll and DNA distributions. J Exp Mar Biol Ecol 68:129–144

Olson RJ, Shalapyonok AA, Sosik HM (2003) An automated submersible flow cytometer for pico- and nanophytoplankton: FlowCytobot. Deep Sea Res I 50:301–315

Olson RJ, Sosik HM (2007) A submersible imaging-in-flow instrument to analyze nano- and microplankton: Imaging FlowCytobot. Limnol Oceanogr Meth 5:195–203

Olson RJ, Sosik HM, Chekalyuk AM (1999) Photosynthetic characteristics of marine phytoplankton from pump-during-probe fluorometry of individual cells at sea. Cytometry 37:1–13

Olson RJ, Sosik HM, Shalapyonok A (2000) Effects of iron enrichment on phytoplankton in the Southern Ocean during late summer: Active fluorescence and flow cytometric analyses. Deep Sea Res II 47:3181–3200

Olson RJ, Vaulot D, Chisholm SW (1985) Marine phytoplankton distributions measured using shipboard flow cytometry. Deep Sea Res 32:1273–1280

Olson RJ, Zettler ER, Anderson OK (1989) Discrimination of eukaryotic phytoplankton cell types from light scatter and autofluorescence properties measured by flow cytometry. Cytometry 10:636–643

Olson RJ, Zettler ER, Chisholm SW (1991) Advances in oceanography through flow cytometry. In: Demers S (ed) Particle analysis in oceanography. Springer, Berlin, pp 351–399

Olson RJ, Zettler ER, DuRand MD (1993) Phytoplankton analysis using flow cytometry. In: Kemp PF, Sherr BF, Sherr EB, Cole JJ (eds) Handbook of Methods in Aquatic Microbial Ecology. Lewis Publishers, Boca Raton, pp 175–186

Ortyn WE, Hall BE, George TC, Frost K, Basiji DA, Perry DJ, Zimmermann CA, Coder D, Morrissey PJ (2006) Sensitivity measurement and compensation in spectral imaging. Cytometry Part A 69A:852–862

Ortyn WE, Perry DJ, Venkatachalam V, Liang L, Hall BE, Frost K, Basiji DA (2007) Extended depth of field imaging for high speed cell analysis. Cytometry Part A 71A:215–231

Palenik B, Haselkorn R (1992) Multiple evolutionary origins of prochlorophytes, the chlorophyll b-containing prokaryotes. Nature 16:265–167

Partensky F, Hess WR, Vaulot D (1999) *Prochlorococcus*, a marine photosynthetic prokaryote of global significance. Microbio Mol Biol Rev 63:106–127

Pel R, Floris V, Gons HJ, Hoogveld HL (2004) Linking flow cytometric cell sorting and compound-specific ^{13}C-analysis to determine population-specific isotopic signtures and growth rates in cyanobacteria-dominated lake plankton. J Phycol 40:857–866

Perry MJ, Porter SM (1989) Determination of the cross-section absorption coefficient of individual phytoplankton cells by analytical flow cytometry. Limnol Oceanogr 34:1727–1738

Platt T (1989) Optimal utlization of flow cytometry. Cytometry 10:500

Podar M, Abulencia CB, Walcher M, Hutchinson D, Zengler K, Garcia JA, Holland T, Cotton D, Hauser L, Keller M (2007) Targeted access to the genomes of low-abundance organisms in complex microbial communities. Appl Env Microbiol 73:3205–3214

Reckermann M (2000) Flow sorting in aquatic ecology. Sci Mar 64:235–246

Reckermann M, Colijn F (2000) Aquatic flow cytometry: achievements and prospects, Forward. Sci Mar 64:119–120

Redfield AC (1958) On the biological control of chemical factors in the environment. Am Sci 46:205–221

Rocap G, Distel D, Waterbury JB, Chisholm SW (2002) Resolution of *Prochlorococcus* and *Synechococcus* ecotypes by using 16S–23S ribosomal DNA internal transcribed spacer sequences. Appl Env Microbiol 68:1180–1191

Rocap G et al (2003) Genome divergence in two *Prochlorococcus* ecotypes reflects oceanic niche differentiation. Nature 424:1042–1047

Rodríguez F, Derelle E, Guillou L, Le Gall F, Vaulot D, Moreau H (2005) Ecotype diversity in the marine picoeukaryotes *Ostreococcus* (Chlorophyta, Prasinophyceae). Environ Microbiol 7:853–859

Rose JM, Caron DA, Sieracki ME, Poulton NJ (2004) Counting heterotrophic nanoplanktonic protists in cultures and in aquatic communities by flow cytometry. Aquat Microb Ecol 34:263–277

Rutten TP, Sandee B, Hofman ART (2005) Phytoplankton monitoring by high performance flow cytometry: A successful approach? Cytometry 64A:16–26

Rynearson TA, Armbrust EV (2005) Maintenance of clonal diversity during a spring bloom of the centric diatom *Ditylum brightwellii*. Mol Ecol 14:1631–1640

Rynearson TA, Newton JA, Armbrust EV (2006) Spring bloom development, genetic variation, and population succession in the planktonic diatom *Ditylum brightwellii*. Limnol Oceanogr 51:1249–1261

Sathyendranath S, Watts L, Devred E, Platt T, Caverhill C, Maass H (2004) Discrimination of diatoms from other phytoplankton using ocean-colour data. Mar Ecol Prog Ser 272:59–68

Sekar R, Fuchs BM, Amann R, Pernthaler J (2004) Flow sorting of marine bacterioplankton after fluorescence *in situ* hybridization. Appl Env Microbiol 70:6210–6219

Shalapyonok A, Olson RJ, Shalapyonok LS (2001) Arabian Sea phytoplankton during Southwest and Northeast Monsoons 1995: composition, size structure and biomass from individual cell properties measured by flow cytometry. Deep-Sea Res II 48:1231–1262

Shalapyonok AA, Olson RJ, Shalapyonok LS (1998) Ultradian growth in *Prochlorococcus* spp. Appl Env Microbiol 64:1066–1069

Shankle AM, Mayali X, Franks PJS (2004) Temporal patterns in population genetic diversity of *Prorocentrum micans* (Dinophyceae). J Phycol 40:238–247

Shapiro HM (2003) Practical flow cytometry, 4th edn. Wiley-Liss, New York

Sieracki CK, Sieracki ME, Yentsch CS (1998) An imaging-in-flow system for automated analysis of marine microplankton. Mar Ecol Prog Ser 168:285–296

Søgaard M, Hansen DS, Fiandaca MJ, Stender H, Schønheyder HC (2007) Peptide nucleic acid fluorescence in situ hybridization for rapid detection of *Klebsiella pneumoniae* from positive blood cultures. J Med Microbiol 56:914–917

Sosik HM, Chisholm SW, Olson RJ (1989) Chlorophyll fluorescence from single cells: Interpretation of flow cytometric signals. Limnol Oceanogr 34:1749–1761

Sosik HM, Green RE, Pegau WS, Roesler CS (2001) Temporal and vertical variability in optical properties of New England shelf waters during late summer and spring. J Geophys Res 106:9455–9472

Sosik HM, Olson RJ (2002) Phytoplankton and iron limitation of photosynthetic efficiency in the Southern Ocean during late summer. Deep Sea Res I 49:1195–1216

Sosik HM, Olson RJ (2007) Automated taxonomic classification of phytoplankton sampled with imaging-in-flow cytometry. Limnol Oceanogr Meth 5:204–216

Sosik HM, Olson RJ, Neubert MG, Shalapyonok AA, Solow AR (2003) Growth rates of coastal phytoplankton from time-series measurements with a submersible flow cytometer. Limnol Oceanogr 48:1756–1765

Stepanauskas R, Sieracki M (2007) Mathcin phylogeny and metabolism in the uncultured marine bacteria, one cell at a time. Proc Nat Acad Sci USA 104:9052–9057

Takabayashi M, Lew K, Johnson A, Marchi AL, Dugdale R, Wilkerson FP (2006) The effect of nutrient availability and temperature on chain length of the diatom, *Skeletonema costatum*. J Plankton Res 28:831–840

Thyssen M, Tarran GA, Zubkov MV, Holland RJ, Gregori G, Burkill PH, Denis M (2008) The emergence of automated high frequency flow cytometry: revealing temporal and spatial phytoplankton variability. J Plank Res 30:333–343

Troussellier M, Courties C, Vaquer A (1993) Recent applications of flow cytometry in aquatic microbial ecology. Biol Cell 78:111–121

Uitz J, Claustre H, Morel A, Hooker SB (2006) Vertical distribution of phytoplankton communities in open ocean: An assessment based on surface chlorophyll. J Geophys Res 111:C08005. doi:10.1029/2005JC003207

Urbach E, Chisholm SW (1998) Genetic diversity in *Prochlorococcus* populations flow cytometrically sorted from the Sargasso Sea and the Gulf Stream. Limnol Oceanogr 43:1615–1630

Vaulot D, Courties C, Partensky F (1989) A simple method to preserved oceanic phytoplankton for flow cytometric analyses. Cytometry 10:629–635

Vaulot D, Marie D (1999) Diel variability of photosynthetic picoplankton in the equatorial Pacific. J Geophys Res 104:3297–3310

Vaulot D, Marie D, Olson RJ, Chisholm SW (1995) Growth of *Prochlorococcus*, a photosynthetic prokaryote, in the Equatorial Pacific Ocean. Science 268:1480–1482

Vaulot D, Romari K, Not F (2002) Are autotrophs less diverse than heterotrophs in marine picoplankon? Trends Microbiol 10:266–267

Veldhuis M, Kraay GW, Timmermans KR (2001) Cell death in phytoplankton: correlation between changes in membrane permeability, photosynthetic activity, pigmentation and growth. Eur J Phycol 36:167–177

Veldhuis MJW, Kraay GW (2000) Application of flow cytometry in marine phytoplankton research: current applications and future perspectives. Sci Mar 64:121–134

Vives-Rego J, Lebaron P, Nebe-von Caron G (2000) Current and future applications of flow cytometry in aquatic microbiology. FEMS Microbiol Rev 24:429–448

Wallner G, Erhart R, Amann R (1995) Flow cytometric analysis of activated sludge with rRNA-targeted probes. Appl Env Microbiol 61:1859–1866

Wallner G, Fuchs B, Spring S, Beisker W, Amann R (1997) Flow sorting of microorganisms for molecular analysis. Appl Env Microbiol 63:4223–4231

Wang X, Chan RKY, Cheng ASK (2005) Underwater cytometer for *in situ* measurement of marine phytoplankton by a technique combining laser-induced fluorescence and laser Doppler velocimetry. Opt Lett 30:1087–1089

West NJ, Scanlan DJ (1999) Niche-partitioning of *Prochlorococcus* populations in a stratified water column in the eastern North Atlantic Ocean. Appl Env Microbiol 65:2585–2591

Wood AM, Horan PK, Muirhead K, Phinney DA, Yentsch CM, Waterbury JB (1985) Discrimination between types of pigments in marine *Synechococcus* spp. by scanning spectroscopy, epifluorescence microscopy, and flow cytometry. Limnol Oceanogr 30:1303–1315

Worden AZ, Binder BB (2003a) Application of dilution experiments for measuring growth and mortality rates among *Prochlorococcus* and *Synechococcus* populations in oligotrophic environments. Aquat Microb Ecol 30:159–174

Worden AZ, Binder BJ (2003b) Growth regulation of rRNA content in *Prochlorococcus* and *Synechococcus* (marine cyanobacteria) measured by whole-cell hybridization of rRNA-targeted peptide nucleic acids. J Phycol 39:527–534

Worden AZ, Chisholm SW, Binder BJ (2000) In situ hybridization of *Prochlorococcus* and *Synechococcus* (marine cyanobacteria) spp. with rRNA-targeted peptide nucleic acid probes. Appl Env Microbiol 66:284–289

Worden AZ, Nolan JK, Palenik B (2004) Assessing the dynamics and ecology of marine picophytoplankton: The importance of the eukaryotic component. Limnol Oceanogr 49: 168–179

Yentsch CM, Horan PK (1989) Cytometry in the aquatic sciences. Cytometry 10:497–499

Yentsch CM et al (1983) Flow cytometry and cell sorting: A technique for analysis and sorting of aquatic particles. Limnol Oceanogr 28:1275–1280

Yentsch CS, Phinney DA (1989) A bridge between ocean optics and microbial ecology. Limnol Oceanogr 34:1694–1705

Zhang H, Bhattacharya D, Maranda L, Lin S (2008) Mitochondrial *cob* and *cox1* and their mRNA editing in *Dinophysis acuminata* from Narragansett Bay, with special reference to the phylogenetic position of *Dinophysis*. Appl Env Microbiol 74:1546–1554

Zhang K, Martiny AC, Reppas NB, Barry KW, Malek J, Chisholm SW, Church GM (2006) Sequencing genomes from single cells by polymerase cloning. Nature Biotech 24:680–686

Zhu F, Massana R, Not F, Marie D, Vaulot D (2005) Mapping of picoeucaryotes in marine ecosystems with quantitative PCR of the 18S rRNA gene. FEMS Microbiol Ecol 52:79–92

Zubkov MV, Burkhill PH (2006) Syringe pumped high speed flow cytometry of oceanic phytoplankton. Cytometry A 69A:1010–1019

Zubkov MV, Burkhill PH, Topping JN (2007) Flow cytometric enumeration of DNA-stained oceanic planktonic protists. J Plankton Res 29:79–86

Zubkov MV, Fuchs BM, Archer SD, Kiene RP, Amann R, Burkill PH (2001) Linking the composition of bacterioplankton to rapid turnover of dissolved dimethylsulphoniopropionate in an algal bloom in the North Sea. Environ Microbiol 3:304–311

Zubkov MV, Fuchs BM, Tarran GA, Burkill PH, Amann R (2003) High rate of uptake of organic nitrogen compounds by *Prochlorococcus cyanobacteria* as a key to their dominance in oligotrophic oceanic waters. Appl Env Microbiol 69:1299–1304

Zwirglmaier K, Jardillier L, Ostrowski M, Mazard S, Garczarek L, Vaulot D, Not F, Massana R, Uloa O, Scanlan DJ (2008) Global phylogeography of marine *Synechococcus* and *Prochlorococcus* reveals a distinct partitioning of lineages among oceanic biomes. Environ Microbiol 10: 147–161

Chapter 9
The Use of the Fluorescence Signal in Studies of Seagrasses and Macroalgae

Susana Enríquez and Michael A. Borowitzka

1 Introduction

Seagrasses and macroalgae (seaweeds) are important primary producers and habitats in estuarine and marine intertidal and shallow benthic areas. Unlike terrestrial plants, aquatic macrophytes also show a wide diversity in photosynthetic and accessory pigment systems and chloroplast structure (Larkum and Vesk 2003; Larkum 2003). Furthermore, aquatic plants are exposed to a more variable light field than terrestrial plants with significant spectral changes with depth and water quality and, in shallow waters, extremely high light flashes due to the lensing effects of waves (Enríquez et al. 2002; Hanelt et al. 2003). Considering that the main structural and most likely functional diversity of photosynthetic mechanisms is present in the wide taxonomic diversity of the aquatic environment, the study of the fluorescence signal on marine macrophytes presents unique challenges and opportunities for the better understanding of how photosynthetic organisms utilize light.

Chlorophyll *a* fluorescence studies have proven to be a powerful tool for the elucidation of fundamental processes in photosynthesis, with important advances made especially between 1960 and the mid-1980s (see Schreiber and Bilger 1993). The ecophysiological application of the broad number of techniques that use the chlorophyll *a* fluorescence signal for the study of plant photosynthetic performance and/or stress is, however, more recent, and the result of a successful collaboration between basic photosynthesis research, applied research and ongoing instrument development (van Kooten and Snel 1990). This application has allowed the description of spatial and temporal patterns in the natural variation of the chlorophyll *a* fluorescence signal, with relevant achievements such as the description of the diurnal down-regulation of the dissipation of light energy absorbed in excess by PSII (e.g. Demmig-Adams and Adams 1992; Schreiber and Bilger 1993; Björkman and Demmig-Adams 1994). However, most of this pioneer applied research has been performed on terrestrial plants and unicellular algae. The application of chlorophyll *a* fluorescence for the study of aquatic macrophyte photosynthesis was, however, limited until the appearance in April 1997 of the first underwater Pulse-Amplitude-Modulated fluorometer (Diving-PAM) developed by Gademann and Schreiber and made commercially available by Heinz Walz GmbH, Germany. From 1995 till now, the number of articles published applying chlorophyll fluorescence to the analysis of the photosynthetic performance of macroalgae and seagrasses, has increased more than five times (source ISI-Thomson Web of Knowledge). It now represents more than 20% of the total literature produced in aquatic sciences, although its application to aquatic multicellular organisms is still far below that on unicellular algae, either free-living in the water column or as endosymbionts within corals. Chlorophyll *a* fluorescence is not only being used in studies of the photosynthetic process and in the ecophysiology of seagrasses and macroalgae,

S. Enríquez (✉)
Laboratorio de Fotobiología, Unidad Académica Puerto Morelos, Instituto de Ciencias del Mar y Limnología, Universidad Nacional Autónoma de México, Apdo. Postal 1152, Cancún, Quintana Roo, México
e-mail: enriquez@cmarl.unam.mx

M.A. Borowitzka
Algae R&D Center, School of Biological Sciences and Biotechnology, Murdoch University, Murdoch 6150, WA, Australia

D.J. Suggett et al. (eds.), *Chlorophyll* a *Fluorescence in Aquatic Sciences: Methods and Applications*, Developments in Applied Phycology 4, DOI 10.1007/978-90-481-9268-7_9, © Springer Science+Business Media B.V. 2011

but is also finding applications in the improvement of the culture of commercially important algae species (Hwang et al. 2007; Pang et al. 2007).

In this chapter, we describe the major achievements in seagrass and macroalgae studies generated by the use of the chlorophyll *a* fluorescence signal. In general, large spatial variability within the same individual has been recognized (e.g., Durako and Kunzelman 2002; Enríquez et al. 2002; Turner and Schwarz 2006; Colombo-Pallota et al. 2006; Strain et al. 2006; Belshe et al. 2007), due to intrinsic components related to the structural complexity of these organisms (i.e., anatomy, self-shading, tissue age, conditions and light history). This large intrinsic variation is illustrated in this first section of the chapter with two contrasting examples: one from the study of Colombo-Pallota et al. (2006) on the giant kelp *Macrocystis pyrifera*; and the other from the study of Enríquez et al. (2002) on the tropical seagrass *Thalassia testudinum*. Despite this large natural variation, chlorophyll fluorescence has proven to be a powerful tool for the analysis of spatial and temporal variation in quantum efficiency and photosynthetic performance, and for the analysis of short-term responses to several environmental factors.

The determination of the rate of non-cyclic electron transport (ETR) between PSII and PSI also has attracted much attention together with the comparison between ETR values and gross photosynthetic rates (Gross Photosynthesis, GPS) measured as oxygen evolution (see also Suggett et al., Chapter 6, this volume). However, results are controversial due, in part, to methodological problems and misconceptions in the interpretation of the variation of the chlorophyll *a* fluorescence signal. Therefore, in this chapter a definition of the most common parameters and their notations is also provided (but see also Cosgrove and Borowitzka, Chapter 1, this volume) for details on terminology). We define direct fluorescence parameters such as maximum (F_v/F_m) and effective ($\Delta F/F_m{'}$) photochemical efficiency of Photosystem II (PSII), photochemical (q_P) and non-photochemical (q_N) quenching coefficients, the photoprotective capacity of the tissue to dissipate the excess of energy absorbed (non-photochemical quenching, NPQ), and, thanks to quenching analysis and the use of chemicals the distinction among its different components (q_E, q_T and q_I). We also define indirect parameters such as the rate of non-cyclic electron transport (ETR) between PSII and PSI, derived from the knowledge of other properties of the tissue such as the functional absorption cross-section of PSII (σ_{PSII}); the functional absorption cross-section of the photosynthetic unit ($\sigma_{PSII+PSI}$ or σ_{PSU}); and the fraction of incident light absorbed by the tissue (Absorptance, A).

The use of new descriptors such as relative ETR (rETR) proposed by Ralph et al. (2002a) is also examined in this chapter as rETR has found increasing acceptance. ETR or rETR are commonly utilized to describe the photosynthetic response to irradiance, as an analogue of the P vs E curves (i.e., ETR vs E curves). A description of this response appears often derived from an instrumental option of the PAM fluorometers commercialized by Heinz Walz GmbH, known as the Rapid Light Curve (RLC). Due to its simplicity and convenience, RLCs have been extensively used for the study of the photosynthesis performance of marine macrophytes. However, this experimental determination (RLC) utilizes very short illumination periods (<1 min) and, consequently, it does not support the definition of ETR, which requires $\Delta F/F_m{'}$ to be determined at the photosynthetic steady-state.

The second section of this chapter ("Protocols used, limitations and specific modifications for aquatic macrophytes") focuses on the specific problems and limitations that the use and interpretation of the fluorescence signal has in seagrass and macroalgal studies. We aim to highlight the major problems and to establish the limits for the interpretation of the chlorophyll *a* fluorescence signal imposed by the methodology used and the current molecular model of the photosynthetic apparatus.

In conclusion, we would like to stress that the use of the chlorophyll *a* fluorescence signal in seagrass and macroalgae studies opens many more methodological research possibilities than the simple estimation of photosynthesis. Overall it is a powerful tool to assess the properties of the photosynthetic apparatus and its response to the environment. Comparative analysis within or among species will allow the description of significant differences among species and/or populations in their sensitivity to light stress, light limitation or any other environmental pressures. Moreover, it will allow the understanding of the broad structural and functional diversity of the photosynthetic apparatus and mechanisms still present in the marine environment. At this time studies on macroalgae are still rather limited compared to those on seagrasses, especially when considering the wide diversity of macroalgae.

Marine ecophysiology and, specifically, the application of chlorophyll fluorescence to the study of macrophyte photosynthesis, has enormous potential to develop a fuller understanding of this natural diversity. This is probably one of the most important challenges of the use of the chlorophyll fluorescence signal in seagrass and macroalgae studies and its main contribution to photosynthesis research.

2 Major Achievements Using the Chlorophyll *a* Fluorescence Signal in Seagrass and Macroalgae Studies

Chlorophyll *a* fluorescence parameters show large variability within the same individual for complex organisms such as marine macrophytes (e.g., Durako and Kunzelman 2002; Enríquez et al. 2002; Turner and Schwarz 2006; Colombo-Pallota et al. 2006; Strain et al. 2006; Belshe et al. 2007b). This spatial variability responds to environmental variation, but also has an intrinsic component due to self-shading within the canopy (Enríquez et al. 2002) and differences within the same individual in tissue age, condition and light history (e.g., Enríquez et al. 2002; Colombo-Pallota et al. 2006; Turner and Schwarz 2006). As Critchley and Russell (1992) have previously recognized for terrestrial plants, structural complexity is usually accompanied by large heterogeneity in light exposure at different levels of plant organization: from the level of the photosynthetic membrane, to the light gradients created within grana stacks, within leaves and within canopies. Similar levels of heterogeneity can affect aquatic macrophyte photosynthesis. However, larger intrinsic variation in the chlorophyll *a* fluorescence signal can be expected in the marine environment considering the large variation in intensity and quality of the light fields, and the broad variability present in life forms, tissue anatomy, absorption cross-section, type of pigmentation and photosynthetic machinery.

The giant kelp *Macrocystis pyrifera* represents one extreme of this variation. This brown alga occupies the entire water column (>18 m depth) and in consequence, the blades are exposed to large changes in irradiance and light quality. Significant variation in the fluorescence signal associated with changes in tissue pigmentation, light absorption capacity and photosynthetic performance has been reported for the blades of *M. pyrifera* (Colombo-Pallota et al. 2006). Similarly, only a few centimeters within the canopy serve to produce significant differences in the photoacclimatory condition of the leaf of the tropical seagrass *Thalassia testudinum* (Enríquez et al. 2002). In their study, Enríquez et al. (2002) found a 19% of reduction in the maximum photochemical yield of PSII (F_v/F_m) along the second youngest leaf, from a maximum value located at 4 cm from the sediment and a progressive reduction towards the minimum observed in the tip at 19 cm from the sediment. This study also found an initial increase in F_v/F_m from 0.8 at the water-sediment interface to the maximum of 0.85 observed at 4 cm from the sediment (Enríquez et al. 2002). The initial increase has been explained as tissue maturation. Most seagrasses have basal leaf meristems, which are very often buried in the sediment. On the other hand, the 19% of progressive reduction in F_v/F_m has been interpreted as a photoacclimatory response of the leaf to the strong light gradient measured in the same study within the canopy (Enríquez et al. 2002). According to these authors, the increment in daily dose of saturating irradiance as the leaf elongates and occupies higher and more illuminated positions within the canopy, could explain the pattern of F_v/F_m variation found along the leaf. Enríquez et al. (2002) reported a strong, negative association between F_v/F_m and the percentage of daylight hours of exposure to saturating light (H_{sat}). In support of this observation, Cayabyab and Enríquez (2007) found a positive and linear association between the estimated H_{sat} for three experimental light treatments and the field-control treatment, and the relative F_v/F_m reduction observed between apical and basal leaf sections of the same seagrass species.

Despite the large natural variation in the fluorescence signal often shown by seagrasses and macroalgae, chlorophyll fluorescence has proven to be a powerful tool for the analysis of diurnal variation in quantum efficiency (e.g., Hanelt 1992; Hanelt et al. 1993; Franklin et al. 1996; Cabello-Pasini et al. 2000; Enríquez et al. 2002; Campbell et al. 2003; Silva and Santos 2003; Lan et al. 2005), and spatial and temporal variation in tissue photosynthetic performance: within the same individual (e.g., Enríquez et al. 2002; Colombo-Pallota et al. 2006; Cayabyab and Enríquez 2007); between populations (e.g., Häder et al. 1998; Durako and Kunzelman 2002; Major and Dunton 2002; Durako et al. 2003; Silva and Santos 2003;

Turner and Schwarz 2006; Strain et al. 2006; Campbell et al. 2003; Belshe et al. 2007b); and among species (e.g., Hanelt 1992; Häder et al. 1998; Campbell et al. 2003; Durako et al. 2003; Silva and Santos 2003; Gómez et al. 2004; Lan et al. 2005). In some species of seagrass (*Zostera capricorni* and *Halophila ovalis*) lateral heterogeneity has also been shown using fluorescence imaging (Ralph et al. 2005).

The utility of chlorophyll *a* fluorescence has also been applied to the analysis of short-term responses to: (1) light intensity and quality (e.g. Flores-Moya et al. 1998, Häder et al. 1998, Ralph 1999, Ralph and Burchett 1995, Campbell et al. 2003, Cayabyab and Enríquez 2007); (2) light limitation (e.g., Lamote and Dunton 2006); (3) UV radiation (e.g., Dawson and Dennison 1996; Aguirre-von-Wobeser et al. 2000; Yakovleva and Titlyanov 2001; Bischof et al. 2002; van de Poll et al. 2002; Apprill and Lesser 2003; Roleda et al. 2006); (4) desiccation (e.g., Björk et al. 1999; Seddon and Cheshire 2000); (5) water flow (Enríquez and Rodríguez-Román 2006); (6) temperature (e.g., van de Poll et al. 2002; Campbell et al. 2006); (7) osmotic stress (e.g., Ralph 1998); (8) nutrients and inorganic carbon (e.g., Flores-Moya et al. 1998; Carr and Björk 2003; Cabello-Pasini and Figueroa 2005); (9) herbicide addition (e.g., Ralph 2000; Haynes et al. 2000; Harrington et al. 2005; Küster and Altenburger 2007); and (10) pathogens (e.g., Ralph and Short 2002).

Several types of fluorometers are commercially available (Huot and Babin, Chapter 3, this volume). Some fluorometers are able to perform rapid measurements, which allow the determination of Q_A-reoxidation kinetics, S-states, effective antenna size and/or the OJIP transients. However, the most popular device applied for the description of the variation in the photosynthetic performance of seagrasses and macroalgae, utilizes Pulse-Amplitude-Modulated (PAM) fluorometry (Schreiber 2004), to measure the variation in the maximum (F_v/F_m) and effective ($\Delta F/F_m'$) photochemical efficiencies of photosystem II (PSII), the most common parameters obtained from the instrument. The notation F_v/F_m refers to the maximum photochemical efficiency of PSII, and requires complete relaxation of the competing mechanisms with the photochemical energy conversion. $\Delta F/F_m'$ is the notation used to describe the variation in the photochemical efficiency of PSII under illuminated conditions. If this parameter is determined at the photosynthetic steady-state at a certain irradiance (E), it may provide information about the ability of this organism to move electrons beyond PSII. Knowledge of the absorption cross-section of PSII may allow determining the rate of non-cyclic electron transport (ETR) between PSII and PSI. Genty et al. (1989) presented a general equation for the estimation of ETR values:

$$\mathrm{ETR} = \Delta F / F'_m \times E_{PAR} \times \sigma_{PSII} \tag{1}$$

where E_{PAR} is the instantaneous incident irradiance in µmol quanta $m^{-2}\ s^{-1}$, and σ_{PSII} is the functional absorption cross-section of PSII. σ_{PSII} can be derived from fluorescence measurements, using the pump-and-probe technique (Falkowski and Raven 1997).

Without direct knowledge of σ_{PSII}, many authors that work with two dimensional structures (2D, i.e., tissues) have substituted this parameter with the product of the fraction of incident light absorbed by the photosynthetic pigments of the organism (Absorptance, A) and the fraction of that light absorbed by PSII ($\sigma_{PSII} / \sigma_{PSU}$) as:

$$\mathrm{ETR} = \Delta F / F'_m \times E_{PAR} \times A \times \frac{\sigma_{PSII}}{\sigma_{PSU}} \tag{2}$$

Therefore, ETR is referred to surface area units. Assuming that a terrestrial leaf generally absorbs 84% of the incident irradiance, and that 50% of these quanta are absorbed by PSII, this equation has been simplified to:

$$\mathrm{ETR} = \Delta F / F'_m \times E_{PAR} \times 0.84 \times 0.5 \tag{3}$$

In mature, healthy terrestrial plant leaves, 0.84 may represent a constant value for absorptance (A, see Gabrielsen 1948; Björkman and Demmig 1987) but certainly this is not the case for the photosynthetic tissues of marine macrophytes, which have shown a large variability for this functional descriptor (see Enríquez et al. 1992; Campbell et al. 2006 and Durako 2007 for seagrasses and Enríquez et al. 1994 for marine macrophytes). Similarly, the assumption that 50% of the total light absorbed by the tissue is absorbed by PSII may be correct in many cases, especially for chlorophytes, but other situations where this assumption is not valid cannot be ignored (see Baker and Oxborough 2004). However, the linear electron flow between PSII and PSI requires a balanced trapping of excitons by both reaction centers. Antenna disconnection, state-transitions and spill-over allow rapid down-regulation responses in order to balance suboptimal situations (Fork and Satoh 1986). Being aware of these restrictions, it is assumed that it is possible to estimate photosynthetic rates as ETR (in µmol e^- m^{-2} s^{-1}) from fluorescence measurements ($\Delta F/F_m'$) knowing the photosynthetically active irradiance value (E_{PAR}, in µmol quanta m^{-2} s^{-1}) for the steady-state of photosynthesis, and the fraction of incident light absorbed by the pigmented tissue (Absorptance, A), and assuming that this energy is equally distributed between both photosystems. The initial equation (1) of Genty et al. (1989) can be rewritten then as:

$$ETR = \Delta F / F'_m \times E_{PAR} \times A \times 0.5 \qquad (4)$$

In contrast to the limited effort invested in the description of tissue absorptance or PSII absorption cross-section, a large effort in ecophysiology of marine macrophytes has been invested in the application of the fluorescence technique for the determination of ETR values. Comparison between fluorescence-based photosynthetic rates (ETR) and gross photosynthetic rates (GPS) measured as oxygen evolution, has also attracted considerable attention (e.g., Osmond et al. 1993; Hanelt et al. 1995; Hanelt and Nultsch 1995; Beer et al. 1998, 2000; Beer and Björk 2000; Cabello-Pasini et al. 2000; Franklin and Badger 2001; Longstaff et al. 2002; Carr and Björk 2003; Silva and Santos 2004; Cabello-Pasini and Figueroa 2005; Colombo-Pallota et al. 2006; Enríquez and Rodríguez-Román 2006; Nielsen and Nielsen 2008). In general, a linear relationship has been found between ETR and GPS at low light levels, however, deviations from this linearity are common dependent on species, light history, irradiance, and other factors such as CO_2 and nitrate levels. One common observation is that, at saturating light, when inorganic carbon availability is low and/or oxygen evolution is limited by any other process, ETR values tend to overestimate O_2 evolution rates (cf. Franklin and Badger 2001; Enríquez and Rodríguez-Román 2006). For the linear relationships reported, deviations from a theoretical value of 4 for the slope between ETR and O_2 evolution rates have been commonly estimated.

The current molecular model (cf. Maxwell and Johnson 2000) establishes a value of four electrons transported beyond PSII for each molecule of oxygen evolved. Higher values are expected when several photoprotective mechanisms contribute to reduce PSII photochemical efficiency, while protecting the photosynthetic apparatus from photoinhibition (defined as a significant decline in the photosynthetic rates measured as oxygen evolution). However, errors in the determinations of the simultaneous GPS and ETR measurements or incorrect assumptions for the Absorptance (A) and/or σ_{PSII} / σ_{PSU} factors, may explain molar ratios < 4 and/or, perhaps, part of the variability so far reported for the ETR/GPS slope. To prevent the use of incorrect A and/or σ_{PSII} / σ_{PSU} values, a new parameter, relative ETR (rETR) has been proposed (Beer et al. 2001; Ralph et al. 2002a) and is receiving increasing acceptance. Relative ETR values result from the multiplication of the effective photochemical efficiency ($\Delta F/F_m'$) by the irradiance (i.e., E_{PAR} or $E_{PAR} \times 0.5$ depending on the author).

Using ETR as a descriptor of photosynthetic rates, the fluorescence signal has been used to describe the photosynthetic response to irradiance, as an analogue to the P *vs* E curves (i.e., ETR *vs* E curves). Different PAM fluorometers have the option of performing this type of analysis from an internal-controlled variation in the actinic light intensity supplied by the instrument. This instrumental option is called the rapid light curve (RLC). Due to their simplicity and convenience, RLCs have been extensively applied in the study of the photosynthesis performance of marine macrophytes (e.g. Beer et al. 1998; Ralph and Short 2002; Longstaff et al. 2002; Campbell et al. 2003; Carr and Björk 2003; Durako et al. 2003; Silva and Santos 2003; Ralph and Gademann 2005; Colombo-Pallota et al. 2006; Belshe et al. 2007a; Saroussi and Beer 2007). They have also been used for comparative analyses of the variation in maximum photosynthetic rates (ETR_{max}), photosynthetic efficiency (i.e., α or the slope of the linear phase at subsaturating light of the ETR *vs* E curves), and saturation irradiance (E_k).

Recently Nielsen and Nielsen (2008) have compared measurements of ETR and gross O_2 evolution in algae with different thallus ticknesses using a conventional PAM and an imaging PAM. They used the thin thallus alga *Ulva lactuca*, the thick thallus alga *Fucus serratus* and 1 to 8 cell thick 'artificial thalli' built from layers of *Enteromorpha intestinalis*. At subsaturating to saturating irradiances imaging ETR/ETR_{max} and P/P_{max} were similar irrespective of thallus thickness, however the conventional ETR/ETR_{max} determinations were unrelated to P/P_{max}. On the other hand, at super-saturating irradiances conventional ETR/ETR_{max} and P/P_{max} values were similar, but imaging ETR/ETR_{max} values were significantly less. These results appear to be a function of the depth to which the saturating light pulse penetrates the tissue (different between the imaging and the conventional PAM), which in turn affects the determination of F_0 and F_m. Therefore, they highlight potential errors related to the fluorometer used and/or to the thickness of the photosynthetic tissue of the organism.

2.1 Quenching Analysis

The most important contribution of PAM-fluorometry has been the development of quenching analysis (Schreiber et al. 1998). Surprisingly, quenching analysis so far has received little attention in the study of aquatic macrophyte photosynthesis. Quenching analysis allows the determination of the photochemical

$$q_P = \frac{(F_m' - F)}{F_m' - F_o'} \quad (5)$$

and non-photochemical quenching coefficients

$$q_N = \frac{1-(F_m' - F_o')}{(F_m - F_o)} \quad (6)$$

As both coefficients are difficult to be estimated (and interpreted) under F_o quenching (F_o'), another parameter has been develop to describe the magnitude of the non-photochemical quenching, NPQ. This parameter is defined as

$$NPQ = \frac{(F_m - F_m')}{F_m'} \quad (7)$$

and describes the photoprotective capacity of the tissue to dissipate the excess of energy absorbed as heat, by means of several non-photochemical quenching processes, and/or to balance the excitation energy between both photosystems via changes in the absorption cross-section of PSII or state-transitions. Diurnal variation in NPQ associated with changes in the instantaneous irradiance and the operation of the xanthophyll cycle has been found to occur in seagrasses (e.g., Ralph et al. 2002b) and macroalgae (e.g., Franklin et al. 1992, 1996; Duval et al. 1992; Urmacher et al. 1995; Schofield et al. 1998; Bischof et al. 2006), but also its lack of operation has been reported for some algal species (Franklin et al. 1996; Franklin and Larkum 1997). The xanthophyll cycle (XC) is one of the most important down-regulation mechanisms in higher plants and several algal Divisions, for the protection of the photosynthetic membranes from an excess of energy absorbed (Franklin et al. 1992, Demmig-Adams and Adams 1992; Niyogi et al. 1998; Rodrigues et al. 2002). The xanthophyll cycle is the conversion of violaxanthin into antheraxanthin and zeaxanthin induced by high light intensity, and the back reactions under low light conditions (Yamamoto et al. 1962). Ralph et al. 2002b found relative low values of NPQ (between 0 and 0.6) for the seagrass *Zostera marina* and that most of the XC pool was in the form of antheraxanthin (> 70%). An unusually high concentration of the intermediate compound, antheraxanthin, was found under wide range of light conditions. The authors interpret this result as due to a partial (incomplete) conversion of violaxanthin to zeaxanthin, which may explain the unusually low NPQ values observed for a higher plant. Other seagrass species, however, have shown: (i) increases in the proportion of zeaxanthin in the total XP pool with increasing irradiances (i.e., *Zostera capricorni*, Flanigan and Critchley 1996); and (ii) higher NPQ values after the peak of irradiance (midday), and for the leaves grown under high light (i.e., NPQ ≈ 3, *Halophila stipulacea*, Sharon and Beer 2008). The increment of zeaxanthin in the total XC pool at higher irradiances is indicative of the presence of a fully functioning photoprotective xanthophyll cycle. The NPQ values estimated for *Halophila stipulacea* are closer to the values observed in higher plants. More studies need to be done to determine if some seagrass species have developed a different functional pattern in NPQ to the pattern evolved by their higher plant ancestors.

NPQ variation in three brown macroalgae *Laminaria abyssalis, L. digitata* (Rodrigues et al. 2002) and *Macrocystis pyrifera* (García-Mendoza and Colombo-Pallota 2007) has been described in two interesting studies. The first study examines the importance of the xanthophyll cycle in the photoprotective ability of the two *Laminaria* species, and concludes that the higher susceptibility to photoinhibition shown by *L. abyssalis* could be explained by its limited de-epoxidation capacity and reduced xanthophyll-cycle pool size. This provides a mechanistic explanation for the differential depth distribution pattern observed: *L. abyssalis* is restricted to deep waters, whereas *L. digitata* grows in the intertidal area. The second study (García-Mendoza and Colombo-Pallota 2007) describes NPQ variation along the blades of the giant kelp *M. pyrifera.* These authors found a significant variation in NPQ between the surface blades (NPQ values > 4) and the deepest blades located at 18 m depth (NPQ < 1). This variation was associated with other important changes in the photoacclimatory condition of the tissue, such as changes in pigment content, absorptance, photosynthetic rates, etc., in response to the exponential light attenuation of the water column (Colombo-Pallota et al. 2006). However, the relevant contribution of this study is the observation that the large NPQ found was related to the amount of zeaxanthin synthesized and, in contrast to higher plants, the transthylakoid proton gradient was not associated with it. A general assumption of the photosynthetic model is that the transthylakoid proton gradient (ΔpH) triggers zeaxanthin synthesis through the xanthophyll cycle (Horton et al. 1991). García-Mendoza and Colombo-Pallota (2007) interpretation of this observation is that the proton gradient only serves to activate the violaxanthin de-epoxidase enzyme, but does not induce NPQ. This observation illustrates the enormous potential that the application of chlorophyll *a* fluorescence has in marine ecophysiology.

So far, most of the information provided by the analysis of the natural variation of the fluorescence signal has been derived from higher plants and green algae. The photosynthetic apparatus of green algae has been the most successful in the colonization of the terrestrial environment, but the main structural and most likely functional diversity of photosynthetic mechanisms is still present in the wide taxonomic diversity of the algae found in the marine environment. The observation reported by García-Mendoza and Colombo-Pallota (2007) is of especial interest as it highlights the important role of marine macrophyte ecophysiology for the full understanding of this natural diversity.

2.2 Analysis of Quenching Components: Use of Chemicals

Quenching analysis is strengthening by the use of different chemicals that inhibit the activity of specific mechanisms. It also allows distinction among different components of NPQ, such as q_E, q_T and q_I (see Schreiber et al. 1994). Although the current knowledge of NPQ in aquatic macrophyte photosynthesis is still limited, even less information is available on the variation of q_E, q_T and q_I. The study of García-Mendoza and Colombo-Pallota (2007) on *Macrocystis pyrifera,* used the inhibitor of zeaxanthin synthesis, dithiothreitol (DTT), and the uncoupler NH_4Cl which disrupts the trans-thylakoid proton gradient (q_E). The slow NPQ relaxation observed upon the addition of the uncoupler took place only in the presence of de-epoxidated XC pigments as a result of zeaxanthin epoxidation. Based on this observation, the authors concluded that the transthylakoid gradient was only related to the activation of the violaxanthin de-epoxidase enzyme, but not to the induction of NPQ, as a fast NPQ relaxation, similar to that reported for higher plants (Horton et al. 1991) and green algae (García-Mendoza et al. 2002), was not observed in the response of *M. pyrifera.*

There is also little information available on the variation of the fluorescence signal of aquatic macrophytes attributed to state transitions (q_T). Seagrasses are higher plants monocotyledons and no significant differences are expected from the variation predicted by the current model. However, large differences may be present in the other taxonomic groups either in the mechanism involved and/or in the importance of state transition for photosynthesis down-regulation, especially in phycobilisome-containing organisms such as red algae. In the red algae fluorescence changes commonly associated with state 2 transition have been reported to be ΔpH-dependent, as low intensities of a light that preferentially activates PSII induce a ΔpH gradient across the membrane (Delphin et al. 1996, 1998; Ritz et al. 1999). This ΔpH gradient can be better observed if the ATPase is inhibited with dicyclohexyl carbodiimide (DCCD) (Delphin et al. 1996). Recent work on cyanobacteria has reported carotenoid-related quenching associated with a carotenoid-binding protein that may be responsible for the energy dissipation through interaction with the phycobilisome core (Rakhimberdieva et al. 2004; Kirilovski 2007). Fluorescence quenching in red algae resembles cyanobacterial quenching, but the role of state-transitions in photoprotection and the presence of a similar mechanism in red algae has not been investigated as yet. More effort is still required in marine research to comprehend the diversity of down-regulatory mechanisms and their specific benefits and/or costs for species survival.

The third component of NPQ, q_I, is the slowest quenching coefficient. This coefficient is associated with photoinhibition (Schreiber 2004). The study of q_I is also the analysis of the dependence of PSII activity on the repair mechanisms, more specifically, on D1 protein turnover, as this protein is considered the primary site of PSII photodamage (Prasil et al. 1992; Long et al. 1994; Ohad et al. 1994). D1 protein expression and concentration can be directly measured using molecular techniques. The indirect analysis of D1 protein turnover requires the use of different inhibitors of chloroplast protein synthesis such as chloramphenicol (CAP), streptomycin or lincomycin. The dependence of the PSII repair mechanism on D1 protein synthesis has been confirmed for macroalgae (e.g., Franklin and Larkum 1997; Schofield et al. 1998; Häder et al. 2002) and seagrasses (e.g., Flanigan and Critchley 1996; Enríquez et al. 2002). For some species such as the filamentous green alga, *Chlorodesmis fastigiata*, the quantum requirement for PSII inactivation was estimated to be 2×10^7 photons (Franklin and Larkum 1997), a value slightly lower, although similar in magnitude, to that reported for higher plants (3×10^7 photons, Park et al. 1995). In seagrasses, differences in the photoacclimatory response of leaves associated with inhibition of D1 protein synthesis have been reported (Flanigan and Critchley 1996; Enríquez et al. 2002). In the seagrass *Thalassia testudinum* leaf segments exposed for most of the day to saturating irradiances as the leaf grows and the segments occupy upper positions in the canopy, showed a photoacclimatory response to high light, and, associated with it, a suppression of the PSII repair mechanism under illumination. In contrast, shaded segments from the basal part of the leaf, maintain higher costs of D1 protein synthesis under illumination to prevent a drastic decline in photosynthesis efficiency (Enríquez et al. 2002). In this study, the variation in the fraction of PSII unable to reduce Q_B was estimated using another type of fluorometer, the Plant Efficiency Analyser (PEA, Hansatech Instruments Ltd, UK). The fast temporal resolution of this type of fluorometer allows the analysis of the OJIP transients (see Strasser et al. 2004). The OJIP sequence represents the sequential reduction of PSII internal electron acceptors (Govindjee 1995). Enríquez et al. (2002) reported a sequential increase in the amplitude of the J transient towards the tip of the leaf. They interpret this observation as the accumulation of closed PSII, known as Q_B non-reducing, in the more illuminated leaf sections. Q_B is the secondary electron acceptor of PSII and, consequently, facilitates the re-oxidation of Q_A. When leaves are unable to reduce Q_B, for instance after infiltrating them with the herbicide 3′-(3,4-dichlorophenyl)-1′1′-dimethyl urea (DCMU), PSII can only undergo a single turnover. This leaf condition leads to the transformation of the sequence OJIP into OJ (Strasser et al. 1995). This observation has facilitated the interpretation of the variation in the amplitude of the J transient in regular, non-poisoned leaves. This amplitude is described as the normalized fluorescence yield

$$V(t) = \frac{(F(t) - F_0)}{F_m - F_0} \quad (8)$$

at the J transient [V(J)], and parameterizes the variation of the fraction of PSII unable to reduce Q_B. In their study Enríquez et al. (2002) interpret the sequential increase in the amplitude of the J transient observed towards the tip of the leaf as the accumulation of Q_B non-reducing PSII in the more illuminated leaf sections. This increment in the amplitude of the J transient was significantly and negatively associated with the 19% of reduction observed in F_v/F_m ($r = -0.97$, $P < 0.01$, Fig. 1).

The progressive accumulation of Q_B non-reducing PSII in the more illuminated apical sections of *T. testudinum* leaves is consistent with acceptor-side

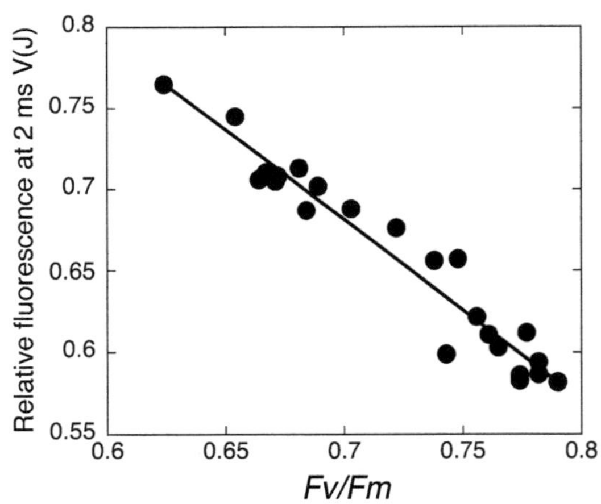

Fig. 1 Negative association between the variation observed in F_v/F_m along the mature section of the second youngest leaf of *Thalassia testudinum* and the changes observed in the normalized induction curve at the J (2 ms) transient (From Enríquez et al. 2002)

photoinhibition (Vass et al. 1992). However, the increment in the fraction of PSII able to generate charge separation but unable to reduce Q_B, may be related to the induction of a photoprotective mechanism (Öquist et al. 1992; van Wijk and van Hasselt 1993). The importance of this alternative mechanism in relation to the regular operation of the xanthophyll cycle in seagrasses needs still to be clarified, but the results of Enríquez et al. (2002) support previous interpretations suggested for *Zostera capricorni* (Flanigan and Critchley 1996) and terrestrial plants (van Wijk and van Hasselt 1993; Critchley and Russell 1994) of the importance of D1 protein turnover regulation in the photoacclimatory response of plants, and the involvement of Q_B non-reducing PSII in photoprotection.

3 Protocols Used, Limitations and Specific Modifications for Aquatic Macrophytes

3.1 Determination of the Variation in F_v / F_m and $\Delta F/F_m'$

The most common parameters obtained from the analysis of the fluorescence signal in marine macrophytes, are F_v/F_m and $\Delta F/F_m'$. F_v is known as variable fluorescence and results from the difference in the minimum (F_o) and maximum (F_m) fluorescence. For an accurate determination of F_v/F_m, a complete relaxation of the competing mechanisms with the photochemical energy conversion (non-photochemical quenching, NPQ) is necessary. As some of these mechanisms have slow kinetics, a long dark-adaptation period of the tissue (6–7 h see Enríquez et al. 2002) for organisms previously exposed to full sun light is recommended. An initial evaluation is required to allow the determination of the minimum number of hours required by a specific organism for complete relaxation of NPQ. The length of this period is highly dependent on the species, photoacclimatory condition and, specially, on the previous light exposure of the tissue. From *in situ* field determinations, knowledge of the diurnal variation in the effective photochemical efficiency ($\Delta F/F_m'$) may allow F_v/F_m determinations from the maximum diurnal value (preferentially at sun-rise, but, sometimes it is possible to measure it a few hours after sun-set).

F_v/F_m and $\Delta F/F_m'$ have a large spatial variation depending on the species, population and/or light environmental fluctuations. *Macrocystis pyrifera* and *Thalassia testudinum* exemplify the large variation in tissue condition (i.e., light exposure, age and light history) often shown by aquatic macrophytes. Any analysis of the natural variation in the fluorescence signal of seagrasses and macroalgae requires knowledge of the magnitude of the intrinsic variability present within the organism examined. This initial analysis will allow for the selection of the appropriate tissue (i.e., age, orientation, position) for the comparison.

To illustrate the importance of this intrinsic source of variability in F_v/F_m, nine different patterns of F_v/F_m variation along the second youngest leaf of *T. testudinum* are shown in Fig. 2. Leaves belong to the same nine meadows examined by Enríquez and Pantoja-Reyes (2005). This comparison uncovers significant differences in the F_v/F_m pattern among sites, and large F_v/F_m variation within the same leaf for some sites (Fig. 2). The magnitude of this variation does not correlate with the magnitude of leaf self-shading within the canopy (Enríquez and Pantoja-Reyes 2005). Site number 5, located at 2.8 m depth, showed the largest F_v/F_m variation: 66.7% of F_v/F_m reduction between the maximum value observed at 2 cm from the basal meristem and the minimum value measured at the leaf tip. This site had much lower canopy attenuation coefficient than site number 7 (canopy K_d = 3.9 and 11.5 m^{-1}, respectively),

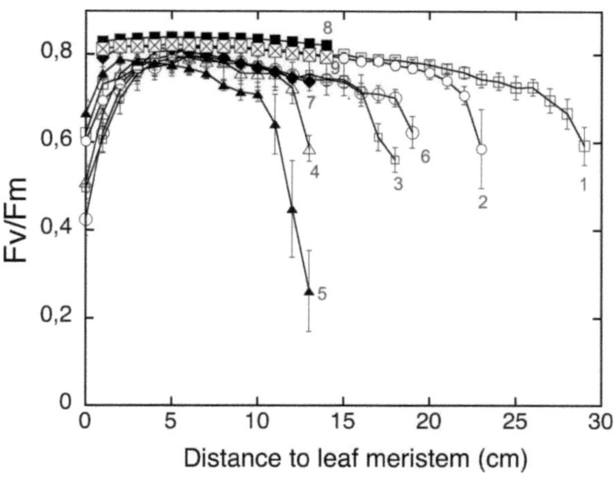

Fig. 2 The patterns of F_v/F_m variation along the second youngest leaf of the seagrass *Thalassia testudinum* are shown for the nine different meadows compared in Enríquez and Pantoja-Reyes (2005). Leaves were dark-adapted for 6–7 h according to Enríquez et al. (2002). Each point represents the average ± SE of 30 leaves. Measurements were taken every 1 cm

which only showed 9.9% of F_v/F_m reduction and was located near the surface at 0.5 m depth. The variation in F_v/F_m was even lower in site number 9 (3.3% of F_v/F_m reduction), located at 4 m depth and with similar leaf self-shading than that estimated for site number 5 ($K_d = 3.3\ m^{-1}$). The nine seagrass meadows studied by Enríquez and Pantoja-Reyes (2005), although located within the same fringing reef lagoon are probably exposed to different environmental stressors, which may explain the large differences found in the pattern of F_v/F_m variation along the leaf. As the leaves also differ in elongation rate and life span, the effect of tissue age, condition and light history cannot be ignored in the interpretation of this F_v/F_m variation (Enríquez et al. 2002).

This comparison serves once again to emphasize the large intrinsic natural variation in tissue light exposure, age and light history of aquatic macrophytes, which strongly affects the variability in the fluorescence signal. This variation has to be considered for any comparative study, and, especially, for the analysis of relative changes in tissue condition and photosynthetic performance.

As stated before, $\Delta F/F_m'$ is the notation used to describe the variation in the photochemical efficiency of PSII under illuminated conditions. The diurnal variation in $\Delta F/F_m'$ is dependent on the environmental light fields (water column and canopy) but is also related to the intrinsic components indicated previously. If this parameter is determined using a leaf clip or any other device that shades the tissue, photochemical quenching relaxes and induces a fast variation in $\Delta F/F_m'$. This variation is not dependent on the current condition of the tissue but on shading. To prevent this variation, it is recommended that the tissue is not shaded while determining $\Delta F/F_m'$, but this is not possible for most macroalgae and seagrass species. Macrophyte photosynthetic tissues move freely in the water column and require attachment to the fiber optics and the instrument. In consequence, it is critical to maintain a similar dark-adaptation period among replicates to allow photochemical quenching (q_P) to completely relax (≈ 1 min, see Iglesias-Prieto et al. 2004). The use of a similar dark-adaptation period will significantly reduce $\Delta F/F_m'$ variability among replicates.

Epiphytic organisms: A very common feature of seagrasses and macroalgae is the presence of epiphytic organisms on the leaf or thallus surface (Bulthuis and Woekerling 1983; Cebrián et al. 1999; Borowitzka et al. 2006). The possible effect of these organisms, especially diatoms and other small algae on the photosynthetic performance of aquatic macrophytes has been little studied. As the presence of these organisms may interfere in the chlorophyll *a* fluorescence signal of the species analysed, most workers avoid this potential problem by selecting areas of the tissue relatively free of epiphytes. However, some species, such as the long-lived leaves of the seagrass of the genus *Posidonia* (Trautman and Borowitzka 1999), may have a highly structured and dense epiphyte community established on the leaves (Ballesteros 1987; Cebrián et al. 1999). The potential effect of the epiphyte community on the photosynthetic performance of seagrass leaves was examined by Ralph and Gademann (1999) who concluded that removing epiphytes from leaves of the seagrass *Posidonia australis* reduced the ETR_{max} by over 70%. Differences in the amount of light absorbed by the leaf before and after removing the epiphyte community, and down-regulation responses associated with a strong increment in the leaf light environment, need to be taken into account before to interpret any direct mechanistic effect of epiphytes on the photosynthetic performance of aquatic macrophytes from variation in the chlorophyll *a* fluorescence signal.

3.2 Limitation of the Use of Rapid Light Curves (RLC)

Determinations of $\Delta F/F_m'$ at the photosynthetic steady-state of a certain irradiance (E) may provide estimations of the photosynthetic rates measured as electron transport rates, ETR. If the instrument or an external source of light provides a succession of actinic light intensities, the photosynthetic response to light (ETR vs E) can be experimentally described. Figure 3 shows the time required for leaf segments of the seagrass *Thalassia testudinum* to achieve the photosynthetic steady-state. The first determination (Figure 3a) was obtained under an irradiance of 452 μmol photons $m^{-2}\ s^{-1}$. The second determination (Fig. 3b) was measured using 115 μmol photons $m^{-2}\ s^{-1}$. In both experiments, the time required to reach the steady-state was longer than 6 min.

The Walz PAM fluorometers have the option of performing rapid light curves (RLC). PAM-CONTROL-type instruments can use an up to 10 min time period

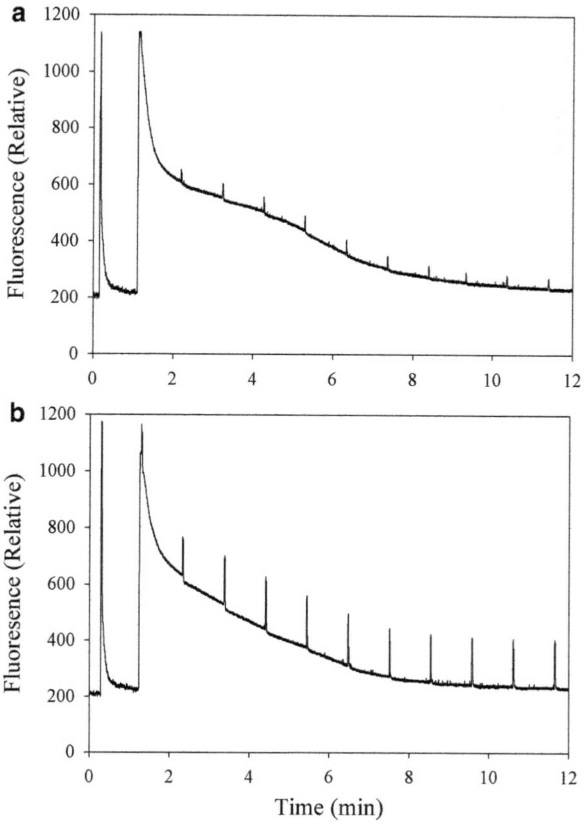

Fig. 3 Variation in the chlorophyll a fluorescence signal of *Thalassia testudinum* leaves after given an initial saturating actinic light pulse followed by the turning on off two different actinic light intensities: (**a**) 452 μmol photons m^{-2} s^{-1}; and (**b**) 115 μmol photons m^{-2} s^{-1}. The time required to reach the steady-state for both irradiances is shown

or a manually controlled time for the illumination period. However, instruments such as MINI-PAM, diving PAM, etc., can only extend this illumination time to 5 min using the manufacturers software. The illumination time for RLCs generally used in the study of macroalgae and seagrass photosynthesis, is 10–20 s (e.g., Durako et al. 2003; Silva and Santos 2003; Ralph and Gademann 2005; Colombo-Pallota et al. 2006; Strain et al. 2006). This type of experimental determination of the RLC, which utilizes a very short illumination period does not support one of the main restrictions for the use of ETR values as photosynthesis descriptors: i.e. $\Delta F/F_m'$ has to be determined at the photosynthetic steady-state. RLCs determined using short light acclimation periods cannot estimate the rate of electron transport, ETR, and consequently, cannot be considered a tool to evaluate the photosynthetic response to light (i.e., ETR *vs* E curve). This type of determination cannot be used either to estimate the equivalent descriptors derived from the classic photosynthetic response curve to irradiance (P *vs* E curve), such as, P_{max}, α, and E_k, as stated in Ralph and Gademann (2005). The physiological information obtained using very short illumination intervals (RLC < 1 min), has a difficult interpretation, as this response takes place under an incomplete induction of NPQ and without a steady flux of electrons in between the two photosystems. However, as long as the experimental approach takes standardized measurements, this type of response may provide useful information for the description of relative changes in photosynthetic activity, but it is important to consider that these determinations are physiologically unclear, and that they cannot describe the photosynthetic rates or the photoacclimatory response of the tissue.

3.3 The Importance of Photosynthesis Induction

ETR values derived from the RLC function are often lower at higher irradiances than determinations using oxygen evolution (Carr and Björk 2003). This common observation may be the consequence of the former considerations but it also may be due to other experimental problems. For example, Colombo-Pallota et al. (2006) used 10 s light exposure intervals to determine RLCs on the blades of *M. pyrifera.* Photosynthetic rates were also measured as oxygen evolution at similar 11 irradiances with light exposures of 3 min. While the oxygen evolution measurements did not show any decline after the maximum photosynthetic rates (P_{max}) were attained, the ETR values estimated from the RLCs showed significant declines at higher irradiances in all blades from all depths (Colombo-Pallota et al. 2006). This type of response has been commonly reported (see Fig. 4a for the response of the green alga *Caulerpa prolifera*) and interpreted as indicative of tissue photoinhibition at high light intensities. However, the oxygen evolution measurements of Colombo-Pallota et al. (2006) did not confirm similar reductions in the photosynthetic rates and, therefore, cannot support this interpretation. Enríquez and Rodríguez-Román (2006) using 1 min of illumination time for

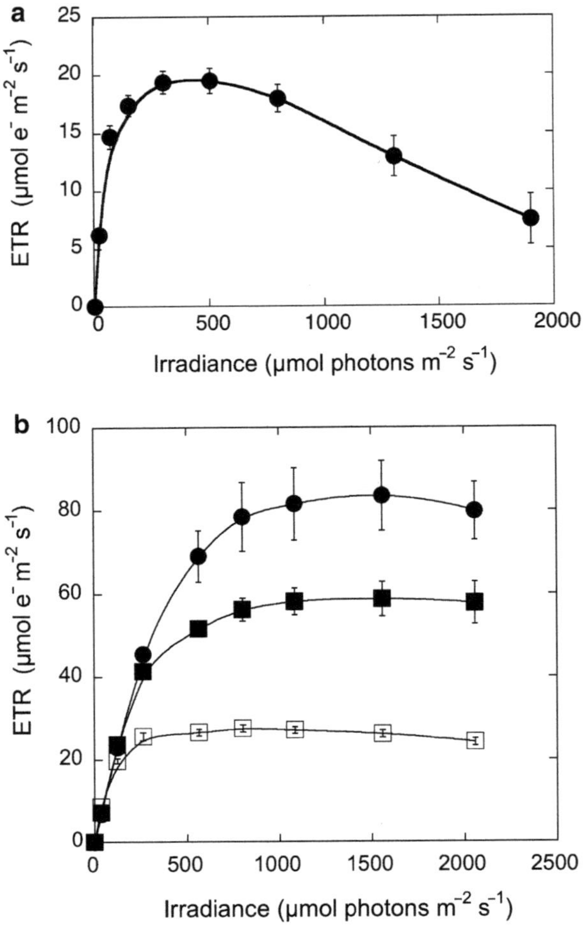

Fig. 4 **(a)** RLC determined with 15 s intervals of illuminated periods on *Caulerpa prolifera*. **(b)** RLC obtained with 1 min interval on *Thalassia testudinum* leaf segments. Open squares are control values obtained before leaf incubation; solid squares are estimated ETR values of leaf segments at 8 cm from the meristem after illuminating the leaf with at 330 μmol photons m^{-2} s^{-1} for 4 h; solid circles represent values obtained at 5 cm from the leaf meristem and after illumination (From Enríquez and Rodríguez-Román 2006)

the RLCs, did not observe a decline in the ETR values of *T. testudinum* leaves at increasing irradiances. Nevertheless, these authors found significant differences in the estimation of the maximum ETR values comparing measurements obtained before and after preliminary tissue incubation at 330 μmol photons m^{-2} s^{-1} and constant temperature (Enríquez and Rodríguez-Román 2006, Fig. 4b). The estimated ETR values obtained after the preliminary illumination were 2.6 times higher, on average, than the values shown by the initial non-incubated control leaves (Fig. 4b). Significant variation in ETR_{max} between basal and apical leaf sections were also observed after the incubation, as maximum ETR values were 3 and 2.2 times higher, respectively, at 5 cm and at 8 cm, than the control values (Fig. 4b). These differences were masked in the initial determinations.

Enríquez and Rodríguez-Román (2006) used the fluorescence signal to assess the effect of water flow on the photosynthesis of three marine macrophytes. They pre-incubated the samples, at similar temperature, before the experimental determinations were performed in a specially designed acrylic chamber. These authors indicate in their study that the previous photosynthesis induction was required to counterbalance the dark inhibition of the enzyme ribulose-1,5-biphosphate carboxylase/oxygenase (Rubisco, Gutteridge et al. 1986). Although it is difficult to discern what variability in the ETR estimations of Enríquez and Rodríguez-Román (2006) was produced by incomplete inductions of the photosynthetic steady-state (RLC were performed with 1 min interval of successive light exposures) or by dark inhibition of Rubisco, this study pointed out the importance of Rubisco regulation and photosynthesis induction in the interpretation of the fluorescence signal. An accurate determination of aquatic macrophyte photosynthesis, especially for P_{max} estimations, requires full photosynthesis induction and the offset of Rubisco dark-inhibition. If RLCs are determined using short time intervals, the total illumination period may not be enough to counterbalance Rubisco dark-inhibition. This may lead to the underestimation of the ETR values, but even worse, to the incorrect conclusion that the organism is strongly photoinhibited at increasing irradiances.

3.4 Determination of Absorptance, PSII Effective Absorption Cross-Section and the Number of Reaction Centers

According to Eq. 4, for an accurate estimation of ETR in macroalgae and seagrass studies, it is necessary to know the ability of the tissue to absorb light. This optical property is called absorptance (A) and describes the fraction of incident light absorbed by the tissue. The critical information for the determination of ETR, is the estimation of the fraction of light absorbed by photosynthetic pigments. Photosynthetic pigments have minimal light absorption in the 725–750 nm range. Subtracting absorptance values at this specific wavelengths from the 400–700 nm estimations, allows correction for non-photosynthetic absorption (Duysens 1956).

Absorptance can be determined using an integrating sphere attached to a radiometer or a spectrophotometer, or using the direct spectrophotometer transmission mode. Both approaches require special attention to the optics of the methodological procedures. The integrating sphere allows direct determinations of the light transmitted (T) and reflected (R) by the surface. Subtracting transmittance and reflectance from one allows the estimation of absorptance. The use of an integrating sphere is not simple and is often associated with significant methodological errors due to partial port closures and non-controlled light losses. The addition of two methodological errors (i.e., Transmittance and Reflectance determinations) if a Taylor-type sphere is used, often reduces the accuracy of absorptance estimations. More accessible methodologies developed for dispersive samples such as leaves, have been available for decades (i.e., Shibata 1959). The Shibata method, known as the opal-glass technique, has provided successfully results for aquatic macrophyte tissues (see Enríquez et al. 1992, 1994, 2005; Enríquez and Sand-Jensen 2003; Cayabyab and Enríquez 2007) although its utility and convenience has not been sufficiently recognized so far. Shibata's method (1959) uses a scattering material such as an opal glass, placed in between the tissue and the photomultiplier in order to ensure that both blank and sample light beams are highly scattered. This method is not appropriate if reflectance is larger than 10–15%, depending on the sensitivity of the photomultiplier, as it assumes that reflectance (R) is negligible, and absorptance is derived from transmittance (T) determinations. Aquatic macrophytes tend to show low reflectance values (Lüning and Dring 1985; Frost-Christensen and Sand-Jensen 1992; Enríquez 2005), although there is significant variation among species. More information is currently available for seagrasses than for macroalgae. Reflectance in seagrass leaves varies between 0.03 and 0.09, with most reported values so far between 0.05 and 0.07 (Runcie and Durako 2004; Enríquez 2005; Thorhaug et al. 2006; Durako 2007). This represents a 6% average error through back-scattering in the spectrophotometric determination. Light absorption expressed as absorptance (A; the fraction of light absorbed no units) can be calculated from the measurement of absorbance (D; units optical density, OD) using the equation:

$$A = 1 - 10^{-D} \tag{9}$$

assuming, as indicated before, that reflectance tends to zero and absorptance can be estimated from the transmittance values (A = 1−T). A further correction for reflectance can be performed subtracting a known reflectance value (R) from the former equation ($A=[1-10^{-D}]-R$) (see Enríquez 2005).

In the spectrophotometric determinations, the distance between the sample and the photomultiplier is also critical. Minimal distance will prevent significant light losses of the transmitted light and absorptance overestimation. Knowledge of the magnitude of this methodological error can be obtained from the absorbance values at 725–750 nm. Two examples are presented in Fig. 5, which shows direct spectrophotometric measurements (in black) and the corrected spectra (in blue lines) after subtracting for residual scattering and non-pigmented absorption (D values between 725 and 750 nm). Figure 5a illustrates an incorrect determination. Absorbance in the 725–750 nm range has a value > 0.1, which represents, approximately, 21% of residual scattering. In contrast, Fig. 5b shows an accurate determination where residual scattering plus non-photosynthetic absorption is reduced to 6% ($D_{725-750} = 0.03$). A comparative study has shown the close compatibility between the use of an integrating sphere and the spectrophotometric opal-glass technique developed by Shibata (Enríquez et al. 1994). Both methodologies require the correction for non-photosynthetic absorption using a white tissue (bleached or etiolated) as reference or baseline (see Enríquez 2005, and the differences observed between Fig. 5a and b without or with the use of this white tissue). The reference or white tissue has to be as similar as possible to the tissue examined in thickness and internal anatomy.

Other more accessible methodologies have been proposed (see Beer and Björk 2000), with wide acceptance in seagrass and macroalgae studies. However, these other options have not been validated yet against a more robust method and, consequently, their suitability for absorptance determination remains undefined.

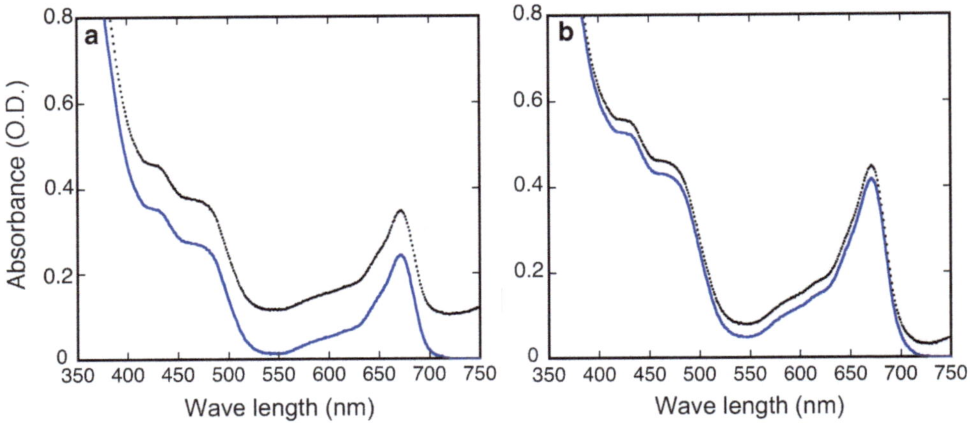

Fig. 5 Light absorption spectra of *Thalassia testudinum* leaves obtained with the opal-glass technique developed by Shibata (1959). Direct spectrophotometric measurements are in black and corrected spectra for residual scattering and non-pigmented absorption, are represented by blue lines. (**a**) Incorrect determination using air as reference or base-line. (**b**) Correct determination using a white leaf as reference or base-line

Direct estimations of the functional absorption cross-section of PSII, σ_{PSII}, can be also derived from fluorescence measurements, using the pump-and-probe technique (Falkowski and Raven 1997). The slope of a flash-intensity saturation curve describes the effective absorption cross-section of PSII. This method does not produce a multiple turnover fluorescence induction curve, as the pump-and-probe combination of flashes induces a single turnover of PSII.

If the effective absorption cross-section of PSII is known, the use of the herbicide DCMU allows the description of the number of reaction centers. Leaves or blades infiltrated with DCMU have a population of PSII that can only undergo a single turnover. This condition allows the description of the number of reaction centers from the estimation of the area over the fluorescence induction curve of the DCMU treated tissues, and knowing PSII absorption cross-section, as this area is proportional to the number of Q_A (see Falkowski and Raven 1997). This possibility has not yet been explored in aquatic macrophyte studies.

3.5 Use of Relative ETR Values (rETR)

As indicated before, to prevent the problems associated with incorrect assumptions of absorptance and/or σ_{PSII} / σ_{PSU} values a new descriptor, relative ETR (rETR) has been proposed (Beer et al. 2001, Ralph et al. 2002). Relative ETR values result from the multiplication of the effective photochemical efficiency ($\Delta F/F_m{}'$) by the irradiance (i.e., E or E × 0.5 depending on the author). This new parameter (rETR) can be considered a good descriptor of relative changes in the photosynthetic rates, if the organisms or tissues examined have similar absorption cross-sections. If significant differences are present, rETR cannot assess the relative contribution of absorption to the variation observed in ETR. To illustrate the possible error introduced in this type of comparison, we show here data from the experimental approach developed by Cayabyab and Enríquez (2007). Figure 6 compares basal and apical segments of leaves of the seagrass *Thalassia testudinum* grown in mesocosms under high light conditions (HL). These leaf segments differ in their capacity of light absorption. Absorptance for the basal segments was estimated to be 0.63 ± 0.07, whereas in the apical segments this value was reduced to 0.51 ± 0.06. Figure 6a shows the variation observed in $\Delta F/F_m{}'$ as a function of irradiance for both type of segments. These measurements were obtained in simultaneous determinations of ETR and oxygen evolution rates, after incubating leaf sections for time intervals between 5 and 10 min. Fluorescence measurements were taken using a PAM 101-102-103 Fluorometer (Heinz Walz, Germany) attached with the octopus fiber optics to the DW3 Hansatech chamber (Hansatech Instruments Ltd, UK), where the oxygen evolution determinations were performed. Significant differences between apical and basal leaf sections were observed using rETR values (Fig. 6b). Interestingly,

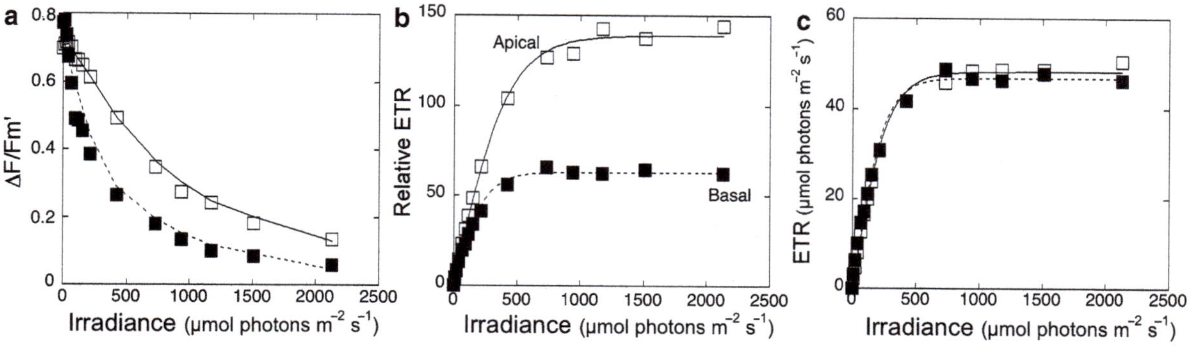

Fig. 6 (**a**) Variation observed in $\Delta F/F_m'$ for apical (open squares) and basal (solid squares) segments of *Thalassia testudinum* leaves, as a function of irradiance; (**b**) variation in the relative ETR values for the same segments; and (**c**) variation in the ETR values after correcting for absorptance (0.51 for the apical segments and 0.63 for the basal segments) (Data from Cayabyab and Enríquez 2007)

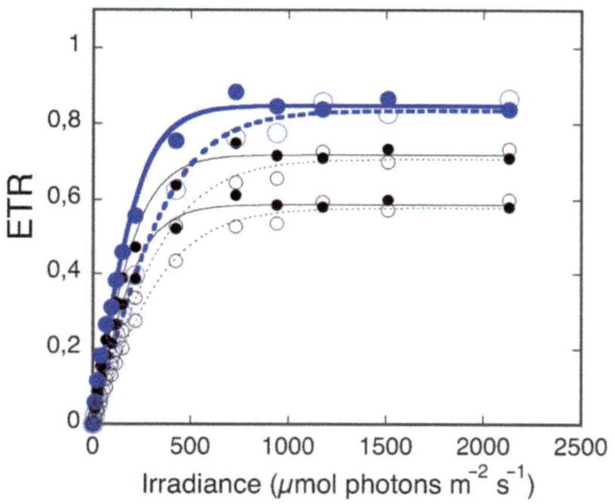

Fig. 7 Simulation of the possible variation in the estimated ETR response to irradiance for the apical (open circles) and basal (solid circles) segments of *Thalassia testudinum* leaves, as a function of the differences reported in leaf absorptance by Enríquez (2005) for the same species. Blue lines and symbols represent the closest estimations to the actual values for the basal and apical leaf segments of the HL treatment reported in Cayabyab and Enríquez (2007)

after correcting rETR by the estimated absorptance (Fig. 6c), the conclusion was the opposite as no significant differences in ETR were distinguishable between both leaf sections.

Aquatic macrophytes display large variation in tissue light absorption (Enríquez et al. 1994) within the same individual, among populations and among species. Changes in light absorption may also occur diurnally due to chloroplast movement (Sharon and Beer 2008). Aquatic photosynthetic tissues are often structurally and optically thinner than terrestrial tissues, which explains these larger differences in comparison with the consistency shown by absorptance in terrestrial leaves (≈ 0.84). For mature leaves of the seagrass *Thalassia testudinum* a 27% of variation in absorptance has been reported (between 0.42 and 0.69, Enríquez 2005). Considering this natural variation, we simulate the resulting differences in ETR in Fig. 7. The error introduced in the estimated maximum ETR values is proportional to the variation in A, and similar for the basal (solid circles) and the apical (open circles) leaf sections, however, an interesting observation is that this error in the ETR estimations is significantly different between leaf sections and irradiances at subsaturating light (Fig. 7).

3.6 The Use of Electron Transport Rates Values (ETR) as Descriptors of Gross Photosynthesis (GPS)

A significant interest in aquatic ecology has been the examination of the suitability of the chlorophyll *a* fluorescence signal as a tool for the estimation of photosynthetic rates. There is a general assumption that a linear association between ETR and gross photosynthesis (GPS) exists, as long as environmental stresses do not impose restrictions on photosynthetic carbon fixation (Genty et al. 1989; Edwards and Baker 1993). The attempts to confirm this linearity in seagrass and macroalgae studies have provided, however, controversial results. In general, the linearity between ETR and GPS has been only confirmed at low light intensities (e.g., Beer et al. 1998; Beer and Björk 2000; Longstaff et al. 2002; Cabello-Pasini and Figueroa 2005; Colombo-Pallota et al. 2006; Enríquez and Rodríguez-Román 2006).

At saturating light, ETR values tend to overestimate O_2 evolution rates (cf. Franklin and Badger 2001, Enríquez and Rodríguez-Román 2006) if photosynthesis has been fully induced. If not, ETR values can also underestimate O_2 evolution rates. Deviations from the linearity have been frequently reported dependent on species, light history, irradiance, tissue thickness, and other factors such as CO_2 and nitrate levels (e.g., Beer et al. 1998; Beer and Björk 2000; Franklin and Badger 2001; Beer and Axelsson 2004; Enríquez and Rodríguez-Román 2006; Nielsen and Nielsen 2008).

In the former example taken from the experimental approach of Cayabyab and Enríquez (2007), apical and basal leaf sections, which differed in the absorption cross-section and pigmentation, did not show significant differences in their photosynthetic rates per unit of area based on the fluorescence measurements (ETR values). However, Cayabyab and Enríquez (2007) did find significant differences in the photosynthetic rates based on the oxygen evolution determinations (Fig. 8).

This result questions again the appropriateness of the use of chlorophyll fluorescence for photosynthetic rates estimations. Oxygen evolution determinations are more intrusive and require more experimental effort than fluorescence measurements. However, they are still better descriptors of photosynthesis than fluorescence-based estimations (ETR). The capacity of ETR to estimate photosynthesis depends on the consistency of the association between ETR and GPS. In their study Cayabyab and Enríquez (2007) analyzed

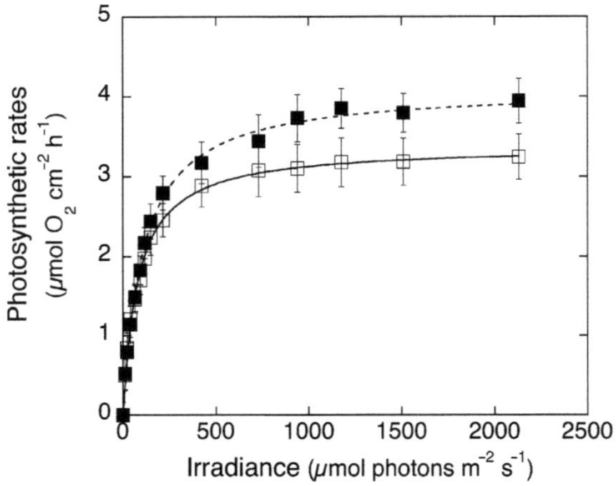

Fig. 8 Variation of the photosynthetic response to irradiance (P vs. E curves) for apical (open squares) and basal (solid squares) leaf segments of the tropical seagrass *Thalassia testudinum* (Data from Cayabyab and Enríquez 2007)

the photoacclimatory plasticity and efficiency of the leaves of the seagrass *Thalassia testudinum*, using simultaneous determinations of ETR and GPS. These determinations were performed on: (1) leaves freshly collected from their natural environment in a shallow reef lagoon (near the laboratory facilities of the Universidad Nacional Autónoma de México in Puerto Morelos, Cancún, Mexico); and (2) leaves grown in mesocosms under three controlled light intensities (HL = 1.5 times the natural light fields in the lagoon, ML = similar light intensities, and LL = 50% of reduction in the natural light fields). In their study, Cayabyab and Enríquez (2007) also compare basal and apical segments from the same leaf and treatment. Figure 9 shows the different associations found in the EPS vs GPS relationships for the four treatments (field control, HL, ML and LL) and the two leaf sections (basal and apical):

A linear relationship is only observed at low irradiances (the 3–7 initial values from the 14 used in this comparison, Fig. 9). The slope of the initial linear phase of the ETR vs GPS association also shows significant differences among treatments and leaf segments (Table 1). Values close to four electrons per oxygen molecule evolved were estimated for the leaves grown in the reef lagoon (field-control treatment) and for the basal sections of all treatments of the experimental leaves (leaf segments grown in mesocosms under HL, ML and LL treatments, Table 1).

The apical sections of the experimental leaves (leaf segments grown in the field but acclimated in mesocosms to HL, ML and LL treatments) showed significant higher values for this slope (between 5,8 and 7,1 e^- delivered per O_2 molecule evolved, Table 1). These values clearly differ from the basal and apical sections of the field growing leaves, but they were not significantly different than the estimated for their respective basal leaf sections (Table 1). Comparing the E_k values determined by Cayabyab and Enríquez (2007) for each treatment and leaf section, and the observed irradiance for the loss of the linearity between ETR and GPS, a linear association was found ($r = 0.71$, $P < 0.05$, Fig. 10). According to this relationship, the loss of linearity in *T. testudinum* leaves appears to be, on average, 80 μmol quanta m^{-2} s^{-1} beyond the saturation irradiance (E_k).

Looking at the whole relationship shown in Fig. 9 (each point represents the average of 4 determinations at 14 different irradiances), it is noteworthy that all of

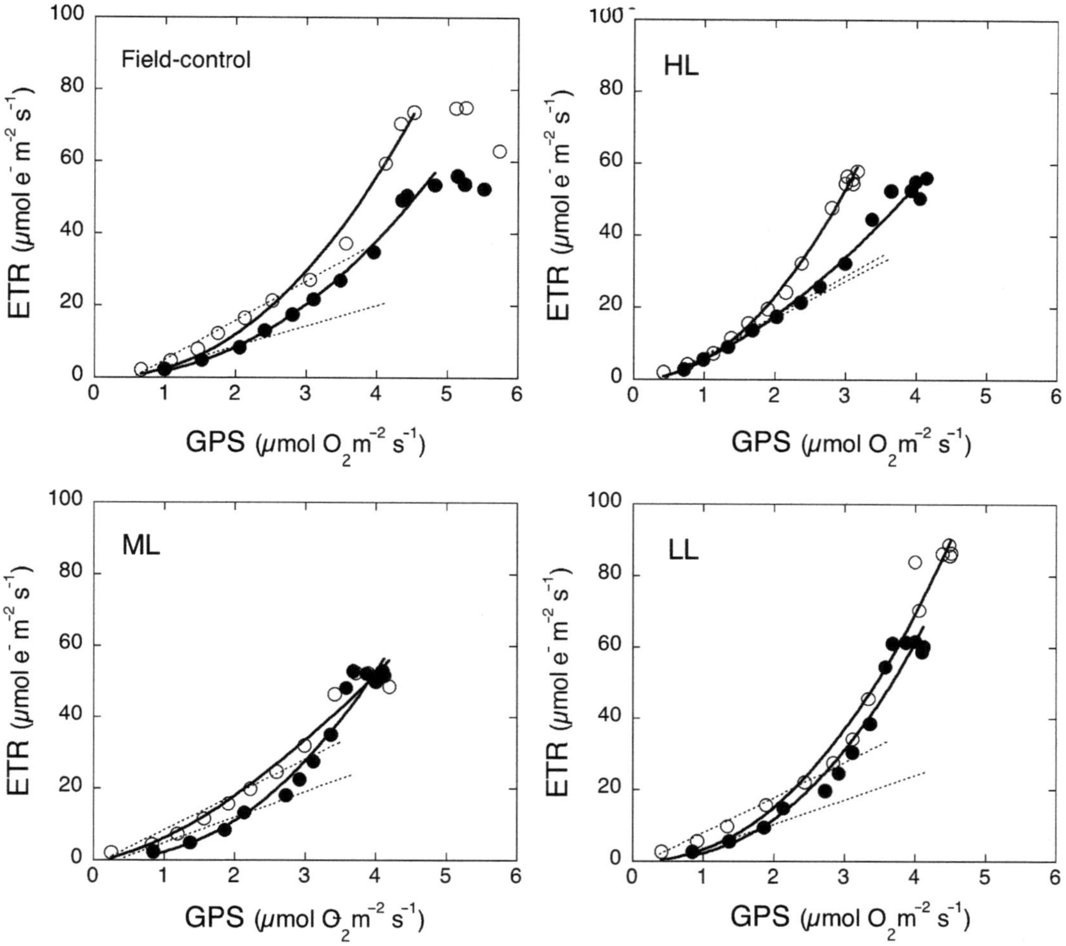

Fig. 9 Variation of the association between fluorescence based photosynthetic rates (ETR values) and Gross photosynthesis measured as oxygen evolution (GPS) for apical (open circles) and basal (solid circles) leaf segments of the tropical seagrass *Thalassia testudinum*. Data from Cayabyab and Enríquez (2007). Data are adjusted to a power function (ETR = a * GPS^b). The initial linearity of these associations is also indicated with dotted lines

Table 1 Slope ± SE for the linear phase of the association between ETR and GPS (Gross Photosynthesis). Correlation coefficient (r) for the power function fit, and parameters estimated from the power function (ETR = a * GPS^b), for the four treatments and two leaf sections of the experimental approach of Cayabyab and Enríquez (2007)

			Parameters from the power function $ETR = a * GPS^b$	
Sample	Slope (linear)	r -power function	a ± SE	b ± SE
CT-field apex	4.25 ± 0,6	0.95	6.96 ± 2.29	1.42 ± 0.21
CT-field base	4.1 ± 0,7	0.99	1.48 ± 0.22	2.36 ± 0.11
HL- apex	6.44 ± 0,6	0.99	5.49 ± 0.46	2.06 ± 0.08
HL- bases	5.3 ± 1.4	0.99	5.68 ± 0.72	1.63 ± 0.10
ML- apex	5.8 ± 0.6	0.99	6.23 ± 1.03	1.53 ± 0.13
ML- bases	4.45 ± 0.7	0.98	2.30 ± 0.73	2.26 ± 0.24
LL- apex	7.12 ± 0.6	0.99	3.29 ± 0.84	2.20 ± 0.18
LL- bases	5.8 ± 0.95	0.98	3.20 ± 0.60	2.17 ± 0.14

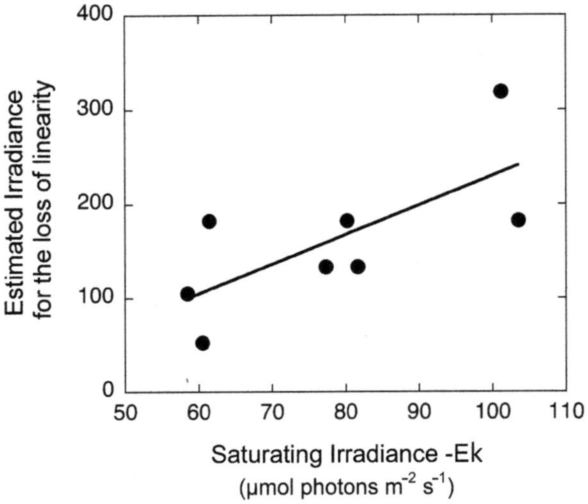

Fig. 10 Relationship between the saturating irradiance E_k, determined for Cayabyab and Enríquez (2007) for each light treatment and leaf section, and the estimated irradiance for the loss of linearity between ETR and GPS values, also from Cayabyab and Enríquez (2007)

them were accurately explained by a power function (i.e., GPS = a* ETR^b, Table 1), except for the last three points of the field-growing leaves, which were excluded from the fitting. This comparison supports the specificity of the association between GPS and ETR values for each photoacclimatory condition.

As the utility of the fluorescence technique for photosynthesis estimation is highly dependent on validation, its capacity to substitute oxygen evolution determinations seems very reduced, as it requires a previous description of the specific association between ETR and gross photosynthesis (GPS) for each photoacclimatory condition. Its utility, however, for the description of relative changes in photosynthesis with respect to an experimental control is not questionable.

4 Final Comments

At the end of this chapter, it is important to reiterate that the use of the fluorescence signal in seagrass and macroalgal studies opens more methodological research possibilities than the simple estimation of photosynthesis. It is, overall, a powerful tool to assess the properties of the photosynthetic apparatus, and, therefore, for the understanding of the broad structural and functional diversity of mechanisms still present in the marine environment. For instance, the rhodophytes do not have a xanthophyll cycle (see Ursi et al. 2003) thus the different carotenoid profiles described for Rhodophyta (Schubert et al. 2006) may be related to differences among species in their photoprotective capacity and sensitivity to light absorbed in excess. This research is currently in progress (Schubert and García-Mendoza 2008). Similarly, there is evidence for xanthophyll cycles other than the violaxanthin/antheraxanthin/zeaxanthin cycle of the Chlorophyta and the de-epoxidation of diadinoxanthin to diatoxanthin in the Phaeophyta, such as a lutein-siphonoxanthin interconversion proposed for the green alga *Caulerpa racemosa* (Ranilello et al. 2006).

Marine ecophysiology and, specifically, the application of chlorophyll *a* fluorescence for the study of macrophyte photosynthesis, has enormous potential to develop a fuller understanding of its natural diversity. This is probably one of the most important challenges of the use of the fluorescence signal in seagrass and macroalgal studies, and its main contribution to photosynthesis research.

References

Aguirre-von-Wobeser E, Figueroa FL, Cabello-Pasini A (2000) Effect of UV radiation on photoinhibition of marine macrophytes in culture systems. J Appl Phycol 12:159–168

Apprill AM, Lesser MP (2003) Effects of ultraviolet radiation on *Laminaria saccharina* in relation to depth and tidal height in the Gulf of Maine. Mar Ecol Prog Ser 256:75–85

Baker NR, Oxborough K (2004) Chlorophyll a fluorescence as a probe of photosynthetic productivity. In: Papageorgiou GC, Govindjee (eds) Chlorophyll *a* fluorescence: a signature of photosynthesis. Springer, Dordrecht, pp 65–82

Ballesteros E (1987) Estructura i dinámica del poblament algal de les fulles de *Posidonia oceanica* (L.) Delile als herbeis de Tossa de Mar (Girona). Bull Inst Cat Hist Nat 54:19–30

Beer S, Axelsson L (2004) Limitations in the use of PAM fluorometry for measuring photosynthetic rates of macroalgae at high irradiances. Eur J Phycol 39:1–7

Beer S, Björk M (2000) Measuring rates of photosynthesis of two tropical seagrasses by pulse amplitude modulated (PAM) fluorometry. Aquat Bot 66:69–76

Beer S, Vilenkin B, Weil A, Veste M, Susel L, Eshel A (1998) Measuring photosynthetic rates in seagrasses by pulse amplitude modulated (PAM) fluorometry. Mar Ecol Progr Ser 174:293–300

Beer S, Larsson C, Poryan O, Axelsson L (2000) Photosynthetic rates of *Ulva* (Chlorophyta) measured by pulse amplitude modulated (PAM) fluorometry. Eur J Phycol 35:69–74

Beer S, Björk M, Gademan R, Ralph P (2001) Measurements of photosynthesis in seagrass. In: Short FT, Coles R (eds) Global seagrass research methods. Elsevier, Amsterdam, pp 183–198

Belshe EF, Durako MJ, Blum JE (2007a) Photosynthetic rapid light curves (RLC) of *Thalassia testudinum* exhibit diurnal variation. J Exp Mar Biol Ecol 342:253–268

Belshe EF, Durako MJ, Blum JE (2007b) Diurnal light curves and landscape-scale variations in photosynthetic characteristics of *Thalassia testudinum* in Florida Bay. Aquat Bot 89:16–22

Bischof K, Peralta G, Kräbs G, van de Poll WH, Pérez-Lloréns JL, Breeman AM (2002) Effects of solar UV-B radiation on canopy structure of *Ulva* communities form southern Spain. J Exp Bot 53:2411–2421

Bischof K, Rautenberger R, Brey L, Pérez-Llorens JL (2006) Physiological acclimation to gradients of solar irradiance within mats of the filamentous green macroalga *Chaetomorpha linum* from southern Spain. Mar Ecol Prog Ser 306:165–175

Björk M, Uku J, Weil A, Beer S (1999) Photosynthetic tolerances to desiccation of tropical intertidal seagrasses. Mar Ecol Progr Ser 191:121–126

Björkman O, Demmig B (1987) Photon yield of O_2 evolution and chlorophyll fluorescence characteristics at 77 K among vascular plants of diverse origins. Planta 170:489–504

Björkman O, Demmig-Adams B (1994) Regulation of photosynthetic light energy capture, conversion and dissipation in leaves of higher plants. In: Schulze E-D, Caldwell MM (eds) Ecophysiology of photosynthesis. Series Ecological Studies, vol 100. Springer, Berlin, pp 17–47

Borowitzka MA, Lavery PS, van Keulen M (2006) Epiphytes of seagrasses. In: Larkum AWD, Orth RJ, Duarte CM (eds) Seagrasses: biology, ecology and conservation. Springer, Dordrecht, pp 441–461

Bulthuis DA, Woekerling WJ (1983) Biomass accumulation and shading effects of epiphytes on leaves of the seagrass *Heterozostera tasmanica* in Victoria, Australia. Aquat Bot 16:137–148

Cabello-Pasini A, Figueroa FL (2005) Effect of nitrate concentration on the relationship between photosynthetic oxygen evolution and electron transport rate in *Ulva rigida* (Chlorophyta). J Phycol 41:1169–1177

Cabello-Pasini A, Aguirre-von-Wobeser E, Figueroa FL (2000) Photoinhibition of photosynthesis in *Macrocystis pyrifera* (Phaeophyceae), *Chondrus crispus* (Rhodophyceae) and *Ulva lactuca* (Chlorphyceae) in outdoor culture systems. J Photochem Photobiol B 57:169–178

Campbell SJ, Miller C, Steven A, Stephens A (2003) Photosynthetic responses of two temperate seagrasses across a water quality gradient using chlorophyll fluorescence. J Exp Mar Biol Ecol 291:57–78

Campbell SJ, McKenzie LJ, Kerville SP (2006) Photosynthetic responses of seven tropical seagrasses to elevated seawater temperature. J Exp Mar Biol Ecol 330:455–468

Carr H, Björk M (2003) A methodological comparison of photosynthetic oxygen evolution and estimated electron transport rate in tropical *Ulva* (Chlorophyceae) species under different light and inorganic carbon conditions. J Phycol 39:1125–1131

Cayabyab NM, Enríquez S (2007) Leaf photoacclimatory responses of the tropical seagrass *Thalassia testudinum* under mesocosm conditions. A mechanistic scaling-up study. New Phytol 176:108–123

Cebrián J, Enríquez S, Fortes M, Agawin N, Vermaat JE, Duarte CM (1999) Epiphyte accrual on *Posidonia oceanica* (L.) Delile leaves: Implications for light absorption. Bot Mar 42:123–128

Colombo-Pallota MF, García-Mendoza E, Ladah LB (2006) Photosynthetic performance, light absorption, and pigment composition of *Macrocystis pyrifera* (Laminariales, Phaeophyceae) blades from different depths. J Phycol 42:1225–1234

Critchley C, Russell AW (1992) Maximizing light interception: the role of photoplast disposition in cells and thylakoid membrane architecture. In: Abrol YP, Mohanty P, Govindjee (eds) Photosynthesis. Photoreaction to plant productivity. Oxford/IBH Publishing, New Delhi, pp 257–276

Critchley C, Russell AW (1994) Photoinhibition of photosynthesis in vivo: the role of protein turnover in photoystem II. Physiol Plant 92:188–196

Dawson SP, Dennison WC (1996) Effects of ultraviolet and photosynthetically active radiation on five seagrass species. Mar Biol 125:629–638

Delphin E, Duval J-C, Etienne A-L, Kirilovsky D (1996) State transitions or ΔpH-dependent quenching of photosystem II fluorescence in red algae. Biochemistry 35:9435–9445

Delphin E, Duval J-C, Etienne A-L, Kirilovsky D (1995) pH-dependent photosystem II fluorescence quenching induced by saturating, multiturnover pulses in red algae. Plant Physiol 118:103–113

Demmig-Adams B, Adams WW III (1992) Photoprotection and other responses of plants to high light stress. Annu Rev Plant Physiol Plant Mol Biol 43:599–626

Durako MJ (2007) Leaf optical properties and photosynthetic leaf absorptances in several Australian seagrasses. Aquat Bot 87:83–89

Durako MJ, Kunzelman JI (2002) Photosynthetic characteristics of *Thalassia testudinum* measured in situ by pulse-amplitude modulated (PAM) fluorometry: methodological and scale-based considerations. Aquat Bot 73:173–185

Durako MJ, Kunzelman JI, Kenworthy WJ, Hammerstrom KK (2003) Depth-related variability in the photobiology of two populations of *Halophila johnsonii* and *Halophila decipiens*. Mar Biol 142:1219–1228

Duval J-C, Harker M, Rousseau B, Young AJ, Britton G, Lemoine Y (1992) Photoinhibition and zeaxanthin formation in the brown algae *Laminaria saccharina* and *Pelvetia canaliculata*. In: Murata N (ed) Research in photosynthesis, vol 14. Kluwer, Dordrecht, pp 581–584

Duysens LMN (1956) The flattening effect of the absorption spectra of suspensions as compared to that of solutions. Biochem Biophys Acta 19:1–12

Edwards GE, Baker NR (1993) Can CO_2 assimilation in maize leaves be predicted accurately from chlorophyll fluorescence analysis? Photosynth Res 37:89–102

Enríquez S (2005) Light absorption efficiency and the package effect in the leaves of the seagrass *Thalassia testudinum*. Mar Ecol Prog Ser 289:141–150

Enríquez S, Pantoja-Reyes NI (2005) Form-function analysis of the effect of canopy morphology on leaf self-shading in the seagrass *Thalassia testudinum*. Oecologia 145:235–243

Enríquez S, Méndez ER, Iglesias-Prieto R (2005) Multiple scattering on coral skeletons enhances light absorption by symbiotic algae. Limnol Oceanogr 50:1025–1032

Enríquez S, Rodríguez-Román A (2006) Effect of water flow on the photosynthesis of three marine macrophytes from a fringing-reef lagoon. Mar Ecol Prog Ser 323:119–132

Enríquez S, Sand-Jensen K (2003) Variation in light absorption properties of *Mentha aquatica* L. as a function of leaf form: implications for plant growth. Int J Plant Sci 164:125–136

Enríquez S, Agustí S, Duarte CM (1992) Light absorption by seagrass *Posidonia oceanica* leaves. Mar Ecol Prog Ser 86:201–204

Enríquez S, Agustí S, Duarte CM (1994) Light absorption by marine macrophytes. Oecologia 98:121–129

Enríquez S, Merino M, Iglesias-Prieto R (2002) Variations in the photosynthetic performance along the leaves of the tropical seagrass *Thalassia testudinum*. Mar Biol 140:891–900

Falkowski PG, Raven JA (1997) Aquatic photosynthesis. Blackwell Science, Malden/Massachussetts, USA, p 375

Flanigan YS, Critchley C (1996) Light response of D1 turnover and photosystem II efficiency in the seagrass *Zostera capricorni*. Planta 198:319–323

Flores-Moya A, Gómez I, Viñegla B, Altamirano M, Pérez-Rodríguez E, Maestre C, Caballero RM, Figueroa FL (1998) Effects of solar radiation on the endemic Mediterranean red alga *Rissoella verruculosa*: photosynthetic performance, pigment content and the activities of enzymes related to nutrient uptake. New Phytol 139:673–683

Fork DC, Satoh K (1986) The control by state transitions of the distribution of excitation energy in photosynthesis. Annu Rev Plant Physiol 37:335–361

Franklin LA, Badger MR (2001) A comparison of photosynthetic electron transport rates in macroalgae measured by pulse amplitude modulated chlorophyll fluorometry and mass spectrometry. J Phycol 37:756–767

Franklin LA, Larkum WD (1997) Multiple strategies for a high light existence in a tropical marine macroalga. Photosynth Res 53:149–159

Franklin LA, Levavasseur G, Osmond CB, Henley WJ, Ramus J (1992) Two components of onset and recovery during photoinhibition of *Ulva rotundata*. Planta 186:399–408

Franklin LA, Seaton CGR, Lovelock C, Larkum AWD (1996) Photoinhibition on a coral reef. Plant Cell Env 19:825–836

Frost-Christensen H, Sand-Jensen K (1992) The quantum efficiency of photosynthesis in macroalgae and submerged angiosperms. Oecologia 91:377–384

Gabrielsen EK (1948) Effects of different chlorophyll concentrations on photosynthesis in foliage leaves. Physiol Plant 1:5–37

García-Mendoza E, Colombo-Pallota MF (2007) The giant kelp *Macrocystis pyrifera* presents a different non-photochemical quenching control than higher plants. New Phytol 173:526–536

García-Mendoza E, Matthijs HCP, Schubert H, Mur L (2002) Non-photochemical quenching of chlorophyll fluorescence in *Chlorella fusca* acclimated to constant and dynamic light conditions. Photosynth Res 74:303–315

Genty B, Briantais J, Baker NR (1989) The relationship between the quantum yield of photosynthetic electron transport and quenching of chlorophyll fluorescence. Biochim Biophys Acta 990:87–92

Gómez I, López-Figueroa F, Ulloa N, Morales V, Lovengreen C, Huovinen P, Hess S (2004) Patterns of photosynthesis in 18 species of intertidal macroalgae from southern Chile. Mar Ecol Prog Ser 270:103–116

Govindjee (1995) Sixty-three years since Kautsky: chlorophyll *a* fluorescence. Aust J Plant Physiol 22:131–160

Gutteridge S, Parry MAJ, Burton S, Keys AJ, Mudd A, Feeney J, Jc S, Pierce J (1986) A nocturnal inhibitor of carboxylation in leaves. Nature 324:274–276

Häder D-P, Lebert M, Figueroa FL, Jiménez C, Viñegla B, Pérez-Rodríguez E (1998) Photoinhibition in Mediterranean macroalgae by solar radiation measured on site by PAM fluorescence. Aquat Bot 61:225–236

Häder D-P, Lebert M, Sinha RP, Barbieri ES, Helbling EW (2002) Role of protective and repair mechanisms in the inhibition of photosynthesis in marine macroalgae. Photochem Photobiol Sci 1:809–814

Hanelt D (1992) Photoinhibition of photosynthesis in marine macrophytes of the South China Sea. Mar Ecol Progr Ser 82:199–206

Hanelt D, Nultsch W (1995) Field studies of photoinhibition show non-correlations between oxygen and fluorescence measurements in the arctic red alga *Palmaria palmata*. J Plant Physiol 145:31–38

Hanelt D, Huppertz K, Nultsch W (1993) Daily course of photosynthesis and photoinhibition in marine macroalgae investigated in the laboratory and in the field. Mar Ecol Progr Ser 97:31–71

Hanelt D, Uhrmacher S, Nultsch W (1995) The effect of photoinhibition on photosynthetic oxygen production in the brown alga *Dictyota dichotoma*. Bot Acta 108:99–105

Hanelt D, Wiencke C, Bischof K (2003) Photosynthesis in marine macroalgae. In: Larkum AWD, Douglas SE, Raven JA (eds) Photosynthesis in algae. Kluwer, Dordrecht, pp 413–435

Harrington L, Fabricius K, Eaglesham G, Negri A (2005) Synergistic effects of diuron and sedimentation on photosynthesis and survival of crustose coralline algae. Mar Poll Bull 51:415–427

Haynes D, Ralph PJ, Prange J, Dennison W (2000) Impact of the herbicide diuron on photoshyntesis in three species of tropical seagrasses. Mar Pollution Bull 41:288–293

Horton P, Ruban A, Rees D, Pascal AA, Noctor G, Young AJ (1991) Control of the light harvesting function of chloroplast membranes by aggregation of the LHCII chlorophyll protein complex. FEBS Lett 292:1–4

Hwang EK, Bak JM, Park CS (2007) Assessment of optimal depth and photon irradiance for cultivation of the brown alga, *Sargassum fulvellum* (Turner) C Agardh. J Appl Phycol 19:787–793

Iglesias-Prieto R, Beltrán VH, LaJeunesse TC, Reyes-Bonilla H, Thomé PE (2004) Different algal symbionts explain the vertical distribution of dominant reef corals in the eastern Pacific. Proc R Soc Lond B 271:1757–1763

Kirilovski D (2007) Photoprotection in cyanobacteria: the orange carotenoid protein (OCP)-related non-photochemical-quenching mechanism. Photosynth Res 93:7–16

Küster A, Altenburger R (2007) Development and validation of a new fluorescence-based bioassay for aquatic macrophyte species. Chemosphere 67:194–201

Lamote M, Dunton KH (2006) Effects of drift macroalgae and light attenuation on chlorophyll fluorescence and sediment sulfides in the seagrass *Thalassia testudinum*. J Exp Mar Biol Ecol 334:174–186

Lan C-Y, Kao W-Y, Lin H-J, Shao K-T (2005) Measurements of chlorophyll fluorescence reveals mechanisms for habitat niche separation of the intertidal seagrasses *Thalassia hemprichii* and *Halodule uninervis*. Mar Biol 148:25–34

Larkum AWD (2003) Light-harvesting systems in algae. In: Larkum AWD, Douglas SE, Raven JA (eds) Photosynthesis in algae. Kluwer, Dordrecht, pp 277–304

Larkum AWD, Vesk M (2003) Algal plastids: their fine structure and properties. In: Larkum AWD, Douglas SE, Raven JA (eds) Photosynthesis in algae. Kluwer, Dordrecht, pp 11–28

Long SP, Humphries S, Falkowski PG (1994) Photoinhibition of photosynthesis in nature. Annu Rev Plant Physiol Plant Mol Biol 45:633–662

Longstaff BJ, Kildea T, Runcie JW, Cheshire A, Dennison WC, Hurd C, Kana T, Raven JA, Larkum AWD (2002) An in situ study of photosynthetic oxygen exchange and electron transport rate in the marine macroalga *Ulva lactuca* (Chlorophyta). Photosynth Res 74:281–293

Lüning K, Dring MJ (1985) Action spectra and spectral quantum yield of photosynthesis in marine macroalgae with thin and thick thalli. Mar Biol 87:119–129

Major KM, Dunton KH (2002) Variations in light-harvesting characteristics of the seagrass, *Thalassia testudinum*: evidence for photoacclimation. J Exp Mar Biol Ecol 275:173–189

Maxwell K, Johnson GN (2000) Chlorophyll fluorescence – a practical guide. J Exp Bot 51:659–668

Nielsen HD, Nielsen SL (2008) Evaluation of imaging and conventional PAM as a measure of photosynthesis in thin- and thick-leaved marine macroalgae. Aquat Biol 3:121–131

Niyogi KK, Grossman AR, Bjorkman O (1998) Arabidopsis mutants define central role for the xhanthophyll cycle in the regulation of photosynthetic energy conversion. The Plant Cell 10:1121–1143

Ohad I, Keren N, Zer H, Gong H, More TS, Gal A, Tal S, Domovich Y (1994) Light-induced degradation of the photosystem II reaction centre D1 protein in vivo: an integrative approach. In: Baker NR, Bowyer JR (eds) Photoinhibition of photosynthesis. From the molecular mechanisms to the field. BIOS Scientific Publ, Oxford, pp 161–177

Öquist G, Anderson JM, McCaffery S, Chow WS (1992) Mechanistic differences in photoinhibition of sun and shade plants. Planta 188:422–431

Osmond CB, Ramus J, Levasseur G, Franklin LA, Henley WJ (1993) Fluorescence quenching during photosynthesis and photoinhibition of *Ulva rotundata* Blid. Planta 190:97–106

Pang SJ, Zhang ZH, Zhao HJ, Sun JZ (2007) Cultivation of the brown alga *Hizikia fusiformis* (Harvey) Okamura: stress resistance of artificially raised young seedlings revealed by chlorophyll fluorescence measurement. J Appl Phycol 19:557–565

Park Y-I, Chow WS, Anderson JM (1995) Light inactivation of functional Photosystem II in leaves of peas grown in moderate light depends on photon exposure. Planta 196:401–411

Prasil O, Adir N, Ohad I (1992) Dynamics of photosystem II: mechanism of photoinhibition and recovery processes. In: Barber J (ed) Topics in photosynthesis. The photosystems: structure, function and molecular biology, vol 11. Elsevier, Amsterdam, pp 295–348

Rakhimberdieva MG, Stadnichuk IN, Elanskaya IV, Karapetyan NV (2004) Carotenoid-induced quenching of the phycobilisome fluorescence in photosystem II-deficient mutant of *Synechocystis* sp. FEBS Lett 574:85–88

Ralph PJ (1998) Photosynthetic responses of *Halophila ovalis* (R. Br.) Hook. f. to osmotic stress. J Exp Mar Biol Ecol 227:203–230

Ralph PJ (1999) Light-induced photoinhibitory stress responses of laboratory-cultured *Halophila ovalis*. Bot Mar 42:11–22

Ralph PJ (2000) Herbicide toxicity of *Halophila ovalis*, assessed by chlorophyll a fluorescence. Aquat Bot 66:141–152

Ralph PJ, Burchett MD (1995) Photosynthetic responses of *Halophila ovalis* (R. Br.) Hook. *f.* to high irradiance stress, using chlorophyll *a* fluorescence. Aquat Bot 51:55–66

Ralph PJ, Gademann R (1999) Photosynthesis of the seagrass *Posidonia australis* Hook. *f* and associated epiphytes, measured by *in situ* fluorescence analysis. In: Walker DI, Wells FE (eds) The seagrass flora and fauna of Rottnest Island, Western Australia. Perth, Western Australian Museum, pp 63–71

Ralph PJ, Gademann R (2005) Rapid light curves: a powerful tool to assess photosynthetic activity. Aquat Bot 82:222–237

Ralph PJ, Short FT (2002) Impact of the wasting disease pathogen, *Labyrinthula zosterae*, on the photobiology of eelgrass *Zostera marina*. Mar Ecol Progr Ser 226:265–271

Ralph PJ, Gademann R, Larkum AWD, Kühl M (2002a) Spatial heterogeneity in active chlorophyll fluorescence and PSII activity of coral tissues. Mar Biol 141:639–646

Ralph PJ, Polk SM, Moore KA, Orth RJ, Smith WO Jr (2002b) Operation of the xanthophyll cycle in the seagrass *Zostera marina* in response to variable irradiance. J Exp Mar Biol Ecol 271:189–207

Ralph PJ, Macinnis-Ng CMO, Frankart C (2005) Fluorescence imaging applications: effect of leaf age on seagrass photokinetics. Aquat Bot 81:69–84

Ranilello R, Lorenti M, Brunet C, Buia MC (2006) Photoacclimation of the invasive alga *Caulerpa racemosa* var. *cylindracea* to depth and daylight patterns and a putative role for siphonaxanthin. Mar Ecol 27:20–30

Ritz M, Neverov VKV, Etienne J-L (1996) pH-dependent fluorescence quenching and its photoprotective role in the unicellular red alga *Rhodella violaceae*. Photosynthetica 37:267–280

Rodrigues MA, Dos Santos CP, Young AJ, Strbac D, Hall DO (2002) A smaller and impaired xanthophyll cycle makes the deep sea macroalgae *Laminaria abyssalis* (Phaeophyceae) highly sensitive to daylight when compared with shallow water *Laminaria digitata*. J Phycol 38:939–947

Roleda M, Wiencke C, Lüder UH (2006) Impact of ultraviolet radiation on cell structure, UV-absorbing compounds, photosynthesis, DNA damage, and germination in zoospores of Arctic *Saccorhiza dermatodea*. J Exp Bot 57:3847–3856

Runcie JW, Durako MJ (2004) Among-shoot variability and leaf-specific absorptance characteristics affect diel estimates of *in situ* electron transport of *Posidonia australis*. Aquat Bot 80:209–220

Saroussi S, Beer S (2007) Alpha and quantum yield of aquatic plants derived from PAM fluorometry: uses and misuses. Aquat Bot 86:89–92

Schofield O, Evens TJ, Millie DF (1998) Photosystem II quantum yields and xanthophyll-cycle pigments of the macroalga *Sargassum natans* (Phaeophyceace): responses under natural sunlight. J Phycol 34:104–112

Schreiber U (2004) Pulse-Amplitude-Modulation (PAM) fluorometry and saturation pulse method: an overview. In: Papageorgiou GC, Govindjee (eds) Chlorophyll *a* fluorescence: a signature of photosynthesis. Springer, Dordrecht, pp 279–319

Schreiber U, Bilger W (1993) Progress in chlorophyll fluorescence research: major developments during the last years in retrospect. Prog Bot 54:151–173

Schreiber U, Bilger W, Neubauer C (1994) Chlorophyll fluorescence as a nonintrusive indicator for rapid assessment of in vivo photosynthesis. In: Schulze E-D, Caldwell MM (eds) Ecophysiology of photosynthesis. Springer, Berlin, pp 49–70

Schreiber U, Bilger W, Hormann H, Neubauer C (1998) Chlorophyll fluorescence as a diagnostic tool: basics and some aspects of practical relevance. In: Raghavendra AS (ed) Photosynthesis. A comprehensive treatise. Cambridge University Press, Cambridge, pp 320–336

Schubert N, García-Mendoza E (2008) Photoinhibition in red algal species with different carotenoid profiles. J Phycol 44:1437–1446

Schubert N, García-Mendoza E, Pacheco-Ruiz I (2006) Carotenoid composition of marine red algae. J Phycol 42:1208–1216

Seddon S, Cheshire A (2000) Photosynthetic responses of *Amphibolis antartica* and *Posidonia australis* to temperature and desiccation using chlorophyll fluorescence. Mar Ecol Prog Ser 220:119–130

Sharon Y, Beer S (2008) Diurnal movement of chloroplasts in *Halophila stipulacea* and their effect on PAM fluorometric measurement of photosynthetic rates. Aquat Bot 88:273–276

Shibata K (1959) Spectrophotometry of translucent biological materials: opal glass transmission method. Methods Biochem Anal 7:77–109

Silva J, Santos R (2003) Daily variation patterns in seagrass photosynthesis along a vertical gradient. Mar Ecol Progr Ser 257:37–44

Silva J, Santos R (2004) Can chlorophyll fluorescence be used to estimate photosynthetic production in the seagrass *Zostera noltii*? J Exp Mar Biol Ecol 307:207–216

Strain E, Beardall J, Thomson R, Roberts S, Light B (2006) Spatio-temporal variability in the photosynthetic characteristics of *Zostera tasmanica* measured by PAM. Aquat Bot 85:21–28

Strasser RJ, Srivastava A, Govinjee (1995) Polyphasic chorophyll *a* fluorescence transient in plants and cyanobacteria. Photochem Photobiol 61:32–42

Strasser RJ, Tsimilli-Michael M, Srivastava A (2004) Analysis of the chlorophyll *a* fluorescence transient. In: Papageorgiou GC, Govindjee (eds) Chlorophyll *a* fluorescence: a signature of photosynthesis. Springer, Dordrecht, pp 321–362

Thorhaug A, Richardson AD, Berlyn GP (2006) Spectral reflectance of *Thalassia testudinum* (Hydrocharitaceae) seagrass: low salinity effects. Am J Bot 93:110–117

Trautman DA, Borowitzka MA (1999) The distribution of epiphytic organisms on *Posidonia australis* and *P. sinuosa*, two seagrasses with differing leaf morphology. Mar Ecol Prog Ser 179:215–229

Turner SJ, Schwarz A-M (2006) Biomass development and photosynthetic potential of intertidal *Zostera capricorni* in New Zealand estuaries. Aquat Bot 85:53–64

Urmacher S, Hanelt D, Nultsch W (1995) Zeaxanthin content and the degree of photoinhibition are linearly correlated in the brown alga *Dictyota dichotoma*. Mar Biol 123:159–165

Ursi S, Pedersén M, Plastino M, Snoeijs P (2003) Intraspecific variation of photosynthesis, respiration and photoprotective carotenoids in *Gracilaria birdiae* Gracilariales (Rhodophyta). Mar Biol 142:1007

van Wijk KJ, van Hasselt PR (1993) Photoinhibition of photosystem II *in vivo* is preceded by down-regulation through light-induced acidification of the lumen: Consequences for the mechanism of photoinhibition *in vivo*. Planta 189:359–368

Van de Poll WH, Eggert A, Buma AGJ, Breeman AM (2002) Temperature dependence of UV radiation effects in Arctic and temperate isolates of three red macrophytes. Eur J Phycol 37:59–68

Van Kooten O, Snel JFH (1990) The use of chlorophyll fluorescence nomenclature in plant stress physiology. Photosynth Res 25:147–150

Vass I, Styring S, Hundal T, Koivuniemi A, Aro E-M, Anderson B (1992) Reversible and irreversible intermediates during photoinhibition of photosystem II: stable reduced QA species promote chlorophyll triplet formation. Proc Nat Acad Sci USA 89:1408–1412

Yakovleva IM, Titlyanov EA (2001) Effect of high visible and UV irradiance on subtidal *Chondrus crispus*: stress, photoinhibition and protective mechanisms. Aquat Bot 71:47–61

Yamamoto HY, Nakayama TOM, Chichester CO (1962) Studies of the light interconversion of the leaf xanthophylls. Arch Biochem Biophys 97:168–173

Chapter 10
Chlorophyll Fluorescence in Reef Building Corals

Mark E. Warner, Michael P. Lesser, and Peter J. Ralph

1 Introduction

The ecological success of reef-building corals throughout the tropics is due in large part to the endosymbiotic dinoflagellates that reside within the gastrodermal cells of these cnidarian hosts. These algae, belonging to the genus *Symbiodinium*, are often referred to by the common term of "zooxanthellae." This mutualism between *Symbiodinium* spp. and tropical and sub-tropical coral species has been a key component to the evolutionary persistence of reef-building corals since the Triassic (Stanely 2003). The importance of these algae in the long-term success of reef-building corals cannot be over emphasized, as they can contribute a significant portion of photosynthetically derived carbon to the host via translocation. The coral metabolizes this carbon, thereby meeting up to 90 percent or more of the animal's daily metabolic demand from the byproducts of photosynthesis by the symbionts (Muscatine 1990). The large proportion of respiratory carbon provided by these endosymbionts contributes directly to the energetic demands of coral, with calcification rates that are typically elevated by three times as much when compared to corals held in the dark or species that do not harbor endosymbionts (i.e. aposymbiotic) (Gattuso 1999). Corals also feed heterotrophically; however this energy source cannot be directly monitored using the techniques described in this chapter and consequently will not be discussed.

Continued degradation of coral reefs throughout the world has spawned much interest in establishing techniques that allow for rapid assessment and understanding of how abiotic and biotic stressors may impact the photobiology of these symbioses. In the past 10–15 years, there has been an explosion of interest in using active chlorophyll *a* fluorescence to assess the photosynthetic condition of *Symbiodinium*, both *in hospite* and *in vitro*. The advent of commercially available instrumentation, some of which is submersible (e.g. the Diving-PAM), and the non-invasive nature of the method, has been met with widespread appeal in the last decade. In this regard, the use of fluorescence methodologies has led to a well-established body of evidence supporting thermal disruption of energy transfer in photosystem II (PSII) that is related to the larger phenomena of coral bleaching (described in detail below). Furthermore, fluorescence-based methods have provided more detailed understanding of both the pathways of photodamage and photoprotection in *Symbiodinium* across multiple temporal and spatial scales in many coral species throughout the world. This information is also quite topical when placed in the context of our current understanding of the massive genetic diversity of the genus *Symbiodinium*, and current efforts are underway to understand the potential links between phylogenetic diversity and physiological differences of these algae, which can originate from adaptive traits and/or physiological plasticity.

M.E. Warner (✉)
College of Earth, Ocean, and Environment, University of Delaware, 700 Pilottown Rd. Lewes, DE 19958, USA
e-mail: mwarner@udel.edu

M.P. Lesser
Department of Molecular, Cellular, and Biomedical Sciences, University of New Hampshire, Durham, NH 03824, USA

P.J. Ralph
Plant Functional Biology and Climate Change Cluster, University of Technology PO Box 123, Broadway, Sydney, NSW, Australia

D.J. Suggett et al. (eds.), *Chlorophyll* a *Fluorescence in Aquatic Sciences: Methods and Applications*, Developments in Applied Phycology 4, DOI 10.1007/978-90-481-9268-7_10, © Springer Science+Business Media B.V. 2011

This chapter is not intended to be an exhaustive review of the existing literature, but as a partial review of the status of the field of chlorophyll fluorescence in reef corals, some of the significant results that have been revealed thus far, and the available instrumentation that one may use to investigate chlorophyll fluorescence in corals. Secondly, we will review specific methods and protocols that one may use with currently available instrumentation, as well as some of the possible caveats that may arise with particular research questions that have been approached to date, and some of the open areas of study that remain to be explored.

2 Natural Patterns of Fluorescence

As with other aquatic systems, the most common use of fluorescence methods in reef corals to date has been to estimate the maximum quantum efficiency of PSII photochemistry (denoted as F_v/F_m) and the PSII photochemical efficiency in the light-adapted state, also known as the effective, or steady state, quantum yield of chlorophyll fluorescence (typically denoted as $\Delta F/F_m'$, $\Delta F'/F_m'$ or Φ_{PSII}; see Chapter 1 (this volume) by Cosgrove and Borowitzka for further discussion concerning appropriate nomenclature and symbols for fluorescence work). Studies in the laboratory and the field have now confirmed a daily, mid-day, decline in PSII photochemical activity in *Symbiodinium* of reef-building corals and other algal-cnidarian symbioses (Ralph et al. 1999; Gorbunov et al. 2001; Lesser and Gorbunov 2001; Jones and Hoegh-Guldberg 2001; Winters et al. 2003; Lesser and Farrell 2004). This decrease in PSII activity is observed as a decrease in the effective quantum yield (Φ_{PSII}) as well as dark-adapted quantum yields of PSII fluorescence (F_v/F_m), and has been shown to correlate with the development of non-photochemical quenching (NPQ) as well as the daily occurrence of peak solar radiation exposure.

Importantly, this down-regulation in PSII photochemistry is viewed as reversible (formerly referred to as "dynamic") and photoprotective (Ralph et al. 1999; Gorbunov et al. 2001) and should not be interpreted as detrimental to the photosynthetic function of PSII in *Symbiodinium* spp. When both productivity by respirometry and chlorophyll fluorescence are measured simultaneously, there is typically no apparent negative impact to total net daily productivity (oxygen production) of the whole coral colony, despite the observed daily decline in PSII activity (Hoegh-Guldberg and Jones 1999; Lesser and Gorbunov 2001). This finding is quite similar to that noted for free living microalgae (Behrenfeld et al. 1998). Thus, while there may be a certain percentage of damaged PSII reaction centers that occurs daily in "healthy" shallow water corals (Gorbunov et al. 2001), electron transport through the remaining centers is adequate to maintain maximal photosynthetic productivity. Interestingly, some studies have noted differences in the effective quantum yield and calculated PSII electron transport rate (ETR) over different portions of the day, wherein ETR values can be higher in the early morning compared to the late afternoon, higher in the afternoon compared to the morning, or there is no change over the day (see later section on measuring and interpreting ETR below). Such daily fluctuations of PSII activity are highly specific for a particular species of coral and are most likely influenced by several factors, including the phylotype and physiology of resident *Symbiodinium*, colony morphology (colour and tissue thickness), animal behavior (polyp extension and contraction), and the light history of the particular location (from depth to within the colony) from which fluorescence measurements were taken (Winters et al. 2003; Levy et al. 2004).

Recent efforts to characterise diel fluctuations have led to the use of a metric known as the excitation pressure over PSII abbreviated as Q_m, and quantified as:

$$Q_m = 1 - [\Delta F/F_m' \text{ at peak sunlight} / F_V/F_m \text{ at dawn or dusk}]$$

Here, it is important to note the proper time for recording F_v/F_m so as to collect the true maximal value that is free from the influence of extended NPQ or transitory diel damage or reversible down-regulation of PSII. Similar to the concept of excitation pressure derived by Maxwell et al. (1995), Iglesias-Prieto et al. (2004) introduced Q_m as a useful metric of symbiont performance at maximal irradiance, such that values near zero are indicative of a high proportion of open PSII reaction centers and possible light limitation, while values near one are indicative of maximal PSII reaction center closure and possible photoinhibition. Iglesias-Prieto et al. (2004) utilized this metric to show the differential use of light by the corals *Pocillopora verrucosa* and *Pavona giantea* under field conditions, which harbor different genetic types of *Symbiodinium* (ITS2 types D1 and C1-c respectively). Both species displayed equivalent Q_m even though they were found at different depths. These results can help explain why *P. gigantea* is not as common at shallow depths at this study site, as its symbionts would experience >30% higher light pressure as compared to *P. verrucosa* and suffer from greater PSII photoinactivation. These results show a significant proximal mechanism (i.e., differences in zooxanthellae genotype) for niche diversification between different species of coral. Recent work has further noted significant differences in excitation pressure between different Caribbean corals residing at the same depth and that differences in Q_m can be more sensitive than gross comparisons of F_v/F_m values between different coral species for such comparisons (Warner et al. 2006) (Fig. 1).

Fewer studies have documented natural trends in active chlorophyll fluorescence in reef corals over time scales greater than a few days–weeks. Warner et al. (2002) followed fluorescence patterns in several species of Caribbean corals over a 5-year period in the Bahamas. This work demonstrated a clear fluctuation in F_v/F_m on a seasonal scale, with the lowest F_v/F_m values typically recorded in the late summer to early fall and the highest values in the late winter to early spring. In addition, these fluctuations were dampened in deeper corals, and the lowest Φ_{PSII} was noted in corals residing in the shallowest habitats. These results were interpreted as a combination of seasonal photoacclimation and possible PSII damage resulting from excessively warm water temperatures during some summers (Warner et al. 2002) when symbiont densities dropped below typical seasonal averages (Fitt et al. 2000). While it is speculated that similar patterns in seasonal changes in F_v/F_m may dominate other coral species, Hill and Ralph (2005) were unable to detect significant differences in F_v/F_m in three species of reef corals collected on the southern Great Barrier Reef during the winter and summer. However, direct comparison of this work to the aforementioned seasonal study in the Caribbean is not easy due to the extended holding times (2 days) of collected corals at low light levels in the latter study prior to fluorescence measurement. Clearly, further work is needed in order to understand the physiological mechanisms that underpin changing patterns of chlorophyll fluorescence on seasonal scales for better long-term monitoring of reef corals, and if such seasonal changes in fluorescence are common, one should take care to understand this inherent variability when using chlorophyll fluorescence for other studies, especially when performed at different times of the year.

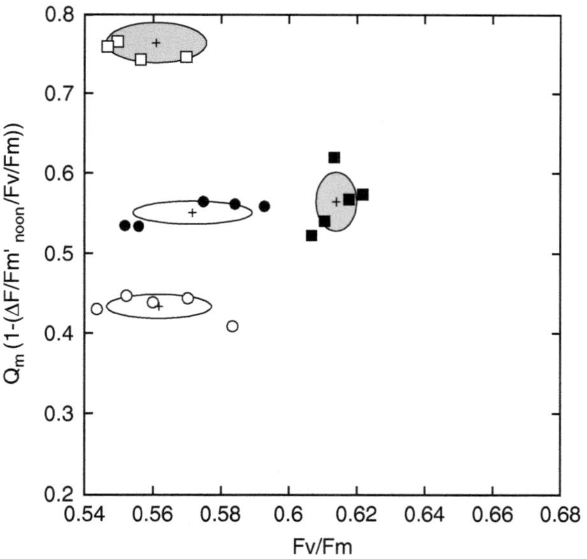

Fig. 1 A comparison between the quantum yield of PSII (F_v/F_m) and the excitation pressure over PSII (Qm) for *Symbiodinium* spp. within the Caribbean corals, *Montastraea faveolata* (●), *Diploria strigossa* (○), *Montastraea cavernosa* (closed squares), and *Montastraea annularis* (□). All corals were held in outdoor mesocosms at 28 °C and acclimated to 50% of incident solar radiation. Note the similarity in F_v/F_m between three coral species, yet the marked difference in Qm. n = 5 for each point, shaded area represents SEM. Unpublished data contributed by R. Iglesias-Prieto

2.1 Multiple and Single Turnover Instrumentation

Importantly, there are fundamental differences between some instrumentation that is commonly used for investigating active chlorophyll fluorescence in corals. In particular, comparison of single and multiple turnover instruments has shown that multiple turnover protocols, such as those employed by the PAM fluorometer, may lead to increases in maximal fluorescence (F_m) of up to 35% leading to a higher estimate of PSII quantum yields (Kolber et al. 1998). While we are not aware of any published studies that have utilized both single and multiple turnover instrumentation with corals, as an example, one can compare the marked difference in F_v/F_m values for the Caribbean coral *Montastraea faveolata* sampled at a similar depth in the Bahamas near Lee Stocking Island (compare Gorbunov et al. 2001 and Warner et al. 2002). The differences in fluorescence yields observed when comparing single and multiple turnover protocols are also supported by studies on algal cells and isolated chloroplasts (Samson and Bruce 1996; Kolber et al. 1998; Vasil'ev and Bruce 1998). Single turnover protocols (such as FRRf) allow for the near simultaneous turnover of all PSII reaction centers through a single actinic exposure of all reaction centers on the timescale of microseconds, which allows for the calculation of additional parameters, such as the effective absorption cross section of PSII (σ_{PSII}). It is therefore important to note that the absolute PSII quantum yield values from PAM and fast repetition rate (FRR) fluorometers may not be the same, but both instruments generally agree when tracking relative changes in quantum yields with variations in the physical environment of a coral (Suggett et al. 2003; authors personal obs.). In addition to the basic fluorescence parameters that can be measured with a variety of active fluorescence methods (PAM, FRRf), the FRR technique measures the effective absorption cross section of PSII (σ_{PSII}) and energy transfer between PSII complexes, expressed as the "connectivity factor" (*p*) (Kolber et al. 1998; Suggett et al. 2003). These parameters help to characterize the size of PSII units, their capacity to absorb and utilize light, and the functional integrity of light-harvesting complexes. In addition to the effective and dark-adapted quantum yield of PSII, these parameters add significantly to the biophysical and physiological interpretation of changes in fluorescence yields in corals and other photoautotrophs.

Another type of fluorometer is now commercially available that employs the Fluorescence-Induction and Relaxation method (see Gorbunov and Falkowski 2004; FIRe, Satlantic, Inc.) which allows simultaneous measurement of fluorescence from single and multiple turnover protocols. However, the current design of this instrument limits its use to bench-top laboratory-based measurements. In addition to the instruments mentioned above, other studies have monitored the photobiology of *Symbiodinium* with multiple turnover fluorometers equipped with faster data acquisition response times (e.g. Plant Efficiency Analyzer; Hansatech Instruments, U.K. and double-modulated fluorometers; Photon System Instruments, Czech Republic), thereby providing a complete recording of fluorescence induction from F_o to $F_{M,}$ also known the classic Kautsky curve. The resulting data are often referred to as O-J-I-P curves, in reference to the polyphasic sequence of transients that may be distinguished by such curves when plotted on a semi-log time scale. While a detailed discussion of these transients is beyond the scope of this chapter (see Strasser et al. 1995, 2004, for more in depth discussion), we do mention some examples where the use of O-J-I-P curves on corals are informative (see section on detecting stress below (Hill et al. 2004a; Hill and Ralph 2005; Ulstrup et al. 2005). Likewise, it is noteworthy that while this method has its own limitations (samples must remain in complete darkness during measurement), this technique provides further detail of electron transport through PSII and the PQ (plastoquinone) pool that is not readily available by many PAM instruments such as the Diving-PAM or FRRf.

2.2 Non-Photochemical Quenching

As previously mentioned, the mid-day decline in PSII activity typically correlates well with significant light energy dissipation via non-photochemical pathways, and a rise in non-photochemical quenching (NPQ) of the fluorescence signal. Fluorescence techniques, in tandem with other measurements such as HPLC of photosynthetic and photoprotective pigments, have led to a better understanding of the possible sources of this quenching signal. NPQ is a common photoprotective mechanism in terrestrial plants and algae that is primarily controlled by the rate of photosynthetic electron transport, and influences the redox state of the plastoquinone pool and ΔpH of the thylakoid membrane. However, for *Symbiodinium* spp. *in hospite*, NPQ typically progresses well before photosynthetic electron transport is saturated (Gorbunov et al. 2001). This is in contrast to many free-living phytoplankton, where NPQ is employed when photosynthetic electron transport is almost fully (~90%) saturated (Falkowski et al. 1986). As irradiance in corals increases beyond saturation (E_k), the quantum yield of NPQ when measured by single turnover (FRRf) approaches 0.7–0.8 (Gorbunov et al. 2001) (note when accounting for the 30–40% difference in maximal dark adapted fluorescence between single and multiple turnover instruments, this value would provide an NPQ value ~ 2.0 when measured with a multiple turnover instrument). These values are as high as observed in plants and phytoplankton, and indicate that up to 80% of excess excitation energy dissipates as heat due to NPQ. NPQ may originate from energy dissipation at the light harvesting complex (antenna), the PSII reaction centers or both.

Results from single-turnover instruments (e.g. the submersible fast repetition rate fluorometer or FRR), have shown that enhanced NPQ correlates well with a significant decline in the effective absorption cross-section for PSII (either σ_{PSII} measured in the dark, or σ_{PSII}' measured in the light activated state), a well-known signature associated with a decrease in light harvesting capacity (Gorbunov et al. 2001; Lesser and Gorbunov 2001; Levy et al. 2004, 2006). Likewise, studies with corals have confirmed that rapidly reversible NPQ in the light-harvesting antennae is associated with the xanthophyll cycle, whereby a rapid conversion of diadinoxanthin to diatoxanthin occurs, which is thought to direct photon energy away from the light harvesting antennae upstream of the PSII reaction centers (Brown et al. 1999; Gorbunov et al. 2001; Warner and Berry-Lowe 2006).

NPQ has been sub-divided into three different classes, based on relaxation kinetics of the quenching signal once a subject is placed in the dark after a defined period of light exposure. These are commonly categorized by the speed of relaxation time in the dark (from fastest to slowest respectively) as energy dependant (qE), state transitional (qT), and photoinhibitory (qI) based quenching (Hill et al. 2005). Under typical light exposures, it was generally assumed that the majority of NPQ was energy dependent, and largely related to antennae based quenching, while qI becomes more prevalent when light levels are in excess of what can be effectively used for photochemistry and become damaging. While documenting such trends are possible in reef corals, the origin of NPQ can be much more variable and complex (described below). State transitions are well described in terrestrial plants and chlorophytes wherein reduction of the plastoquinone (PQ) pool leads to a signal cascade that results in phosphorylation of PSII light harvesting complexes, which subsequently migrate from PSII to PSI. This leads to a diversion in excitation flux away from PSII in what is commonly referred to as a State-1 to State-2 transition (Allen et al. 1981; Wollman 2001; Bellafiore et al. 2005). A similar phenomenon (although mechanistically different) is noted to occur in many types of cyanobacteria where changes in phycobilisome coupling between PSII and PSI drive state transitions (Allen and Mullineaux 2004). A few studies have suggested the possibility of state transitions and qT quenching in corals, especially during periods of excess light and temperature (Jones and Hoegh-Guldberg 2001; Hill and Ralph 2005). However if such a mechanism does exist, it is quite different from the traditional concept of LHC detachment and migration via redox driven kinase/phsophorylase activity developed from work in other systems. To date, supporting evidence for state transitions, either by phosphorylation via phosphothreonine pathways (Rintamäki et al. 1997; Cardol et al. 2003) or detection of enhanced energy transfer to PSI by 77K spectroscopy, has not been found in freshly isolated or cultured *Symbiodinium* (Warner, Iglesias-Prieto 2001; 2007, personal observations). Nevertheless, other photoprotective pathways, such as uncoupling light harvesting to PSII reaction centers or photosystem I cyclic electron flow are other strong possibilities (McCabe-Reynolds et al. 2008) that require further exploration in *Symbiodinium*.

While thermal dissipation in the pigment bed attributed to the xanthophyll cycle can be a primary photoprotective mechanism *in vivo*, continued thermal dissipation in the reaction centers is stimulated by changes in the pH gradient across the thylakoid membrane as well (Weis and Berry 1987; Müller et al. 2001). Thus, NPQ in some corals is associated with a co-occurring decline in the functional absorption cross-section of PSII as well as the quantum yield of photochemistry (q_p) (Gorbunov et al. 2001). This same irradiance-induced decline in σ_{PSII} has been observed in both phytoplankton (Olaizola et al. 1994) and terrestrial plants and supports the overall interpretation that NPQ is associated with thermal energy dissipation in the pigment bed and not the PSII reaction centre.

In order to delineate possible origins of NPQ, single-turnover instrumentation, such as the FRRF holds a clear advantage over multiple-turnover instruments such as the PAM fluorometer, as one may quantify the recovery kinetics of F_v/F_m, NPQ, and σ_{PSII} simultaneously. For example, using a custom-built FRR fluorometer, Gorbunov et al. (2001) noted a biphasic

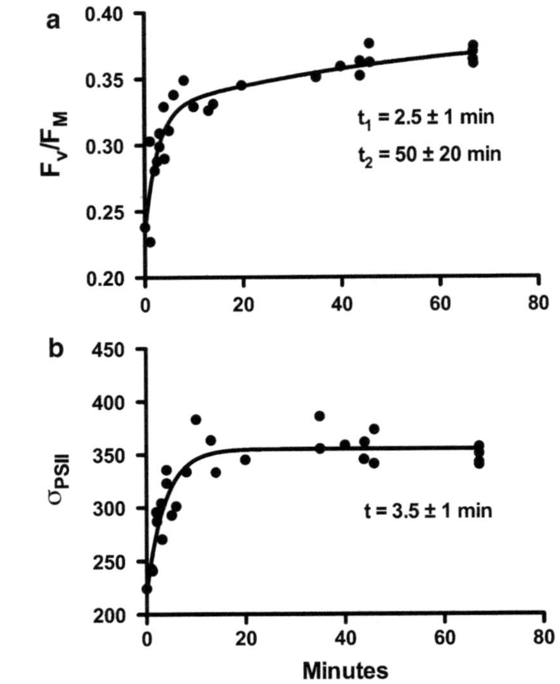

Fig. 2 Dark recovery of the quantum yield of PSII (F_v/F_m) (**a**) and the functional absorption cross section for PSII (σ_{PSII}) in *Symbiodinium* within the reef building Caribbean coral, *Montastraea faveolata*, following exposure to mid-day PAR of 1,500 μmol photons m^{-2} s^{-1}. Recovery kinetics were evaluated by a two exponential fit to F_v/F_m and a single exponential fit to σ_{PSII} (Reproduced from Gorbunov et al. 2001. With permission. Copyright (2008) by the American Society of Limnology and Oceanography, Inc.)

recovery pattern in F_v/F_m in the Caribbean coral *Montastraea faveolata* following exposure to a dose of mid-day light of 1,500 μmol photons m^{-2} s^{-1} (Fig. 2). As the time constant for σ_{PSII} recovery closely matched the more rapid phase of recovery in F_v/F_m, one can infer that this portion of the recovery in PSII activity is directly related to down-regulation in energy transfer at the LHC, while the slower portion of recovery phase is attributed to down-regulated PSII reaction centers. The origin of NPQ can vary significantly between coral species. Gorbunov et al. (2001) noted a significant rise in NPQ originating from the PSII reaction center in the Caribbean coral *Montastraea cavernosa* while light levels were well below the saturation point for maximal photosynthesis (E_k), while Levy et al. (2006) noted a good correlation between NPQ and the daily changes in PSII functional absorption cross section. While *in situ* and laboratory studies have shown a central role for NPQ in the photoprotection of the *Symbiodinium* photosynthetic apparatus, there is evidence for variability in NPQ capabilities between different genotypes of *Symbiodinium* (Warner and Berry-Lowe 2006) which has the potential to directly affect their ecological distribution and whole-colony organismal performance.

3 Detecting Stress

Active chlorophyll fluorescence has become the method of choice for detecting photoinactivation in *Symbiodinium* within reef-corals subjected to many forms of abiotic stress, and the initial onset of coral bleaching in particular. Coral bleaching is broadly defined as the loss of symbionts and/or the loss of algal pigments, and while many stressors may elicit a bleaching response (e.g. excessive changes in salinity, temperature, or light), elevated temperature combined with high irradiance are two factors that are considered to be the causative agents of many large-scale bleaching events that have significantly impacted coral reefs over the past 2 decades (e.g., Hoegh-Guldberg 1999; Lesser 2004; Brown and Dunne 2008). Early estimates of PSII inactivation by thermal stress were made by quantifying changes in F_v/F_m in cultured *Symbiodinium* using the DCMU fluorescence enhancement method (Samuelsson and Öquist 1977; Iglesias-Prieto et al. 1992; Lesser 1996) and in whole corals by PAM fluorometry (Warner et al. 1996). Initial PAM fluorometry work provided the first evidence for differences in PSII inactivation between different coral species, and that elevated NPQ in some species provided evidence of greater photoprotection in heat-resistant corals, and importantly provided a non-invasive proxy for stress prior to colony paling (Warner et al. 1996). Simultaneous efforts by Iglesias-Prieto (1997) utilized induction kinetics of cultured *Symbiodinium* to show PSII damage as well.

Subsequent work on this topic in many laboratories has led to a much greater understanding of the temporal nature and fine detail of changes in PSII fluorescence in reef corals under thermal and light stress (Hoegh-Guldberg and Jones 1999; Jones et al. 1999; Warner et al. 1999; Gorbunov et al. 2001; Bhagooli and Hidaka 2004; Hill et al. 2004a; Lesser and Farrell 2004; Hill and Ralph 2006), and this work has added to several proposed models to further describe a cascade of cellular pathways that are most likely involved in the disruption of photosynthesis in *Symbiodinium* during coral bleaching (see Lesser 2004, 2006 and Smith et al. 2005). Not surprisingly, this work has also led to several competing (yet not necessarily mutually exclusive) hypotheses concerning the initial points of breakdown in the photosynthetic machinery of *Symbiodinium* when corals are exposed to bleaching conditions. In particular, fluorescence data collected from several types of instruments has been used to infer direct damage to the PSII reaction center (Warner et al. 1996, 1999; Iglesias-Prieto 1997; Hill et al. 2004a; Hill and Ralph 2006) as well as some point in carbon fixation downstream of PSII (Jones et al. 1998; Yakovleva and Hidaka 2004).

While the breakdown of PSII integrity is substantiated by direct loss in the D1 reaction center protein (Warner et al. 1999; Lesser and Farrell 2004; Robison and Warner 2006) as well as *de novo* synthesis of light harvesting proteins (Takahashi et al. 2008), testing specific details concerning hypothesized degradation of the dark reactions have been elusive due to the unstable nature of the form II Rubisco in *Symbiodinium*, however carbonic anhydrase activity would appear not to be heat labile (Leggat et al. 2004). Despite this instability in Rubisco, Lesser (1996) showed a significant decline in the Rubisco activity (via ^{14}C uptake) when cultured *Symbiodinium* was exposed to elevated temperatures. This decline in Rubisco activity occurred simultaneously with significant declines in the maximum quantum yield of PSII fluorescence (Lesser 1996). Recent use of the electron donor diphenyl carbazide (DPC), coupled with PAM fluorometry and oxygen electrodes has provided evidence for the integrity of the oxygen evolving complex well above bleaching temperatures (Hill and Ralph 2008b), while both an acceleration (Tchernov et al. 2004) and decline (Robison and Warner 2006) in the rate of electron transport on the acceptor side of PSII has been documented in heat treated *Symbiodinium*. While significant advances have been made in this arena, it is doubtful that one prevailing point of damage or mechanism of disruption will ever be elucidated given the diversity in physiological responses between different types of *Symbiodinium*, the corals studied thus far, and the tremendous variability in experimental designs and instrumentation used in bleaching experiments. Moreover, a complete understanding of the bleaching response will require a better understanding of the cellular and biochemical responses of the cnidarian host, and to what extent the alga and host are impacted by temperature manipulation remains an open and important question (Baird et al. 2009).

In many cases, photoinactivation is initially detected as a significant drop in F_v/F_m as well as effective quantum yield, and an initial loss in the return to dark acclimated F_v/F_m values recorded prior to the application of the stressor, and in some cases these patterns are directly attributable to the genetic type of *Symbiodinium* present (Berkelmans and van Oppen 2006; Warner et al. 2006; Ulstrup et al. 2006). As PSII activity continues to decline, a significant loss in *Symbiodinium* number becomes apparent. The continued decline in PSII activity is an important marker, and is in contrast to F_v/F_m or $\Delta F/F_m'$ values that decrease upon the initial stress and then subsequently stabilize at a new level, where PSII photoinactivation becomes less significant. The latter example may be better characterized as acclimation to present conditions and is presently a topic for further investigation. Similar to the discussion of natural patterns of fluorescence in reef corals above, different types of instrumentation have provided various levels of detail concerning alteration to photochemical charge separation and energy transfer during experimental bleaching studies. Single turnover fluorescence methods have shown both no change (Lesser and Farrell 2004) as well as a significant increase in σ_{PSII} (Suggett et al. 2008) upon heating. Minimal change in σ_{PSII} was interpreted as greater NPQ originating from (possibly damaged) PSII reaction centers in corals exposed to elevated temperature and light (Lesser and Farrell 2004), while a rise in σ_{PSII} with thermal stress may indicate that light harvesting antennae are becoming energetically isolated while there is a greater loss in PSII reaction center proteins, similar to results seen in other microalgae during iron limitation (Greene et al. 1991).

In addition to the extensive use of fluorescence methods to study coral bleaching related to temperature and light, active chlorophyll fluorescence has also been used to investigate the impact of specific toxins that are common in some reef environments (e.g. cyanide used (illegally) for capturing ornamental reef fish) (Jones et al. 1999; Jones and Kerswell 2003) as well as the isolated impacts of salinity and turbidity stress on reef corals (Kerswell and Jones 2003; Philipp and Fabricius 2003).

4 Protocols and Pitfalls

As most users of fluorescence instrumentation have often noted, the ability to rapidly gather large amounts of data is not a typical hindrance, yet proper instrument verification, protocol design and downstream data analyses are key to maximizing the use of active chlorophyll fluorescence and reporting valid, and comparable, results. Some elements regarding the use of active fluorescence on reef corals specifically have been discussed in a previous review (Fitt et al. 2001), and while this was written with a focus on PAM fluorometry in general, much of the advice is applicable for any type of instrumentation. Several of the elements covered below, while simple in principle, can lead to significant errors in data collection if not adhered to. Naturally, the most important question one should have when considering the use of active chlorophyll fluorescence in their investigation is, "is this method appropriate for my specific research questions," rather than vice-versa (Platt 1964).

4.1 Dark Acclimation, Sample Area and Related Matters

While there is no one appropriate set time period for dark acclimation prior to recording F_v/F_m, many studies have shown that a range from 10–30 min is typically enough time for full relaxation of NPQ, and that each user must empirically test dark acclimation times with their particular instrument and corals. Importantly, *Symbiodinium* in some corals have a propensity toward chlororespiration when held in the dark for extended time periods. This dark-reduction of PSII will give artificially low F_v/F_m values. There is mounting evidence for chlororespiration in *Symbiodinium* when corals are held in the dark and this could explain the dark reduction of the PQ pool (Jones and Hoegh-Guldberg 2001; Hill and Ralph 2005). While our current understanding of chlororespiration in most algal groups is still quite weak, work with terrestrial plants has provided a working pathway wherein a plastid encoded NAD(P)H-deydrogenase (Ndh) complex transfers electrons to plastoquinone, followed by PQ oxidation with oxygen and a terminal oxidase (PTOX) (Peltier and Cournac 2002). Further, oxygen tensions within coral tissues typically follow a diel swing from hyperoxic to hypoxic conditions between day and night respectively (Dykens and Schick 1982, Kühl et al. 1995, Gardella and Edmunds 1999). The rapid depletion of oxygen in the dark maintains the PQ pool and PSII in a partially reduced state until the dim early morning light in the proceeding day can initiate oxygen production once again and alleviate the temporary PSII reduction (Jones and Hoegh-Guldberg 2001; Hill and Ralph 2008a). Correcting this phenomenon is ideally accomplished by providing a short application of far-red light to the coral or by allowing the coral to sit in very dim light (e.g. <5 μmol photons m^{-2} s^{-1}) instead of complete darkness prior to dark measurement (Hill and Ralph 2008a). Recent evidence suggests that dark reduction of PSII can vary between corals with different morphologies (e.g. thick versus thin tissue corals) (Hill and Ralph 2008a) and one should pay close attention for possible differences when working with a particular coral species.

In addition, when a PAM fluorometer is used, one should ensure that the measuring beam of the fluorometer is not so bright as to create an actinic effect, which will artificially raise the F_o level, and can easily be detected by monitoring the baseline fluorescence for a few minutes prior to applying a saturation light pulse. Therefore, it is useful to carefully consider the settings of the fluorometer (if published), when evaluating results. Similarly, when continually monitoring coral fluorescence over a longer time interval (i.e. several hours to days) from a fixed point, multiple turnover flashes which are spaced close together (e.g. every 5 min) may also induce an artificial reduction of PSII in the dark in some corals (authors personal observation). Conversely, a common mistake when using single turnover instruments such as the FIRe fluorometer is not providing a saturation flash that is long enough to ensure complete closure of all PSII reaction centers, which is easily characterized by a true plateau in the fluorescence trace following a single turnover flash. Beyond single turnover flash length, it is recommended that one ensure that flash intensity is high enough to complete saturation as well. If possible, this could entail test runs with increasing flash intensity to ensure that F_v/F_m quanta^{-1} does not change.

Another point of concern is how to physically dark acclimate a coral fragment without creating secondary effects. One example of this is the use of dark "holders" or "clips" that may be used to hold the fiber sensors of the fluorometer to a particular place on the surface of the coral. Given that reduced flow and hypoxic conditions may further reduce PSII (Ulstrup et al. 2005), one should always verify that such devices are not creating further artifacts in their particular system. Probe placement for light-acclimated readings should be equally scrutinized as well. Small changes in light intensity can significantly impact changes in effective quantum yield, and one should ensure that the probe and/or probe holder are not shading the immediate area of measurement if true $\Delta F/F_m'$ values are desired. Animal behavior (e.g. polyp contraction from physical contact) can contribute to measurements that are not representative of the ambient conditions, and therefore standard measurement times should be chosen when one is attempting to collect similar data from several colonies (Levy et al. 2003). Equally important is the light measurement taken at the time of F and F_M measurement, and in many cases it is desirable that this is done at the same relative plane as the coral surface (Fig. 3).

Photochemical quenching (qP) has been used in a few coral and cultured *Symbiodinium* studies to infer the relative proportion of 'open' PSII reaction centers (Warner et al. 1996; Jones et al. 1998; Robison and Warner 2006). Such an approach is not entirely correct due to dynamic nature of reaction center connectivity. If qP is to provide an accurate index of the redox state of PSII, one must assume negligible energy transfer between PSII complexes (often termed the pool or puddle model), which in many cases is not correct (Lavergne and Trissl 1995; Baker 2008). Measurement of qP by traditional PAM fluorometry protocols is complicated by many factors, including: the rapid quenching of the fluorescence signal in many corals (personal observation), the need to accurately measure F_o' which can be quite difficult without the use of a far-red light source and/or careful analysis of the fluorescence signal during the light acclimated to dark transition. Considering that a far-red light source is not available on the Diving-PAM fluorometer, nor is the complete fluorescence trace recorded when used remotely *in situ*, one should approach the measurement of photochemical quenching and subsequent interpretation of the data collected with great care (see Baker 2008 for further discussion).

Some studies have documented differences in $\Delta F/F_m'$ (and by extension relative electron transport rate or ETR) collected by PAM fluorometers when using different size fiber diameters. Ralph et al. (2002) noted differences between both fiber sizes (from 1–8 mm diameter) as well as intracolony location (polyp vs. coenosarc). Beyond the changes in field of view between different fiber diameters, fluorescence signal heterogeneity may also arise due to the light microclimate, polyp structure, tissue thickness, and number of resident symbionts at the point of sampling (Hill et al. 2004b). Not only will the light history of symbionts be naturally different at different polyp locations in some corals, but one should also consider the light fields created by any artificial actinic light source from the fluorometer itself (see ETR section below for further discussion on the importance of internal light fields).

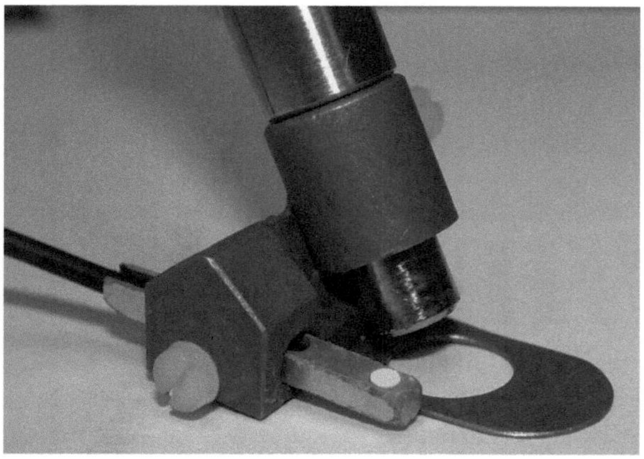

Fig. 3 An example of a modified holder designed to allow simultaneous measurement of chlorophyll fluorescence and light measurement at the same location on a coral colony. Photo by M. Warner

4.2 Electron Transport Rate

Of all of the fluorescence parameters currently under use with corals, the electron transport rate or ETR is most likely the most misunderstood and misused variable found in the literature. This confusion stems from general guidelines that are often overlooked and are largely applicable to all microalgae when attempting to use ETR to infer productivity as well as some special circumstances that are applicable to corals for PSII ETR calculations. For a more in depth discussion of using electron transport to infer productivity, we direct the reader to Suggett et al., Chapter 6, this volume.

Similar to many studies that have attempted to correlate ETR by PAM fluorometry to primary productivity (measured by indirect respirometry or ^{14}C uptake) in algae (Geel et al. 1997; Flameling and Kromkamp 1998; Franklin and Badger 2001; Hancke et al. 2008a, b), ETR calculated from corals can significantly overestimate photosynthetic productivity over a broad range of light intensities (Hoogenboom et al. 2006), including those well below the level for saturation of photosynthesis (E_k). There are several mechanistic reasons for this non-linearity. First, it is important to note the many other pathways for PSII electron transport in addition to carbon fixation that may contribute to this disconnect. These include the Mehler reaction, Rubisco oxygenase (i.e., photorespiration), nitrate assimilation, mitochondrial respiration, and alternative oxidase activity and PSII cyclic electron transport. The importance of many of these pathways is still uncertain in *Symbiodinium* and a better understanding of how such pathways may play a dominant role in this system is still needed (Suggett et al. 2008). Secondly, the multiple turnover saturation pulses employed by the PAM fluorometer reduce the plastoquinone (PQ) pool, which in turn increases the chance of over estimating the effective quantum yield which will directly impact down-stream ETR calculations (Samson et al. 1999; Suggett et al. 2003). This point is especially important at low light levels, given that PQ pool fluorescence quenching is proportional to the actinic light intensity used, thus, estimates of Φ_{PSII} between multiple and single turnover instrumentation tend to converge at higher light levels (Suggett et al. 2003).

Pre-configured protocols designed for measuring ETR by PAM fluorometry (e.g. use of "light curve" settings) were originally designed to allow rapid assessment of the light adapted state of benthic material whilst constrained by the use of SCUBA. This required a rapid assessment of the sample without allowing sufficient time for steady state conditions to be established, as the light steps are usually less than 30s. This fundamental difference with traditional oxygen-based photosynthesis to irradiance (P-E) curves must be recognized and RLC curves should not be interpreted as P-E curves. Although the curves have a similar shape and can be paramaterised using similar curve fitting routines, the RLC may be used to infer a sample's capacity to acclimate to a series of short light steps (Ralph and Gademann 2005). Furthermore, RLC's do not always provide sufficient measurement steps at low irradiance (leading to erroneous slope or α calculations) and are typically quite fast (10–20 s per light step), thereby allowing for greater uncertainty of the light history of the sample and therefore the true biological meaning of the data. While ETR to E plots look strikingly similar to P-E curves, interpreting the results of non-linear curve fitting of the former using similar methods as the latter should be approached with caution and often does not provide synonymous data (e.g. in many cases ETR_{max} is not equivalent to traditional P_{max}). With these cautionary notes notwithstanding, if one attempts to use PSII ETR derived values to infer total photosynthetic productivity, a linear relationship between photosynthesis (measured by another method such as respirometry or ^{14}C uptake) and ETR under the planned experimental conditions must be confirmed prior to drawing such a conclusion.

Despite the limitations of using ETR to infer productivity mentioned above, there is still tremendous utility in understanding PSII ETR in itself. However, one should be aware that several of the assumptions built into the often used and simple calculations, which were originally derived for working with terrestrial plants or suspensions of microalgae, are not equally applicable to reef corals. In particular, the general equation for "relative" ETR:

$$rETR = \frac{\Delta F}{F_m'} \times E$$

makes no assumption of the percentage of light energy transfer between PSII and PSI (this is typically assumed to be equivalent, yet see Hancke et al. 2008b for further discussion) and, more importantly, does not account for the absorptance (i.e. the fraction of incident light absorbed) of the coral or *Symbiodinium*. Recent efforts by Enriquez et al. (2005) have shown the tremendous importance of understanding light scattering and absorptance in reef corals. Coral skeletons are incredibly efficient at scattering sunlight, and typical PAR levels measured within the tissues of a coral are substantially higher than those measured at the outer surface of the animal and not easy to obtain without sophisticated instrumentation (Kühl et al. 1995). Despite the fact that this can lead to an underestimation of the incident PAR *in hospite*, one solution toward a better approximation of ETR is to record the spectral reflectance of the coral surface where fluorescence measurements are recorded and then convert the absorbance band at 675 nm (the wavelength for Chl *a* absorption that is free from interference from accessory algal and animal pigments, see Rodriguez-Román et al. 2006 to absorptance (Enriquez et al. 2005). Thus, ETR may be calculated by the modified equation:

$$\mathrm{ETR} = \frac{\Delta \mathrm{F}}{\mathrm{F}_{\mathrm{m}}'} \times \mathrm{E} \times a_{675}$$

where *a* is the calculated absorptance. While absorptance is typically quite high (~90 %) in normally pigmented corals, a reduction in symbiont density can lead to a significant enhancement in reflected light and thus a substantial reduction in absorptance. Not accounting for this change can lead to significant errors in ETR calculation, such as when comparing bleached vs. normally pigmented coral colonies (pers. obs.). Naturally, one should also consider other sources of fluorescence such as the endolithic green algae, which under specific conditions, can dominate the spectral signal in a bleached coral (Rodriguez-Román et al. 2006). Understanding how light will influence photosynthesis in reef corals at this level is largely unexplored and much more work is warranted in understanding how light absorption by the animal and algal pigments may influence the photosynthetically usable radiation (PUR) that is truly available for photosynthesis.

5 Conclusion

With the increased use of fluorescence in coral reef monitoring efforts, it is our hope that further development will continue in the advancement of new instrumentation that can provide more temporal detail of the fluorescence signal from reef building corals. In particular, the need for submersible, diver operated, single turnover instrumentation that allows for as much end user control and customization in flash protocol design and data acquisition as possible should be considered a high priority for manufacturers.

The development of, commercially available fluorescence instrumentation has clearly led to a significant rise in the use of this technique in coral reef biology. While some of the protocols and methods using chlorophyll fluorescence that have been established from many years of research in plant physiology are applicable to studying corals, one should take care to recognize that these symbioses are composed of a diverse group of dinoflagellates, and as such, have very different physiologies relative to terrestrial plants. Similarly, while chlorophyll fluorescence is commonly considered a valuable proxy for interpreting the relative "health" of a reef building coral, one must not loose sight of the possible caveats that may arise if instrument and protocol limitations, as well as natural variability in fluorescence across numerous temporal and spatial scales are not fully considered.

Acknowledgements Support to M. Warner was provided by the National Science Foundation (IOB 544765) and from the Australian Research Council for P. Ralph. Elements of this chapter were a direct result of several fruitful discussions among many colleagues who attended the AquaFluo, Chlorophyll Fluorescence in Aquatic Sciences meeting in 2007, as well as the UNESCO-IOC-World Bank Coral Reef Targeted Research Group workshop, "Understanding the stress response of corals and *Symbiodinium* in a rapidly changing environment," in 2005 in Puerto Morelos, Mexico.

References

Allen JF, Bennett J, Steinback KE, Arntzen CJ (1981) Chloroplast protein phosphorylation couples plastoquinone redox state to distribution of excitation energy between photosystems. Nature 291:25–29

Allen JF, Mullineaux CW (2004) Probing the mechanism of state transitions in oxygenic photosynthesis by chlorophyll fluorescence spectroscopy, kinetics and imaging. In: Papageorgiou GC, Govindjee (eds) Chlorophyll *a* fluorescence: a signature of photosynthesis. Springer, Dordrecht 2:447–461

Baird A, Bhagooli R, Ralph P, Takahashi S (2009) Coral bleaching: the role of the host. Trends Ecol Evol 24:16–20

Baker NR (2008) Chlorophyll fluorescence: a probe of photosynthesis in vivo. Annu Rev Plant Biol 59:89–113

Berkelmans R, van Oppen MJH (2006) The role of zooxanthellae in the thermal tolerance of corals: a 'nugget of hope' for coral reefs in an era of climate change. Proc Roy Soc London B 273:2305–2312

Bhagooli R, Hidaka M (2004) Photoinhibition, bleaching susceptibility and mortality in two scleractinian corals, *Platygyra ryukyuensis* and *Stylophora pistillata*, in response to thermal and light stresses. Comp Biochem Physiol A 137:547–555

Behrenfeld M, Prasil O, Kolber Z, Babin M, Falkowski PG (1998) Compensatory changes in photosystem II electron turnover rates protect photosynthesis from photoinhibition. Photosynth Res 58:259–268

Bellafiore S, Barneche F, Peltier G, Rochaix JD (2005) State transitions and light adaptation require chloroplast thylakoid protein kinase STN7. Nature 433:892–895

Brown B, Ambarsari I, Warner ME, Fitt WK, Dunne RP, Gibb SW, Cummings DG (1999) Diurnal changes in photochemical efficiency and xanthophyll concentrations in shallow water reef corals: evidence for photoinhibition and photoprotection. Coral Reefs 18:99–105

Brown BE, Dunne RP (2008) Solar radiation modulates bleaching and damage protection in a shallow water coral. Mar Ecol Prog Ser 362:99–107

Cardol P, Gloire G, Havaux M, Remacle C, Matagne R, Franck F (2003) Photosynthesis and state transitions in mitochondrial mutants of *Chlamydomonas reinhardtii* affected in respiration. Plant Physiol 133:2010–2020

Dykens JA, Schick JM (1982) Oxygen production by endosymbiotic algae controls superoxide dismutasae activity in their animal host. Nature 297:579–580

Enriquez S, Méndez ER, Iglesias-Prieto R (2005) Multiple scattering on coral skeletons enhances light absorption by symbiotic algae. Limnol Oceanogr 50:1025–1032

Falkowski PG, Wyman K, Ley AC, Mauzerall D (1986) Relationship of steady state photosynthesis to fluorescence in eukaryotic algae. Biochem Biophys Acta 849:183–192

Fitt WK, McFarland FK, Warner ME, Chilcoat GC (2000) Seasonal patterns of tissue biomass and densities of symbiotic dinoflagellates in reef corals and relation to coral bleaching. Limnol Oceanogr 45:677–685

Fitt WK, Brown BE, Warner ME, Dune RP (2001) Coral bleaching: interpretation of thermal tolerance limits and thermal thresholds in tropical corals. Coral Reefs 20:51–65

Flameling IA, Kromkamp J (1998) Light dependence of quantum yields for PSII charge separation and oxygen evolution in eukaryotic algae. Limnol Oceanogr 43:284–297

Franklin LA, Badger MR (2001) A comparison of photosynthetic electron transport rates in macroalgae measured by pulse amplitude modulated chlorophyll fluorometry and mass spectrometry. J Phycol 37:756–767

Gardella DJ, Edmunds PJ (1999) The oxygen environment adjacent to the tissue of the scleractinian *Dichocoenia stokesii* and its effects on symbiont metabolism. Mar Biol 135:289–295

Gattuso JP (1999) Photosynthesis and calcification at cellular, organismal and community levels in coral reefs: a review on interactions and control by carbonate chemistry. Am Zool 39:160–183

Geel C, Versluis W, Snel JFH (1997) Estimation of oxygen evolution by marine phytoplankton from measurement of the efficiency of Photosystem II electron flow. Photosynth Res 51:61–70

Gorbunov MY, Kolber ZS, Lesser MP, Falkowski PG (2001) Photosynthesis and photoprotection in symbiotic corals. Limnol Oceanogr 46:75–85

Gorbunov MY, Falkowski PG (2004) Fluorescence induction and relaxation (FIRe) technique and instrumentation for monitoring photosynthetic processes and primary production in aquatic ecosystems. In: van der Est A, Bruce D (eds) Proceedings of the 13th International Congress of Photosynthesis, vol. 2. Allen Press, pp 1029–1031

Greene RM, Geider RJ, Falkowski PG (1991) Effect of iron limitation on photosynthesis in a marine diatom. Limnol Oceanogr 36:1772–1782

Hancke K, Hancke TB, Olsen LM, Johnsen G, Glud RN (2008a) Temperature effects on microalgal photosynthesis-light responses measured by O_2 production, pulse-amplitude-modulated fluorescence, and ^{14}C assimilation. J Phycol 44:501–514

Hancke TB, Hancke K, Johnsen G, Sakshaug E (2008b) Rate of O_2 production derived from pulse-amplitude-modulated fluorescence: testing three biooptical approaches against measured O_2-production rate. J Phycol 44:803–813

Hill R, Frankart C, Ralph PJ (2005) Impact of bleaching conditions on the components of non-photochemical quenching in the zooxanthellae of a coral. J Exp Mar Biol Ecol 322:83–92

Hill R, Larkum AWD, Frankart C, Kuhl M, Ralph PJ (2004a) Loss of functional photosystem II reaction centers in zooxanthellae of corals exposed to bleaching conditions: using fluorescence rise kinetics. Photosynth Res 82:59–72

Hill R, Schreiber U, Gademann R, Larkum AWD, Kühl M, Ralph PJ (2004b) Spatial heterogeneity of photosynthesis and the effect of temperature-induced bleaching conditions in three species of corals. Mar Biol 144:633–640

Hill R, Ralph PJ (2005) Diel Seasonal changes in fluorescence rise kinetics of three scleractinian corals. Funct Plant Biol 32:549–559

Hill R, Ralph PJ (2006) Photosystem II heterogeneity of *in hospite* zooxanthellae in scleractinian corals exposed to bleaching conditions. Photochem Photobiol 82:1577–1585

Hill R, Ralph PJ (2008a) Dark-induced reduction of the plastoquinone pool in zooxanthellae of scleractinian corals and implications for measurements of chlorophyll *a* fluorescence. Symbiosis 46:45–56

Hill R, Ralph PJ (2008b) Impact of bleaching stress on the function of the oxygen evolving complex of zooxanthellae from scleractinian corals. J Phycol 44:299–310

Hoegh-Guldberg O, Jones RJ (1999) Photoinhibition and photoprotection in symbiotic dinoflagellates from reef-building corals. Mar Ecol Prog Ser 183:73–86

Hoegh-Guldberg O (1999) Climate change, coral bleaching and the future of the world's coral reefs. Mar Fresh Res 50:839–866

Hoogenboom MO, Anthony KRN, Connolly SR (2006) Energetic cost of photoinhibition in corals. Mar Ecol Prog Ser 313:1–12

Iglesias-Prieto R, Matta JL, Robins WA, Trench RK (1992) Photosynthetic response to elevated temperature in the symbiotic dinoflagellate *Symbiodinium microadriaticum* in culture. Proc Natl Acad Sci 89:10302–10305

Iglesias-Prieto R (1997) Temperature-dependent inactivation of photosystem II in symbiotic dinoflagellates. In: Lessios HA, Macintyre IG (eds) Proc 8th Int Coral Reef Symp 2:1313–1318

Iglesias-Prieto R, Beltran VH, LaJeunesse TC, Reyes-Bonilla H, Thome PE (2004) Different algal symbionts explain the vertical distribution of dominant reef corals in the eastern pacific. Proc Roy Soc Lond Ser B 271:1757–1763

Jones RJ, Hoegh-Guldberg O, Larkum AWD, Schreiber U (1998) Temperature-induced bleaching of corals begins with impairment of the CO_2 fixation mechanism in zooxanthellae. Plant Cell Environ 21:1219–1230

Jones RJ, Kildea T, Hoegh-Guldberg O (1999) PAM chlorophyll fluorometry: a new *in situ* technique for stress assessment in scleractinian corals, used to examine the effects of cyanide fishing. Mar Poll Bull 38:864–874

Jones RJ, Hoegh-Guldberg O (2001) Diurnal changes in photochemical efficiency of the symbiotic dinoflagellates (Dinophyceae) of corals: photoprotection, photoinactivation, and the relationship to coral bleaching. Plant Cell Environ 24:89–99

Jones RJ, Kerswell (2003) Phytotoxicity of photosystem II (PSII) herbicides to coral. Mar Ecol Prog Ser 261:149–159

Kerswell AP, Jones RJ (2003) Effects of hypo-osmosis on the coral *Stylophora pistillata*: Nature and cause of 'low salinity bleaching. Mar Ecol Prog Ser 253:145–154

Kolber Z, Prasil O, Falkowski PG (1998) Measurements of variable chlorophyll fluorescence using fast repetition rate techniques: defining methodology and experimental protocols. Biochem Biophys Acta 1367:88–106

Kühl M, Yehuda C, Dalsgaard T, Jørgensen BB, Revsbech NP (1995) Microenvironment and photosynthesis of zooxanthellae in scleractinian corals studied with microsensors for O_2, pH and light. Mar Ecol Prog Ser 117:159–172

Lavergne J, Trissl HW (1995) Theory of fluorescence induction in photosystem II: Derivation of analytical expressions in a model including excitation-radical-pair equilibrium and restricted energy transfer between photosynthetic units. Biophys J 68:2474–2492

Leggat W, Whitney S, Yellowlees D (2004) Is coral bleaching due to the instability of the zooxanthellae dark reactions? Symbiosis 37:137–153

Levy O, Dubinsky Z, Schneider K, Achituv Y, Zakai D, Gorbunov M (2004) Diurnal hysteresis in coral photosynthesis. Mar Ecol Prog Ser 268:105–117

Levy O, Achituv Y, Yacobi YZ, Dubinsky Z, Stambler N (2006) Diel 'tuning' of coral metabolism: physiological responses to light cues. J Exp Biol 209:273–283

Lesser MP (1996) Elevated temperatures and ultraviolet radiation cause oxidative stress and inhibit photosynthesis in symbiotic dinoflagellates. Limnol Oceanogr 41:271–283

Lesser MP, Gorbunov MY (2001) Diurnal and bathymetric changes in chlorophyll fluorescence yields of reef corals measured in situ with a fast repetition rate fluorometer. Mar Ecol Prog Ser 212:69–77

Lesser MP, Farrell JH (2004) Exposure to solar radiation increases damage to both host tissue and algal symbionts of corals during thermal stress. Coral Reefs 16:187–192

Lesser MP (2004) Experimental biology of coral reef ecosystems. J Exp Mar Biol Ecol 300:217–252

Lesser MP (2006) Oxidative stress in marine environments: biochemistry and physiological ecology. Annu Rev Physiol 68:253–278

Levy O, Dubinsky Z, Achituv Y (2003) Photobehavior of stony corals: responses to light spectra and intensity. J Exp Biol 206:4041–4049

Maxwell DP, Falk S, Huner NPA (1995) Photosystem II excitation pressure and development of resistance to photoinhibition. I. Light-harvesting complex II abundance and zeaxanthin content in *Chlorella vulgaris*. Plant Physiol 107:687–694

McCabe-Reynolds JC, Bruns BU, Fitt WK, Schmidt GW (2008) Enhanced photoprotection pathways in symbiotic dinoflagellates of shallow-water corals and other cnidarians. Proc Natl Acad Sci USA 105:13674–13678

Müller P, Li XP, Niyogi KK (2001) Non-photochemical quenching. A response to excess light energy. Plant Physiol 125:1558–1566

Muscatine L (1990) The role of symbiotic algae in carbon and energy flux in reef corals. In: Dubinsky Z (ed) Ecosystems of the world: coral reefs. Elsevier, Amsterdam, pp 1–9

Olaizola M, LaRouche J, Kolber Z, Falkowski PG (1994) Non-photochemical fluorescence quenching and the diadinoxanthin cycle in a marine diatom. Photosynthesis Research 41:357–370

Peltier G, Cournac L (2002) Chlororespiration. Annu Rev Plant Biol 53:523–550

Philipp E, Fabricius K (2003) Photophysiological stress in scleractinian corals in response to short-term sedimentation. J Exp Mar Biol Ecol 299:57–78

Platt JR (1964) Strong inference. Science 146:347–353

Ralph PJ, Gademann R, Larkum AWD, Schreiber U (1999) *In situ* underwater measurements of photosynthetic activity of coral zooxanthellae and other reef-dwelling dinoflagellate endosymbionts. Mar Ecol Prog Ser 180:139–147

Ralph PJ, Gademann R, Larkum AWD, Kühl M (2002) Spatial heterogeneity in active chlorophyll fluorescence and PSII activity of coral tissues. Mar Biol 141:639–646

Ralph PJ, Gademann R (2005) Rapid light curves: a powerful tool for the assessment of photosynthetic activity. Aquat Bot 82:222–237

Rintamäki E, Salonen M, Suoranta U-M, Carlberg I, Andersson B, Aro E-M (1997) Phosphorylation of light-harvesting complex II and photosystem II core proteins shows different irradiance-dependent regulation in vivo. J Biol Chem 272:30476–30482

Robison JD, Warner ME (2006) Differential impacts of photoacclimation and thermal stress on the photobiology of four different phylotypes of *Symbiodinium* (Pyrrhophyta). J Phycol 42:568–579

Rodriguez-Román A, Hernánedez-Pech X, Thomé PE, Enriquez S, Iglesias-Prieto R (2006) Photosynthesis and light utilization in the Caribbean coral *Montastraea faveolata* recovering from a bleaching event. Limnol Oceanogr 51:2702–2710

Samson G, Bruce D (1996) Origins of the low yield of chlorophyll *a* fluorescence induced by single turnover flash in spinach thylakoids. Biochem Biophys Acta 1276:147–153

Samson G, Prasil O, Yaakoubd B (1999) Photochemical and thermal phases of chlorophyll *a* fluorescence. Photosynthetica 37:163–182

Samuelsson G, Öquist G (1977) A method for studying photosynthetic capacities of unicellular algae based on *in vivo* chlorophyll fluorescence. Plant Physiol 40:315–319

Smith DJ, Suggett DJ, Baker NR (2005) Is photoinhibition of zooxanthellae photosynthesis the primary cause of thermal bleaching in corals? Global Change Biol 11:1–11

Stanely GD (2003) The evolution of modern corals and their early history. Earth Sci Rev 60:195–225

Strasser RJ, Srivastava A, Govindjee (1995) Polyphasic chlorophyll *a* fluorescence transient in plants and cyanobacteria. Photochem Photobiol 61:32–42

Strasser RJ, Tsimilli-Michael M, Srivastava A (2004) Analysis of the chlorophyll *a* fluorescence transient. In: Papageorgiou GC, Govindjee (eds) Chlorophyll *a* fluorescence: a signature of photosynthesis. Springer, Dordrecht, pp 321–362

Suggett DG, Oxborough K, Baker NR, Macintyre H, Kana TM, Geider RJ (2003) Fast repetition rate and pulse amplitude modulation chlorophyll a fluorescence measurements for assessment of photosynthetic electron transport in marine phytoplankton. Eur J Phycol 38:371–384

Suggett DG, Warner ME, Smith DJ, Davey P, Hennige S, Baker NR (2008) Photosynthesis and production of hydrogen peroxide by *Symbiodinium* (Pyrrhophyta) phylotypes with different thermal tolerances. J Phycol 44:948–956

Takahashi S, Whitney S, Itoh S, Maruyama T, Badger M (2008) Heat stress causes inhibition of the *de novo* synthesis of antenna proteins and photobleaching in cultured *Symbiodinium*. Proc Natl Acad Sci 105:4203–4208

Tchernov D, Gorbunov MY, de Vargas C, Yadav SN, Milligan AJ, Haggblom M, Falkowski PG (2004) Membrane lipids of symbiotic algae are diagnostic of sensitivity to thermal bleaching in corals. Proc Nat Acad Sci USA 101:13531–13535

Ulstrup KE, Hill R, Ralph PJ (2005) Photosynthetic impact of hypoxia on *in hospite* zooxanthellae in the scleractinian coral *Pocillopora damicornis*. Mar Ecol Prog Ser 286:125–132

Ulstrup KE, Berkelmans R, Ralph PJ, van Oppen MJH (2006) Variation in bleaching sensitivity of two coral species across a latitudinal gradient on the Great Barrier Reef: the role of zooxanthellae. Mar Ecol Prog Ser 314:135–148

Vasil'ev S, Bruce D (1998) Non-photochemical quenching of excitation energy in photosystem II. A picosecond time-resolved study of the low yield of chlorophyll a fluorescence induced by single-turnover flash in isolated spinach thylakoids. Biochem 37:11046–11054

Warner ME, Fitt WK, Schmidt GW (1996) The effects of elevated temperature on the photosynthetic efficiency of zooxanthellae *in hospite* from four different species of reef coral: a novel approach. Plant Cell Environ 19:291–299

Warner ME, Fitt WK, Schmidt GW (1999) Damage to photosystem II in symbiotic dinoflagellates: a determinant of coral bleaching. Proc Natl Acad Sci USA 96:283–292

Warner ME, Chilcoat GC, McFarland FK, Fitt WK (2002) Seasonal fluctuations in the photosynthetic capacity of photosystem II in symbiotic dinoflagellates in the Caribbean reef-building coral *Montastraea*. Mar Biol 141: 31–38

Warner ME, Berry-Lowe S (2006) Differential xanthophyll cycling and photochemical activity in symbiotic dinoflagellates in multiple locations of three species of Caribbean coral. J Exp Mar Biol Ecol 339:86–95

Warner ME, LaJeunesse TC, Robison JD, Thur RM (2006) The ecological distribution and comparative photobiology of symbiotic dinoflagellates from reef corals in Belize: potential implications for coral bleaching. Limnol Oceanogr 51:1887–1897

Weis E, Berry JA (1987) Quantum efficiency of photosystem II in relation to energy-dependent quenching of chlorophyll fluorescence. Biochem Biophys Acta 894:198–208

Winters G, Loya Y, Rottgers R, Beer S (2003) Photoinhibition in shallow-water colonies in the coral *Stylophora pistillata in situ*. Limnol Oceanogr 48:1388–1393

Wollman FA (2001) State transitions reveal the dynamics and flexibility of the photosynthetic apparatus. EMBO J 20:3623–3630

Yakovleva I, Hidaka M (2004) Differential recovery of PSII function and electron transport rate in symbiotic dinoflagellates as a possible determinant of bleaching susceptibility of corals. Mar Ecol Prog Ser 43:43–53

Chapter 11
Assessing Nutrient Status of Microalgae Using Chlorophyll *a* Fluorescence

Kirsten Shelly, Daryl Holland, and John Beardall

1 Introduction

Phytoplankton require carbon, nitrogen and phosphorus in the approximate elemental ratio 106:16:1 (Redfield 1958), although this can vary among different taxa and growth conditions (Geider and La Roche 2002; Ho et al. 2003; Sterner et al. 2008). Other nutrients, including iron, calcium, manganese, copper and zinc are also required in smaller and varying amounts (Morel and Hudson 1985). Silica can also be limiting to the growth of diatoms (Carrick and Lowe 2007; Shipe et al. 2007). If the concentration of a particular nutrient drops in real terms or in relation to the other key nutrients, it may limit the growth rate, maximum biomass and/or species composition of a phytoplankton community.

Considering that oceanic phytoplankton account for up to 50% of the total global primary production (Behrenfeld et al. 2001), an understanding of the limitations to the growth of phytoplankton is of considerable importance, and increasingly so, as carbon cycling dominates scientific and political discussions. The availability of nutrients is one potential limitation, with light being the other major limiting factor. Excepting carbon (C), nitrogen (N) and phosphorus (P) are the principal nutrients likely to be in limiting supply (Birch and Gordon 1981; Lean and Pick 1981; Schindler 1977; Sundareshwar et al. 2003; Wynne and Berman 1980), and are traditionally thought to be limiting in marine and freshwater ecosystems respectively, because they are found in ratios below what is typically found in nutrient replete phytoplankton cells (Hecky and Kilham 1988). More recent studies have found, however, that areas such as the North Pacific Sub-tropical Gyre and regions of the Mediterranean Sea are in fact P-limited (Karl 1999; Krom et al. 1991), while freshwater systems such as Lake Victoria may be N-limited (Guildford and Hecky 2000). In areas such as the equatorial Pacific and Southern Ocean, which are deemed 'high nutrient, low chlorophyll' (HNLC) regions, iron is a significant limiting resource. (Behrenfeld and Bale 1996; Boyd et al. 1999; Martin and Fitzwater 1988; Timmermans et al. 1998). Davey et al. (2008) have also noted that P-depletion is occurring in the North Atlantic, where aeolian dust inputs are relieving Fe-limitation and thus allowing nitrogen fixation.

Anthropogenic activities have become a major source of nutrients in many of the world's aquatic systems and frequently lead to toxic algal blooms and cultural eutrophication (Fisher et al. 1999; Nixon 1995; Wood and Oliver 1995; Smith 2003). The identification of nutrients limiting phytoplankton growth in water bodies is therefore of considerable importance to our understanding of the ecology of aquatic systems and to water management practices. It enables managers to draw up appropriate nutrient loading budgets for catchments and respond to possible perturbations on an informed basis (Beardall et al. 2001b).

Traditional methods for measuring nutrient limitation, such as nutrient enrichment bioassays and certain chlorophyll *a* fluorescence techniques have inherent problems that limit their applicability to natural populations, thus a rapid and reliable technique for measuring nutrient limitation in natural populations would greatly improve our understanding of these complex systems. While there are alternative approaches to determining nutrient limitation (Beardall et al. 2001b;

K. Shelly, D. Holland, and J. Beardall (✉)
Monash University, School of Biological Sciences, Clayton, Victoria 3800, Australia
e-mail: john.beardall@monash.edu

D.J. Suggett et al. (eds.), *Chlorophyll* a *Fluorescence in Aquatic Sciences: Methods and Applications*, Developments in Applied Phycology 4, DOI 10.1007/978-90-481-9268-7_11, © Springer Science+Business Media B.V. 2011

Van Mooy and Devol 2008), this present chapter will concentrate on the use of chlorophyll fluorescence as a tool for the rapid indication of nutrient status.

2 Defining Nutrient Limitation

The term 'nutrient limitation' can have several meanings, and it is therefore important to define precisely which kind of limitation is being measured (Beardall et al. 2001b). For example, many 'bioassay' or 'nutrient enrichment' studies fail to differentiate between two kinds of limitation; Liebig limitation, i.e., limitation of the maximum biomass achievable (Beardall et al. 2001b; Debaar 1994); and the limitation of instantaneous growth rate, or Blackman limitation (Holland et al. 2004; Singh and Lal 1935). Measurements of Liebig limitation do not necessarily reflect limitation by a particular nutrient at the time of sampling and at the prevailing cell density and species composition of the population. Rather they specify simply that should growth continue, one particular nutrient will eventually become the limiting factor. This distinction has major implications for the way that we approach studies of nutrient limitation in natural populations (see Cullen et al. 1992). For example, in oligotrophic waters, a low biomass of rapidly dividing phytoplankton can be maintained by heavy grazing pressure combined with efficient recycling of nutrients in the water column. It follows that the growth rate of these individual cells is not limited by nutrient availability; the maximum population size, however, is determined by the overall nutrient concentration and other biotic and abiotic factors (Beardall et al. 2001b; Moore et al. 2008).

3 The Effects of Nutrient Limitation on Phytoplankton

3.1 Nitrogen

Denitrification in sediments is the main process involved in removing available nitrogen from aquatic ecosystems (Knowles 1982; Seitzinger 1988). Previously it was believed that nutrient fluxes from marine sediments were depleted in nitrogen, comparatively to phosphorus, because denitrification is higher in marine sediments than in freshwater sediments. Although this may be true, as previously stated there are known marine regions described as phosphorus limited. Most nitrogen is present as N_2 and therefore biologically unavailable (Howarth 1988). However, nitrogen fixation, performed chiefly by cyanobacteria, converts N_2 into nitrogen compounds such as ammonia, nitrate and nitrogen oxide. Nitrogen fixation is therefore a process that can alleviate nitrogen limitation, but energetic constraints and limitation of nitrogen fixers by another nutrient, such as phosphorus, molybdenum or iron, are just two ways in which this process can come undone and lead to nitrogen limitation (Vitousek and Howarth 1991).

Nitrogen limitation can affect a number of processes in phytoplankton, including photosynthesis, photochemical energy conversion, and protein synthesis (Berges et al. 1996; Jansen et al. 1996; Masi and Melis 1997). Photosynthesis can be affected through a reduction in the energy-collecting efficiency because of a reduction in chlorophyll *a* (which contains nitrogen). N-limitation likewise reduces protein synthesis, which ultimately affects the protein content of the PSI and PSII reaction centres, leading to a reduction in photochemical energy conversion. PSII is thought to be affected more than PSI because of the rapid turnover of the D1 and D2 proteins, whereas PSI proteins are relatively stable, and proteins with a rapid turnover rate are likely to be affected by decreases in protein synthesis attributed to the lack of nitrogen (Berges et al. 1996). Although PSII is generally affected more by nitrogen limitation than PSI, this may vary between species (Behrenfeld et al. 1994; Berges et al. 1996). Given that nitrogen limitation leads to impairment of protein synthesis, it is conceivable that the D1 repair cycle may be inhibited and the repair capacity of the organism compromised (see Fig. 1 and Behrenfeld et al. 1994).

3.2 Phosphorus

Phosphorus is essential for phytoplankton growth and division as it is a principal constituent of various biomolecules (such as nucleic acids and lipids), as well as being directly involved in energy transfer and cellular metabolic transfer (Frost and Xenopoulos 2002;

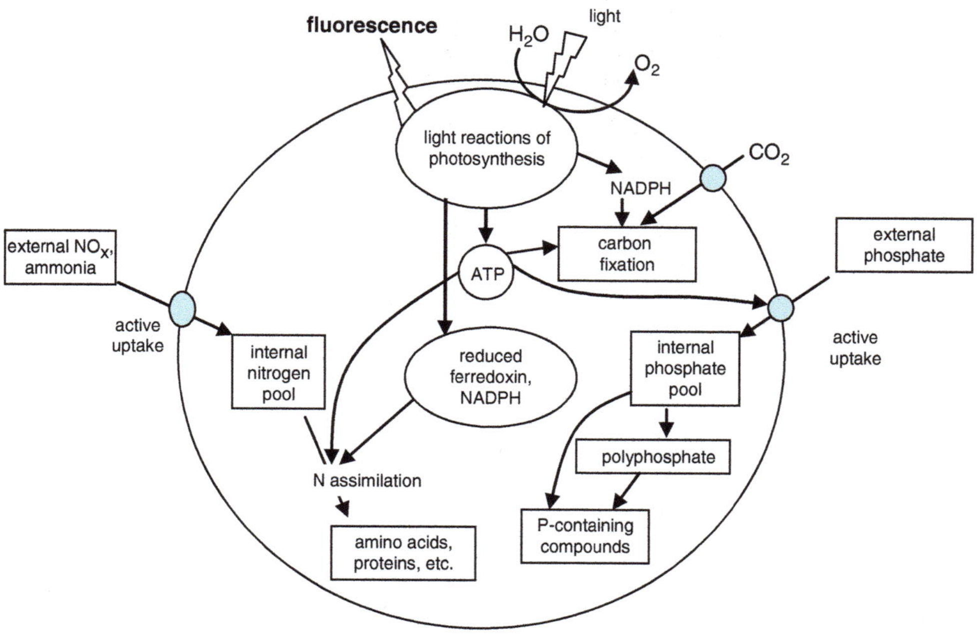

Fig. 1 A schematic view of the interactions between uptake and assimilation of nutrients and photosynthetic energy flow. Processes such as uptake and assimilation will act as energy dissipation processes and ultimately decrease the flow of energy into chlorophyll fluorescence. Active transport processes as depicted require a source of energy (usually ATP)

Iglesias et al. 1993). Phosphorus exchange between the sediment and water-column has been proposed to be controlled, at least in part, by sulphate concentrations (Caraco et al. 1990, 1989). Sulphide (the reduced form of sulphate, which is present in anoxic sediments) regulates dissolved phosphorus levels by preferentially binding to iron in the sediment – when levels of sulphate are low (e.g. in freshwater), phosphorus is readily adsorbed to the sediment, but at high sulphate levels (e.g. seawater) phosphorus tends to be released (Blomqvist et al. 2004). A growing number of marine environments are now thought to be phosphorus limited (Bertilsson et al. 2003), particularly in the tropics, where phosphorus adsorption by calcareous sediments is a contributing factor (Morse et al. 1985). Phosphorus limitation in freshwaters on the other hand is primarily due to a lack of supply from catchment sources, and because mineral forms of phosphorus (from geological sources, Brooks and Edington 1994) do not readily release biologically available phosphates in freshwaters.

P-limitation has been shown to induce a range of metabolic changes, including reduced photosynthetic capacity, increased phosphate uptake capacity, a shift in protein, lipid and carbohydrate production, and an increase in cell biomass (Ganf et al. 1986; Gotham and Rhee 1981; Graziano et al. 1996; Titman and Kilham 1976). P-limitation in phytoplankton can also strongly affect pelagic grazer communities and their ecological activities, presumably because of decreased nutritional quality of their food source (Frost and Xenopoulos 2002).

3.3 Iron

Iron limitation of phytoplankton growth can occur in marine, freshwater, and terrestrial systems (Weger et al. 2002). In high-nutrient low-chlorophyll (HNCL) marine zones, iron limitation results from low total iron, compounded by complexation of these low levels by organic matter (Weger et al. 2002). Concentrations of dissolved iron in oceanic waters have been measured in the range of 1nM down to 20 pM in HNCL areas (Lewandowska and Kosakowska 2004). Iron is far less soluble than other nutrients such as nitrogen and phosphorus and its concentration is constrained by the insolubility of Fe hydroxide which, once formed, causes further Fe inputs to increase the levels of Fe hydroxide rather than other, more biologically accessible forms. Fe hydroxides are not only less soluble than Fe ions, but they also tend to aggregate and sink (Sunda and Huntsman 1997). Soluble Fe (II) is the form common in freshwater, but this is generally oxidised to the less soluble Fe (III) once it enters the sea (Geider and LaRoche 1994; Gerringa et al. 2000). Exacerbating the situation, the vast majority (99.9%) of dissolved iron is thought to be organically bound and the concentration of chelators (that readily bind iron) far exceeds the amount of available iron. Iron limitation in marine systems therefore comes as no surprise (Raven et al. 1999).

Iron is required by cells *inter alia* for the functioning of PSI (Raven et al. 1999), and iron limitation is therefore likely to cause a decline in photosynthetic function. Iron limitation may also cause the down-regulation of nitrogenase activity, indirectly leading to nitrogen limitation through a reduction in fixation (Kupper et al. 2008). Iron was readily available 2.5 billion years ago when the photosynthetic apparatus evolved, but since iron has became a limiting resource in some areas, some phytoplankton taxa have evolved to require less iron for PSI functionality. Organisms such as the cyanobacterium *Prochlorococcus* and the open-ocean diatom *Thalassiosira* are examples of low-iron-requiring species (see Raven et al. 2005).

4 Measuring Nutrient Limitation

4.1 Nutrient Enrichment Bioassays

An increase in phytoplankton cell density is used as a measure of the capacity of a water sample to support growth in bioassay experiments. Water from a given site is filtered and inoculated with the test species; in freshwater, the test species used is often *Selenastrum capricornutum* (Miller et al., 1974, 1978; a.k.a. *Pseudokirchneriella subcapitata*), and marine species include, but are not restricted to, *Thalassiosira psuedonana* (Hayes et al. 1984). Growth of the test organism is measured in the presence of specific added nutrients and if addition of a particular nutrient leads to enhanced growth, it is assumed that the nutrient was limiting in the original sample. Alternatively, unfiltered site water is supplemented with nutrients and the growth of the native population is measured. This method can also be used to determine physiological responses to nutrient enrichment (see Guildford et al. 2003; McAndrew et al. 2007).

Bioassays are easy to set up and measurements of biomass over time are relatively simple to make, rendering them attractive in laboratories with few resources, and in the field, where *in situ* light and temperature can be used. Nevertheless, there are some inherent problems with this approach. Ideally, the growth rate is the measured parameter, however bioassays frequently rely on only a final biomass or yield, attained after a period of many days (see Hecky and Kilham 1988). This only provides information about the maximum biomass that the water body could sustain, and which nutrient will first become limiting if the population was to increase. These types of assays also tend to reflect the nutrient requirements for growth of one specific test organism, which may bear little resemblance to the responses of the species found in the water body under investigation. Filtration of the water sample prior to inoculation with the test organism may also remove colloids and organic complexes that are frequently a source of nutrients, especially phosphorus (Wood and Oliver 1995).

As bioassays are performed over several days in artificial incubations, this isolates the phytoplankton from sources of nutrients that would occur *in situ* (e.g. benthic fluxes and terrestrial run-off), and may lead to changes in species composition (Elser and Kimmel 1986; Wood and Oliver 1995). Most bioassay techniques remove other potential sources of limitation such as light and carbon, which may artificially induce limitation that was not present *in situ*. Hence the results of bioassays, while accepted by most researchers as valid, are not wholly applicable to natural systems. If done correctly, however, with adequate control of all growth factors and careful monitoring of population dynamics, bioassays provide a useful measure of Leibig and/or Blackman limitation, and will therefore continue to find applications (Holland et al. 2004; Moore et al. 2008).

During bioassay incubations, phytoplankton biomass can be measured directly by cell counts, or indirectly through chlorophyll extractions or chlorophyll *a* fluorescence measures. Chlorophyll *a* fluorescence is generally the quickest, easiest and least destructive way to measure growth, but one must always account for different outputs of different organisms, and be aware of the fact that the nutrient-limitation status of a population may itself cause changes to the fluorescence output, independent of biomass (see below).

4.2 Chlorophyll *a* Fluorescence as a Measure of Nutrient Stress

Nutrients such as carbon, nitrogen and phosphorus have clear direct or indirect interactions with photosynthesis. Some of these are summarized in Fig. 1. Evidently inorganic carbon (as CO_2 or bicarbonate) can be assimilated and C fixation acts as a way of dissipating photosynthetic electron flow. Similarly, nitrogen assimilation requires ATP and reductant that originates from photosynthetic electron transport (Fig. 1). In addition, active transport of these nutrients into cells can involve expenditure of energy as ATP (originating either from respiration or from photosynthesis). Thus uptake and assimilation of nutrients such as N, P and C can impact on the overall energetics of cellular processes and on the demand of cells for reductant and ATP. Such a demand can act to dissipate electron flow in photosynthesis and can thus impact on chlorophyll *a* fluorescence output (Fig.1).

The effect of environmental stress, such as nutrient limitation, on algal photosynthesis has been measured extensively using variable chlorophyll *a* fluorescence. Nutrient deficiency will often result in a major impairment of photochemical efficiency. Declines in photochemical efficiency – measured as the variable chlorophyll *a* fluorescence divided by the maximum fluorescence, or F_v/F_m (see Maxwell and Johnson 2000; Chapter 1 by Cosgrove and Borowitzka, this volume) – have been observed over a range of algal taxa for N-, P- or Fe-starved cells (e.g. Berges et al. 1996; Greene et al. 1992; Kolber et al. 1988; LaRoche et al. 1993; McKay et al. 1997). Lippemeier et al. (1999) noted that changes in variable chlorophyll *a* fluorescence could be seen in response to Si-limitation and re-supply in diatoms. F_v/F_m has been known to recover following nutrient re-addition following starvation and can be used as a measure of phytoplankton 'health' for bioassays on natural populations (Boyd et al. 1999; Geider and LaRoche 1994; LaRoche et al. 1993). Behrenfeld and Bale (1996) reported increases in F_v/F_m from 0.25 to 0.55 over a 24 h period following *in situ* Fe enrichment, and this increase persisted for 8 days in a HNLC region of the equatorial Pacific; this has been reported extensively and will be discussed in more detail later in this chapter. The measurement of F_v/F_m may therefore be a useful index of nutrient status, although, Moore et al. (2008), working in the North Atlantic, found no increase in F_v/F_m after 48 h bioassays, even though carbon uptake and chlorophyll *a* clearly indicated nitrogen limitation. As with all bioassays, care must be taken in interpretation of results carried out on populations isolated in containers.

The applicability of F_v/F_m as a proxy for nutrient limitation in the open ocean has recently come into question. Cullen and Davis (2003) found that the way in which commercial instruments are calibrated can lead to substantial changes in F_v/F_m and potentially lead to erroneous results, especially in the open ocean where fluorescence signals may be very low. In many cases it would be possible to assess this uncertainty and determine the likely impact of any error in the calibration, and this has been incorporated into some recent studies (Moore et al. 2008).

4.3 Natural Population Enrichments and Chlorophyll *a* Fluorescence

In situ experiments and molecular probes have provided evidence of iron limitation in high nutrient low chlorophyll (HNCL) regions of the ocean. These methods, however, are not only labour intensive but expensive and have spatial and temporal limitations (Behrenfeld and Kolber 1999). Behrenfeld and Kolber (1999) discovered diel fluorescence patterns in South Pacific Ocean phytoplankton samples, and, upon supplying the area with iron, these patterns were rapidly lost. They carried out laboratory tests on a marine *Synechococcus* sp. which indicated that the patterns seen in the ocean experiments were associated with iron limitation and state transitions. Greene et al. (1994) also suggested that iron availability was the principle mechanism that controlled phytoplankton photosynthesis in the equatorial Pacific. Behrenfeld and Kolber (1999) suggest that the mechanism behind this chlorophyll *a* fluorescence change is a re-arrangement of the light harvesting complexes; a state change from State I to State II or vice versa. The physiological explanation for this state change is not known, but regardless, it appears that chlorophyll *a* fluorescence measurements can provide a simple and reliable physiological indicator of iron limitation.

5 NIFTS

5.1 What is a NIFT?

Measurable changes in variable chlorophyll *a* fluorescence in response to nutrient limitation or re-supply require several days or hours respectively. In contrast, resupply of pulses of the limiting nutrient has been shown to cause transient fluctuations in chlorophyll *a* fluorescence on a much shorter timescale (Beardall et al. 2001a; 1996). Recent studies – spanning marine and freshwater organisms, and both field assemblages and laboratory cultures of algae – have shown that transient changes in steady state chlorophyll *a* fluorescence and $\Phi PSII_{e\text{-}max}$ following nutrient addition could potentially serve as a diagnostic tool for assessing nutrient limitation in phytoplankton. Turpin and Weger (1988) were the first to report that if nitrogen is added to phytoplankton cultures limited in that particular nutrient, a rapid but transient change (within minutes) in chlorophyll *a* fluorescence is seen. This change in chlorophyll *a* fluorescence has been termed a nutrient-induced fluorescence transient, or NIFT (Wood and Oliver 1995). Inorganic carbon (Miller et al. 1991), nitrogen (Holmes et al. 1989; Wood and Oliver 1995; Young and Beardall 2003) and phosphorus (Beardall et al. 2001b; Gauthier and Turpin 1997; Roberts 1998; Wood and Oliver 1995) additions have all produced NIFTs in cultures deficient in the supplied nutrient. There is also some evidence that this phenomenon may occur with silicate (Lippemeier et al. 2001). The chlorophyll *a* fluorescence response is generally not observed if a non-limiting nutrient or distilled water is added to the cells (Beardall et al. 2001a, b; Holland et al. 2004). Although the precise physiology behind NIFTs is still poorly understood, it is believed that the transient change in chlorophyll *a* fluorescence is, in part at least, attributable to a reallocation of energy from photosynthesis to nutrient uptake (discussed below).

NIFTs have been demonstrated in some natural populations, for example the Murrumbidgee River, New South Wales (Wood and Oliver 1995), Lake Lucerne in Switzerland (Beardall et al. 2001b) and the Derwent River, Tasmania (Holland et al. 2004). In these studies, most of the samples showed no NIFT response, indicating that nutrient limitation of the kind identified by NIFTs may be relatively uncommon in natural systems. For example, Wood and Oliver (1995) identified NIFTs on 6 out of 19 occasions, and Beardall et al. (2001b) saw a NIFT response to phosphate in some Lake Lucerne samples, but not in any Lake Zürich samples, in line with the P-status of these two water bodies. The Derwent populations sampled by Holland et al. (2004) exhibited NIFTS in response to P-resupply, in keeping with the apparent P-status of those waters, but only after being isolated from the river for at least 24 h. NIFT analyses show promise as a rapid and sensitive means of detecting limitation of phytoplankton by specific nutrients.

5.2 How to Measure NIFTs

A NIFT response can be determined quickly and easily in any adequately equipped fluorescence spectrophotometer. Good results have been obtained with a spectrofluorometer equipped with a 100 µmol photons m^{-2} s^{-1} actinic excitation beam at 436 nm, measuring emission at 686 nm (Roberts et al. 2008). Other studies have successfully used actinic light of 90 µmol photons m^{-2} s^{-1} (Holland et al. 2004) and 150 µmol photons m^{-2} s^{-1} (Young and Beardall 2003).

A typical NIFT determination occurs as follows:

1. A sub-sample (2–3 mL) of the phytoplankton suspension of interest is transferred to a cuvette and placed in the fluorescence spectrometer, and is constantly mixed with a magnetic stirrer to ensure even dispersion.
2. The actinic light is turned on, and fluorescence emission is logged at regular (30 s or less) intervals. The output is monitored until the signal stabilises. This usually takes a few minutes, but may take longer.
3. A concentrated nutrient sample (10–20 µL) is pipetted into the cuvette (the relevant concentration is discussed below).
4. A NIFT response (either a rise or fall in fluorescence output) will begin to appear almost immediately, if present. The output can then be logged until fluorescence either reaches a new steady state or returns to the previous steady value. This typically takes 5–30 min, but may vary with nutrient concentration, physiological state of the cells etc.

There are several ways to track a NIFT response. The most common parameter used is F_t, the steady state chlorophyll *a* fluorescence under actinic light.

NIFT responses can also be identified in the chlorophyll *a* fluorescence parameters F_m (the maximum fluorescence measured after a saturating flash) and $\Phi PSII_{e\text{-}max}$, the effective quantum yield of PSII; the latter two parameters require the use of a PAM fluorometer. These parameters have been used to determine mechanisms behind the NIFT response (Beardall et al. 2001a): when measuring for nutrient limitation, however, it is sufficient to only measure F_t. An excitation wavelength of 436 nm and an emission wavelength of 686 nm are typically used, although these may be varied, and it is conceivable that multi-channel fluorescence spectrometers (such as the PhytoPAM) may be used to distinguish different NIFT responses in mixed populations (D. Holland unpublished data).

5.3 The Characteristics of the NIFT Response are Dependent on the Limiting Nutrient

The precise shape of the response depends on the limiting nutrient. For phosphorus (and inorganic carbon; Miller et al. 1991), the NIFT responses follow the pattern shown in Fig. 2a, where the dominant trend is for a drop in chlorophyll *a* fluorescence followed by recovery. With NH_4^+ additions to N-limited cells, a small rise prior to the main drop in chlorophyll *a* fluorescence is often observed (Holland et al. 2004; Young and Beardall 2003). With NO_3^- additions to N-limited cells, the major response is a chlorophyll *a* fluorescence

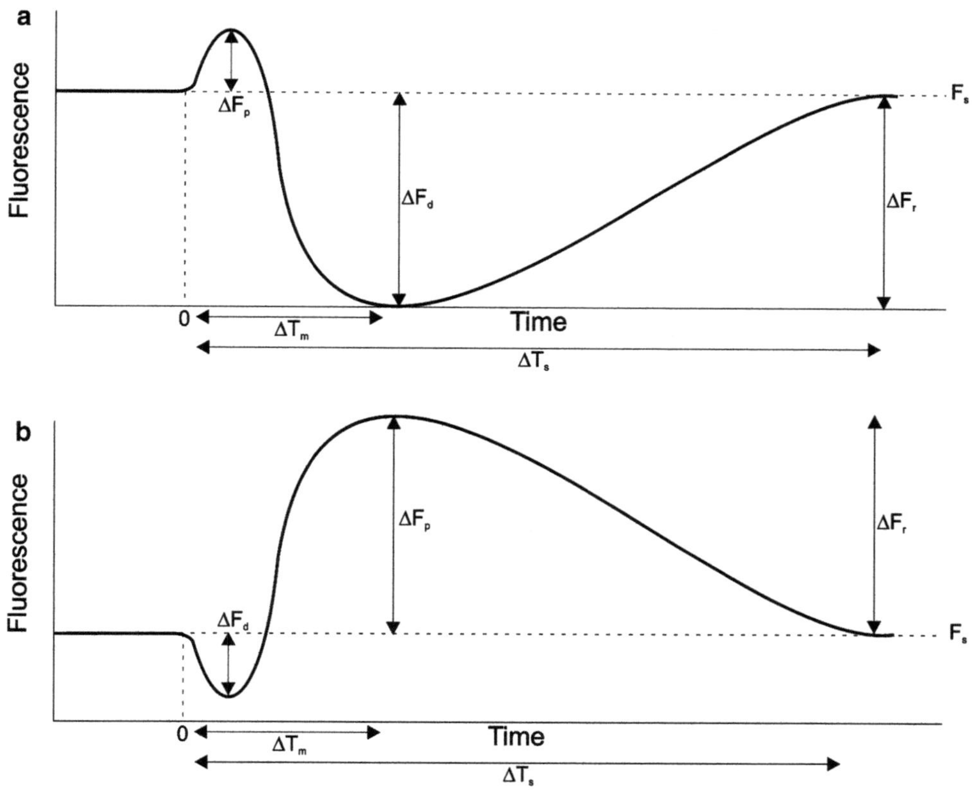

Fig. 2 A schematic view of chlorophyll fluorescence output during a NIFT following addition of a limiting nutrient at time T = 0. The horizontal line represents the steady state fluorescence prior to nutrient addition. Several parameters can be calculated from the NIFT curve. These include the time to minimum fluorescence; the difference between the minimum and steady state fluorescence (ΔF_d); the height of the rise in fluorescence to a peak (ΔF_p)the time to minimal fluorescence (ΔT_m) or to recovery of steady state (ΔT_s); and the integrated area above or below (depending on whether fluorescence drops or rises) the quenching curve. Typical curves are shown fro addition of **a**) phosphorus, inorganic carbon or NH_4^+ and **b**) NO_3^-. The initial dip or rise is not always present. See text for details

rise to a peak, followed by recovery (Fig. 2b). The shape of the drop and/or rise in fluorescence, as well as its magnitude, is also dependent on algal species and the degree of limitation (Holland et al. 2004; Roberts et al 2008; Young and Beardall 2003). In steady-state P- or N-limited cultures, the magnitude and rapidity of the fluorescence response (maximal change in fluorescence) increases slightly with growth rate (Young and Beardall 2003; Roberts et al. 2008). In P-starved *D. tertiolecta* cultures, the shape of the NIFT response also changes significantly as starvation increases and the ability to show a NIFT response can be much reduced in moribund cells (Roberts 1998). Similar data showing an initial increase to a peak, then a decrease, in the NIFT response with the progression of N-starvation are presented by Young and Beardall (2003). Changes in the NIFT response to nutrient starvation have also been reported in natural populations isolated in containers (Holland et al. 2004).

Several parameters can be calculated from the NIFT curve (Fig. 2). These include the time to minimum chlorophyll *a* fluorescence; the difference between the minimum and steady state chlorophyll *a* fluorescence (ΔF); the time to recovery of steady state; and the integrated area above or below (depending on whether chlorophyll *a* fluorescence drops or rises) the quenching curve. Roberts et al. (2008) studied the NIFT response of *D. tertiolecta* grown at different phosphorus concentrations, while Young and Beardall (2003) studied NIFTs in the same species under nitrogen limitation. The studies cited above found that the size of the three parameters depended on the growth rate, the cell concentration, the extent of limitation and the amount of the limiting nutrient. Interestingly, the half-saturation constant for the NIFT response as a function of limiting nutrient concentration was approximately equal to that for uptake. Absolute values of, and the relationship between, these parameters, have the potential to be diagnostic of cell phosphate and nitrogen status, in *D. tertiolecta* at least. Although there is good evidence for co-limitation of phytoplankton communities (North et al. 2007; Saito et al. 2008), there have been no studies of NIFTs in co-limited cultures. Whether such co-limitation could be detected by NIFT methods is yet to be determined.

When measuring NIFTs, the ideal concentration of the nutrient spike is one that gives the maximum ΔF while minimising the time to recovery. In a nutrient-deficient laboratory culture with a cell concentration in the order of 10^6 mL^{-1}, the following nutrient additions have been successfully used to induce NIFTs (note: no NIFT studies have been performed on Fe limitation):

- For P-limitation – 10 μM PO_4^{3-} (Holland et al. 2004; Roberts et al. 2008)
- For N-limitation – 100 μM NH_4^+, 10 μM NO_3^- (Holland et al. 2004; Young and Beardall 2003)

Roberts et al. (2008) showed that the time to recovery correlates with the time taken for the nutrient spike to be depleted. Higher cell concentrations therefore have faster recovery times for a given nutrient spike: it is therefore likely that field samples, having typically lower cell biomass values than cultured samples, will show a prolonged recovery phase; the initial kinetics of percentage decline in chlorophyll *a* fluorescence, however, would likely remain similar (Roberts et al. 2008).

5.4 NIFT Responses of Different Taxa

The shape of a NIFT response curve has been found to differ across species of phytoplankton and with limiting nutrient. Re-supply of ammonium to the green alga *Selenastrum minutum* when ammonium-limited caused an immediate decrease in chlorophyll *a* fluorescence (Holmes et al. 1989; Turpin and Weger 1988). In contrast, *Dunaliella tertiolecta* displays an increase in chlorophyll *a* fluorescence after ammonium re-supply, with only a small magnitude, short, drop initially (Young and Beardall 2003). Some of the differences in response could potentially be a result of different techniques (Kromkamp and Forster 2003), including varying instrumentation (Cullen and Davis 2003) and illumination conditions as well as laboratory vs natural field populations.

A variety of taxa have been tested for NIFT responses. These responses are summarised in Table 1. However, we do not have enough data to ascertain whether these responses are all governed by the same mechanisms.

5.5 Mechanisms Behind NIFTs

Transient changes in chlorophyll *a* fluorescence following nutrient supply are indicative of a reallocation of light energy from carbon fixation to nutrient uptake (Roberts et al. 2008). Apart from the early

Table 1 Species which have been tested for NIFTs

Species	Presence/Absence	Nutrient limitation	Reference
Green algae			
Chlorella emersonii	Present	Phosphorus	(Holland et al. 2004)
Dunaliella tertiolecta	Present	Phosphorus and nitrogen	(Roberts et al. 2008; Young and Beardall 2003)
Selenastrum minutum	Present	Nitrogen	(Turpin and Weger 1988)
Red Algae			
Porphyridium	Present	Nitrogen	Shelly, unpublished data
Cyanobacteria			
Oscillatoria sp.	Present	Phosphorus and nitrogen	(Holland et al. 2004)
Synechococcus sp.	Absent	Phosphorus, nitrogen	K. Shelly, unpublished data
Microcystis aeruginosa Kutz UTEX LB2063	Present	Phosphorus and nitrogen	(Wood and Oliver 1995)
Diatoms			
Phaeodactylum tricornutum	Absent	Phosphorus, nitrogen, silica	Shelly, unpublished data
Thalassiosira weissflogii	Present	Phosphorus(?), silica*	(Lippemeier et al. 2001)
Dinoflagellates			
Woloszynskia	Absent	Phosphorus, nitrogen	Shelly, unpublished data

* Although not interpreted as NIFTS in the original paper, these data are compatible with the time frame of NIFT processes and show an initial drop in fluorescence, followed by recovery over a time course in which cell numbers and growth cannot account for any changes in fluorescence intensity

work carried out by Turpin and Weger (1988), most of the studies that have examined the mechanisms behind NIFTs in detail did so using phosphorus, although the general conclusions are likely to be applicable to NIFTs induced in nitrogen-limited cells, or any other limiting nutrient that is actively taken up by cells.

The simplest interpretation of the P-induced drop in chlorophyll *a* fluorescence is that providing additional phosphate stimulates energy-dependent uptake processes and demand for electrons from the photosynthetic electron transport chain. This energy dissipation relieves pressure on photosynthesis and hence chlorophyll *a* fluorescence is quenched. However, the mechanisms behind NIFT responses appear to be more complex than can be accounted for by this simple explanation. For instance, NO_3^- addition produces a *rise* in chlorophyll *a* fluorescence. Drops in chlorophyll *a* fluorescence during P-induced and NH_4^+-induced NIFTS correlate with increases in non-photochemical quenching (NPQ) (Petrou et al. 2008; Young and Beardall 2003). Conversely, the rise in chlorophyll *a* fluorescence seen during a NO_3^--induced NIFT is paralleled by a drop in NPQ (Young and Beardall 2003).

The study carried out by Petrou et al. (2008) demonstrated that the NIFT response in P-starved *D. tertiolecta* is associated with changes in energy distribution between photosystems I and II as well as light-stress-induced non-photochemical quenching. Following the addition of phosphate to phosphorus starved *Dunaliella tertiolecta* cells, a state transition from state 1 to 2 was evident and is consistent with the findings of Gauthier and Turpin (1997), which also showed a decreased PSII/PSI ratio during Pi-uptake in *Selenastrum minutum*. It was also found that surplus light energy absorbed during a NIFT was re-directed through non-photochemical quenching to the xanthophyll cycle, contributing to the decline in chlorophyll *a* fluorescence signal (NIFT response) upon nutrient addition. Both state transitions and energy-dependent quenching are equally important for P-starved cells to effectively take up phosphate. P-starved cells can, via state transitions, produce additional ATP within the chloroplast. State 2 conditions are favoured by the increased ATP demand which favours cyclic electron flow and as a result, increased ATP production (Delosme 1991; Delphin et al. 1996). Synthesis of ATP, however, also requires phosphate so the capacity for ATP synthesis in P-starved cells would be restricted, even with improved cyclic photophosphorylation, unless phosphates were diverted from other cellular processes. Adoption of both of these mechanisms by P-starved *D. tertiolecta* cells was evident in the experiments by Petrou et al. (2008) which showed both a state transition and increased total NPQ during the NIFT.

The study by Petrou et al. (2008) confirms that state transitions are involved in the phosphorus NIFT response in *D. tertiolecta*. State transitions involve a change in light-harvesting pigment protein complexes (LHCP) between PS I and PS II. Since chlorophyll *a* fluorescence at room temperature essentially all comes from PS II, moving LHCP to PS I makes that chlorophyll become "silent" and hence leads to a drop in fluorescence intensity. The use of inhibitors in the study of Petrou et al. (2008), however, must always be viewed with some caution as these are not always as specific as intended, and further work using state transition mutants such as have been found for *Chlamydomonas* (Kruse et al. 1999) would be informative. The study by Petrou et al. (2008) also revealed that P-starved *D. tertiolecta* had a decreased capacity for carbon fixation due to down-regulated photosynthetic electron transport efficiency (decline in Φ_{PSII}) and re-direction of light energy to NPQ during the NIFT. It appears that the rapid chlorophyll *a* fluorescence quenching of a phosphorus induced NIFT results from a combination of a) a state transition, which has the consequence that cells shift from linear electron flow to cyclic electron flow around PS I, thereby generating additional ATP for Pi uptake, and b) an increase in energy-dependent quenching, which dissipates excess light energy through enhanced xanthophyll cycle activity.

In some cases, however, a rise in chlorophyll *a* fluorescence is seen during a NIFT (e.g. addition of NO_3^-, Young and Beardall 2003). The mechanism behind these changes is yet to be ascertained, but seem to result in a *decrease* in NPQ, the opposite to what is seen with phosphorus-stimulated NIFTS, and so may reflect the additional requirement of NO_3^- assimilation for reducing equivalents (and hence linear electron transport) as well as ATP. Whether this involves state transitions from State 2 to State 1 is yet to be determined.

If state transitions do play a significant role in NIFTs, this may explain why we have been unable to obtain reliable NIFT responses from diatoms, in which the state transition phenomenon is apparently not found (Owens 1986; Ting and Owens 1994). Only the paper by Lippemeier et al. (2001) contains data (for *Thalassiosira weissflogii*) that can be interpreted as evidence of possible NIFT responses to Si addition and (less convincingly) to phosphorus addition to the appropriately nutrient starved cells

6 Conclusion

There are many questions that remain unanswered in the case of the NIFT, for example if a state transition is involved in the NIFT phenomena, then it is doubtful that a NIFT will be seen in diatoms, rendering this technique of little use in areas such as the southern ocean where diatoms dominate. Unpublished studies have also indicated that the time of the onset of NIFTs can be affected by light or dark treatments. Investigations into possible processes which allow for P uptake in the dark as well processes involving PSI as a source of ATP for uptake in the light are needed to truly understand the mechanisms behind the NIFT. Further work is also needed in order to evaluate the applicability of the technique to a broad range of algal taxa, as well as to help gain further insight into the true mechanisms behind the NIFT response.

Fluorescence techniques can be used to provide novel information about the nutrient status of phytoplankton, in both the laboratory and the field. While the application of fluorescence can be as simple as measuring fluorescence as a proxy for chlorophyll during growth experiments, or using F_v/F_m as an indicator of stress, recent insights into the relationship between fluorescence quenching and nutrient uptake provide a new and rapid technique – the NIFT – that has the potential to provide what are essentially instantaneous, *in situ* measures of nutrient limitation, in both laboratory and natural populations. The NIFT phenomenon, however, is a relatively new development, and has yet to be thoroughly evaluated over a broad range of algal taxa, and at the low cell numbers characteristic of natural populations. There is, however, little doubt that as these techniques continue to be refined, wide-ranging applications will be found. Recently, delayed chlorophyll fluorescence has been advocated as a possible indicator of nutrient stress. How sensitive this might be to a range of nutrients and degree of limitation is still uncertain, but further investigation of this phenomenon is called for.

References

Beardall J, Young EB, Roberts SC (1996) Interactions between photosynthesis and the acquisition of N and P in *Dunaliella*. In: Mathis P (ed) Photosynthesis: from light to biosphere, vol V. Kluwer, Dordrecht, pp 435–438

Beardall J, Berman T, Heraud P, Omo Kadiri M, Light BR, Patterson G, Roberts S, Sulzberger B, Sahan E, Uehlinger U, Wood B (2001a) A comparison of methods for detection of phosphate limitation in microalgae. Aquat Sci 63:107–121

Beardall J, Young EB, Roberts SC (2001b) Approaches for determining phytoplankton nutrient limitation. Aquat Sci 63:44–69

Behrenfeld MJ, Bale AJ (1996) Confirmation of iron limitation of phytoplankton photosynthesis in the equatorial Pacific Ocean. Nature 383:508–511

Behrenfeld MJ, Kolber ZS (1999) Widespread iron limitation of phytoplankton in the South Pacific Ocean. Sci 283:840–843

Behrenfeld MJ, Lee H, Small LF (1994) Interactions between nutritional status and long-term responses to ultraviolet-B radiation stress in a marine diatom. Mar Biol 118:523–530

Behrenfeld MJ, Randerson JT, McClain CR, Feldman GC, Los OS, Tucker CJ, Falkowski PG, Field CB, Frouin R, Esaias WE, Kolber DD, Pollack NH (2001) Biospheric primary production during an ENSO transition. Science 291: 2594–2597

Berden-Zrimec M, Drinovec L, Molinari I, Zrimec A, Fonda Umani S, Monti M (2008) Delayed fluorescence as a measure of nutrient limitation in *Dunaliella tertiolecta*. J Photochem Photobiol B: Biol 92:13–18

Berges JA, Charlebois DA, Mauzerall DC, Falkowski PG (1996) Differential effects of nitrogen limitation on photosynthetic efficiency of Photosystem I and II in microalgae. Plant Physiol 110:689–696

Bertilsson S, Berglund O, Karl DM, Chisholm SW (2003) Elemental composition of marine *Prochlorococcus* and *Synechococcus*: implications for the ecological stoichiometry of the sea. Limnol Oceanogr 48:1721–1731

Birch PB, Gordon DM (1981) Nitrogen and phosphorus nutrition of *Cladophora* in the Peel-Harvey estuarine system, Western Australia. Bot Mar 24:381–387

Blomqvist S, Gunnars A, Elmgren R (2004) Why the limiting nutrient differs between temperate coastal seas and freshwater lakes: a matter of salt. Limnol Oceanogr 49:2236–2241

Boyd PW, LaRoche J, Gall M, Frew R, McKay RM (1999) Role of iron, light, and silicate in controlling algal biomass in subantarctic waters SE of New Zealand. J Geophys Res 104:13391–13404

Brooks AS, Edington DN (1994) Biogeochemical control of phosphorus cycling and primary production in Lake Michigan. Limnol Oceanogr 39:961–968

Caraco NF, Cole JJ, Likens GE (1989) Evidence for sulphate-controlled phosphorus release from sediments of aquatic systems. Nature 341:316–318

Caraco N, Cole JJ, Likens GE (1990) A comparison of phosphorus immobilization in sediments of freshwater and coastal marine systems. Biogeochemistry 9:277–290

Carrick HJ, Lowe RL (2007) Nutrient limitation of benthic algae in Lake Michigan: the role of silica. J Phycol 43:228–34

Cullen JJ, Davis RF (2003) The blank can make a big difference in oceanographic measurements. Limnol Oceanogr Bull 12:29–35

Cullen JJ, Neale PJ, Lesser MP (1992) Biological weighting function for the inhibition of phytoplankton photosynthesis by ultraviolet radiation. Science 258:646–650

Davey M, Tarran GA, Mills MM, Ridame C, Geider RJ, La Roche J (2008) Nutrient limitation of picophytoplankton photosynthesis and growth in the eastern tropical North Atlantic. Limnol Oceanogr 42:1722–1733

Debaar HJW (1994) Von Liebig law of the minimum and plankton ecology (1899–1991). Prog Oceanogr 33:347–386

Delosme R (1991) Electron transfer from cytochrome *f* to photosystem I in green algae. Photosynthesis Res 29:45–54

Delphin E, Duval J, Etienne A, Kirilovsky D (1996) State transitions or ΔpH-dependent quenching of photosystem II fluorescence in red algae. Biochemistry 35:9435–9445

Elser JJ, Kimmel BL (1986) Alteration of phytoplankton phosphorus status during enrichment experiments - implications for interpreting nutrient enrichment bioassay results. Hydrobiologia 133:217–222

Fisher TR, Gustafson AB, Sellner K, Lacuture R, Haas LW, Margnien R, Karrh R, Michael B (1999) Spatial and temporal variation in resource limitation in Chesapeake Bay. Mar Biol 133:763–778

Frost PC, Xenopoulos MA (2002) Ambient solar ultraviolet radiation and its effects on phosphorus flux into boreal lake phytoplankton communities. Can J Fish Aquat Sci 59: 1090–1095

Ganf GS, Stone SJL, Oliver RL (1986) Use of protein to carbohydrate ratios to analyse for nutrient deficiency in phytoplankton. Aust J Mar Freshwater Res 37:183–197

Gauthier DA, Turpin DH (1997) Interactions between inorganic phosphate (Pi) assimilation, photosynthesis and respiration in the Pi-limited green alga *Selenastrum minutum*. Plant Cell Env 20:12–24

Geider RJ, La Roche J (2002) Redfield revisited: variability in the N:P ratio of phytoplankton and its biochemical basis. Eur J Phycol 37:1–17

Geider RJ, LaRoche J (1994) The role of iron in phytoplankton photosynthesis, and the potential for iron-limitation of primary productivity in the sea. Photosynth Res 39:275–301

Gerringa LJA, de Baar HJW, Timmermans KR (2000) A comparison of iron limitation of phytoplankton in natural oceanic waters and laboratory media conditioned with EDTA. Mar Chem 68:335–346

Gotham IJ, Rhee G (1981) Comparative kinetic studies of phosphate-limited growth and phosphate uptake in phytoplankton in continuous culture. J Phycol 17:257–265

Graziano LM, LaRoche J, Geider RJ (1996) Physiological response to phosphorus limitation in batch and steady state cultures of *Dunaliella tertiolecta* (Chlorophyta): a unique stress protein as an indicator of phosphate deficiency. J Phycol 32:825–838

Greene RM, Geider RJ, Kolber ZS, Falkowski PG (1992) Iron-induced changes in light harvesting and photochemical energy conversion processes in eukaryotic marine algae. Plant Physiol 100:565–575

Greene RM, Kolber ZS, Swift DG, Tindale NW, Falkowski PG (1994) Physiological limitation of phytoplankton photosynthesis in the eastern equatorial Pacific determined from variablility in the quantum yield of fluorescence. Limnol Oceanogr 29:1061–1074

Guildford SJ, Hecky RE (2000) Total nitrogen, total phosphorus, and nutrient limitation in lakes and oceans: Is there a common relationship? Limnol Oceanogr 45:1213–1223

Guildford SJ, Hecky RE, Taylor WD, Mugidde R, Bootsma HA (2003) Nutrient enrichment experiments in tropical Great Lakes Malawi/Nyasa and Victoria. J Great Lakes Res 29:89–106

Hayes PK, Whitaker TM, Fogg GE (1984) The distribution and nutrient status of phytoplankton in the Southern Ocean between 20° and 70°W. Polar Biol 3:153–165

Hecky RE, Kilham P (1988) Nutrient limitation of phytoplankton in freshwater and marine environments: A review of recent evidence on the effects of enrichment. Limnol Oceanogr 33:796–822

Ho TY, Quigg A, Finkel ZV, Milligan AJ, Wyman K, Falkowski PG, Morel FMM (2003) The elemental composition of some marine phytoplankton. J Phycol 39:1145–1159

Holland DP, Roberts SC, Beardall J (2004) Assessment of the nutrient status of phyoplankton: A comparison between conventional bioassays and nutrient-induced fluorescence transients (NIFTs). Ecol Indic 4:149–159

Holmes JJ, Weger HG, Turpin DH (1989) Chlorophyll *a* fluorescence predicts total photosynthetic electron flow to CO_2 or NO_3^-/NO_2^- under transient conditions. Plant Physiol 91:331–337

Howarth RW (1988) Nutrient limitation of net primary production in marine ecosystems. Annu Rev Ecol Syst 19:898–910

Iglesias AA, Plaxton WC, Podesta FE (1993) The role of inorganic phosphate in the regulation of C4 photosynthesis. Photosynth Res 35:205–211

Jansen MAK, Greenberg BM, Edelman M, Matto AK, Gaba V (1996) Accelerated degradation of the D2 protein of Photosystem II under ultraviolet radiation. Photochem Photobiol 63:814–817

Karl DM (1999) A sea of change: biogoechemical variability in the North Pacific subtropical gyre. Ecosystems 2:181–214

Knowles R (1982) Denitrification. Microbiol Rev 46:43–70

Kolber ZS, Zehr J, Falkowski PG (1988) Effects of growth irradience and nitrogen limitation on photosynthetic energy conversion in Photosystem II. Plant Physiol 88:923–929

Krom MD, Kress N, Brenner S, Gordon LI (1991) Phosphorus limitation of primary productivity in the eastern Mediterranean Sea. Limnol Oceanogr 36:424–432

Kromkamp JC, Forster RM (2003) The use of variable fluorescence measurements in aquatic ecosystems: differences between multiple and single turnover measuring protocols and suggested terminology. Eur J Phycol 38:103–112

Kruse O, Nixon PJ, Schmid GH, Mullineaux CW (1999) Isolation of state transition mutants of *Chlamydomonas reinhardtii* by fluorescence video imaging. Photosynth Res 61:43–51

Kupper H, Setlik I, Seibert S, Prasil O, Setlikova E, Strittmatter M, Levitan O, Lohscheider J, Adamska I, Berman-Frank I (2008) Iron limitation in the marine cyanobacterium *Trichodesmium* reveals new insights into regulation of photosynthesis and nitrogen fixation. New Phytol 179:784–798

LaRoche J, Geider RJ, Graziano LM, Murray H, Lewis K (1993) Induction of specific proteins in eukaryotic algae grown under iron-, phosphorus-, or nitrogen-deficient conditions. J Phycol 29:767–777

Lean DRS, Pick FR (1981) Photosynthetic response of lake plankton to nutrient enrichment: a test for nutrient limitation. Limnol Oceanogr 26:1001–1019

Lewandowska J, Kosakowska A (2004) Effect of iron limitation on cells of the diatom *Cylotella meneghiniana* Kützing. Oceanol 46:269–287

Lippemeier S, Hartig P, Colijn F (1999) Direct impact of silicate on the photosynthetic performance of the diatom *Thalassiosira weissflogii* assessed by on- and off-line PAM fluorescence measurements. J Plankton Res 21:269–283

Lippemeier S, Hintze R, Vanselow KH, Hartig P, Coljin F (2001) In-line recording of PAM fluorescence of phytoplankton cultures as a new tool for studying effects of fluctuating nutrient supply of photosynthesis. Eur J Phycol 36:89–100

Martin JH, Fitzwater SE (1988) Iron deficiency limits phytoplankton growth in the North-East Pacific Subartic. Nature 331:341–343

Masi A, Melis A (1997) Morphological and molecular changes in the unicellular green alga *Dunaliella salina* grown under supplemental UV-B radiation: cell characterstics and Photosytem II damage and repair properties. Biochim Biophy Acta 1321:183–193

Maxwell K, Johnson GN (2000) Chlorophyll fluorescence – a practical guide. J Exp Bot 51:659–668

McAndrew PM, Bjorkman KM, Church MJ, Morris PJ, Jachowski N, Williams PJL, Karl DM (2007) Metabolic response of oligotrophic plankton communities to deep water nutrient enrichment. Mar Ecol Prog Ser 332:63–75

McKay RML, Geider RJ, LaRoche J (1997) Physiological and biochemical response of the photosynthetic apparatus of two marine diatoms to Fe stress. Plant Physiol 114:615–622

Miller WE, Maloney TE, Greene EJC (1974) Algal productivity in 49 lake waters, as determined by algal assays. Water Res 8:667–679

Miller WE, Greene EJC, Shiroyama T (1978) The *Selenastrum capricornutum* Printz algal assay: bottle test. Experimental design, application and data interpretation protocol: EPA (USA)

Miller AG, Espie GS, Canvin DT (1991) The effects of inorganic carbon and oxygen on fluorescence in the cyanobacterium *Synechococcus* UTEX 625. Can J Bot 69:1151–1160

Moore CM, Mills MM, Languois R, Milne A, Achterberg EP, La Roche J, Geider RJ (2008) Relative influence of nitrogen and phosphorous availability on phytoplankton physiology and productivity in the oligotrophic sub-tropical North Atlantic Ocean. Limnol Oceanogr 53:291–305

Morel FMM, Hudson RJM (1985) The geobiological cycle of trace elements in aquatic systems: Redfield revisited. In: Stumm W (ed) Chemical processes in lakes. Wiley, New York, pp 251–281

Morse JW, Zullig JJ, Bernstein LD, Miller FJ, Milne P, Mucci A, Choppin GR (1985) Chemistry of calcium carbonate-rich shallow water sediments in the Bahamas. Am J Sci 285:147–185

Nixon SW (1995) Coastal marine eutrophication – a definition, social causes, and future concerns. Ophelia 41:199–219

North RL, Guildford SJ, Smith REH, Havens SM, Twiss MR (2007) Evidence for phosphorus, nitrogen, and iron colimitation of phytoplankton communities in Lake Erie. Limnol Oceanogr 52:315–28

Owens TG (1986) Light harvesting function in the diatom *Phaeodactylum tricornutum*. Plant Physiol 80:739–746

Petrou K, Doblin MA, Smith RA, Ralph PJ, Shelly K, Beardall J (2008) State transitions and nonphotochemical quenching during a nutrient-induced fluorescence transient in phosphorus-starved, *Dunaliella tertiolecta*. J Phycol 44:1204–1211

Raven JA, Evans MCW, Korb RE (1999) The role of trace metals in photosynthetic electron transport in O_2-evolving organisms. Photosynth Res 60:111–149

Raven JA, Brown K, Mackay M, Beardall J, Giordano M, Granum E, Leegood RG, Kilminster K, Walker DI (2005) Iron, nitrogen, phosphorus and zinc cycling and consequences for primary productivity in the oceans. In: Gadd GM, Semple KT, Lappin-Scott HM (eds) Society for General Microbiology Symposium 65: Micro-organisms and earth systems – advances in geomicrobiology. Cambridge University Press, Cambridge, pp 247–272

Redfield AC (1958) The biological control of chemical factors in the environment. Am Sci 46:205–221

Roberts S (1998) Physiological effects of phosphorus limitation on photosynthesis in two green algae Ph.D. thesis, Monash University, Melbourne 116 pp

Roberts S, Shelly K, Beardall J (2008) Interactions among phosphate uptake, photosynthesis, and chlorophyll fluorescence in nutrient-limited cultures of the chlorophyte microalga *Dunaliella tertiolecta*. J Phycol 44:662–669

Saito MA, Goepfert TJ, Ritt JT (2008) Some thoughts on the concept of colimitation: three definitions and the importance of bioavailability. Limnol Oceanogr 53:276–90

Schindler DW (1977) Evolution of phosphorus limitation in lakes. Science 195:260–262

Seitzinger SP (1988) Denitrification in fresh-water and coastal marine ecosystems: ecological and geochemical significance. Limnol Oceanogr 33:702–724

Shipe RF, Carpenter EJ, Govil SR, Capone DG (2007) Limitation of phytoplankton production by Si and N in the western Atlantic Ocean. Mar Ecol Prog Ser 338:33–45

Singh BN, Lal KN (1935) Limitations of Blackman's law of limiting factors and Harder's concept of relative minimum as applied to photosynthesis. Plant Physiol 10:245–268

Smith VH (2003) Eutrophication of freshwater and coastal marine ecosystems – A global problem. Environ Sci Pollut R 10:126–139

Sterner RW, Anderson TR, Elser JJ, Hessen DO, Hood JM, McCauley E, Urabe J (2008) Scale-dependent carbon: nitrogen: phosphorus seston stoichiometry in marine and freshwaters. Limnol Oceanogr 53:1169–1180

Sunda WG, Huntsman SA (1997) Interrelated influence of iron, light and cell size on marine phytoplankton growth. Nature 390:389–392

Sundareshwar PV, Morris JT, Koepfler EK, Fornwalt B (2003) Phosphorus limitation of coastal ecosystem processes. Science 299:563–565

Timmermans KR, van Leeuwe MA, de Jong JTM, McKay RML, Nolting RF, Witte HJ, van Ooyen J, Swagerman MJW, Kloosterhuis H, de Baar HJW (1998) Iron stress in the Pacific region of the Southern Ocean: evidence from enrichment bioassays. Mar Ecol Prog Ser 166:27–41

Ting CS, Owens TG (1994) The effects of excess irradiance on photosynthesis in the marine diatom *Phaeodactylum tricornutum*. Plant Physiol 106:763–770

Titman D, Kilham P (1976) Sinking in freshwater phytoplankton- some ecological implications of cell nutrient status and physical mixing processes. Limnol Oceanogr 21:409–417

Turpin DH, Weger HG (1988) Steady-state chlorophyll *a* fluorescence transients during ammonium assimilation by the N-limited green alga *Selenastrum minutum*. Plant Physiol 88:97–101

Van Mooy BAS, Devol AH (2008) Assessing nutrient limitation of *Prochlorococcus* in the North Pacific subtropical gyre by using an RNA capture method. Limnol Oceanogr 53: 78–88

Vitousek PM, Howarth RW (1991) Nitrogen limitation on land and in the sea: how can it occur? Biogeochemistry 13:87–115

Weger HG, Middlemiss JK, Petterson CD (2002) Ferric chelate reductase activity as affected by the iron-limited growth rate in four species of unicellular green algae (Chlorophyta). J Phycol 38:513–519

Wood MD, Oliver RL (1995) Fluorescence transients in response to nutrient enrichment of nitrogen- and phosphorus-limited *Microcystis aeruginosa* cultures and natural phytoplankton populations: a measure of nutrient limitation. Aust J Plant Physiol 22:331–340

Wynne D, Berman T (1980) Hot water extractable phosphorus – an indicator of nutritional status of *Peridinium cinctum* (Dinophyceae) from Lake Kinneret (Israel). J Phycol 16:40–46

Young EB, Beardall J (2003) Rapid ammonium- and nitrate-induced perturbations to chl *a* fluorescence in nitrogen-stressed *Dunaliella tertiolecta* (Chlorophyta). J Phycol 39:332–342

Chapter 12
The Application of Variable Chlorophyll Fluorescence to Microphytobenthic Biofilms

R.G. Perkins, J.C. Kromkamp, J. Serôdio, J. Lavaud, B. Jesus, J.L. Mouget, S. Lefebvre, and R.M. Forster

1 Introduction to Benthic Biofilms

Community assemblages of diatoms, green algae and cyanobacteria comprise the microphytobenthos (MPB), which inhabit benthic sediment ecosystems (Admiraal 1984; Underwood and Kromkamp 1999; Consalvey et al. 2004). Particular attention has been paid to the analysis of intertidal soft sediment systems, e.g. cohesive mudflat and sandy substrata typical of estuarine habitats. Variable chlorophyll fluorescence has been applied to these systems since the 1990s, in an attempt to investigate the primary productivity and photophysiology of the integrated biofilms, when viewed as a "black box system", and also at the species level (Sections 5, 6 and 7). These transient (i.e. temporary) biofilms are not confined to such soft sediment habitats however, and more recently application of fluorescence methodologies has been applied to biofilms inhabiting rocky shores and stromatolite systems (Kromkamp et al. 2007; Perkins et al. 2007). However the large majority of published work has centred upon benthic soft-sediment biofilms, due to their important ecosystem functions of carbon flow and sediment stability (Underwood and Kromkamp 1999). In the former their high magnitude of productivity fuels carbon flow through invertebrate and bacterial food webs to support important trophic levels of anthropogenically exploited taxa, including coastal fish and shell fisheries and coastal avifauna. In the case of sediment stability, biogenic exopolymers, usually referred to as extracellular polymeric substances (EPS), produced by the MPB in part to facilitate mobility, may contribute significantly to sediment stability, hence increasing the sediment resistance to hydrodynamic stresses and thus resistance to coastal erosion (e.g. Underwood and Kromkamp 1999 and citations there-in). Finally, the photosynthetic production of oxygen can be regarded as an important ecosystem function.

One attribute of MPB ecology has in particular contributed to confounding the application of fluorescence methodology to benthic biofilms. This is the behavioural adaptation of the MPB to migrate vertically in their sediment matrix habitat in response to

R.G. Perkins (✉)
School of Earth and Ocean Sciences, Cardiff University, Main Building, Park Place, Cardiff, CF10 3YE, UK
e-mail: PerkinsR@cf.ac.uk

J.C. Kromkamp
NIOO-KNAW Centre for Estuarine and Marine Ecology, P.O. Box 140, 4400 AC Yerseke, The Netherlands

J. Serôdio
Departamento de Biologia and CESAM – Centro de Estudos do Ambiente e do Mar, Universidade de Aveiro, Campus de Santiago, 3810-193, Aveiro, Portugal

J. Lavaud
CNRS UMR6250 'LIENSs', Institute for Coastal and Environmental Research (ILE), University of La Rochelle, 2 rue Olympe de Gouges, 17042 La Rochelle cedex, France

B. Jesus
Centro de Oceanografia, Faculdade de Ciências da Universidade de Lisboa, Campo Grande, 1749-016 Lisboa, Portugal

J.L. Mouget
Laboratoire de Physiologie et de Biochimie végétales, MMS EA 2160, Université du Maine, Av. O. Messiaen, 72085 Le Mans Cedex 9, France

S. Lefebvre
Laboratoire de Biologie et Biotechnologies Marine, Université de Caen, Esplanade de la Paix, 14032 Caen cedex, France

R.M. Forster
Ecosystem Interactions, Centre for Environment, Fisheries and Aquaculture Science (CEFAS), Pakefield Road, Lowestoft, Suffolk NR33 0HT, UK

D.J. Suggett et al. (eds.), *Chlorophyll* a *Fluorescence in Aquatic Sciences: Methods and Applications*, Developments in Applied Phycology 4, DOI 10.1007/978-90-481-9268-7_12, © Springer Science+Business Media B.V. 2011

environmental stimuli, as well as part of endogenous tidal and diel rhythms (Sections 2, 3, 4, 5 and 6). Consalvey et al. (2004) reviewed this migration in MPB biofilms and Consalvey et al. (2005) discussed the effect of this as part of a review of fluorescence methodology applied to MPB biofilms. Vertical migration appears to follow tidal and diel rhythms such that cells migrate to the surface of the sediment to coincide with daylight emersion periods, whilst migrating back in to the sediment for immersion periods (Consalvey et al. 2004 and citations there-in). In some instances this appears to be modulated by light environment such that low turbidity in the overlaying water column enables the MPB to remain at the sediment surface after the onset of immersion, and during bright moonlit emersion periods, MPB will migrate to the surface (Perkins, Kromkamp, Serôdio and Jesus, personal observations).

Light is undoubtedly a major stimulus in MPB vertical migration. Several studies have shown that cells within MPB biofilms show negative or positive phototaxis. For example, cells will migrate down to avoid potentially harmful high light environments and migrate up to optimise the light environment when ambient light levels are low (Sections 2, 3 and 6). Cells also show "microcycling" such that there can be a constant turnover of taxa at the sediment surface, resulting in a reduction in photodose integrated over time (Section 2, 5, 6 and 7). As well as affecting productivity, such "behavioural" down regulation of photosynthesis may act to make interpretation of fluorescence measurements difficult. For example, cells will migrate to different positions within the sediment matrix making the distance between the cells and the fluorometer probe variable and unknown. This leads to a variable attenuation of applied actinic light, as well as the fluorescence yields used to study aspects of photoacclimation. Furthermore, cells may migrate in response to darkness applied for measurement of dark adapted fluorescence parameters, again altering fluorescence yields (Sections 2, 3 and 6). These effects, and others discussed in this chapter, lead to potential errors in the application of variable chlorophyll fluorescence to migratory MPB biofilms. As a result great care is needed in the interpretation of fluorescence data obtained. Benthic diatom taxa, which can comprise the majority of MPB biomass, also show differences in photophysiology from higher plants, making conventional interpretation of fluorescence data incorrect. For example, diatoms exhibit high levels of down regulation through diadinoxanthin/diatoxanthin xanthophyll cycling, non-photochemical quenching (NPQ), induced as a result of the trans-thylakoid proton gradient resulting from light induced electron transport (Section 3). Thus MPB cells can show behavioural and physiological down regulation. However chlororespiration during dark periods leads to retention of this proton gradient, retaining NPQ in the dark (Section 3). This can suppress maximum fluorescence yield in the dark (F_m) such that it is lower than operational maximum yield in low light (F_m'), making conventional calculation of NPQ from the difference in these yields problematic (Sections 3 and 5).

In this review, the authors have summarised the main areas of fluorescence research as applied to benthic biofilms, principally concentrating on those of intertidal soft sediment ecosystems. Fluorescence is now widely applied in the field of benthic biofilm research, being relatively non-invasive, although obviously the application of the measurement, e.g. actinic light during a light curve, is inherently intrusive to some extent. Fluorescence also has the distinct advantage of being non-destructive when compared to other measurements of, for example, productivity using slurries or intact biofilms with radioactive labels added. The methodology is summarised, along with potential problems, as well as ways to minimise error and achieve correct interpretation of data obtained. The review covers subsurface signal as a result of vertical integration of fluorescence measurements (Section 2), non-photochemical quenching and the xanthophyll cycle (Section 3), measurement and calculation of fluorescence derived electron transport rate (Section 5), methodology used to obtain light curve parameters of photophysiology, e.g. photoacclimation (Section 6) and concludes with a section comparing fluorescence to other methods including radio-labelled carbon uptake and oxymetry (Section 7).

2 The Effects of Subsurface Signal

2.1 *Microphytobenthic Biofilms on Soft Sediments*

Sediments colonised by microphytobenthos are optically dense, causing both downwelling light (including light generated by a fluorometer) as well as upwelling fluorescence to be attenuated within the microalgal biofilm/sediment matrix. As a result, the fluorescence levels F_s and F_m' measured from just above the surface represent the integration of the fluorescence signals emitted at various depths, distorting the relationship between F_o and *chl a* as this relationship will thus not only depend upon the total biomass, but also on the shape of the vertical biomass profile. Of more importance, this effect causes the measurement of the effective quantum yield, $\Delta F/F_m'$, at the surface to differ significantly from the true, inherent value of the microalgae composing the biofilm, because the relationship between F_s and F_m' varies with irradiance within the depth range over which the fluorometer signal is integrated (Forster and Kromkamp 2004; Serôdio 2004).

The depth-integration effect was first predicted from observations of changes in $\Delta F/F_m'$ in undisturbed biofilms under constant irradiance (Underwood and Kromkamp, 1999; Perkins et al. 2001, 2002), and was studied in detail using numerical simulation models (Forster and Kromkamp 2004; Serôdio 2004). Studies showed that fluorescence measurements taken non-invasively at the surface of the sediment result in a substantial light-dependent overestimation of the inherent value of $\Delta F/F_m'$ and relative electron transport rate, rETR, particularly important under supersaturating irradiances (Fig. 1). As a consequence, light curves derived from depth-integrated measurements are likely to appear to saturate at higher irradiances, or to be less photoinhibited when compared to the true physiological response of the biofilm-forming microalgae. The contribution of sub-surface fluorescence signals also affect the determination of the light response of non-photochemical coefficient NPQ, which is expected to be underestimated under high light when measured non-invasively in intact biofilms (Forster and Kromkamp 2004; Serôdio 2004).

While this effect may occur, although to a lesser extent, in other optically dense samples such as thick leaves or macroalgal thalli (Forster and Kromkamp

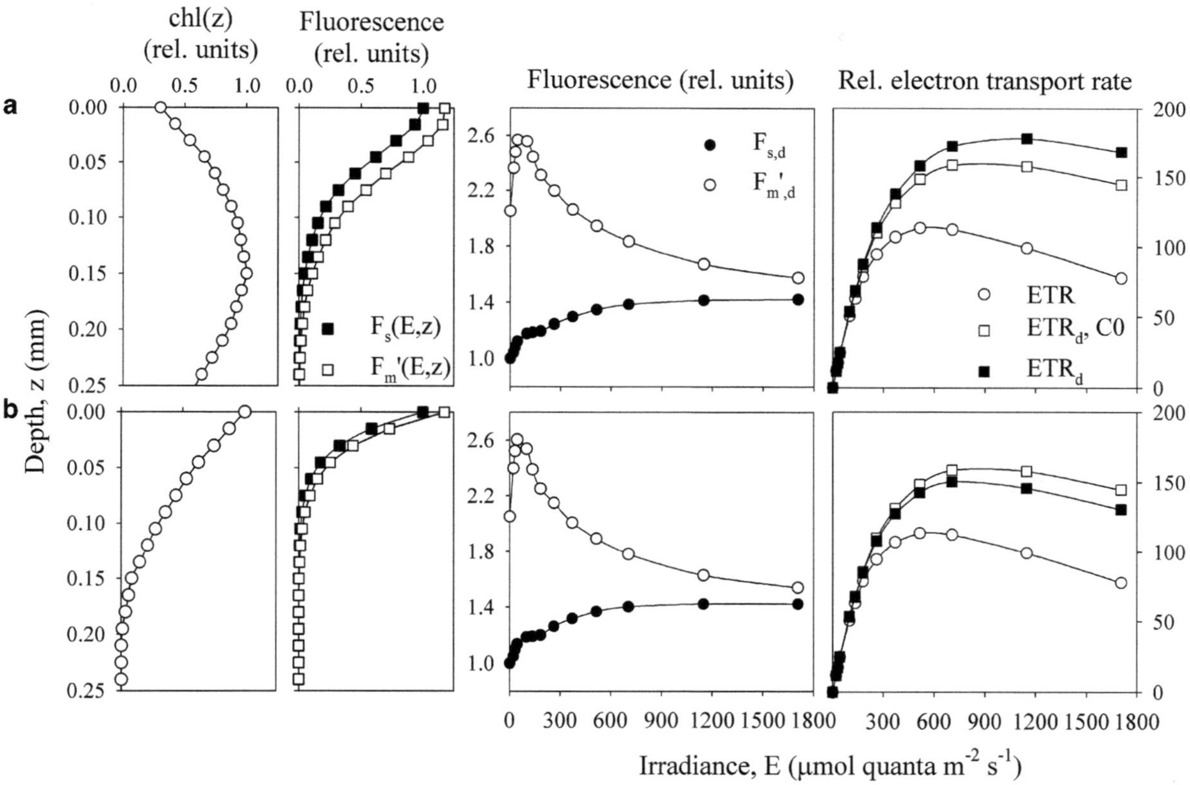

Fig. 1 Effects of vertical distribution of microalgal biomass, chl(z), on the profiles of fluorescence emission, F(z) and $F_m'(z)$, and on fluorescence light-response curves of ETR based on depth-integrated parameters (F_d, $F_{m\,d}'$), as compared to the case of homogeneous photic zone (C0). (**a**) Biomass profile with a subsurface maximum. (**b**) Biomass profile describing accumulation at the surface. Fluorescence levels normalised to surface values (Adapted from Serôdio 2004)

2004; Serôdio 2004; Sušila et al. 2004), in the case of microphytobenthos the problem is further complicated by the occurrence of comparatively large scale vertical migration by motile diatoms. In this case, it becomes highly difficult to interpret fluorescence yields from biofilms where the cells move vertically within the sediment matrix, since the subsurface signal emanates not only from cells at unknown depth, but at a variable depth in the sediment. Additionally, signal strength is proportional to the distance between the fluorometer probe and the cells themselves, thus vertical migration may increase or decrease measured yields, so making interpretation of changes in yield or calculation of photophysiological parameters difficult. For example, a decrease in both F_m' and F can be due to induction of NPQ down regulation on exposure to increasing PAR or to downward migration (negative phototaxis). It is often not possible to differentiate between the two processes using fluorescence methods. For this reason it can be prudent to use subsurface spectral reflectance for measurements such as biomass, see below, (Kromkamp et al. 2006; Morris et al. 2008) as the reflectance spectra are not influenced by NPQ.

It is hard to avoid artefacts of tidally induced vertical migration and positive or negative phototaxis, but the effects can be minimized by not taking measurements during the first and last hour of the emersion period when vertical migration is maximal. In addition it is better to minimize the effects of phototaxis by keeping the duration of measurements as short as possible (e.g. minimise the duration of light steps during rapid light curves, RLCs, see later). PSII quantum efficiency often stabilizes before true steady state fluorescence is reached (Perkins, Kromkamp, Serôdio and Jesus, personal observations).

The issue of the subsurface signal from the cells within the subsurface sediment has been investigated in considerable depth (Kromkamp et al. 1998; Perkins et al. 2002; Forster and Kromkamp 2004; Serôdio 2004; Jesus et al. 2006a, b). Kromkamp et al. (1998) discussed the issue of microcycling of cells such that the fluorescence yield was obtained from a varied surface community over time and that this could explain the persistent high $\Delta F/F_m'$ at high incident irradiance. Perkins et al. (2002) reported over estimation of ETR in sediments due to subsurface signal from cells exposed to a lower light level than that applied at the surface and showed that the community composition could change during a light curve. In addition it appeared that the PSII signal was contaminated by PSI-fluorescence, shown by using a 680 nm bandpass filter compared to the more conventional 695 nm longpass filter. Underwood et al. (2005) used high resolution fluorescence imaging to report diel patterns in $\Delta F/F_m'$ of several benthic diatom species which showed microcycling over an artificially extended emersion period (Fig. 2). Serôdio (2004) and Forster and Kromkamp (2004) also demonstrated similar effects on measurements due to integration of the fluorescence signal over sediment depth. To make matters more complex, the degree of overestimation of $\Delta F/F_m'$, which manifests the greatest at light levels above the light saturation coefficient, E_k, depends also on the shape of the biomass profile, and subsurface biomass maxima, resulting in the highest degree of overestimation of the true $\Delta F/F_m'$, and hence resulting in an overestimate of the maximum electron transport rate, $rETR_{max}$, often by as much as of 60% (Forster and Kromkamp 2004). The effect of this on estimates of depth integrated primary production will be discussed later when comparing fluorescence with other methods used to estimate primary production.

Vertical migration does not only influence the apparent value of $\Delta F/F_m'$, but it also influences the measured minimal fluorescence yield, F_o (Forster and Kromkamp 2004, Perkins et al. 2001; Jesus et al. 2006a); which is used as a proxy of biomass (Barranguet and Kromkamp 2000; Honeywill et al. 2002). It is thus not surprising that Jesus et al. (2006a, b) observed that tidally induced vertical migration resulted in either

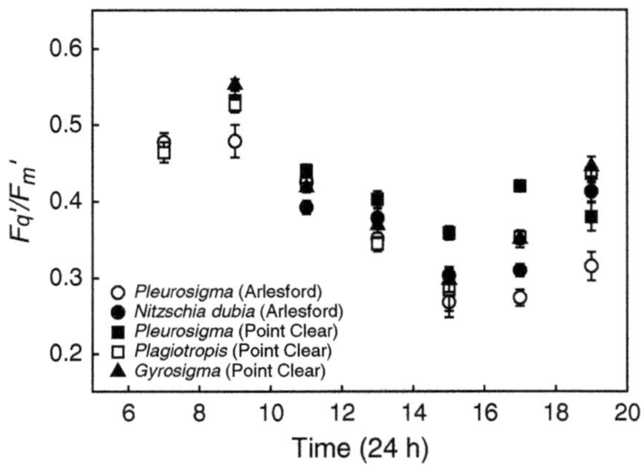

Fig. 2 Quantum efficiency under constant light level of individual cells of five benthic diatom species obtained using high resolution fluorescence imaging. By this method, only cells at the surface of the sediment were analysed, reducing any effect from sub-surface cells

an over estimation or under estimation of biomass dependent upon the time the measurements were made within the emersion period. Cells either migrated up or away from the sediment surface, thereby enriching or depleting the biomass at the sediment surface contributing to the F_o signal measured.

Recent work using migration inhibitors (Cartaxana et al. 2008; Perkins et al. in preparation) and engineered non-migratory biofilms (Jesus et al. 2006a, b; Mouget et al. 2008) have investigated these issues further and confirm that vertical migration and the concomitant "deep layer fluorescence" lead to erroneous estimates of quenching coefficients and overestimation of $\Delta F/F_m'$.

2.2 Stromatolites – the effect of "layered" biofilms

Stromatolites are perhaps an extreme example of a layered biofilm where sub-surface cells "interfere" with the fluorescence signal from surface cells and vice versa. These biogenic organosedimentary structures consist of layers of sand grains cemented together through microorganism (primarily cyanobacterial photosynthesis and bacterial mineralisation) metabolic processes and physicochemical reactions within the matrix of the stromatolite (e.g. Reid et al. 2000). The measurements become an integrated measurement of the sub surface cyanobacteria (Fig. 3) and the surface cells, where diatom epiphytes can be present (Fig. 4, Perkins et al. 2007). Stromatolites consist of a microbial consortium which trap ooids (sand grain particles) by extracellular polymers, mainly of cyanobacterial origin (Visscher et al. 2000; Reid et al. 2000). Reid et al. (2000) described three different developmental stages where the pioneering type 1 stromatolite has the lowest cyanobacterial diversity with *Schizotrix gebeleinii* as the dominant form. Type 2 is characterised by a micritic crust at the surface (often lacking diatom epiphytes, Kromkamp, personal observation) and in climax state type 3 the endolythic *Solentia* spp. bores into the ooids and fuses them together, giving the stromatolite its structure. Types 2 and 3 have the highest cyanobacterial diversity (Baumgartner et al. 2007). Figure 3 shows a typical example of a type 1 stromatolite (Kromkamp et al. 2007) with a clear subsurface layer, but the layer above also contain many cyanobacteria.

Fig. 4 Surface mixed eukaryotic microalgae and cyanobacteria mat on a stromatolite, note the fluffy three dimensional structure of the mat. The area viewed is approximately 60 cm wide (Reproduced courtesy of P. Reid)

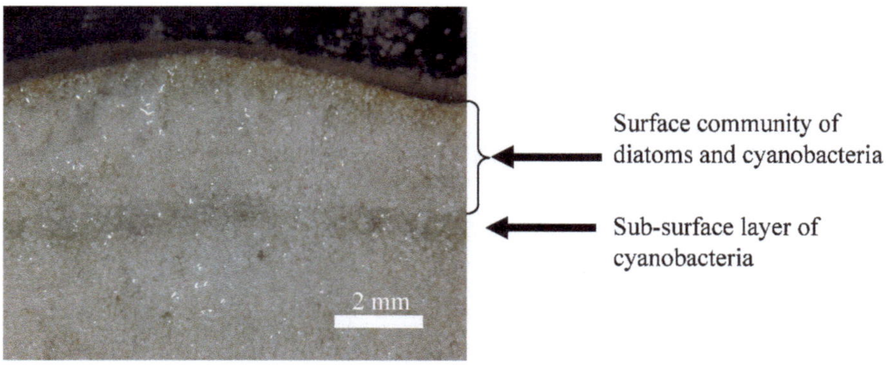

Fig. 3 Cross sectioned sample from a stromatolite showing the surface mixed community and sub-surface layer of cyanobacteria (Reproduced courtesy of P. Reid)

This clearly highlights the difficulty when applying surface measurements using PAM fluorometry: the depth and thickness of the subsurface layer varies as well as the distance of the cells from the stromatolite surface. This means that quantitative analyses of both cyanobacterial biomass as well as effective quantum efficiency is very difficult as the apparent $\Delta F/F_m'$ and F_o is integrated over an unknown depth interval with varying biomass distribution. This problem can be partly circumvented by measuring on a sample cross section, by putting the PAM fiberoptics parallel to the mat surface (Perkins et al. 2007). However, this will disrupt chemical and light gradients, especially of oxygen and light, and hence greatly influence photosynthetic activity (Kromkamp et al. 2007).

The following example demonstrates the problems with working with cyanobacteria (Fig. 5). RLCs were made on two sections of the stromatolite where the depth of the subsurface layer was respectively at 4 and 1 mm below the surface. The layer at the surface contained cyanobacteria (cf. *Schizotrix* sp.) but the cell density (measured by sub sample cell counts and examination using microscopy) was substantially lower than in the subsurface layer. After the RLCs were performed, the surface layer was removed until the subsurface layer was exposed and new RLCs were made. In both cases the measured rETR was much higher when the surface layers were not removed. Several explanations are possible: the cyanobacteria in the surface layer have a higher photosynthetic activity than cyanobacteria in the subsurface biomass maximum. Deeper in the stromatolite the oxygen concentration might be low, causing rapid inactivation of the stromatolite (Kromkamp et al. 2007; Perkins et al. 2007). This might be the case for the layer 4 mm below the surface but is unlikely for the layer 1 mm below the surface. Alternatively, fluorescence from the deeper layers, where the irradiance is lower and where $\Delta F/F_m'$ will thus be higher than at the stromatolite surface, will cause an overestimation of true $\Delta F/F_m'$ of the cells at the surface. The degree of overestimation depends on the depth (and thus the irradiance) of the subsurface layer (which is unknown without destructive sampling) and on the biomass in the subsurface maximum relative to the cyanobacterial biomass in the upper layers. Because the RLCs on the exposed subsurface layers are rather similar, the contribution of "deep layer fluorescence" seems the most likely explanation for the higher rates of rETR measured at the stromatolite surface. A comparison with measurements on stromatolite sample cross sections (Mouget et al. in preparation) seems to corroborate this conclusion.

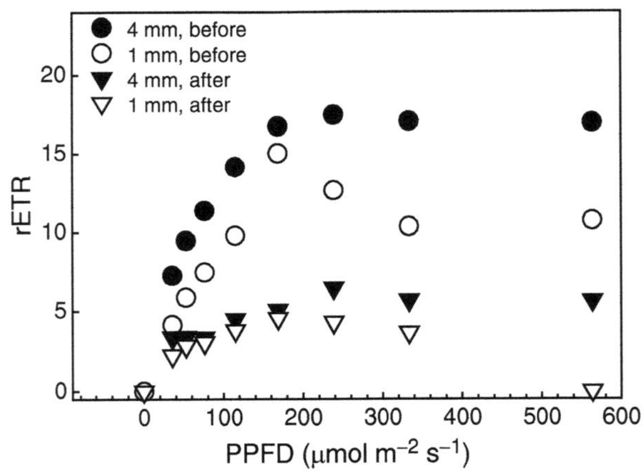

Fig. 5 Light response curves for two stromatolite samples using a fluorimeter probe in the top-down position. The first sample (closed symbols) refers to a layer of cyanobacteria measured at 4 mm depth in the ooids matrix, data clearly show a large decrease in rETR post scraping away of the surface ooids. The second sample (open symbols) refers to a layer 1 mm below the stromatolite surface, again rETR decreased significantly post scraping

2.3 Deconvolution of Depth Integrated Signals

The inherent photophysiological status of the microalgae comprising a biofilm can be estimated from depth integrated fluorescence measurements made on undisturbed samples. A method has been proposed to estimate the inherent light response through the deconvolution of light-response curves based on depth-integrated measurements (Serôdio 2004). This approach is based on the relationship existing between the depth-integrated fluorescence levels measured at consecutive light levels of a light curve and the depth attenuation of the fluorescence signal, which results in a set of recursive equations to be applied to depth-integrated light curves. This approach has assumptions and limitations (Serôdio 2004; Jesus et al. 2006b, c): (i) it assumes an homogeneous photophysiological light response for all the microalgae; (ii) assumes an exponential vertical attenuation of incident actinic light and emitted fluorescence, which implies a vertically homogeneous photic zone; (iii) it requires the attenuation coefficients for downwelling actinic light and upwelling fluorescence to be known; (iv) it does not account for changes in surface biomass during the construction of the light curve on intact biofilms. Nevertheless, numerical simulations using published values for actinic light and fluorescence have shown a significant reduction in the differences between depth-independent light curves and those deconvoluted from depth-integrated curves (Serôdio 2004).

3 Down Regulation Through Non-photochemical Quenching

3.1 NPQ and the Xanthophyll Cycle in Diatoms

Diatoms of MPB biofilms are submitted to an extreme light environment which includes exposure to both high visible and UV irradiances, and fast and unpredictable light fluctuations. Additionally, other environmental pressures (nutrient limitation, extremes in temperature and salinity, etc.) can slow down the photosynthetic machinery and create a situation where the photosynthetically converted light energy cannot be entirely used for metabolic purposes. In order to maintain their photosynthetic productivity at an optimal level by preventing photoinhibition (i.e. decrease in quantum efficiencies F_v/F_m and $\Delta F/F_m'$), diatoms need to acclimate to environmental changes through fast regulation of their photosynthetic activity (Lavaud 2007). Together with the PS II electron cycle (Lavaud et al. 2002c), the non-photochemical quenching of chlorophyll fluorescence (NPQ) is believed to be one of the most important of these ‘photoprotective’ (or ‘photoacclimative’) mechanisms in diatoms (Lavaud 2007). Most of the diatoms of the MPB are capable of vertical migration into the sediment as a behavioral photoprotective strategy in order to avoid excess light exposure at the surface. However, vertical migration and NPQ appear to be complementary, NPQ induction may occur before the onset of the migratory response (Serôdio et al. 2008; Underwood et al. 2005), although recent work using chemical inhibitors of migration and NPQ suggests that migration may occur instead of NPQ induction (Perkins et al. in preparation).

The NPQ mechanism is described in Chapter 1 of this volume. qE, the energy-dependent quenching, which is regulated by the build-up of a transthylakoid proton gradient (ΔpH) and the operation of the xanthophyll cycle (XC) (Lavaud et al. 2002b; Lavaud and Kroth 2006), remains the best known component and plays a major role in the regulation of diatom NPQ. The machinery triggering and controlling NPQ amplitude and kinetics is now well known (Goss et al. 2006; Lavaud 2007). The major characteristic of NPQ in diatoms is its amplitude (Lavaud et al. 2002a; Ruban et al. 2004) such that it can account for up to 90% of energy dissipation (Lavaud et al. 2002a). Estuarine species of the MPB diatoms show a higher (up to five times higher than plankton species, Fig. 6) and faster (10 s induction) switch on/off of NPQ (Serôdio et al. 2005, 2006a, 2008; Herlory et al. 2007; Lavaud et al. 2007; Cruz and Serôdio 2008).

The mechanism of NPQ in diatoms shows other specificities regarding the pigments and proteins that are involved as well as their spatial organization (see Lavaud 2007). In particular the XC, which consists of the enzymatic de-epoxidation/epoxidation of the couple DD-DT, is employed (Fig. 7, Lavaud 2007). Accumulation of photoprotective diatoxanthin, DT, depends on the concomitant activity of two enzymes, a

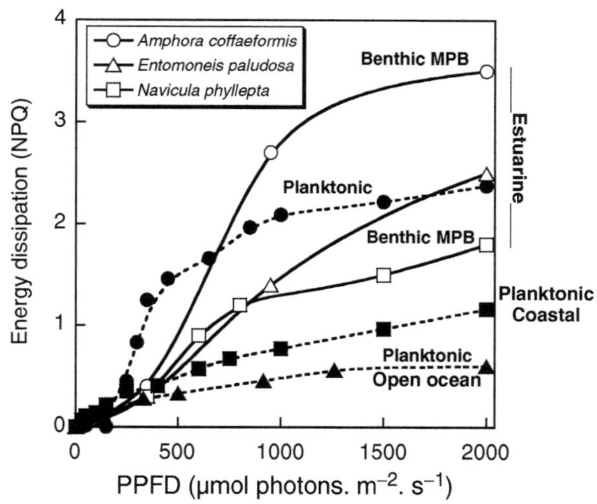

Fig. 6 NPQ versus irradiance (PPFD, Photosynthetic Photon Flux Density) in different species of diatoms : in black, planktonic species from the open ocean (*Thalassiosira oceanica*), coastal (*Skeletonema costatum*) and estuarine (*Phaeodactylum tricornutum*) habitats, in white three benthic species isolated from the microphytobenthos (MPB) in Baie de Bourgneuf (Atlantic coast, France). Adapted from Lavaud et al. (2007) (planktonic species) and unpublished data (benthic species)

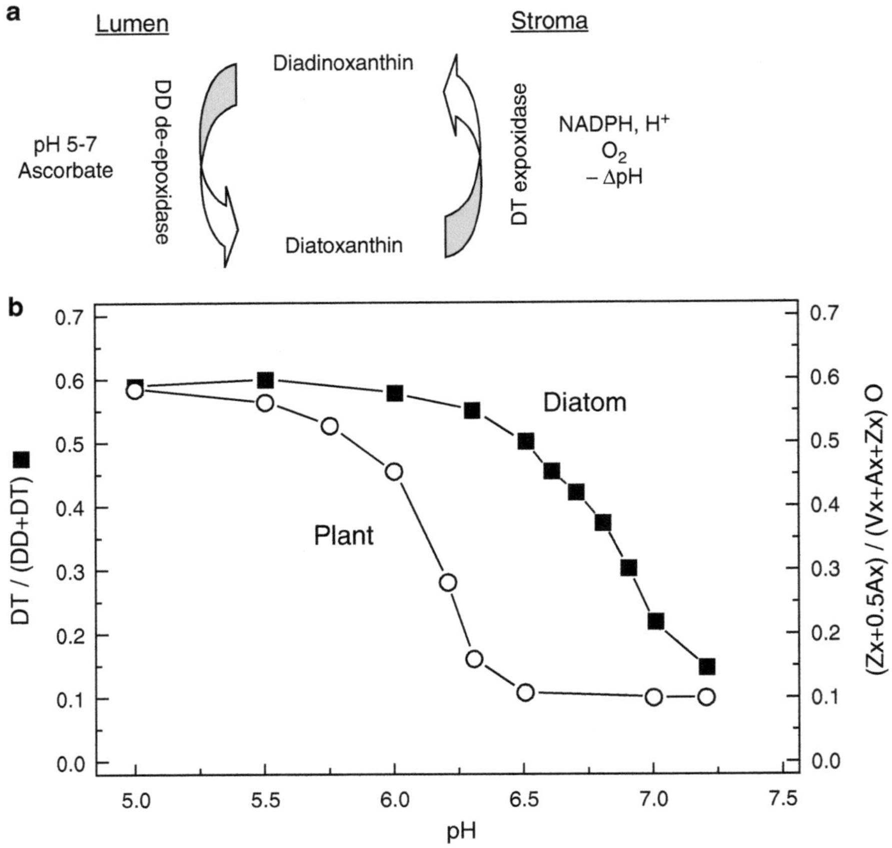

Fig. 7 (**a**) Simplified scheme of the xanthophyll cycle (XC) regulation. Co-factor requirement for the enzymes is shown as well as the pH optimum. – ΔpH means that the DT epoxidase is inhibited by high ΔpH under high light (Goss et al. 2006). DD, diadinoxanthin; DT, diatoxanthin (adapted from Lavaud 2007). (**b**) pH dependence of DD and VX de-epoxidation in isolated thylakoid membranes of the diatom *Phaeodactylum tricornutum* (open circles) and spinach (closed squares). De-epoxidation is calculated as [DT/(DD+DT)] for the diatom and as [(ZX+0.5AX)/(VX+AX+ZX)] for the plant. VX, violaxanthin; AX, antheraxanthin; ZX, zeaxanthin (Adapted from Jakob et al. 2001)

de-epoxidase and an epoxidase, the activity of which depends on the light intensity via the build-up of the transthylakoid ΔpH and the availability of co-factors (Fig. 7a).

The regulation of the XC in diatoms shows some striking peculiarities, the main one being the triggering of the diadinoxanthin, DD, de-epoxidase by a weak ΔpH (and thus a rather high lumen pH, Fig. 7b) (Jakob and Wilhelm 2001) so that the DD de-epoxidation occurs at lower irradiances and/or shorter illumination periods than in higher plants (Fig. 7b). Additionally, MPB diatoms can react to high light stress by accumulating large amounts of DD-DT (Rech et al. 2005; van de Poll et al. 2006; Schumann et al. 2007) in order to increase their photoprotection capacity via NPQ (Perkins et al. 2006; Schumann et al. 2007; Cruz and Serôdio 2008). All together, these differences in the regulatory components and mechanistic components of the NPQ process in diatoms have been suggested to ensure more flexibility and thus quicker response to the light environment (Lavaud 2007).

Field experiments have shown that the photoprotection ensured by both the XC and NPQ are essential for the MPB diatoms to maintain an optimal photosynthetic activity (Serôdio et al. 2005), even if not 100 % efficient (Serôdio et al. 2008). Additionally, the differential ability of species to cope (including NPQ) with prolonged high light and UV (Waring et al. 2006) is believed to potentially control spatial distribution (Fig. 6) (Lavaud et al. 2007) as well as species succession within MPB biofilms (Tuji 2000; Serôdio et al. 2005a; Underwood et al. 2005).

As described later in this chapter, RLCs have become a powerful approach to estimate the photosynthetic productivity and the photophysiology of MPB assemblages isolated from the field or directly *in situ*. Nevertheless, the assessment of NPQ on MPB via RLCs is rendered problematic by the confounding effects on F_m and F_m' levels of (1) downward vertical migration of the cells in the sediment, (2) the light and fluorescence attenuation in the upper layers of the sediment, (3) the contribution of fluorescence originating in "deeper" layers with a concomitant lower irradiance, and (4) a sustained NPQ under prolonged darkness (Serôdio et al. 2005; Jesus et al. 2006b). In order to solve this problem, Serôdio and co-workers proposed to assess NPQ following $\Delta\alpha_{RLC}$, which is the variation of the initial slope of RLC under high light exposure. α_{RLC} indeed linearly correlates with NPQ irrespective of measurement conditions, e.g. *ex situ*/*in situ*, summer/winter, low light/high light acclimation, etc. (Serôdio et al. 2005, 2006a, 2008; Cruz and Serôdio 2008) making it a good alternative to the classic calculation of NPQ as $(F_m-F_m')/F_m'$ (or $(F_0-F_0')/F_0'$, see Fig. 8a) (discussed in Serôdio et al. 2006a). Additionally, it has recently been reported that RLC construction itself can generate rapid endogenous changes of the photosynthetic activity which in turn modulates chl *a* fluorescence emission, and can potentially affect the ETR measurement and subsequent derivation of photophysiological parameters (Perkins et al. 2006; Cruz and Serôdio 2008). Together with the redox state of Q_A, NPQ has been shown to be one the main endogenous mechanisms potentially disturbing chl *a* fluorescence emission (Perkins et al. 2006; Jesus et al. 2006b; Herlory et al. 2007).

There are a number of physiological/technical features which can strongly influence NPQ amplitude and kinetics, and subsequently the shape of RLCs in MPB communities dominated by diatoms: (1) the species composition (Perkins et al. 2002; Serôdio et al. 2005a; Underwood et al. 2005); (2) the light/dark past history of the cells and hence the cells photoacclimation state (Perkins et al. 2006; Cruz and Serôdio 2008), (3) the accumulated light dose during the RLCs (Perkins et al. 2006; Herlory et al. 2007; Cruz and Serôdio 2008), (4) the fluorometer used (Perkins et al. 2006; Cruz and Serôdio 2008). Figure 8 shows that, in the MPB diatom species *Navicula phyllepta,* the length of each irradiance step (10, 30, 60 or 140 s) can influence

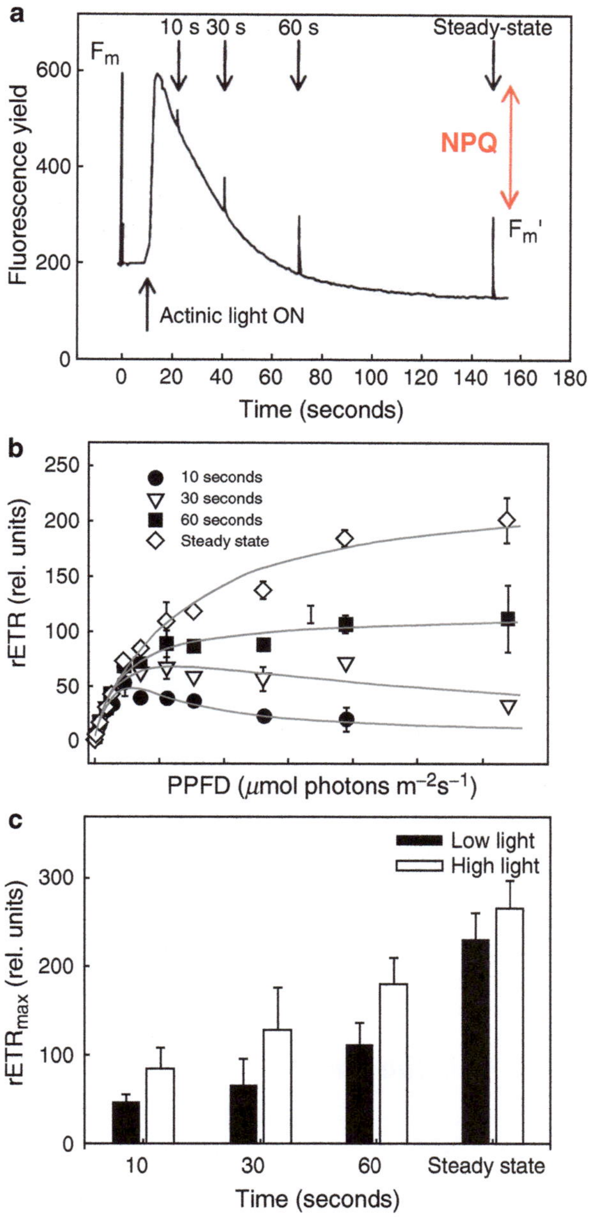

Fig. 8 (**a**) Fluorescence induction kinetics in a *Navicula phyllepta* low light culture (25 µmol photons m^{-2} s^{-1}) under a 370 µmol photons m^{-2} s^{-1} actinic light; application of saturating pulses to measure F_m' are indicated by downward arrows after 10, 30, 60 s and when fluorescence reached 'steady-state' (140 s); NPQ = $(F_m-F_m')/F_m'$. (**b**) Rapid Light Response Curves (RLCs) of the same low light culture acclimated at 100 µmol photons m^{-2} s^{-1} for 1 h. (**c**) Maximum Electron Transport Rate (rETR max) parameter as extracted from the RLCs built on N. *phyllepta* low (100 µmol photons m^{-2} s^{-1} for 1h) and high (400 µmol photons m^{-2} s^{-1} for 1 h) light acclimated cultures. PPFD, Photosynthetic Photon Flux Density (Adapted from Perkins et al. 2006)

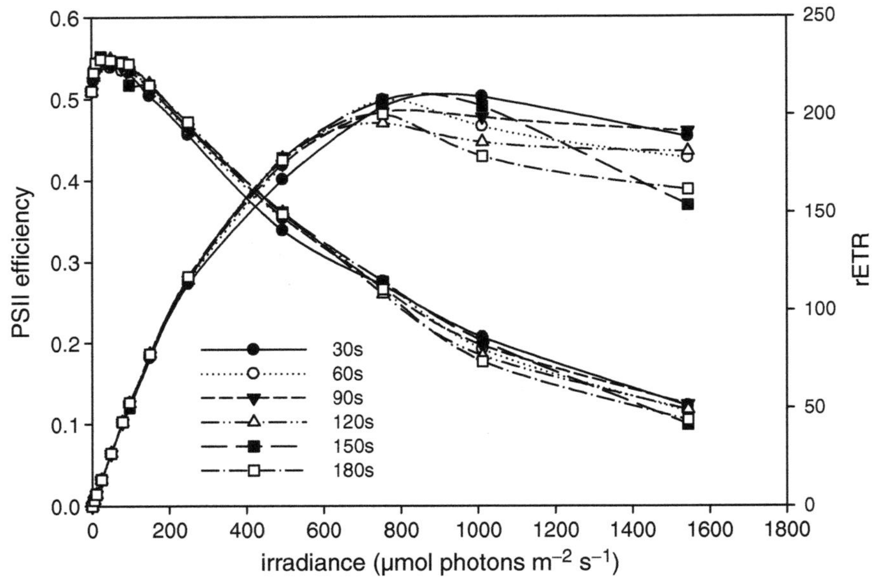

Fig. 9 Rapid Light Response Curves (RLCs) made in situ on an intact microphytobenthos (MPB) community (after 15 min dark adaptation) in the Eden estuary (Scotland). Each light step lasted 3 min and the PS II efficiency ($\Delta F/F_m'$) was measured at 30 s intervals. Despite the fact that F and F_m' only stabilized near the end of the 3 min exposure at each light step, $\Delta F/F_m'$ stabilized already in less than 30 s and rETR measurement was only disturbed for the highest intensities (over 800 μmol photons.m^{-2}.s^{-1}) whatever the length of the light steps (from 30 to 180 s) (Kromkamp J., unpublished data)

partly the level of NPQ induction (Fig. 8a), the profile of RLCs (Fig. 8b) and the subsequent determination of ETR_{max} as a function of the light history of the cells (Fig. 8c). However, it should be mentioned that this phenomenon is not always observed on intact MPB communities *in situ* as shown in Fig. 9.

3.2 NPQ in the Dark

In diatoms, NPQ occurs not only during light exposure but also during prolonged darkness (several hours; Jakob et al. 1999; Consalvey et al. 2004; Serôdio et al. 2005). Dark NPQ is due to chlororespiration, the amplitude of which is especially high in diatoms (Dijkman and Kroon 2002; Lavaud et al. 2002c). This pathway allows electrons to flow from the NADPH, H^+ to O_2, both synthesized in light, via the plastoquinone (PQ) pool: it is the respiratory chain of the plastids (Kuntz 2004). In diatoms, the chlororespiratory pathway is switched on very rapidly after the onset of darkness (Dijkman and Kroon 2002). Its amplitude and duration directly relates to the irradiance and duration of the former illumination through accumulation of reducing equivalents (Lavaud et al. 2002c), in other words it depends on the past light history of the cells (Cruz and Serôdio 2008). By transferring electrons through PQ, the chlororespiratory pathway generates, in the dark, the build-up of a transthylakoid ΔpH (Ting and Owens 1993). This dark ΔpH has been shown to be sufficient to activate the DD de-epoxidase and to drive the synthesis of DT, and the development of a large NPQ (up to 2) (Jakob et al. 1999; Jakob and Wilhelm 2001; Serôdio et al. 2005, 2006b). This is only possible because in diatoms, as specified above (Fig. 7b), the DD de-epoxidase needs only a weak acidification of the lumen to be activated, even though the rate of chlororespiration and subsequent change in the ΔpH are low (Jakob and Wilhelm 2001). Such a process would allow the cells to prevent photoinhibition during a subsequent sudden exposure to high light by keeping activated the dissipative function of the LHC system; an obviously adaptive advantage for the diatom species growing in a fluctuating light environment. Dark NPQ is even more physiologically relevant for diatoms of the MPB which can spend more than 18 h per day in the dark in the sediment before migrating to the surface and experiencing high light exposure (Serôdio et al. 2005).

As well as the NPQ which develops in light, the dark NPQ generates a methodological problem due to our inability to instantaneously measure the 'true' F_m (and F_0) level during dark-adaption of the cells *in situ* where their past light/dark history is usually unknown. Indeed, to achieve correct measurement of fluorescence parameters, complete Q_A oxidation and NPQ relaxation are required, which is usually reached after a short (15 min) dark-adaptation in controlled laboratory conditions but which might not be enough *in situ* (Perkins et al. 2001; Consalvey et al. 2004; Jesus et al. 2006a). Dark NPQ can easily quench F_m by at least 10–15% depending on the diatom species (Jakob and Wilhelm 2001). When the cells are further exposed to a low irradiance (below 150 µmol photons m^{-2} s^{-1}, Mouget and Tremblin 2002) during the measurement following dark-adaptation (typically at the beginning of RLCs acquisition), the whole photosynthetic machinery is fully activated which (1) reoxidizes Q_A and the PS II, (2) dissipate the ΔpH and subsequently change the equilibrium of the XC, hence relaxing the fluorescence quenching (Consalvey et al. 2004; Serôdio et al. 2005, 2006a). Hence, it is rather common to observe F_m' level transiently higher than the dark F_m level (Mouget and Tremblin 2002; Consalvey et al. 2004; Serôdio et al. 2005, 2006a). As a consequence it significantly affects the measurement and calculation of many fluorescence parameters. It can also significantly perturb the use of F_0 and its changes as a proxy for the dynamics of MPB diatom biomass at the surface and within the sediment (Consalvey et al. 2004; Jesus et al. 2006a). Solutions have been proposed to rule this problem out: (1) measurement of the 'true' F_m level in the presence of DCMU which is only applicable in controlled laboratory conditions (see Chapter 12, paragraph 12.4), (2) short exposure to low dose of far-red or low light instead of dark-adaptation in order to reoxidize PS II and dissipate the ΔpH and NPQ (Consalvey et al. 2004; Jesus et al. 2006a), (3) use $F_m'{}_m$, the maximum F_m' value measured as the 'true' F_m, instead of the dark F_m level (Serôdio et al. 2006a, 2008; Cruz and Serôdio 2008).

4 The Quantification of the Microalgal Biomass Using Fluorescence

The quantification of the microalgal biomass of microphytobenthic biofilms is a methodological challenge, due to the thinness of the sediment photic zone, the large horizontal and vertical heterogeneity, and the rapid changes in microalgae near the surface due to vertical migratory movements. Serôdio et al. (1997) was the first study to investigate the possibility of using in vivo chlorophyll fluorescence to non-destructively quantify the microalgal biomass of microphytobenthic biofilms. They experimentally established a linear relationship between sediment *chl a* content and minimum fluorescence yield F_o, showing that large F_o variations in the dark represented changes in the amount of microalgae at the sediment surface, as the result of cell vertical migratory movements. The use of F_o as a biomass proxy was shown to be preferable to other fluorescence parameters such as F_m or F_m' because it varied the least with previous light history, temperature and microalgal group. F_o was shown to allow the estimation of the miroalgal biomass present in the photic zone of the sediment (defined as 'photosynthetically active biomass'; Guarini et al. 2000; Honeywill et al. 2002) but also of the microalgal biomass in the photic zone weighted by its contribution to depth-integrated photosynthesis (defines as 'productive biomass'; Serôdio et al. 2001). This method overcomes the operational difficulties with previously used methods, based on destructive and time-consuming procedures, and introduced considerable operational advantages, including the possibility to obtain repeated measurements in the same sample over time, without any physical disturbance of the sediment-air or sediment-water interface, and allowing the concurrent measurement of other variables in the same sample (Serôdio et al. 1997).

However, the determination of F_o in MPB biofilms is not without problems and has been the source of significant discussion (e.g. Consalvey et al. 2004; Jesus et al. 2006b, c). These problems are related to MPB vertical movements during the dark adaptation (DA) period and to problems concerning the presence of NPQ in the dark (exhibited by diatom dominated biofilms). The determination of fluorescence parameters requiring DA (F_o, as well other parameters and indices such as F_m or F_v/F_m) requires that PSII reaction centres and Q_A be in their fully oxidised form (Schreiber et al. 1986). To achieve this state it is conventional to place samples in full darkness until a DA steady state is reached. However, dark adapting microphytobenthic biofilms presents a number of specific problems unique to this type of community.

Microphytobenthic biofilms are know to exhibit a behavioural photo-regulation mechanism where microalgae migrate vertically within the sediment matrix as a response to changes in ambient light in order to maintain an optimum light environment and avoid photo-inhibition (e.g. Perkins et al. 2002; Jesus et al. 2006b, c). This response to changes in ambient light can hinder the determination of F_0 if vertical migration occurs during the DA period necessary for Q_A and PSII re-oxidation. This has been recognized as a potential problem since the first studies using fluorescence (e.g. Serôdio et al. 1997) and short adaptation periods have been suggested as alternatives. Several forms of determining the minimum fluorescence yield have been tested, depending on the length of DA used (F', F_0', F_0^2, F_0^5, F_0^{10}, F_0^{15}; the superscript number denoting the duration of the dark period in minutes). By far, the most commonly used parameter is F_0^{15} (Kromkamp et al. 1998; Underwood et al. 1999; Barranguet and Kromkamp 2000; Perkins et al. 2001; Hagerthey et al. 2002; Honeywill et al. 2002; Perkins et al. 2002; Consalvey et al. 2004; Defew et al. 2004; Jesus et al. 2006b, c), followed by 5 min (Serôdio et al. 1997, 2001, Serôdio 2003; Serôdio and Catarino 2000; Jesus et al. 2006b, c) and 2 min (Serôdio et al. 2006b, 2007, 2008). However, the 15 min of DA can be excessive in some biofilms inducing significant downward migration during that period (e.g. Jesus et al. 2006a, b, c), suggesting that shorter time periods might be preferable. In fact, Jesus et al. (2006c) compared the relationship between chl a with F_0', F_0^5 and F_0^{15} and found no significant differences from F_0' to F_0^{15} in muddy sediment assemblages, suggesting that light history had little or no effect in the epipelic assemblages and that no DA was necessary for a good relationship between chl a and fluorescence. However, in sandy sediments there was evidence of a light history effect and a 5 min DA period was necessary to remove this effect. Other authors (e.g. Serôdio et al. 2007) have found that shorter DA times (2 min) might be even better than 5 min for the determination of photosynthesis fluorescence indices that require the input of a minimum fluorescence yield parameter.

Intertidal biofilms exhibit the additional problem of not showing an homogeneous behavioural response to light stimulus throughout the tidal cycle, i.e. biofilms close to the beginning of the emersion period will tend to increase F_0 values during the DA period as a result of cells migrating to the sediment surface, and measurements close to the end of the emersion period will be very sensitive to darkness and cells will migrate downwards quickly over the DA period. Although, this does not seem to be the case in all estuaries (e.g. Kromkamp et al. 1998; Barranguet and Kromkamp 2000; Hagerthey et al. 2002; Honeywill et al. 2002). To reduce this problem it was proposed that measurements are taken closer to the middle of the emersion period and that a low light or far-red treatment is used instead to the dark treatment (Jesus et al. 2006b). It is not clear why low light and far-red light work similarly but both treatments have to be applied at a reduced photon flux quantity to work properly. Thus, it is possible that this reduced photon flux promotes the dissipation of the chlororespiration trans-thylakoid proton gradient (ΔpH) exhibited by diatoms in the dark (Ting and Owens 1993; Dijkman and Kroon 2002c; Lavaud et al. 2002c). The reduction in the ΔpH promotes the epoxidation of diatoxanthin into diadinoxanthin, therefore, decreasing the NPQ caused by diatoxanthin presence in the dark (Jakob et al. 1999; Jakob and Wilhelm 2001). Another advantage of low light is that it seems to promote the presence of cells at the surface, thus reducing the problem of migration downwards that occurs during the DA period (Jesus et al. 2006b).

5 Calculation of Electron Transport Rate: ETR v rETR

5.1 *Multiple and Single Turnover Methods*

The rate of electron transport by PSII depends on the amount of light absorbed by the antenna of PSII and the efficiency at which the absorbed light by PSII is used by the reaction centers (RCII) to drive charge separation. Basically two methods are used which are both based on the light doubling method originally proposed by Bradbury and Baker (1981) which we will call the multiple turnover (MT) method and the single turnover (ST) method and which relate to the description of the pulse-amplitude modulation principle by Schreiber et al. (1986) and the pump and probe method (Falkowski et al. 1986b) respectively. The differences between these two approaches have recently been reviewed by Kromkamp and Forster (2003).

5.2 *The MT-method*

The MT-method is usually used by scientists using the PAM family of fluorometers which uses a multiple turnover flash to measure the quantum efficiency of PSII. Because the flash duration used to measure the maximum fluorescence (F_m) is relatively long (50 ms – 1 s) it allows for multiple charge separations during the flash and will thus lead to fully reduced Q_A, Q_B and PQ-pool. When the PSII effective quantum efficiency ($\Delta F/F_m'$) is measured the absolute rate of electron transport per unit area (ETR^A) can be calculated as follows:

$$ETR^A = E \bullet A_{II} \bullet \Phi_{RC} \bullet \Delta F / F_m' \quad (1)$$

A_{II} is the fraction of the incident light (E) which is absorbed by PSII. Φ_{RC} is yield (in electrons) of reduced Q_A per trapped photon, i.e. the maximum quantum yield of photochemistry within PSII, and it is usually assumed to be equal to 1 (Kolber and Falkowski 1993). Note that ETR^A is expressed in μmol electrons s^{-1} m^{-2} of surface area. When working with higher plants it is assumed that leaves absorb approx. 85% of incident light, and about half of this is partitioned to PSII. Assuming that Φ_{RC} is close to 1, Eq. 1 can be rewritten as:

$$ETR^A \approx 0.43 \bullet E \bullet \Delta F \ / F_m' \quad (2)$$

Often the fraction of absorbed light is determined by measuring the transmittance of light through a piece of macroalgal thallus using a light sensor (Beer et al. 1998; Longstaff et al. 2002). Implicit in this assumption is that the measured signal is equivalent to the integrated fluorescence yield over the entire path length, an assumption which might not be met, certainly when several cell layers are involved. The artefacts associated with this assumption were discussed above where the effects of "deep layer" fluorescence are described.

When working with optically thin suspensions of phytoplankton or unicellular algal cultures a slightly different approach is taken: in this case the ETR is usually calculated per mg chl a (μmol e^- (mg chl a)$^{-1}$$s^{-1}$):

$$ETR = E \bullet a'^*_{PSII} \bullet \Delta F \ / F_m' \bullet \Phi_{RC} \quad (3a)$$

Where a'^*_{PSII} is the optical absorption cross section of PSII (here expressed in m^2 (mg chl a)$^{-1}$, which is the product of the cross section of a single PSII unit and the number of PSII per mg chl ($a'^*_{PSII} = a^*_{PSII} \bullet n_{PSII}$)). As it is rather difficult to measure the absorption cross section of PSII it is normally assumed that a'^*_{PSII} is half the optical cross section of the cells (a^* m^2 (mg chl a)$^{-1}$, i.e. it is assumed that 50% of the absorbed light is funneled to PSII and the other half to PSI). Below we will discuss this assumption. Thus, assuming Φ_{RC} equals 1 we can rewrite Eq. 3a as:

$$ETR = E \bullet a^* \bullet 0.5 \bullet \Delta F / F_m' \quad (3b)$$

The optical cross section can be easily determined using a spectrophotometer equipped with an integrating sphere, i.e. by using the filterpad method for natural phytoplankton which requires concentration on a filter before the absorption measurements can be made (Tassan and Ferrari 1998; Simis et al. 2005).

5.3 The ST-method

Researchers using a saturating single turnover flash to measure F_m' usually take the approach developed by Falkowski et al. (1986b) and Kolber and Falkowski (1993), based on the development of the pump and probe fluorometer, which was followed up by the Fast Repetition Rate Fluorometer (FRRF) (Kolber et al. 1998). Here the rate of photosynthetic electron transport is described as follows:

$$ETR = E \bullet a^*_{PSII} \bullet n_{PSII} \bullet q_p \bullet \Phi_{tm} \bullet \Phi_{RC} \bullet f \quad (4)$$

Here a^*_{PSII} is the optical cross section of a single PSII unit (m^2/mol PSII) and n_{PSII} is the number of PSII units per mg chl a, usually assumed equal to 0.002 (thus assuming 500 chl a molecules per PSII (Falkowski 1981; Kolber and Falkowski 1993). The photochemical quenching coefficient q_P (moles electrons transferred per mole photons absorbed by PSII) reflects the proportion of oxidized PSII centers and is often used as a proxy of the number of open reaction centers. This is, however, only true when the PSII centers are not connected (Kramer et al. 2004). The trapping efficiency Φ_{tm} is the efficiency at which trapped photons in the pigment bed are transferred to an open RCII and f is the fraction of functional PSII centers. The product of the optical PSII cross section and the trapping efficiency equals the effective PSII cross section (σ_{PSII}, units m^{-2} (mol PSII)$^{-1}$):

$$\sigma_{PSII} = a^*_{PSII} \bullet \Phi_{tm} \quad (5)$$

As the FRRF can measure the functional cross section σ_{PSII} from the rise of F to F_m during the induction flashlet sequence, Eq. 4 can be rewritten as

$$ETR = E \bullet \sigma_{PSII} \bullet n_{PSII} \bullet q_p \bullet \Phi_{tm} \bullet f \quad (6)$$

Equation 6 is the one that is proposed by Kolber and Falkowski (1993) and Kolber et al. (1998) for use with the pump and probe and FRR fluorometer. The fraction of inactive centers (f) was measured as $\Phi_{sat}/0.65$, where Φ_{sat} is the maximum PSII efficiency measured and 0.65 the assumed maximum PSII efficiency of healthy cells without non-functional PSII centers. Kromkamp and Forster (2003) argued that this factor should be omitted because non-functional PSII centers will affect the level of F_o', and this is already incorporated in the value of q_P.

5.4 Assumptions and Uncertainties

Fraction of light absorbed by PSII. Both the MT as well as the ST method usually use a number of a-priori assumptions. When using the MT method to calculate ETR it is necessary to know which fraction of the absorbed light is funneled to PSII. Using a combination of optical and biophysical techniques Suggett et al. (2004) tested the hypothesis that about 50% of the light is absorbed by PSII. For a large range of taxa (diatoms, green alga, haptophytes and a cryptophyte) the fraction of light absorbed by PSII varied between 0.48 and 0.58, justifying this assumption. However, in the pelagophytin diatom *Aureococcus anophagefferens* the PSII-antenna absorbed about 36% of the light and in two cyanobacterial *Synechoccocus* species PSII absorbed 25–32% of the total light. The fraction of light absorbed by PSII was independent of the growth irradiance. A similar approach was taken by Johnsen and Sakshaug (2007), but they used a slightly different scaling procedure to match the fluorescence excitation spectra to the absorption spectra in order to arrive at the fraction of light absorbed by PSII: their results show that 48–88% of the light absorbed by the photosynthetic pigments was absorbed by PSII, which is generally higher than the estimates obtained by Suggett et al. (2004). Whether this difference is entirely due to methodological question or to growth conditions and different algal species remains an open question, but clearly this topic requires further research.

Estimates of the number of PSII. When using the ST protocol the absorption is quantified by multiplying the measured functional cross section σ_{PSII} with the n_{PSII}. This latter factor is also rather difficult to measure, certainly using field material, and for this reason it is assumed that n_{PSII} equals 0.002, i.e. a PSII contains 500 mol chl a (mol PSII)$^{-1}$ (Kolber and Falkowski 1993), a value based on the determinations by the photosynthetic unit (PSU) size by Mauzerall and Greenbaum (1989). Assuming that a PSU contains 4 RCII it can be calculated from the data presented by the references in Table 1 that generally n_{PSII} varies between 500–725 mol chl a per mol PSII, although lower values have been reported for *Isochrysis galbana.*

Photoacclimation generally results in more n_{PSII} when the cells are grown in low irradiances (Table 1). Kromkamp and Limbeek (1993) observed that when

Table 1 Number of PSII (nPSII, mol chla (mol PSII)$^{-1}$) in different algae grown in low light (LL) or high light (HL)

Species	Growth conditions[a]	n_{PSII}	Source
Thalassiosira weisflogii	LLHL	723552	(Dubinsky et al. 1986)
Isochrysis galbana	LLHL	636264	(Dubinsky et al. 1986)
Prorocentrum micans	HLLL	725513	(Dubinsky et al. 1986)
Dunaliella tertiolecta	LLHL	710833	(Falkowski et al. 1981)
Skeletonema costatum	HLLL	590605	(Falkowski et al. 1981)
Skeletonema costatum	LL-sinusoidalML-sinusoidal	467376	(Kromkamp and Limbeek 1993)
Skeletonema costatum	LL-fluctuatingML-fluctuating	290237	(Kromkamp and Limbeek 1993)
Chlorella pyrenodiosa	LLHL	588390	(Myers and Graham 1971)
Emiliania huxleyi B11	LLHL	665488	(Suggett et al. 2007)
Emiliania huxleyi B92	LLHL	720528	(Suggett et al. 2007)

[a] LL= low light, ML = medium light and HL = high light growth conditions

the marine diatom *Skeletonema costatum* was grown in fluctuating light simulating vertical mixing, this resulted in smaller but more n_{PSII}. This makes sense as it allows the cells to both harvest the same amount of light as with large PSU, but it will result in a higher rate of maximal photosynthesis. Recently Suggett et al. (2004) compared estimates of n_{PSII} obtained using the ST turnover oxygen measurements with those obtained from a biooptical: $n_{PSII} = \sigma_{PSII}/a^*_{PSII}$: they observed a linear 1:1 relationship between both methods and for most eukaryotic algae n_{PSII} varied between 500–600 mol chl a per mol PSII. The pelagophyte *A. anophagefferens* had a larger n_{PSII} and cyanobacteria seem to contain smaller n_{PSII} (240–280 mol chl a per mol PSII).

Uncertainties in σ_{PSII}. Although the FRRF technique allows estimation of the measurement of the functional PSII cross section, it is necessary to stress that LED's (mainly blue) are used to induce the fluorescence induction curve from which σ_{PSII} is estimated. Without spectral correction of the effective absorption of the FRRF in relation to the underwater light field, this may lead to an overestimation of σ_{PSII} (Suggett et al. 2001).

5.5 Calculation of ETR in Microphytobenthos Studies

Most research measuring ETR on MPB or macroalgae have used a PAM-type fluorometer because a commercially available FRR-type fluorometer (such as Chelsea's Fast^{Tracka} or Satlantic's FIRe) that are able to measure σ_{PSII}, have not been available (RandD versions of both instruments have been used for coral reef research). As is clear from the above section on the MT-protocol, calculation of absolute rates of PSII electron transport requires knowledge of incident irradiance and the optical absorption cross section, and this is exceedingly difficult when working on benthic biofilms. For this reason the relative rate of ETR (rETR) is often calculated as $\Delta F/F_m' \bullet E$ as opposed to the absolute ETR calculated as $\Delta F/F_m' \bullet E \bullet a^*$. Because the value of a* (the proportion of light absorbed) will change as a result of photoacclimation (time scales of change are hours-days because they relate to de-novo synthesis or break-down of pigments) it is not always possible to compare rates of rETR between publications.

Morris and Kromkamp (2003) cultured the benthic diatom *Cylindrotheca closterium* at two different growth rates and compared the relationship between ETR and oxygen evolution at a range of different temperatures during steady state growth. In general they observed that the relationship between ETR and oxygen evolution was not very sensitive to short-term changes in temperatures. However the relationship of rETR versus oxygen evolution was rather different between low and high growth rate, but when they examined the relationship on the basis of absolute ETR, the differences were minor. This was due to the a* values of the two cultures being different between the two growth rates. Nevertheless, often changes in rETR reflect changes in absolute ETR (Fig. 10): three out of the four different algal species showed a similar pattern in change in $rETR_{max}$ and absolute ETR_{max} (expressed per cell), despite large changes occurring in the photosynthetic physiology after transfer from replete to a P-free medium. The exception was *Emiliania huxleyi* which showed an unexpected increase per cell, because the optical absorption cross section per cell increased. If the absolute ETR was expressed per unit chl-*a* all cells showed a good correlation between the changes in rETR and ETR (data not shown). This suggests that with some care it may be possible to deduce changes in photophysiology from relative rates of ETR, but definitely more research is needed to confirm this.

A possible way to obtain MPB cells in order to measure a* spectrophotometrically would be to use the lens tissue method (Eaton and Moss 1966): however, this will only select for a fraction of the migrating species, and the observed a* value might not be representative for the total MPB community. In order to avoid this problem, Morris et al. (2008) reconstructed a* from HPLC pigment analyses using the procedure of Bidigare (1990) and compared the absorption values to the ones measured using a filter pad method (collecting cells by filtration). From this it was concluded that the package effect reduced the maximum absorption spectra obtained from the HPLC data by about 30%, in line with other studies on MPB (Mercado et al. 2004) and phytoplankton (Berner et al. 1989; Johnsen and Sakshaug 2007) although Nelson et al. (1993) reported that less than 25% of the phytoplankton in Californian coastal waters showed a measurable package effect. Using the reconstructed and measured a* values, Morris et al. (2008) demonstrated that the

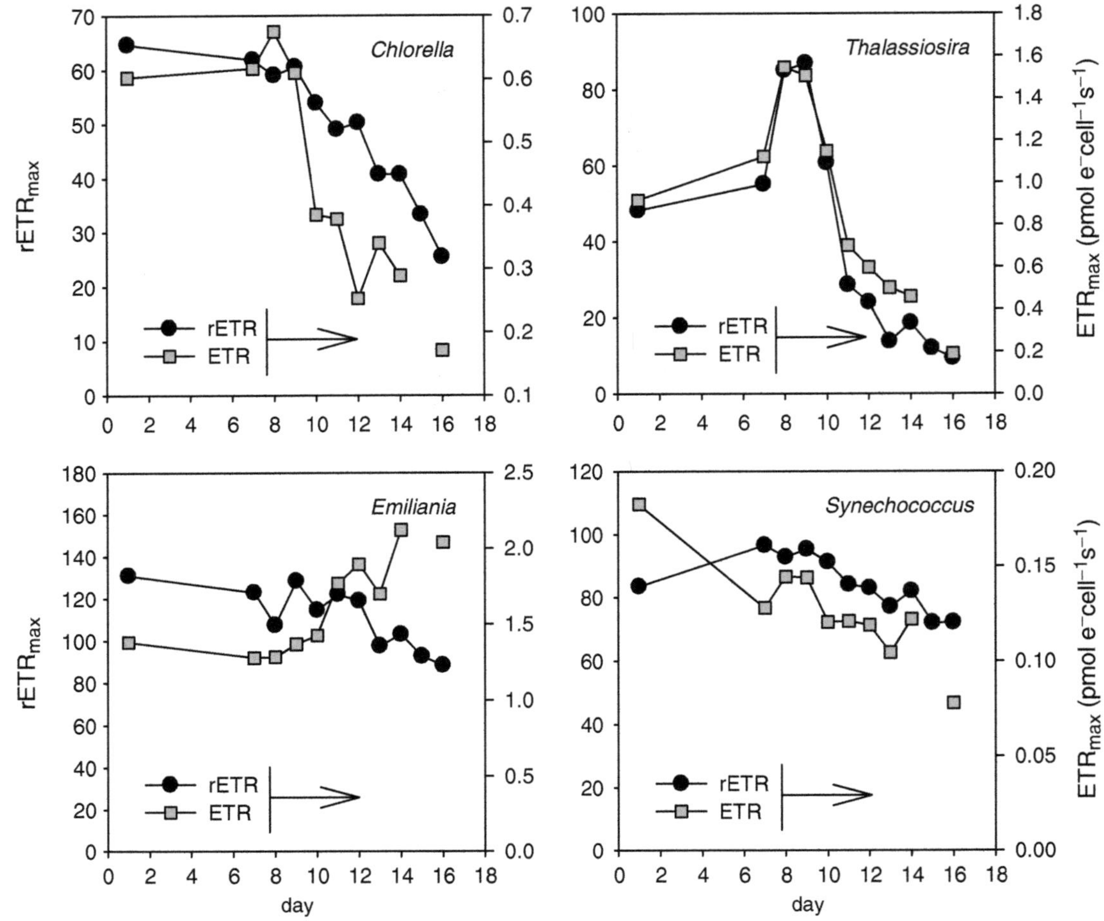

Fig. 10 Comparison between $rETR_{max}$ (left Y-axis) and absolute ETR_{max} (right Y-axis, expressed per cell) for four different algal species: *C. autotrophica, T. pseudonana, E. huxleyi* and *Synechococcus* CCY9502. On day 7 (arrow) the cells were resuspended in phosphate-free medium and were thus starved for P (Kromkamp J., unpublished data)

PAM derived quantum efficiencies for C-fixation (Φ_C), on both intact diatom biofilms as on sediment slurries with defined optical conditions, matched the measured Φ_C very well. Only when growth rate of the diatom biofilm slowed down when the biomass reached an apparent steady state did the PAM-derived Φ_C of the biofilm overestimate the measured Φ_C of the sediment slurry, whereas Φ_C of the slurry matched the measured Φ_C. Most likely the overestimate of the true Φ_C was caused by fluorescence derived from deeper layers (see above). This suggests that quantification of ETR on MPB biofilms is possible using reconstructed a* values from HPLC derived pigments, provided the diatoms in the biofilm are actively growing and the biofilm does not reach a high biomass (i.e., a particularly thick biofilm).

6 Light Response Curves

6.1 A Brief Overview of Methodology

Light response curves (the photosynthesis – irradiance curve following Falkowski and Raven (1997) – are used to determine a range of photophysiological and productivity parameters. Incremental increases or decreases in the actinic light environment (PAR, usually applied by the flourometer in use) are applied, with measurement of the quantum efficiency ($\Delta F/F_m'$). The well known and accepted equations for calculation of ETR are applied and the resulting values plotted as a function of PAR. The result is a saturating curvilinear response, encompassing a near linear light limited phase, a light saturated phase and then, in some instances, a third phase which is not always present and often attributed to down regulation/photoinhibition. Various methods of curve fitting have been applied to this curve, from which physiological parameters can be derived, including the method of Eilers and Peeters (1988) followed by curve fitting such as that following the Nelder-Mead model (Press et al. 2003). Parameters derived are usually:-

1. $rETR_{max}$ – the maximum relative electron transport rate when light becomes saturating
2. α – the maximum light use coefficient
3. E_k – the light saturation coefficient, calculated as $rETR_{max} / \alpha$
4. β – the coefficient of down regulation / photoinhibition

These parameters, except β, along with the PAR at which saturation occurs (E_s) can be calculated using the parameters derived from various models such Eilers and Peeters (1988).

The application of the methodology to benthic biofilms has been reviewed in Consalvey et al. (2005), but here it will be discussed principally to compare the three main types of light response curves: steady state light curves (SS), rapid light curves (RLC) and non-sequential light curves (NSLC). Light curve parameters ($rETR_{max}$, α etc.) are mostly interpreted in similar ways for all light curves, except for the coefficient β. In SS light curves the decline in ETR indicated by the value of β is attributed to photoinhibition, whereas in RLCs β is a measurement of down regulation as the short duration of the light curve is not thought sufficient to induce photoinhibition (White and Critchley 1999). The other parameter worth mentioning here is $rETR_{max}$, where the letter (r) preceding ETR denotes whether the measurement is of "relative" electron transport rate, i.e. in the absence of any measurement of the light absorption coefficient a* (Sakshaug et al. 1997; Beer et al. 2000; Forster and Kromkamp 2004; Perkins et al. 2006). In most instances, measurement of a functional value of a* is highly problematic for benthic biofilms, due to interference from the sediment matrix and the time required for measurement. As a result the majority of work has used rETR and hence determined $rETR_{max}$. Data are often considered to be comparative as a result of this relative measurement, as opposed to absolute values for comparison to other measurements of productivity (e.g. ^{14}C, O_2 described later in this chapter).

6.2 Steady State Light Curves

Steady state (SS) light curves are those where the operational fluorescence yield (F or F′) is allowed to reach a stable value after each incremental increase / decrease in PAR (Falkowski and Raven 1997). Thus the effects of Q_A reduction/oxidation and NPQ induction/reversal are allowed to reach completion for the light environment applied. Once this has been achieved the saturating pulse is applied to determine the rise in fluorescence yield to F_m' and hence calculation of $\Delta F/F_m'$. SS light curves are often considered to be a measurement of the potential photophysiology of the cells at the time of measurement, as opposed to their actual operational photophysiology at that time. Hence, they reflect the capacity for photosynthesis and for this reason allow comparison between SS light curves of different species and under different environmental conditions. SS light curves can thus be used for examination of phenotypic (light history exposure) and genotypic (species specific variation) adaptation. For example, the effects of light dose history prior to the light curve (phenotypic adaptation) are greatly changed, or possibly negated, due to the photoacclimation during each step of the light curve. Obviously some aspects of prior photoacclimation will remain, such as changes in pigment bed which take longer to be modified by the cells. In fact the time to reach steady state is a function of light history; e.g. the rate and magnitude of NPQ induction / reversal is a function of previous light dose history. This time period is therefore highly variable, and can be in the order of several minutes or longer. This has several drawbacks when applied to benthic biofilms:

1. The long duration of the light curves (up to 40 min for an 8 step light curve[1]) often makes replication of measurements impossible.
2. Investigation of temporal or spatial variation is prohibited by the long time period spent obtaining a single set of measurements.
3. Changes in biofilm surface community may occur over this period of time due to microcycling (e.g. Kromkamp et al. 1998) and positive or negative

[1] Walz fluorometers such as the Diving-PAM and Water-PAM are programmed to have 8 step light curves. However other manufacturer's fluorometers differ and can allow the number of steps to be dictated by the user.

Fig. 11 Selected HRFI fluorescence images (at *F*′) used in the construction of light response curves Perkins et al. (2002). Image (**a**) was obtained near the start of the light curve, at 380 µmol photons m^{-2} s^{-1} PAR and image (**b**) at 1150 µmol photons m^{-2} s^{-1} PAR. Note the change in surface biofilm from a mixed diatom and euglenid community to just *Euglena* spp. in less than 8 min. Scale bar is 100 µm

phototaxis (e.g. Serôdio et al. 2006a) induced by the applied PAR, as well as possible diel and tidal patterns in vertical migration (see elsewhere in this chapter), hence measurements at incremental light steps may be made on different cells.

As a result, SS light curves are usually only used with benthic biofilms to determine light parameters in theoretical investigations, rather than in determination of absolute values of light curve parameters. However even in these situations the issue mentioned in Point 3 above can make the light curve largely uninformative, or at least difficult to interpret due to the rapid changes in community structure between light steps. Perkins et al. (2002) demonstrated the complete change in surface community in a benthic biofilm using high resolution fluorescence imaging (HRFI) in an 8 min light curve, changes being observed in surface diatom taxa and finally a complete change to a *Euglena* spp. dominated surface biofilm (Fig. 11). As a result SS light curves are often limited to work using cultures of cells or engineered biofilms; in both cases migration is inhibited, or preferably prevented (e.g. Jesus et al. 2005, Perkins et al. 2006; Mouget et al. 2008).

6.3 Rapid Light Curves

The definition of a RLC is somewhat difficult: when does a light curve become “rapid”? For example Kromkamp et al. (1998) used light curves that were not SS curves, but had light steps of 2 min duration. This followed on from the first “rapid” light curve work by Schreiber et al. (1994) and Schreiber et al. (1997), the former using 5 min steps, the latter using truly rapid light curves of 10 s. It is the opinion of the authors that to be a true RLC, light steps should be of less than 60 s duration. Using this definition, RLCs were not applied to benthic biofilms until the work of Serôdio et al. (2005b), with work on seagrass reviewing the use of RLCs by Ralph and Gademann (2005) in the same year. The methodology was further investigated using benthic diatom cultures and microphytobenthos suspensions (Perkins et al. 2006; Serôdio et al. 2006a; Herlory et al. 2007; Cruz and Serôdio 2008) and is now an accepted method for field and laboratory measurements of operational photophysiological state.

In some cases RLCs can be considered to be a compromise between SS light curves and practical limitations of data acquisition and replication. This however is an error as RLCs measure an entirely different photophysiological state, namely the operational photophysiology at that time. The duration of light steps used is often a compromise between practical time limitations and obtaining the required measurements. The longer the duration of each light step, obviously the longer the cells exposed to this new PAR will have to photoacclimate and the measurement becomes an intermediate between a RLC and a SS light curve. Indeed many workers now prefer RLCs between 10 and 30 s at each light step, with 60 s considered too long due to effects of photoacclimation during the light curve. Also an important consideration, as discussed above, is that the longer each light step, the longer the cells will have to migrate in response to the applied PAR or as part of diel and tidal patterns of vertical migration, hence altering the community structure being investigated during the light response curve itself.

Typically RLCs are of 10, 20 or 30 s at each light step and often have 8 incremental steps[1] after an initial measurement in darkness (dark adaptation prior to the light curve is discussed below). Thus RLCs may be as short as 80–240 s. However migration and photoacclimation will still occur to some extent, with cells migrating vertically at speeds such as 0.17 – 0.28 μm s^{-1} (Consalvey et al. 2004). Serôdio et al. (1997) estimated the effective measurement depth using fluorescence to be 270 μm and Kromkamp et al. (1998) estimated detectable fluorescence measurements from only 150 μm sediment depth. Both these estimates are well within the depths to which benthic diatoms migrate, demonstrating the ability of the surface community under investigation to change over even a RLC. Furthermore, Perkins et al. (2006) reported detectable photoacclimation by cultures of benthic diatoms in RLCs with light steps of 10, 30 and 60 s duration. In this case, the level of photoacclimation increased as a function of the increase in light step duration, especially for light curves with incremental increases in, as opposed to decreases in, PAR (Fig. 12). They further concluded that RLCs with decreasing PAR light steps were preferable to minimise this effect and that light steps of 20–30 s duration were preferable. However if samples have been dark adapted prior to the light curve, then back pressure on the electron transport chain due to the time required to reactivate dark reactions, e.g. RUBISCO activation, could result in under estimation of $\Delta F/F_m'$. It should also be noted that decreasing light steps are not appropriate for SS light curves in which reversal of NPQ is required.

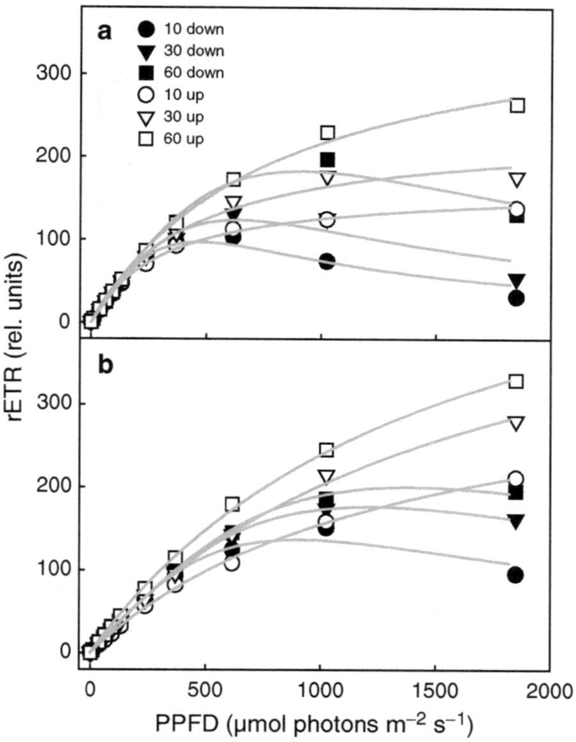

Fig. 12 Rapid light curves for cultures of *N. phyllepta* exposed to 1 h light acclimation period of low (**a**: 25 μmol photons m^{-2} s^{-1}) or high (**b**: 400 μmol photons m^{-2} s^{-1}) light. Light response curves were run with light step durations of 10, 30 and 60 s and with either increasing (up). Note the increase in $rETR_{max}$, lack of saturation or down regulation for light curves with incremental increases in light level, as opposed to curves with decreasing light levels where saturation and down regulation occurred (From Perkins et al. 2006)

Perkins et al. (2006) indicated that 10 s was not thought to be an appropriate duration due to limitations of measurements by the flourimeter used (Walz Diving-PAM control unit). When cells had been exposed to high light, their rapid induction of NPQ to quench the fluorescence yield after the incremental increase in PAR was such that there was a considerable change in operational fluorescence yield (F′) between measurement of F′ and application of the saturating pulse and subsequent measurement of F_m'. Where this happened, F_m' was under estimated relative to F' causing under estimation of $\Delta F/F_m'$ and hence the calculated value of ETR.

While this question demands further research, it is clear that the most appropriate light step to be used in each case will depend on the relative benefit of avoiding the mentioned artefact when applying 10 s, and of the confounding errors introduced by increasing the light step to 20s, namely short-term photoacclimation and migratory responses during the construction of the light curve. For example, the detection of diel rhythms in the photoacclimation status of microphytobenthos using RLCs has been shown to be possible when applying short light steps of 10 or 20 s, whilst becoming almost undetectable when using light steps of 40 s due to the rapid light-activation of dark-acclimated community (Serôdio et al. 2005).

Despite the mentioned potential problems, incremental RLCs using 10 s light steps were shown to allow characterization of the steady-state photoacclimation status of samples acclimated to a wide range of ambient irradiances, through the estimation of the light-saturation parameter E_k from RLC parameters (Serôdio et al. 2006a; Cruz and Serôdio 2008). These studies have also demonstrated that such short RLCs could effectively 'capture' the steady-state photoacclimation status formed prior to the RLC. RLC parameters responded to ambient irradiance (Fig. 13) in a well-defined way, with the initial slope of RLCs

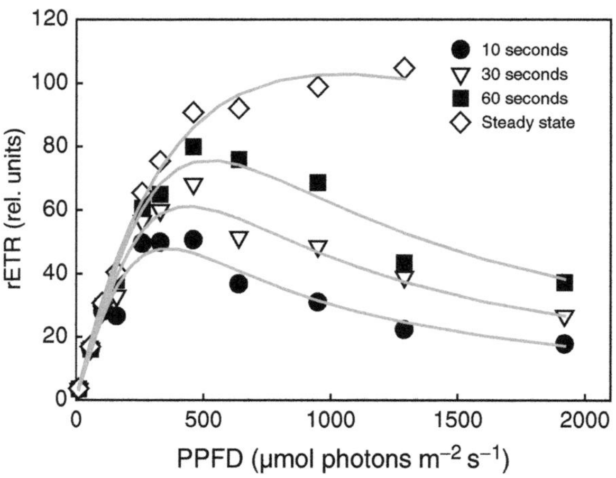

Fig. 13 Non-sequential Light Curves for cultures of *Navicula phyllepta* run with light step durations of 10, 30, 60 s and reaching steady state (SS). Note the increase in $rETR_{max}$ as a function of the duration of light step and the lack of saturation for steady state light curves (From Perkins et al. 2006)

decreasing with ambient irradiance (related to the build-up of NPQ; see section on XC), and maximum ETR increasing with ambient irradiance. As a consequence, RLCs could be related to SS light curve parameters, with significant correlations being found between the initial slopes of both light curves for samples acclimated to lower ambient irradiances, and between the maximum ETR of both curves for samples acclimated to higher ambient irradiances (Serôdio et al. 2006a).

In their application to benthic biofilms, RLCs have distinct benefits over traditional SS light curves. They minimise changes in community structure due to migration, as well as reducing methodologically induced photoacclimation to modify the photoacclimation state of the cells under investigation. However the authors suggest that care needs to be taken in their interpretation as they are comparative measurements that should only be compared between studies using the same methodologies. Data are only comparable so long as the duration of the incremental light steps used are the same. It should also be noted that the magnitude of light level used for each light step will also result in variation between studies; photoacclimation will be proportional to the total light dose experienced during the light curve and measurements with higher light levels will induce more photoacclimation over the duration of the light curve.

It is worth mentioning at this point the issue of dark adaptation prior to the light curve itself. In many studies light curves are preceded by a short period of dark adaptation. This is usually to obtain an approximate measurement of the maximum fluorescence yield F_m for calculation of parameters such as NPQ (where NPQ $= (F_m - F_m')/F_m'$, see elsewhere in this chapter). However it can also be argued that such dark acclimation modifies the photoacclimation state of the cells investigated and, if of sufficient duration, dark acclimation can induce vertical migration (Jesus et al. 2006a, b, c). However work by Kromkamp et al. (2008) indicated that a dark adaptation period of 1 min prior to the light curve did not affect the RLC data obtained in a series of light curves logged over an emersion period. It is probable that the effect of the dark period is proportional to the photoacclimation state of the cells, and hence on the photodose (light history effect) experienced by the cells, as well as the propensity for the cells to migrate as a function of changes in light environment. It is suggested that, to standardise methodologies and hence increase comparative capability, dark adaptation should be avoided prior to light curves. Comparative values of F_m and hence calculation of parameters such as NPQ may still be achieved as in most cases fluorometer programming results in the first light curve steps being in darkness and at low light. The former replaces the dark adapted measurement, and Jesus et al. (2006a) indicated that, for benthic diatoms which retain NPQ in the dark, low light reverses NPQ more efficiently than darkness, hence giving a value of F_m' which exceeds F_m in some instances (see elsewhere in this chapter). However, as with most methodologies, the use of dark adaptation will depend on what is being investigated, and in some studies reversal of NPQ prior to the light curve may be required.

6.4 Non-sequential Light Curves

The last form of light curve to be discussed here are NSLCs, in which each light step is applied to a separate culture sample or biofilm sample/area that has not been used for any previous measurements as part of the light curve (Perkins et al. 2006; Herlory et al. 2007). For example, the fluorometer probe is placed over one (sub)sample for the selected light step duration, at the end of which $\Delta F/F_m'$ is obtained. The probe is then moved to the next (sub)sample for the next light step of the light curve. There is no requirement here for the measurements to be sequential, but for this methodology dark adaptation prior to the light curve is not possible. In most cases this is a theoretical light curve as it requires an extensive set of replicate samples (e.g. cultures) or, in the case of field measurements, a substantially large area of biofilm which is considered to be homogeneous enough to allow a large number of replicate measurements. This method should therefore almost totally remove the photoacclimation induced by the light curve and, with enough replication, can be considered to be measuring the true operational photoacclimation state of the cells. The disadvantage is the number of replicates required.

Perkins et al. (2006) applied this method to cultures of benthic diatom cells, principally the species *Navicula phyllepta*. Despite using different culture sub samples for each light step, photoacclimation still occurred, with $rETR_{max}$ proportional to the increasing length of each light step. Notable however was that down regulation was observed at irradiances above the point where the curve saturated when light steps were of short duration (e.g. 10 and 30 s), and also that SS light curves did not saturate at all and showed the highest $rETR_{max}$ values. This demonstrates the rapid rate of induction of photoacclimation processes during the light curve steps themselves. Also reported by Perkins et al. (2006) was the effect of light history, when cultures incubated at high light level (400 μmol photons $m^{-2}\ s^{-1}$ PAR) prior to the light curve measurements did not saturate or show down regulation, but cultures incubated at low light (25 μmol photons $m^{-2}\ s^{-1}$ PAR) showed both saturation and down regulation for 10 and 30 s light curves, saturation for 60 s curves, but no saturation for steady state curves. The rate and level of photoacclimation processes was therefore proportional to light history and the length of the light curve steps used. An important point that should be mentioned at this point is that the light curves referred to in Perkins et al. (2006) used rETR. It would be expected that as photoacclimation increased, a higher level of NPQ induction would result in a lower effective ETR. Thus if the cross section of PSII had been measured to determine the "true" ETR as opposed to rETR, the ETR_{max} may decrease as a function of increasing light step duration. This clearly needs further investigation, although the measurement of PSII cross section is difficult for benthic biofilms during light curves.

6.5 Light Curves Summary

The use of steady state light curves has been a powerful tool in the use of fluorescence to investigate photophysiology of photoautotrophs, but due to the long duration required for measurement to be achieved, is not practical for use with benthic biofilms, especially those exhibiting rapid migratory activity in response to changes in light environment or as part of diel and tidal rhythms. It is suggested here that RLCs of 10–30 s duration with no prior dark adaptation step should be utilised, the particular light step to be applied being dependent on the previous evaluation of the relative magnitude of the errors introduced by (i) photoacclimation and (ii) migration during the RLC, and (iii) fluorometer software, factors varying with sample composition and light history, light levels applied during the RLC, and instrument characteristics. Furthermore, RLCs starting at high light and proceeding to lower irradiance may represent an advantageous alternative to minimize the effect of cumulative light dose on the PSII quantum efficiency, although systematic comparative studies are still missing. Having said this, this would require a change in the setup of the automated LC used to be made with Walz-PAM fluorometers, as currently the models mostly used in the field (mini-PAM, Diving-PAM etc.) can only make LC from low to high light. Obviously this does not rule out further work in which different durations of light step are used to investigate photoacclimation etc., indeed this is highly recommended by the authors, especially with regard to the use of RLCs *in situ*.

7 Comparison of Fluorescence with Other Methodologies

In Section 5 we described how absolute rates of electron transport can be calculated. However, ETR is not the common currency in primary production studies. Photosynthetic rate measurements in aquatic organisms usually imply recording of either O_2 emission (e.g., oxymetry, mass spectrometry) or CO_2/HCO_3^- uptake (e.g., ^{14}C labelling, infra-red gas analysis) and in the marine environment as well as in ecosystem models C is the most often used. A comparison between ETR and the rate of oxygen evolution or C-fixation (P^B) can be obtained from a comparison of both methods: when P^B is plotted as a function of ETR the regression coefficient (providing a linear relationship is obtained) is the conversion factor needed to convert ETR into P^B and thus equals the efficiency electron yield (Φ_e) for O_2 evolution or C-fixation (mol O_2 produced or CO_2 fixed per mol electron produced. Thus:

$$P^B = \Phi_e \bullet \mathrm{ETR} \tag{7a}$$

Theoretically the optimal value for Φ_e is 0.25, which means that in the case of O_2-evolution four absorbed photons give rise to four charge separations and thus the production of four electrons. In the case of C-fixation we need to know the photosynthetic quotient (PQ) if we want to convert rates of O_2-evolution to rates of C-fixation. Thus:

$$P^B(\mu\mathrm{mol}\ O_2\,(\mathrm{mg\,chl}a)^{-1}\mathrm{s}^{-1}) = \Phi_e \bullet \mathrm{ETR} \tag{7b}$$

$$P^B(\mu\mathrm{mol}\ C\,(\mathrm{mg\,chl}a)^{-1}\mathrm{s}^{-1}) = \Phi_e \bullet \mathrm{ETR} \bullet \mathrm{PQ}^{-1} \tag{7c}$$

As explained above, two different methods of active fluorescence have been developed: the Pulse Amplitude Modulated (PAM) principle which uses a MT-flash to measure F_m and the Fast Repetition rate Fluorometry (FRRF).[2] It should be noted that when the MT method is used (Eq. 1, 3) with a fluorometer giving a ST-flash, the F_m (F_{m-ST}) measured is lower than that measured with a MT-flash (F_{m-MT}). As a consequence a lower PSII efficiency is obtained and thus a lower rate of ETR, because a ST flash leads to a single reduction of all Q_A, whereas an MT flash also reduces the Q_B and PQ-pool, leading to a higher F_m (Schreiber et al. 1995; Koblizek et al. 2001; Suggett et al. 2003; Kromkamp and Forster 2003). ETR measured using a PAM fluorometer can be up to ca. 35% higher than ETR measured using a FRRF (Suggett et al. 2003; Kromkamp and Forster 2003).

At room temperature, most of the fluorescence emission originates from PSII and because charge separation and the splitting of water in the oxygen evolving complex both take place in RCII it might be expected that ETR and the rate of gross O_2-evolution show a tight coupling. As carbon fixation (the "dark reactions") takes place 'later' and is a 'more distant' step of the photosynthetic process, a more loose relationship might be obtained. Some diatoms seem also be able to use a C4-like C-fixation process using phosphoenolpyruvate carboxykinase (PEPCKase) or phosphoenolpyruvate carboxylase (PEPCase) (Cabello-Pasini et al. 2001; McGinn and Morel 2008) but whether MPB diatoms also possess this capacity has not been investigated yet.

Table 2 presents a compilation of studies where rates of ETR are compared to rates of photosynthesis obtained using more classical methods. Only a small number (12%) of these studies deal with biofilms. Most of these studies (85%) concern marine organisms, and ca. 70% relate to field studies or use natural samples photoacclimated in the laboratory for different periods of time. Natural phytoplankton studies mainly used the ST-approach; five studies used the

[2] Several manufacturers make PAM-like fluorometers (e.g. Walz, PSI, Hansatech, Moldaenke). Fluorometers using the ST-protocol are made by Chelsea Instruments, Satlantic, PSI (and some Walz models). The former 2 use the FRRF protocol and can apply MT and ST flashes.

Table 2 Comparison of photosynthetic parameters in aquatic organisms obtained with fluorescence-based methods and with carbon- and oxygen-based methods

Authors (year)	Methods		Organisms	Growth irradiance / photoacclimation	Observed relationship	Hypothesis for NL	Remarks
	Fluorescence	O_2 / ^{14}C			Linear*/Non Linear, Low-, Medium-, High-Irradiance		
Babin et al. (1996)	FRRF	^{14}C	Phytoplankton Natural samples	Ambient conditions	NL	Nutrient limitations	
Barranguet and Kromkamp (2000)	PAM-101	^{14}C	MPB Natural samples	Ambient conditions	Mostly L, sometimes deviations at HINo significant correlation for P_m, E_k or α between PAM and ^{14}C data	Migration, change in a* and light absorption by sediment	Measurements in parallel
Beer et al. (1998)	Diving-PAM (LC, RLC)	O_2	Seagrasses (*Zostera marina, Cymodocea nodosa, Halophila stipulacea*)	Ambient conditions	L (+, =, –) or NL (deviation at HI), depending on the species	Photorespiration and other O_2 consuming reactions	Simultaneous measure-ments (LC)
Beer and Bjork (2000)	Diving-PAM	O_2	Seagrasses (*Halophila ovalis, Halodule wrightii*) Natural samples	Ambient conditions	L (+, =) or NL (deviation at HI), depending on the species	Photorespiration, estimation of light absorbed by PSII	Simultaneous measure-ments
Beer et al. (2000)	Diving-PAM, PAM-101	O_2	Macroalgae (*Ulva lactuca, U. fasciata*) Natural samples	*U. lactuca* : 1700 μmol photons m^{-2} s^{-1} *U. fasciata* : ambient conditions or 50 or 150 μmol photons m^{-2} s^{-1}	L (=)		Simultaneous measure-ments
Beer and Axelsson (2004)	PAM-101, Diving-PAM	O_2	Macroalgae (*Ulva lactuca, Fucus serratus, Laminaria saccharina, Palmaria palmata, Porphyra umbilicalis*)	Acclimation to 400 μmol photons m^{-2} s^{-1}*L. saccharina* : 100, 400 μmol photons m^{-2} s^{-1} *U. lactuca* : 200, 1700 μmol photons m^{-2} s^{-1}	Mostly LNL at HI	Φ_{PSII} lower than 0.1	Simultaneous measure-ments
Boyd et al. (1997)	PPF	^{14}C	Phytoplankton assemblages (field), dominance of coccolithophorids	Ambient conditions	NL for αL for E_k Partially L (–) for P_m	Limitations in the determination of some parameters (esp. bio-optical), differences in light sources and measurement duration	Significant correlations observed, but parameter values differed

(continued)

Table 2 (continued)

Authors (year)	Methods		Organisms	Growth irradiance / photoacclimation	Observed relationship	Hypothesis for NL	Remarks
Cabello-Pasini and Figueroa (2005)	Diving-PAM	O_2	*Ulva rigida*	Different DIC, irradiances and temperatures	Mostly L (=)	Lower correlation at HI and high nitrate concentration	Simultaneous measure-ments
Carr and Bjork (2003))	Diving-PAM, PAM-2000 (LC, RLC)	O_2	*Ulva reticulata, U. fasciata*	50 μmol photons m^{-2} s^{-1}	NL, especially at HI or high DICL with fresh sample at each light step P_{mf}/P_m changing with I	Light history, changes in σ_{PSII}	Simultaneous measure-ments
Carr and Bjork (2007)	PAM-2000	O_2	*Ulva fasciata*	80 μmol photons m^{-2} s^{-1} Nutrient limitation	Stable ratio (=) P_{mf}/P_m		Simultaneous measurements at one saturating non-inhibitory irradiance level
Corno et al. (2005)	FRRF (FAST^{tracka})	^{14}C	Phytoplankton assemblages (field)	Ambient irradiance	L (+) Divergence in primary production estimates	O_2 consumption processes (mito-chondrial, photorespiration, Mehler reaction), species composition, vertical distribution	
Estevez-Blanco et al. (2006)	FRRF (FAST^{tracka})	^{14}C	Phytoplankton assemblages (field)	Ambient irradiance	L (+)		
Falkowski et al. (1986a))	PPF	O_2	*Thalasiossira weissflogii, Skeletonema costatum, Isochrysis galbana, Dunaliella tertiolecta, Chlorella vulgaris*		L at MI,deviations at LI and HI	Quenching phenomenon (LI), cyclic transport of electrons around PSII (HI)	Measurements of steady-state O_2 evolution
Falkowski et al. (1986b)	PPF	O_2	*Chlorella pyrenoidosa*		L (=) when Φ_{PSII} >0.2Deviation at HI	Cyclic transport of electrons around PSII	Simultaneous measure-ments
Falkowski et al. (1991)	PPF	^{14}C	Phytoplankton Natural samples	Ambient conditions	L (mostly =)		
Flameling and Kromkamp (1998)	PAM-101	O_2	*Emiliana huxleyi, Scenedesmus protuber-ans, Phaeocystis globosa, Phaeodactylum tricornutum*		Mostly NL (deviation at HI) Sp. dependent	O_2 consumption processes or cyclic/pseudo cyclic transport of electrons,	Simultaneous measure-ments at steady state

Franklin and Badger (2001)	PAM-101	$^{18}O_2$ (mass spectrometry)	*Porphyra columbina, Ulva australis, Zonaria crenata* Natural samples	100 µmol photons m^{-2} s^{-1}	L (+) at LINL at HI Sp. dependent	Quenching phenomena, cyclic transport of electrons around PSII, not photorespiration	Simultaneous measurements at steady state
Fujiki et al. (2007)	FRRF (FAST*tracka*)	^{14}C	*Dunaliella tertiolecta*	Different light histories	L (+) at LI and MINL at HI	Cyclic (PSII) or pseudo-cyclic transport of electrons	
Fujiki et al. (2008)	FRRF (Diving Flash)	O_2	Phytoplankton Natural samples	Ambient conditions	Good correlation		FRRF- and O_2-data from different field campaigns
Geel et al. (1997)	PAM-101, Xenon-PAM	O_2	*Phaeodactylum tricornutum, D. tertiolecta, Tetraselmis sp., Isochrysis sp., Rhodomonas sp.*	100 µmol photons m^{-2} s^{-1}	L (+) at LINL above light saturation	Species dependent O_2 consumption process (Mehler-type, not photorespiration)	Simultaneous measurements at steady state
Gilbert et al. (2000a)	Xenon-PAM	^{14}C	Freshwater natural samples		Primary production overestimated by fluorescence	Duration of measurements package effect, Mehler reaction, N_2 reduction photorespiration	
Gilbert et al. (2000b)	PAM-101	O_2	Freshwater species, *Chlorella vulgaris, Cryptomonas ovata, Cyclotella meneghiniana, Synechococcus leopoliensis*	30 µmol photons m^{-2} s^{-1}	L (=) below E_k / NL above E_K, depending on the species	Package effect, Mehler reaction, cyclic transport of electrons (PSII), not photorespiration	Simultaneous measurements
Glud et al. (2002)	Diving-PAM	O_2, ^{14}C	Sea-ice algae		Mostly NL	Self-shading	RLC
Hader et al. (1997)	PAM-2000	O_2	*Cladophora prolifera, C. pellucida* Natural samples	Ambient conditions	Good correlation		Follow-up of diel photoinhibition and recovery
Hancke et al. (2008a)	PAM-101	O_2, ^{14}C	*Prorocentrum minimum, Prymnesium parvum, Phaeodactylum tricornutum*	Continuous light 80 µmol photons m^{-2} s^{-1}	Mostly L (+, =)α, E_k, to a lesser extent P_m	Species-specific differences (especially for P_m), Mehler reaction	Measurements in parallelFluorescence at steady state

(continued)

Table 2 (continued)

Authors (year)	Methods		Organisms	Growth irradiance / photoacclimation	Observed relationship	Hypothesis for NL	Remarks
Hancke et al. (2008b)	PAM-101	O_2	*Prorocentrum minimum, Prymnesium parvum, Phaeodactylum tricornutum*	Continuous light 80 µmol photons m^{-2} s^{-1}	L (−, =, +), depending on the methodHigh correlation for P_m, weak for α and E_k	Differences in slope coefficient in function of the estimation of light absorbed by PSII	Simultaneous measurements at steady state. Different approaches to estimate quanta absorbed by PSII
Hanelt and Nultsch (1995)	PAM-2000	O_2	*Palmaria palmata* Natural samples		NL	Differences in the role of pigments	Simultaneous measure-ments
Hanelt et al. (1995)	PAM-101	O_2	*Dictyota dichotoma* Natural samples		L		Simultaneous measure-ments
Hartig et al. (1998)	PAM-101	^{14}C	MPB (concentrated cell suspensions; aliquots for both methods)	Collection of motile fraction from natural sample	L (mostly -) at MINL at LI, HI	Differences in light sources (LI), alternative electron sinks (Mehler reaction, nitrogen reduction) (HI)	Measurements in parallel. High correlation for Pm, not for α and Ek. Measure of a* problematical
Heinze et al. (1996)	PAM-101	O_2	*Scenedesmus obliquus* Freshwater species	100 µmol photons m^{-2} s^{-1}	L (=)		Simultaneous measure-ments at steady state
Kaiblinger and Dokulil (2006)	FRRF (FAST^{tracka})	^{14}C	Freshwater phytoplankton Dominant organisms : *Planktothrix rubescens* (summer) or diatoms		L (=) at subsaturating I	Influence of high surface irradiance	Vertical profiles
Koblizek et al. (1999)	PAM-101, PEA, PSI	O_2	*Spongiochloris sp* Freshwater species.	20 µmol photons m^{-2} s^{-1}	Mostly L (=)		Exposure for 120 min to 1000 µmol photons m^{-2} s^{-1}, recovery for 90 min in the dark

Kolber and Falkowski (1993)	PPF	^{14}C	Phytoplankton Natural samples		L (mostly =)		Measurements in parallel for fluorescence day profiles
Kromkamp et al. (2001)	PAM-101	O_2	*Planktothrix rubescens* Natural freshwater samples		L (generally +) or curvilinear. Deviation mostly at HI	Electron sinks (light respiration, Mehler reaction)	Simultaneous measure-mentsSteady state ?
Kromkamp et al. (2008)	Water-PAM and FRRF (FAST^{tracka})	^{14}C	*Lake phytoplankton*	Ambient conditions	L (+)		Both FRRF and PAM overestimate C-fixation
Kroon et al. (1993)	PAM-101	O_2, ^{14}C	*Heterocapsa pygmeaea* Marine species	Low irradiances of different qualities, 9-19 µmol photons m^{-2} s^{-1}	Mostly L (=) Close coupling with O_2 data, weaker with ^{14}C	Changes in linear or cyclic electron transport rates, or enzymatic processes, depending on chromatic adaptation	Simultaneous measure-ments at steady state (O_2)
Kroon (1994)	PAM-101	O_2	*Chlorella pyrenoidosa*- Freshwater species	Fluctuating growth irradiance, total daily dose 10.5 µmol photons m^{-2} d^{-1}	L (−)		Simultaneous measure-ments at steady state
Kühl et al. (2001)	Diving-PAM (LC, RLC)	O_2	Brown macroalgae, coralline red algae Natural samples		Good correspondence		
Lefebvre et al. (2007)	FMS1	O_2	*Skeletonema costatum*	Simulated seasonal conditions (5 combinations of irradiances, photoperiods and temperatures)	L (+)	The slope of the relationship varied according to the seasonal treatment	Measurements in parallel, at steady state
Longstaff et al. (2002)	Diving-PAM (ETR measured at diel ambient irradiance, and by RLC)	O_2	*Ulva lactuca*	Diel ambient irradianceNutrient enrichment experiments	L at LI (=) andMI (+) depending on ETR determination NL at HI	Cyclic transport of electrons around PSII, NPQ, Mehler reaction, photorespiration Accuracy of photon absorbance by thalli	Simultaneous measure-ments at steady state (diel ambient irradiance)

(continued)

Table 2 (continued)

Authors (year)	Methods		Organisms	Growth irradiance / photoacclimation	Observed relationship	Hypothesis for NL	Remarks
Masojidek et al. (2001)	PAM-101	O_2	Freshwater cyanobacteria Natural samples	Diel ambient light	Mostly L (+) at LI and NL at HI Diel changes	Photorespiration, Mehler reaction	Simultaneous measure-ments possibly at steady state
Melrose et al. (2006)	FRRF (FAST^{tracka})	^{14}C	Coastal phytoplankton	Ambient conditions	L (–)		FRRF mainly underesti-mated C-fixation
Migné et al. (2007)	Diving-PAM	CO_2 (IRGA)	MPB. Natural assemblages	Diel ambient light,screened by the Perspex dome for benthic chambers	Mostly LNL at HI, depending on season and station	Migration, relative contribution of surface/subsurface cells	Measurements in parallel on different samples
Moore et al. (2003)	FRRF(FAST^{tracka})	^{14}C	Phytoplankton. Natural assemblages	Ambient light regimes	L (+) between ^{14C} and FRRF-modelled productivity estimates NL (α)	O_2 consumption processes or non linear transports of electrons	Spectral differences between light source (underwater light field *in situ*, incubators)
Morris and Kromkamp (2003)	PAM-101	O_2	*Cylindrotheca closterium*	Two growth rates, 7 temperatures (short-term change)	Mostly LNL at LI and HI	Alternative electron sinks (HI), variation in respiration rates (LI)	ETR/O_2 relationship hardly influenced by temperature
Morris et al. (2008)	MiniPAM, Water-PAM	^{14}C	MPB in tidal mesocosms Measurements made on biofilms (ETR) and suspensions (ETR, ^{14}C)	200 μmol photons m^{-2} s^{-1}	Good correlation (=) ETR/P_m for suspensions, weaker for biofilms	Changes with growth phase, diel rhythm, migration, absorption cross-section of MPB,	Measurements in parallel (suspensions)
Perkins et al. (2001)	Diving-PAM	^{14}C	MPB. Natural assemblages	Ambient irradiance (100% and 50%)	L at LI, MINL at HI	Vertical migration and relative contribution of surface/ subsurface cells	Measurements in parallel in sample cores
Perkins et al. (2002)	Xenon-PAM, High resolution imaging system	^{14}C	MPB. Natural assemblages	Ambient irradiance220 μmol photons m^{-2} s^{-1} (Imaging system)	NL	PSI contribution, vertical migration, relative contribution of surface/subsurface cells	Measurements in parallel on sample cores
Prasil et al. (1996)	FRRF	O_2	*Chlorella pyrenoidosa*-Freshwater species	50 μmol photons m^{-2} s^{-1}	NL at HI		Redox state of PQ, cyclic transport around PSII

Raateoja and Seppala (2001)	FRRF (FAST^{tracka})	^{14}C, O_2	*Nannochloris sp.* Marine species	39 and 314 μmol photons m^{-2} s^{-1}	NL	Pre-set parameters to convert electron flows into primary productivity, nitrogen reduction	Influence of growth irradiance, diel cycle and nutrient uptake
Raateoja (2004)	FRRF (FAST^{tracka})	^{14}C	Phytoplankton. Natural assemblages	Ambient irradiance	L at LINL at HI	Methodological discrepancy (measurement periods)	Deviation from linearity at irradiance >200 μmol photons m^{-2} s^{-1}
Raateoja et al. (2004)	FRRF (FAST^{tracka})	^{14}C	Phytoplankton. Natural assemblages	Ambient irradiance (primary production), fluorescent tubes (P/E curves)	NL	Differences in light sources, inadequacy of FRRF for near-surface measurements	
Rech (2004)	FMS1	O_2	*Amphora coffeaeformis, Haslea ostrearia, Entomoneis paludosa, Phaeodactylum tricornutum, Porphyridium cruentum,Skeletonema costatum, H. ostrearia, Dunaliella tertiolecta, P. cruentum, Spirulina platensis*	100 μmol photons m^{-2} s^{-1} 75 and 350 μmol photons m^{-2} s^{-1}	L at LI and MINL at HI L at LI and MI NL at HI		Species-dependent relationRelation varying with the species, and the growth irradiance
Rees et al. (1992)	PAM-101 ?	O_2	*Dunaliella sp.* Marine species	40 μmol photons m^{-2} s^{-1}	NL	Cyclic transport around PSI and PSII, Mehler reaction, etc.	
Rysgaard et al. (2001)	Diving-PAM (RLC)	O_2, ^{14}C	Sea ice algae. Natural communities	Ambient irradiance	L		Calibration ^{14}C/rETR (in situ RLCs, and parallel measurements on ice cores)
Sagert and Schubert (2000)	PAM-2000	O_2	*Palmaria palmata*	7 to 568 μmol photons m^{-2} s^{-1}	Significant correlation for P_m and E_K for lab measurements		Simultaneous measurements

(continued)

Table 2 (continued)

Authors (year)	Methods		Organisms	Growth irradiance / photoacclimation	Observed relationship	Hypothesis for NL	Remarks
Sarma et al. (2005)	FRRF (FAST^{tracka})	O_2 (triple oxygen isotopes, light-dark bottle incubation)	Phytoplankton. Natural assemblages	Ambient irradiance	Weak correlation between gross oxygen production estimated by$^{17}\Delta$ anomaly, light-dark bottle incubation, and FRRF	Differences in time resolution of in situ changes in irradiance	Calibration ^{14}C/FRRF Comparisons with seasons
Sarma et al. (2006)	FRRF (FAST^{tracka})	O_2(triple oxygen isotopes, light-dark bottle incubation)	Phytoplankton. Natural assemblages	Ambient irradiance	Good correlation (ca. =) between daily integrated production estimated by$^{17}\Delta$ anomaly, light-dark bottle incubation, and FRRF		Short-term variations (hourly/daily changes)
Schofield et al. (1995)	PAM-101	^{14}C	Antarctic ice algae Natural communities	Ambient irradiance	Weak agreement between UV-B photoinhibition estimated by fluorescence and ^{14}C fixation	Correlation between ^{14}C- and fluorescence-based estimates changed with time of the day	Experiments on UV-B inhibition of carbon fixation ratesMea-surements in parallel
Schreiber et al. (1995)	Xenon-PAM	O_2	*Ankistrodesmus braunii*- Freshwater species	80 µmol photons m^{-2} s^{-1}	L (ca. =) at HI and MI NL at LI	Cyclic PSII transport rate	
Serôdio et al. (1998)	PAM-101	O_2	*Chlorella vulgarisPhae-odactylum tricornutum Spirulina maxima*	50 µmol photons m^{-2} s^{-1}	L, slope varying with species		
Serôdio (2003)	PAM-101	O_2	MPB. Natural sediment cores	Ambient irradiance	L		Changes in biomass profile due to vertical migration
Serôdio et al. (2007)	WATER-PAM	O_2	MPB. Natural sediment cores	Ambient irradiance	L		Depth-integration of subsurface fluorescence
Silva et al. (1998)	PAM-2000	O_2	*Gelidium sesquipedale*	Ambient irradiance	L		Good correlation whatever the depth

Smyth et al. (2004)	FRRF (FAST*tracka*)	^{14}C (Daily primary production, LC)	Phytoplankton assemblages. Natural communities	Ambient irradiance	NL (α usually higher, P_m and E_k lower, when FRRF-estimated)	Integration with time and depth of instantaneous fluorescence data	
Suggett et al. (2001)	FRRF	^{14}C, O_2	Phytoplankton assemblages. Natural communities	Ambient irradiance	Good correlation (α, P_m usually higher, E_k lower, when FRRF-estimated)	FRRF estimation of photosynthetic unit size, respiration, alternative electron sinks	
Suggett et al. (2003)	Xenon-PAM, FRRF (Chelsea) FRRF (Chelsea)	O_2 (flash yield) $^{18}O_2$ (mass spectrometry)	*Chaetoceros muelleri, Prorocentrum minimum, Dunaliella tertiolecta, Emiliana huxley-Prochlorococcus marinus* *D. tertiolecta, Thalasiossira weissflogii, Nannochloris atomus*	50 or 75 μmol photons m^{-2} s^{-1} (LI), 400 or 375 μmol photons m^{-2} s^{-1} (HI) 5–10 μmol photons m^{-2} s^{-1}	NL (PAM vs FRRF), differences being lower at HIL (–, =) depending on the species	Differences between PAM and FRRF measurements explained by distinct approaches (single- vs multiple turn-over)	Simultaneous measurements

*When the relation is linear, ETR estimated photosynthesis can be equal (=), over- (+), or under-estimated in comparison with carbon- or oxygen-based method 'pump and probe' and 18 used the FRRF techniques (mostly Chelsea's FAST^{tracka}), whereas only six studies used a PAM-type fluorometer in natural systems. The MT-PAM fluorometer is the apparent method of choice for seagrasses and macroalgal studies (16 out of 16) and both types are used for culture studies of unicellular algae. Fluorescence estimates of photosynthesis (from which ca. 20% are expressed as rETR, see above) have been compared mainly with oxygen-evolution data (39 out of 66 references, including all the macroalgal/seagrass studies), and four dealing with $^{18}O_2$ and $^{17}O_2$). In 27 out of the 67 references cited in Table 2 fluorescence was compared to a ^{14}C-based method, the majority reporting on phytoplankton productivity. A limited number (6) compared ETR to both oxygen evolution and carbon fixation. Thirty-five percent of the studies combined the different photosynthesis measurements simultaneously on the same sample, and 11 out of the 67 studies ran experiments in parallel (same time, but on different subsamples).

A global survey of Table 2 shows that in more than 75% of the studies cited, a good correlation was observed between rETR and other methods to estimate photosynthesis. Usually fluorescence methods gave higher estimates of photosynthesis, i.e., the value of Φ_e and/or PQ was overestimated (Eq. 7). The PQ value depends on the N-source for growth and is close to 1 when ammonia is the sole nitrogen source and 1.3 when nitrate is the N-source for growth (Williams 1993). This is because the reduction of nitrate requires photosynthetically produced electrons. Many studies, however, show a large variation in PQ, which is at least partly due to methodological errors (Williams and Robertson 1991). Another reason for the overestimate of C-fixation by ETR is that often an a priori assumption is made of Φ_e, which is assumed to equal 0.25 (= 1 mol oxygen per 4 mol photons absorbed). However, culture work shows that this optimal value is often not reached, and that it is closer to 6–7 mol oxygen per 4 mol photons absorbed (Flameling and Kromkamp 1998). For this reason Raateoja et al. (2004) used a value of 0.18 for Φ_e and assuming a PQ of 1.5, Kromkamp et al. (2008) calculated that Φ_e equalled 0.15. Another reason for an overestimation might be that the chosen value for n_{PSII} when using the ST-method might be overestimated. This is discussed above in Section 5 on rETR vs ETR.

MPB hardly features in the table (12%) and no direct estimates of C-fixation are made using a-priori assumptions for Φ_e. This is because the optical conditions in the MPB are so complex that absolute estimates of ETR are very difficult to make, related to the fact that quantification of rETR is hindered by difficulties in measuring the absorption coefficient of MPB, the presence of vertical migration in muddy sediments and the artefacts caused by deep layer fluorescence. These issues are discussed in the section on rETR vs ETR. To circumvent this problem several approaches have been tried: Barranguet and Kromkamp (2000) measured rETR at the sediment surface and compared the photosynthetic parameters obtained with those obtained from ^{14}C-fixation made on sediment slurries. Assuming no vertical migration and a homogenous chl *a* distribution with depth they calculated a conversion factor (called EE) and estimated depth integrated primary production. Their results demonstrated that with the assumptions mentioned, a single conversion factor could be used to estimate primary production throughout the year at different stations. At high light non-linearity was often observed, which could be due to physiological factors (see below) but which could also be due to deep layer fluorescence, although that was not realized by the authors at the time. The EE factor calculated by Barranguet and Kromkamp (2000) corresponded with the conversion factor obtained from a culture study with the benthic diatom *Cylindrotheca closterium* (Morris and Kromkamp 2003). Because the approach taken by Barranguet and Kromkamp (2000) did not take different vertical biomass distribution profiles in consideration, Serôdio (2003) developed a chlorophyll fluorescence based index to estimate photosynthesis:

$$P_{fluo} = E\frac{F_o}{F_{o,sed}}\Delta F/F_m' \tag{8a}$$

Basically this equation approximates to:-

$$P_{fluo} = F_o \bullet rETR \tag{8b}$$

where F_0 in Eq. 8a is corrected for background fluorescence of the sediment (and which should always be done as well as possible). The rationale of this approach is that F_0 is a good proxy for the [chl *a*] in the photic zone of the sediment, and thus reflects the vertical biomass profile. The author compared the photosynthesis estimates of the fluorescence index with those obtained from microelectrode mechanisms (gross oxygen production using the light dark shift technique, Revsbech and Jørgensen 1983) and found in general a linear relationship between both methods and the slopes did not differ significantly, suggesting that this index might be a powerful method to estimate MPB primary production. The validity of this model was corroborated using a model by Forster and Kromkamp (2004) (Fig. 14) who showed that the slope between the modelled and measured production depended little on the shape of the vertical biomass profile of the MPB. Recently Serôdio et al. (2007) confirmed the validity of the model results by demonstrating a linear relationship between the fluorescence index shown in Eq. 8a and measurements obtained using oxygen microelectrodes. As in the model study by Forster and Kromkamp (2004), and also demonstrated in a laboratory study with light and temperature treatments simulating seasons (Lefebvre et al. 2007), the slope of the regression between both methods varied throughout the year (Fig. 15), thus requiring a regular calibration of the conversion factor. If we consider only the studies for which the measurements were made simultaneously, in all but two studies (Flameling and Kromkamp 1998; Hanelt and Nultsch 1995), was the relation between fluorescence-based and oxygen-based estimates of photosynthesis linear. In most cases, ETR closely matched oxygen- or carbon-based estimates of photosynthesis.

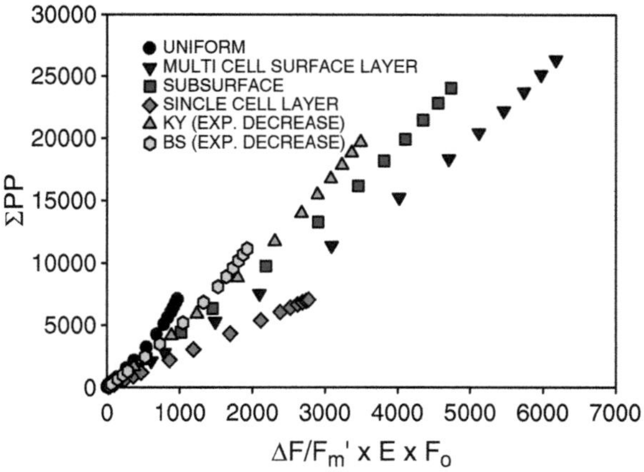

Fig. 14 Relationship between modeled primary production (Y-axis) and a fluorescence index (X-axis, slightly modified after Serodio, 2003). For the calculations one set a single set of photosynthetic parameters was chosen and the results were due to differently shaped depth profiles and thus a corresponding different contribution of fluorescence originating from layers within the photic zone

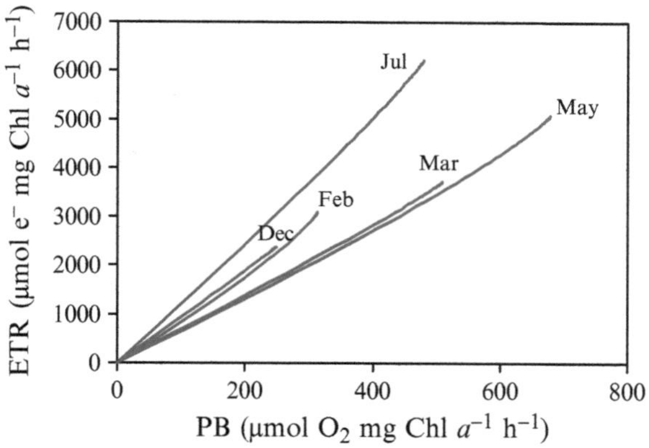

Fig. 15 Productivity measured as electron transport rate (ETR) as a function of that measured by oxygen electrodes (P^B) for microphytobenthic biofilms measured at five different times of the year. Note the near linear, but differing relationship between the two methods at different time points (From Lefebvre et al. 2007)

On the other hand, in less than 25% of the references cited, a non linear, or a weak relation was observed. Usually, the non-linearity between P^B and ETR was observed at high irradiances above E_k. This is a general pattern, independent of the taxonomic level, as it can be observed in seagrasses, macroalgae, microalgae and in cyanobacteria. This non-linearity where ETR became relatively higher than P^B was attributed to light dependent O_2 consuming processes (photorespiration, chlororespiration, mitochondrial respiration in light, Mehler reaction, Flameling and Kromkamp 1998) or other alternative electron sinks (e.g. thioredoxin, nitrogen reduction, Lomas and Glibert 1999), alternative electron pathways (cyclic transport around PSII and/or PSI, Lavaud et al. 2002c), transfer of reductants from chloroplasts to mitochondria (OAA-malate and DHAP-PGA shuttles, Behrenfeld et al. 2004), different time-scale between fluorescence and oxygen or carbon-based measurements, methodological constraints in determination of the fraction of light absorbed by PSII (especially in biofilms, see section on rETR vs ETR), difference in apparatus sensitivity, vertical migration and contribution of subsurface cells in the case of MPB. Moreover, non-linearity could also be light dependent, varying according to the light level experienced by the cells or to their light history, which in turn greatly influences the light absorption (or utilization) characteristics needed to estimate ETR.

Thus, in spite of an increasing number of studies dedicated to investigate the variation of photosynthetic parameters in microalgae using fluorometry and to compare them with other methods, many uncertainties still remain. This calls for a more in depth study of how the different parameters needed to calculate absolute ETR (a^*_{PSII}, n_{PSII}) and convert these to rates of C-fixation (Φ_e, PQ) vary and depend on nutrient status, light histories and growth rate (e.g., see discussion in Morris and Kromkamp (2003)), and this applies especially to organisms producing biofilms, with their specific constraints.

8 General Summary

The application of variable chlorophyll fluorescence to microphytobenthic (MPB) biofilms in soft sediment systems is complex as a result of the signal emanating form subsurface cells, cellular vertical migration within the sediment matrix, a high capacity for down regulation, chlororespiration in the dark and the effects of the physical structure of the sediment/biofilm matrix (light attenuation by the sediment matrix) itself. Despite these factors, fluorescence has yielded a large and valuable range of information on the ecology, physiology and productivity of MPB biofilms and indeed at the species level of diatoms, cyanobacteria and other taxa comprising them. Care in design of methods, including timing measurements away from periods of maximal migration and minimising measurement duration, as well as the use of numerical simulations to deconvolute depth-integrated measurements (Serôdio 2004) have enabled correct interpretation of data, enhancing our knowledge of biofilm functioning and (photo)physiology. Fluorescence has provided valuable information on down regulatory measurements including xanthophyll cycling in the form of non photochemical quenching and PS II electron cycling (Perkins et al. 2006; Serôdio et al. 2005, 2006b; Herlory et al. 2007; Lavaud et al. 2007; Cruz and Serôdio 2008). We now have a far better understanding of the use of fluorescence as a biomass proxy (Honeywill et al. 2002; Jesus et al. 2006a, b), in the calculation of ETR and the construction of light response curves (Perkins et al. 2006; Serôdio et al. 2006a; Herlory et al. 2007; Cruz and Serôdio 2008) and particularly encouraging is the correlation between fluorescence methodology and alternative methods such as oxymetry and carbon uptake (Table 1) for measurement of productivity (Barranguet and Kromkamp 2000; Forster and Kromkamp 2004; Kromkamp and Forster 2003; Morris and Kromkamp 2003; Morris et al. 2008; Serôdio et al. 2005, 2006b, 2007). The application of fluorimetry to the MPB is still comparatively at an early stage and there is much more to be done to investigate the complex photophysiology of the species comprising the biofilms as well as in developing applied outputs of models of productivity and other ecosystem functions of these important ecosystems.

References

Admiraal W (1984) The ecology of estuarine sediment inhabiting diatoms. Prog Phycol Res 3:269–322

Babin M, Morel A, Claustre H, Bricaud A, Kolber Z, Falkowski PG (1996) Nitrogen- and irradiance-dependent variations of the maximum quantum yield of carbon fixation in eutrophic, mesotrophic and oligotrophic marine systems. Deep-Sea Res I 43:1241–1272

Barranguet C, Kromkamp J (2000) Estimating primary production rates from photosynthetic electron transport in estuarine microphytobenthos. Mar Ecol Prog Ser 204: 39–52

Baumgartner LK, Spear JR, Buckley DH, Pace NR, Reid RP, Dupraz C, Visscher P (2007) Microbial diversity in modern marine stromatolites, Highborne Cay, Bahamas. Environ Microbiol 11:2710–2719.

Beer S, Axelsson L (2004) Limitation in the use of PAM fluorometry for measuring photosynthtic rates of macroalgae at high irradiances. Eur J Phycol 39:1–7

Beer S, Bjork M (2000) Measuring rates of photosynthesis of two tropical seagrasses by pulse amplitude modulated (PAM) fluorometry. Aquat Bot 66:69–76

Beer S, Vilenkin B, Weil A, Veste M, Susel L, Eshel A (1998) Measuring photosynthetic rates in seagrasses by pulse amplitude modulated (PAM) fluorometry. Mar Ecol Prog Ser 174:293–300

Beer S, Larson C, Poryan O, Axelsson L (2000) Photosynthetic rates of *Ulva lactuca* (Chlorophyta) measured by pulse amplitude modulated (PAM) fluorometry. Eur J Phycol 35:69–74

Behrenfeld MJ, Prasil O, Babin M, Bruyant F (2004) In search of physiological basis for covariations in light-limited and light-saturated photosynthesis. J Phycol 40:4–25

Berner T, Dubinsky Z, Wyman K, Falskowski PG (1989) Photoadaptation and the "package" effect in *Dunaliella tertiolecta* (Chlorophyceae). J Phycol 25:70–78

Bidigare RR (1990) In vivo absorption properties of algal pigments. In: SPIE Ocean Optics X, pp 290–302

Boyd PW, Aiken J, Kolber Z (1997) Comparison of radiocarbon and fluorescence based (pump and probe) measurements of phytoplankton photosynthetic characteristics in the Northeast Atlantic Ocean. Mar Ecol Prog Ser 149:215–226

Bradbury M, Baker NR (1981) Analysis of the slow phases of the *in vivo* chlorophyll fluorescence induction curve – Changes in the redox state of Photosystem-II electron-acceptors and fluorescence emission from Photosystem-I and Photosystem-II. Biochim BiophysActa 635:542–551

Cabello-Pasini A, Figueroa FL (2005) Effect of nitrate concentration on the relationship between photosynthetic oxygen evolution and electron transport rate in *Ulva rigida* (Chlorophyta). J Phycol 41:1169–1177

Cabello-Pasini A, Swift H, Smith GJ, Alberte RS (2001) Phosphoenolpyruvate carboxykinase from the marine diatom *Skeletonema costatum* and the phaeophyte *Laminaria setchellii*. II. Immunological characterization and subcellular localization. Bot Mar 44:199–207

Carr H, Bjork M (2003) A methodological comparison of photosynthetic oxygen evolution and estimated electron transport rate in tropical *Ulva* (Chlorophyceae) species under different light and inorganic carbon conditions. J Phycol 39:1125–1131

Carr H, Bjork M (2007) Parallel changes in non-photochemical quenching properties, photosynthesis and D1 levels at sudden, prolonged irradiance exposures in *Ulva fasciata* Delile. J Photochem Photobiol B 87:18–26

Cartaxana P, Brotas V, Serôdio J (2008) Effects of two motility inhibitors on the photosynthetic activity of the diatoms *Cylindrotheca closterium* and *Pleurosigma angulatum*. Diatom Res 23:65–74

Consalvey MC, Jesus B, Perkins RG, Brotas V, Underwood GJC, Paterson DM (2004) Monitoring migration and measuring biomass in benthic biofilms: the effects of dark/far-red adaptation and vertical migration on fluorescence measurements. Photosynth Res 81:91–101

Consalvey M, Perkins RG, Paterson DM, Underwood GJC (2005) PAM Fluorescence: A beginners guide for benthic diatomists. Diatom Res 20:1–22.

Corno G, Letelier RM, Abbott MR, Karl DM (2005) Assessing primary production variability in the North Pacific subtropical gyre: a comparison of Fast Repetition Fate Fluorometry and ^{14}C measurements. J Phycol 42:51–60

Cruz S, Serôdio J (2008) Relationship of rapid light curves of variable fluorescence to photoacclimation and non-photochemical quenching in a benthic diatom. Aquat Bot 88:256–264

Defew EC, Perkins RG, Paterson DM (2004) The influence of light and temperature interactions on a natural estuarine microphytobenthic assemblage. Biofilms 1:21–30.

Dijkman NA, Kroon BMA (2002) Indications for chlororespiration in relation to light regime in the marine diatom *Thalassiosira weissflogii*. J Photochem Photobiol B 66: 179–187

Dubinsky Z, Falkowski PG, Wyman K (1986) Light harvesting and utilization by phytoplankton. Cell Physiol 27: 1335–1349

Eaton JW, Moss B (1966) The estimation of numbers and pigment content in epipelic algal populations. Limnol Oceanogr 11:584–595

Estevez-Blanco P, Cermeno P, Espineira M, Fernandez E (2006) Phytoplankton photosynthetic efficiency and primary production rates estimated from fast repetition rate fluorometry at coastal embayments affected by upwelling (Rias Baixas, NW of Spain). J Plankton Res 28:1153–1165

Falkowski PG (1981) Light-shade adaptation and assimilation numbers. J Plankton Res 3:203–216

Falkowski PA, Raven JF (1997) Aquatic Photosynthesis. Blackwell Science, London, UK

Falkowski PG, Owens TG, Ley AC, Mauzerall DC (1981) Effects of growth irradiance levels on the ratio of reaction centers in two species of marine phytoplankton. Plant Physiol 68:969–973

Falkowski PG, Fujita Y, Ley A, Mauzerall D (1986a) Evidence for cyclic electron flow around Photosystem-II in *Chlorella pyrenoidosa*. Plant Physiol 81:310–312

Falkowski PG, Wyman K, Ley AC, Mauzerall DC (1986b) Relationship of steady-state photosynthesis to fluorescence in eucaryotic algae. Biochim Biophys Acta 849:183–192

Falkowski PG, Ziemann D, Kolber Z, Bienfang PK (1991) Role of eddy pumping in enhancing primary production in the ocean. Nature 352:55–58

Flameling IA, Kromkamp J (1998) Light dependence of quantum yields for PSII charge separation and oxygen evolution in eucaryotic algae. Limnol Oceanogr 43:284–297

Forster RM, Kromkamp JC (2004) Modelling the effects of chlorophyll fluorescence from subsurface layers on photosynthetic efficiency measurement in microphytobenthic algae. Mar Ecol Prog Ser 284:9–22

Franklin AL, Badger MR (2001) A comparison of photosynthetic electron transport rates in macroalgae measured by pulse amplitude modulated chlorophyll fluorometry and mass spectrometry. J Phycol 37:756–767

Fujiki T, Suzue T, Kimoto H, Saino T (2007) Photosynthetic electron transport in *Dunaliella tertiolecta* (Chlorophyceae) measured by fast repetition rate fluorometry: relation to carbon assimilation. J Plankton Res 29:199–208

Fujiki T, Hosaka T, Kimoto H, Ishimaru T, Saino T (2008) *In situ* observation of phytoplankton productivity by an underwater profiling buoy system: use of fast repetition rate fluorometry. Mar Ecol Prog Ser 353:81–88

Geel C, Versluis W, Snel JFH (1997) Estimation of oxygen evolution by marine phytoplankton from measurement of the efficiency of photosystem II electron flow. Photosynth Res 51:61–70

Gilbert M, Wilhelm C, Richter M (2000a) Bio-optical modelling of oxygen evolution using *in vivo* fluorescence: comparison of measured and calculated photosynthesis/irradiance (P-I) curves in four representative phytoplankton species. J Plant Physiol 157:307–314

Gilbert M, Domin A, Becker A, Wilhelm C (2000b) Estimation of primary productivity by Chlorophyll *a in vivo* fluorescence in freshwater phytoplankton. Photosynthetica 38:111–126

Glud RN, Rysgaard S, Kuhl M (2002) A laboratory study on O_2 dynamics and photosynthesis in ice algal communities: quantification by microsensors, O_2 exchange rates, ^{14}C incubations and a PAM fluorometer. Aquat Microb Ecol 27:301–311

Goss RE, Pinto A, Wilhelm C, Richter M (2006) The importance of a highly active and delta-pH-regulated diatoxanthin epoxidase for the regulation of the PS II antenna function in diadinoxanthin cycle containing algae. J Plant Physiol 163:1008–1021

Guarini J-M, Blanchard GF, Gros PH, Gouleau D, Bacher C (2000) Dynamic model of the short-term variability of microphytobenthic biomass on temperate intertidal mudflats. Mar Ecol Prog Ser 95:291–303.

Hader DP, Herrmann H, Schafer J, Santas R (1997) Photosynthetic fluorescence induction and oxygen production in two Mediterranean *Cladophora* species measured on site. Aquat Bot 56:253–264

Hagerthey SE, Defew EC, Paterson DM (2002) Influence of *Corophium volutator* and *Hydrobia ulvae* on intertidal benthic diatom assemblages under different nutrient and temperature regimes. Mar Ecol Prog Ser 245:47–59

Hancke K, Hancke TB, Olsen LM, Johnsen G, Glud RN (2008a) Temperature effects on microalgal photosynthesis-light responses measured by O_2 production, pulse-amplitude-modulated fluorescence, and ^{14}C assimilation. J Phycol 44:501–514

Hancke TB, Hancke K, Johnsen G, Sakshaug E (2008b) Rate of O_2 production derived from pulse-amplitude-modulated fluorescence: testing three biooptical approaches against measured O_2-production rate. J Phycol 44:803–813

Hanelt D, Nultsch W (1995) Field studies of photoinhibition show non-correlations between oxygen and fluorescence measurements in the arctic red alga *Palmaria palmata*. J Plant Physiol 145:31–38

Hanelt D, Uhrmacher S, Nultsch W (1995) The effect of photoinhibition on photosynthetic oxygen production in the brown alga *Dictyota dichotoma*. Bot Acta 108:99–105

Hartig P, Wolfstein K, Lippemeier S, Colijn F (1998) Photosynthetic activity of natural microphytobenthos populations measured by fluorescence (PAM) and ^{14}C-tracer methods: a comparison. Mar Ecol Prog Ser 166:53–62

Heinze I, Dau H, Senger H (1996) The relation between the photochemical yield and variable fluorescence of photosystem II in the green alga *Scenedesmus obliquus*. J Photochem Photobiol B 32:89–95

Herlory O, Richard P, Blanchard G (2007) Methodology of light response curves: application of chlorophyll fluorescence to microphytobenthic biofilms. Mar Biol 153:91–101

Honeywill C, Hagerthey SE, Paterson DM (2002) Determination of microphytobenthic biomass using pulse-amplitude modulated minimum fluorescence. Eur J Phycol 37:485–492

Jakob T, Wilhelm C (2001) Unusual pH-dependence of diadinoxanthin de-epoxidase activation causes chlororespiratory induced accumulation of diatoxanthin in the diatom *Phaeodactylum tricornutum*. J Plant Physiol 158:383–390

Jakob T, Goss R, Wilhelm C (1999) Activation of diadinoxanthin de-epoxidase due to a chlororespiratory proton gradient in the dark in the diatom *Phaeodactylum tricornutum*. Plant Biol 1:76–82

Jesus B, Perkins RG, Mendes CR, Brotas V, Paterson DM (2006a) Chlorophyll fluorescence as a proxy for microphytobenthic biomass: alternatives to the current methodology. Mar Biol 150:17–28

Jesus B, Perkins RG, Consalvey MC, Brotas V, Paterson DM (2006b) Effects of vertical migration by benthic microalgae on fluorescence measurements of photophysiology. Mar Ecol Prog Ser 315:55–66

Jesus B, Mendes C, Brotas V, Paterson DM (2006c) Effect of sediment type on microphytobenthos vertical distribution: modelling the productive biomass and improving ground truth measurements. J Exp Mar Biol Ecol 332:60–74

Johnsen G, Sakshaug E (2007) Biooptical characteristics of PSII and PSI in 33 species (13 pigment groups) of marine phytoplankton, and the relevance for pulse-amplitude-modulated and fast-repetition-rate fluorometry. J Phycol 43:1236–1251

Kaiblinger C, Dokulil MT (2006) Application of fast repetition rate fluorometry to phytoplankton photosynthetic parameters in freshwaters. Photosynth Res 88:19–30

Koblizek M, Ciscato M, Komenda J, Kopecky J, Siffel P, Masojidek J (1999) Photoadaptation in the green alga *Spongiochloris* sp. A three-fluorometer study. Photosynthetica 37:307–323

Koblizek M, Kaftan D, Nedbal L (2001) On the relationship between the non-photochemical quenching of the chlorophyll fluorescence and the photosystem II light harvesting efficiency. A repetitive flash fluorescence induction study. Photosynth Res 68:141–152

Kolber Z, Falkowski PG (1993) Use of active fluorescence to estimate phytoplankton photosynthesis in situ. Limnol Oceanogr 38:1646–1665

Kolber ZS, Prásil O, Falkowski PG (1998) Measurements of variable chlorophyll fluorescence using fast repetition rate techniques: defining methodology and experimental protocols. Biochim Biophys Acta 1367:88–106

Kramer DM, Johnson G, Kiirats O, Edwards GE (2004) New fluorescence parameters for the determination of Q(A) redox state and excitation energy fluxes. Photosynth Res 79:209–218

Kromkamp JC, Forster RM (2003) The use of variable fluorescence measurements in aquatic ecosystems: differences between multiple and single turnover measuring protocols and suggested terminology. Eur J Phycol 38:103–112

Kromkamp J, Limbeek M (1993) Effect of short-term variation in irradiance on light-harvesting and photosynthesis of the marine diatom *Skeletonema costatum*: a laboratory study simulating vertical mixing. J Gen Microbiol 139:2277–2284

Kromkamp J, Barranguet C, Peene J (1998) Determination of microphytobenthos PSII quantum efficiency and photosynthetic activity by means of variable chlorophyll fluorescence. Mar Ecol Prog Ser 162:45–55

Kromkamp JC, Domin A, Dubinsky Z, Lehmann F, Schanz F (2001) Changes in photosynthetic properties measured by oxygen evolution and variable chlorophyll fluorescence in a simulated entrainment experiment with the cyanobacterium *Planktothrix rubescens*. Aquat Sci 63:363–382

Kromkamp JC, Morris EP, Forster RM, Honeywill C, Hagerthey S, Paterson DM (2006) Relationship of intertidal surface sediment chlorophyll concentration to hyperspectral reflectance and chlorophyll fluorescence. Est Coasts 29: 183–196

Kromkamp JC, Perkins R, Dijkman N, Consalvey M, Andres M, Reid RP (2007) The effect of burial on stromatolite photosynthetic activity: the role of oxygen. Aquat Microb Ecol 48:123–230

Kromkamp JC, Dijkman NA, Peene J, Simis SGH, Gons HJ (2008) Estimating phytoplankton primary production in Lake Ijsselmeer (The Netherlands) using variable fluorescence (PAM-FRRF) and C-uptake techniques. Eur J Phycol 43:327–344

Kroon BMA (1994) Variability of photosystem II quantum yield and related processes in *Chlorella pyrenoidosa* (Chlorophyta) acclimated to an oscillating light regime simulating a mixed photic zone. J Phycol 30:841–852

Kroon B, Prezelin BB, Schofield O (1993) Chromatic regulation of quantum yields for photosystem II charge separation, oxygen evolution, and carbon fixation in *Heterocapsa pygmaea* (Pyrrophyta). J Phycol 29:453–462

Kühl M, Glud RN, Borum J, Roberts R, Rysgaard S (2001) Photosynthetic performance of surface-associated algae below sea ice as measured with a pulse-amplified-modulated (PAM) fluorometer and O_2 microsensors. Mar Ecol Prog Ser 223:1–14

Kuntz M (2004) Plastid terminal oxidase and its biological significance. Planta 218:896–899

Lavaud J, Rousseau B, Van Gorkom H, Etienne A-L (2002) Influence of the diadinoxanthin pool size on photoprotection in the marine planktonic diatom *Phaeodactylum tricornutum*. Plant Physiol 129: 1398-1406.

Lavaud J (2007) Fast regulation of photosynthesis in diatoms: mechanisms, evolution and ecophysiology. Functl Plant Sci Biotechnol 1:267–287

Lavaud J, Kroth P (2006) In diatoms, the transthylakoid proton gradient regulates the photoprotective non-photochemical fluorescence quenching beyond its control on the xanthophyll cycle. Plant Cell Physiol 47:1010–1016

Lavaud J, Rousseau B, Van Gorkom H, Etienne A-L (2002b) Influence of the diadinoxanthin pool size on photoprotection in the marine planktonic diatom *Phaeodactylum tricornutum*. Plant Physiol 129:1398–1406

Lavaud J, Van Gorkom H, Etienne A-L (2002c) Photosystem II electron transfer cycle and chlororespiration in planktonic diatoms. Photosynth Res 74:49–57

Lavaud J, Strzepek RF, Kroth PG (2007) Photoprotection capacity differs among diatoms: possible consequences on the spatial distribution of diatoms related to fluctuations in the underwater light climate. Limnol Oceanogr 52:1188–1194

Lefebvre S, Mouget JL, Loret P, Tremblin G (2007) Comparison between fluorimetry and oxymetry techniques to measure photosynthesis in the diatom *Skeletonema costatum* cultivated under simulated seasonal conditions. J Photochem Photobiol: B 86:131–139

Lomas MW, Glibert PM (1999) Temperature regulation of nitrate uptake: a novel hypothesis about nitrate uptake and reduction in cool-water diatoms. Limnol Oceanogr 44:556–572

Longstaff BJ, Kildea T, Runcie JW, Chesshire A, Dennison WC, Hurd C, Kana TM, Raven JA, Larkum AWD (2002) An *in situ* study of photosynthetic oxygen exchange and electron transport rate in the marine macroalga *Ulva lactuca* (Chlorophyta). Photosynth Res 74:281–293

Masojidek J, Grobbelaar JU, Pechar L, Koblizek M (2001) Photosystem II electron transport rates and oxygen production in natural waterblooms of freshwater cyanobacteria during a diel cycle. J Plankton Res 23:57–66

Mauzerall D, Greenbaum NL (1989) The absolute size of a photosynthetic unit. Biochim Biophys Acta 974:119–140

McGinn PJ, Morel FMM (2008) Expression and inhibition of the carboxylating and decarboxylating enzymes in the photosynthetic C-4 pathway of marine diatoms. Plant Physiol 146:300–309

Melrose DC, Oviatt CA, O'Reilly JE, Berman MS (2006) Comparisons of fast repetition rate fluorescence estimated primary production and ^{14}C uptake by phytoplankton. Mar Ecol Prog Ser 311:37–46

Mercado JM, Sanchez-Saavedra MD, Correa-Reyes G, Lubian L, Montero O, Figueroa FL (2004) Blue light effect on growth, light absorption characteristics and photosynthesis of five benthic diatom strains. Aquat Bot 78:265–277

Migné A, Gevaert F, Creach A, Spilmont N, Chevalier E, Davoult D (2007) Photosynthetic activity of intertidal microphytobenthic communities during emersion: in situ measurements of chlorophyll fluorescence (PAM) and CO_2 flux (IRGA). J Phycol 43:864–873

Moore CM, Suggett D, Holligan PM, Sharples J, Abraham ER, Lucas MI, Rippeth TP, Fisher NR, Simpson JH, Hydes DJ (2003) Physical controls on phytoplankton physiology and production at a shelf sea front: a fast repetition-rate fluorometer based field study. Mar Ecol Prog Ser 259:29–45

Morris EP, Kromkamp JC (2003) Influence of temperature on the relationship between oxygen- and fluorescence-based estimates of photosynthetic parameters in a marine benthic diatom (*Cylindrotheca closterium*). Eur J Phycol 38:133–142

Morris EP, Forster RM, Peene J, Kromkamp JC (2008) Coupling between Photosystem II electron transport and carbon fixation in microphytobenthos. Aquat Microb Ecol 50:301–311

Mouget J-L, Tremblin G (2002) Suitability of the Fluorescence Monitoring System (FMS, Hansatech) for measurement of photosynthetic characteristics in algae. Aquat Bot 74:219–231

Mouget J-L, Perkins RG, Lefebvre S (2008) Migration or photoacclimation to prevent photoinhibition and UV-B damage in marine microphytobenthic communities? Aquat Microb Ecol 52:223–232

Myers J, Graham JR (1971) Photosynthetic unit in *Chlorella* measured by repetitive short flashes. Plant Physiol 48: 282–286

Nelson NB, Prezelin BB, Bidigare RR (1993) Phytoplankton light-absorption and the package effect in California coastal waters. Mar Ecol Prog Ser 94:217–227

Pemberton STJ, KL AJ, Geider RJ (2004) A methodology to determine primary production and phytoplankton photosynthetic parameters from Fast Repetition Rate Fluorometry. J Plankton Res 26:1337–1350

Perkins RG, Underwood GJC, Brotas V, Snow GC, Jesus B, Ribeiro L (2001) Responses of microphytobenthos to light: primary production and carbohydrate allocation over an emersion period. Mar Ecol Prog Ser 223:101–112

Perkins RG, Oxborough K, Hanlon ARM, Underwood GJC, Baker NR (2002) Can chlorophyll fluorescence be used to estimate the rate of photosynthetic electron transport within microphytobenthic biofilms? Mar Ecol Prog Ser 228:47–56

Perkins RG, Mouget J-L, Lefebvre S, Lavaud J (2006) Light response curve methodology and possible implications in the application of chlorophyll fluorescence to benthic diatoms. Mar Biol 149:703–712

Perkins RG, Kromkamp J, Reid RP (2007) How do stromatolite photosynthetic communities tolerate natural sand burial events? The roles of light and oxygen in photochemical reactivation. Mar Ecol Prog Ser 349:23–32

Prasil O, Kolber Z, Berry JA, Falkowski PG (1996) Cyclic electron flow around photosystem II in vivo. Photosynth Res 48:395–410

Raateoja MP (2004) Fast repetition rate fluorometry (FRRF) measuring phytoplankton productivity: a case study at the entrance to the Gulf of Finland, Baltic Sea. Boreal Env Res 9:263–276

Raateoja MP, Seppala J (2001) Light utilization and photosynthetic efficiency of Nannochloris sp (Chlorophyceae) approached by spectral absorption characteristics and Fast Repetition Rate Fluorometry (FRRF). Boreal Env Res 6, 205-220.

Raateoja M, Seppala J, Kuosa H (2004) Bio-optical modelling of primary production in the SW Finnish coastal zone, Baltic Sea: fast repetition rate fluorometry in Case 2 waters. Mar Ecol Prog Ser 267:9–26

Rech M (2004) Effets de l'éclairement visible et ultraviolet sur la croissance et la photosynthèse de microalgues: incidences sur l'écophysiologie du phytoplancton des claires ostréicoles. Ph.D. thesis, Le Mans University, 200 p

Rech M, Mouget J-L, Morant-Manceau A, Rosa P, Tremblin G (2005) Long-term acclimation to UV radiation: effects on growth, photosynthesis and carbonic anhydrase activity in marine diatoms. Bot Mar 48:407–420

Rees D, Lee CB, Gilmour DJ, Horton P (1992) Mechanisms for controlling balance between light input and utilization in the salt tolerant alga *Dunaliella* C9AA. Photosynth Res 32:181–191

Reid P, Visscher PT, Decho AW, Stolz JF, Bebout BM, Dupraz C, Macintyre LG, Paerl HW, Pinckney L, Prufert-Bebout L, Steppe F, DesMarais DJ (2000) The role of microbes in accretion, lamination and early lithification of modern marine stromatolites. Nature 406:989–992

Revsbech NP, Jørgensen BB (1983) Photosynthesis of benthic microflora measured with high spatial resolution by the oxygen microprofile method: capabilities and limitations of the method. Limnol Oceanogr 28:749–756

Ruban VA, Lavaud J, Rousseau B, Guglielmi G, Horton P, Etienne A-L (2004) The super-excess energy dissipation in diatom algae: comparative analysis with higher plants. Photosynth Res 82:165–175

Rysgaard S, Kuhl M, Glud RN, Hansen JW (2001) Biomass, production and horizontal patchiness of sea ice algae in a high-Arctic fjord (Young Sound, NE Greenland). Mar Ecol Prog Ser 223:15–26

Sagert S, Schubert H (2000) Acclimation of *Palmaria palmata* (Rhodophyta) to light intensity: comparison between artificial and natural light fields. J Phycol 36:1119–1128

Sarma VVSS, Abe O, Hashimoto S, Hinuma A, Saino T (2005) Seasonal variations in triple oxygen isotopes and gross oxygen production in the Sagami Bay, central Japan. Limnol Oceanogr 50:544–552

Sarma VVSS, Abe O, Hashimoto S, Hinuma A, Saino T (2006) Short-term variation of triple oxygen isotopes and gross oxygen production in the Sagami Bay, central Japan. Limnol Oceanogr 51:1432–1442

Schofield O, Kroon BMA, Prezelin BB (1995) Impact of Ultraviolet-B radiation on photosystem-II activity and its relationship to the inhibition of carbon fixation rates for Antarctic ice algae communities. J Phycol 31:703–715

Schreiber U, Schliwa U, Bilger W (1986) Continuous recording of photochemical and non-photochemical chlorophyll fluorescence quenching with a new type of modulation fluorometer. Photosynth Res 10:51–62

Schreiber U, Hormann H, Neubauer C, Klughammer C (1995) Assessment of photosystem II photochemical quantum yield by chlorophyll fluorescence quenching analysis. Aust J Plant Physiol 22:209–220

Schumann A, Goss R, Torsten J, Wilhelm C (2007) Investigation of the quenching efficiency of diatoxanthin in cells of *Phaeodactylum tricornutum* (Bacillariophyceae) with different pool sizes of xanthophyll cycle pigments. Phycologia 46:113–117

Sakshaug E, Bricaud A, Dandonneau Y, Falkowski P, Keifer D, Legendre L, Morel A, Parslow J, Takahashi M (1997) Parameters of photosynthesis: definitions, theory and interpretation of results. J Plankton Res 19:1637–1670

Schreiber U, Bilger W, Neubauer C (1994) Chlorophyll fluorescence as a nonintrusive indicator for rapid assessment of in vivo photosynthesis. Plant Phys. And Biochem 100: 49–70

Serôdio J (2003) A chlorophyll fluorescence index to estimate short-term rates of photosynthesis by intertidal microphytobenthos. J Phycol 39:33–46

Serôdio J (2004) Analysis of variable chlorophyll fluorescence in microphytobenthos assemblages: implications of the use of depth-integrated measurements. Aquat Microb Ecol 36:137–152

Schreiber U, Gademann R, Ralph PJ, Larkum AWD (1997) Assessment of Photosynthetic Performance of *Prochloron* in *Lissoclinum patella* in hospite by chlorophyll fluorescence measurements. Plant Cell Phys 38:945–951

Serôdio J, Catarino F (2000) Modelling the primary productivity of intertidal microphytobenthos: time scales of variability and effects of migratory rhythms. Mar Ecol Prog Ser 192:13–30

Serôdio J, Marques da Silva J, Catarino F (1997) Non destructive tracing of migratory rhythms of intertidal benthic microalgae using in vivo Chlorophyll *a* fluorescence. J Phycol 33:542–553

Serôdio J, da Silva JM, Catarino F (1998) Relationship between chlorophyll fluorescence quenching and O_2 evolution in microalgae. In: Garab G (ed) Photosynthesis: mechanisms and effects, vol V. Kluwer, Dordrecht, pp 4109–4112

Serôdio J, Marques da Silva J, Catarino F (2001) Use of in vivo chlorophyll a fluorescence to quantify short-term variations in the productive biomass of intertidal microphytobenthos. Mar Ecol Prog Ser 218:45–61

Serôdio J, Cruz S, Vieira S, Brotas V (2005) Non-photochemical quenching of chlorophyll fluorescence and operation of the xanthophyll cycle in estuarine microphytobenthos. J Exp Mar Biol Ecol 326:157–169

Serôdio J, Vieira S, Cruz S, Coelho H (2006a) Rapid light-response curves of chlorophyll fluorescence in microalgae: relationship to steady-state light curves and non-photochemical quenching in benthic diatom-dominated assemblages. Photosynth Res 90:29–43

Serôdio J, Coelho H, Vieira S, Cruz S (2006b) Microphytobenthos vertical migratory photoresponse as characterised by light-response curves of surface biomass. Est Coast Shelf Sci 68:547–556

Serôdio J, Vieira S, Barroso F (2007) Relationship of variable chlorophyll fluorescence indices to photosynthetic rates in microphytobenthos. Aquat Microb Ecol 49:71–85

Serôdio J, Vieira S, Cruz (2008) Photosynthetic activity, photoprotection and photoinhibition in intertidal microphytobenthos as studied in situ using variable chlorophyll fluorescence. Cont Shelf Res 28:1363–1375

Silva J, Santos R, Serôdio J, Melo RA (1998) Light response curves for *Gelidium sesquipedale* from different depths, determined by two methods: O_2 evolution and chlorophyll fluorescence. J Appl Phycol 10:295–301

Simis SGH, Tijdens M, Hoogveld HL, Gons HJ (2005) Optical changes associated with cyanobacterial bloom termination by viral lysis. J Plankt Res 27:937–949

Suggett D, Kraay G, Holligan P, Davey M, Aiken J, Geider R (2001) Assessment of photosynthesis in a spring cyanobacterial bloom by use of a fast repetition rate fluorometer. Limnol Oceanogr 46:802–810

Suggett DJ, Oxborough K, Baker NR, MacIntyre HL, Kana TD, Geider RJ (2003) Fast repetition rate and pulse amplitude modulation chlorophyll *a* fluorescence measurements for assessment of photosynthetic electron transport in marine phytoplankton. Eur J Phycol 38:371–384

Suggett DJ, MacIntyre HL, Geider RJ (2004) Evaluation of biophysical and optical determinations of light absorption by photosystem II in phytoplankton. Limnol Oceanogr Meth 2:316–332

Suggett DJ, Le Floc'H E, Harris GN, Leonardos N, Geider RJ (2007) Different strategies of photoacclimation by two strains of *Emiliania huxleyi* (Haptophyta). J Phycol 43: 1209–1222

Sušila P, Lazár D, Ilík P, Tomek P, Nauš J (2004) The gradient of exciting radiation within a sample affects the relative height of steps in the fast chlorophyll *a* fluorescence rise. Photosynthetica 42:161–172

Tassan S, Ferrari GM (1998) Measurement of light absorption by aquatic particles retained on filters: determination of the optical path length amplification by the 'transmittance-reflectance' method. J Plankton Res 29:1699–1709

Ting CS, Owens TG (1993) Photochemical and non-photochemical fluorescence quenching processes in the diatom *Phaeodactylum tricornutum*. Plant Physiol 101:1323–1330

Tuji A (2000) The effect of irradiance on the growth of different forms of freshwater diatoms: implications for succession in attached diatom communities. J Phycol 36:659–661

Underwood GJC, Kromkamp J (1999) Primary production by phytoplankton and microphytobenthos in estuaries. Adv Ecol Res 29:93–153

Underwood GJC, Perkins RG, Consalvey MC, Hanlon A, Oxborough K, Baker N, Paterson DM (2005) Patterns in microphytobenthic primary productivity: species-specific variation in migratory rhythms and photosynthetic efficiency in mixed-species biofilms. Limnol Oceanogr 50:755–767

Van De Poll WH, Van Leeuwe MA, Roggeveld J, Buma AGJ (2006) Nutrient limitation and high irradiance acclimation reduce PAR and UV-induced viability loss in the Antarctic diatom *Chaetoceros brevis* (Bacillariophyceae). J Phycol 41:840–850

Visscher PT, Reid RP, Bebout BM (2000) Microscale observations of sulfate reduction: Correlation of microbial activity with lithified micritic laminae in modern marine stromatolites. Geology 28:919–922

Waring J, Underwood GJC, Baker NR (2006) Impact of elevated UV-B radiation on photosynthetic electron transport, primary productivity and carbon allocation in estuarine epipelic diatoms. Plant Cell Env 29:521–534

White AJ, Critchley C (1999) Rapid light curves: A new fluorescence method to assess the state of the photosynthetic apparatus. Photosynth Res 59:63–72

Williams PJB (1993) Chemical and tracer methods of measuring plankton production. In: Li WKW, Maestrini SY (eds) Measurement of primary production from the molecular to the global scale. ICES, La Rochelle, pp 21–36

Williams PJL, Robertson JE (1991) Overall planktonic oxygen and carbon dioxide metabolisms: the problem of reconciling observations and calculations of photosynthetic quotients. J Plankton Res 13:153–169

Chapter 13
Chlorophyll Fluorescence Applications in Microalgal Mass Cultures

Jiří Masojídek, Avigad Vonshak, and Giuseppe Torzillo

1 Preface

Chlorophyll (Chl) fluorescence has become one of the most common and useful techniques in photosynthetic field research. Its non-invasiveness, sensitivity and the wide availability of reliable instruments, also makes it a convenient and suitable technique in microalgal biotechnology to monitor a culture's photosynthetic performance. Experimentally, homogenous microalgal cultures in suspension have also been ideal objects in photosynthetic studies. For the purpose of this book we summarised results of experiments since the 1990s that have pioneered the practical use of Chl fluorescence methods to monitor the physiological status of fast-growing microalgal mass cultures, optimising and estimating biomass productivity or finding marker processes of certain compound synthesis.

In their biomass, microalgae produce various valuable bioactive substances such as pigments, polyunsaturated fatty acids, antioxidants, essential amino acids or immunologically-effective, virostatic and cytostatic compounds. Therefore, microalgae are cultivated commercially for biomass as food and feed additives, as a source for pharmacology and cosmetics, or, on a small-scale, for research of diagnostic products.

2 Historical Overview of Using Chl Fluorescence in Microalgal Mass Cultures

Microalgal cultures, outdoors, are exposed to changes in environmental conditions, particularly irradiance. To cope with variable light intensity in combination with other stresses during the day, quickly-growing microalgae have developed fast and prompt regulation mechanisms, usually operating on a time-scale of seconds to minutes. Outdoor culture performance can be monitored through the assay of dry weight, or photosynthesis measurements carried out in the laboratory, but the time required to complete these measurements is rather long. In algal biotechnology warning signals must be recognized as soon as possible in order to prevent a significant reduction in daily productivity, or to avoid situations which, in a few days, may culminate in culture loss. Since unfavourable environmental conditions and their synergisms affect the function of Photosystem II (PSII), directly or indirectly, chlorophyll fluorescence represents a useful tool in microalgal biotechnology – giving rapid evidence of stress affecting the photosynthetic activity of the culture and certain quantification on productivity.

In the 1990s, Chl fluorescence measurements were employed to examine the photosynthetic performance of microalgal mass cultures (e.g. Vonshak et al. 1994; Torzillo et al. 1996). In particular, questions were studied associated with the relationship of fluorescence-based measures of PSII photobiochemical activity as a means to estimate primary productivity (Genty et al. 1989;

J. Masojídek (✉)
Institute of Microbiology, Academy of Science, Opatovický mlýn, CZ-37981 Třeboň, Czech Republic
and
Institute of Physical Biology, University of South Bohemia, Zámek 136, CZ-37333 Nové Hrady, Czech Republic
e-mail: masojidek@alga.cz

A. Vonshak
Ben Gurion University J. Blaustein Institutes for Desert Research, Sede Boqer 84990, Israel

G. Torzillo
Istituto per lo Studio degli Ecosistemi, CNR, Via Madonna del Piano 10, 50019 Sesto Fiorentino (Firenze), Italy

D.J. Suggett et al. (eds.), *Chlorophyll* a *Fluorescence in Aquatic Sciences: Methods and Applications*, Developments in Applied Phycology 4, DOI 10.1007/978-90-481-9268-7_13, © Springer Science+Business Media B.V. 2011

Torzillo et al. 1996; Schreiber et al. 1998; Baker 2008; Suggett et al., Chapter 6, this volume). It was essential to determine the fluorescence quenching that results from both photochemical and non-photochemical processes (Bradbury and Baker 1984; Schreiber et al. 1986).

The aim of this chapter is to provide a simple, practical guide to the use of Chl fluorescence for algal biotechnologists who wish to apply this technique in microalgal mass cultures, both in field and laboratory studies. Whilst the principles behind the measurements will be mentioned only briefly, the emphasis will be given to the applications and limitations of fluorescence technique for *in situ* monitoring and its implications.

3 Microalgae Grown for Commercial Purposes and Cultivation Systems

In applied phycology, the term *microalgae* is usually used in its broadest sense to mean both prokaryotic cyanobacteria and eukaryotic algae – unicellular or filamentous photosynthetic microorganisms. Microalgae represent important CO_2 consumers and primary producers – being the basis of the food chain in aquatic environments. Furthermore, they are one of the most efficient converters of solar energy to biomass. Dense, well-mixed mass cultures of microalgae (> 0.5 g biomass per litre) with sufficient nutrition and gas exchange represent artificial systems which are completely different from optically-thin natural phytoplankton populations with low biomass density, often limited by nutrient and carbon supply.

Many cultivation systems have been designed and built to grow microalgae using natural or artificial light. Two basic approaches to microalgal mass production are used: the first applies to cultivation in open reservoirs large in area, while the second are closed vessels – photobioreactors (for review see Pulz et al. 2001; Torzillo et al. 2003; Tredici 2004). The first type – open cultivation systems – is represented by natural or artificial ponds, raceways (ponds akin to race-tracks) and cascades (i.e. inclined-surface systems) (see example in Fig. 1, upper panel). In the second type – photobioreactors (closed or semi-closed systems with natural or artificial illumination) consist of glass or transparent plastic tubes, or panels, positioned horizontally or vertically, arranged as serpentine loops, flexible coils, manifold rows or 'fences', in which the microalgal suspension is continuously circulated (see example in Fig. 1, bottom panel).

In every cultivation system, several basic principles must be considered that affect productivity in mass microalgal cultures, such as: (i) culture depth, or optical cross-section (thicker culture layers would progressively absorb penetrating light more, rendering it unavailable for photosynthesis); (ii) turbulence where various means and techniques could be used; (iii) nutrient content and supply, including gas exchange (CO_2 supply and O_2 removal); (iv) cultivation procedure (batch, continuous or semi-continuous, or multistage processes); (v) biomass concentration and areal density; and (vi) acclimation state of microalgae (Richmond 2004; Grobbelaar 2007). The choice of a suitable cultivation system and the adjustment of the cultivation regime must be worked out for each individual productive microalgal strain.

Myriads of microalgae have been isolated from natural habitats and are kept in numerous culture collections around the world. However, to date, only a few microalgal strains, mostly of aquatic origin, have been cultivated in large-scale production systems of hundreds to thousands of litres.

Arthrospira (Spirulina) platensis is a planktonic filamentous cyanobacterium composed of individual cells (about 8 μm in diameter), which grows in subtropical alkaline lakes with a temperature optimum of about 35°C. In productive cultures, *Arthrospira* is cultivated in shallow mixed ponds or semi-closed tubular photobioreactors. The growth medium contains inorganic salts with a high concentration of bicarbonate, keeping the pH value between 9 and 10. This cyanobacterium is the most cultivated photosynthetic prokaryote since its biomass is widely used as a health food, feed supplement and as a source of fine chemicals. It contains proteins, polyunsaturated fatty acids, phycobiliproteins, carotenoids, polysaccharides, vitamins and minerals.

The microalga *Chlorella* (green alga, Chlorophyta) is a cosmopolitan genus with small globular cells (3–8 μm in diameter), including strains with a high temperature tolerance since some can grow between 15°C and 40°C. *Chlorella* grows autotrophically in an inorganic medium, as well as in mixotrophic and heterotrophic conditions (for example, with an addition of acetic acid or glucose). At present, autotrophic production of *Chlorella* is carried out in open ponds, semi-closed tubular photobioreactors, or inclined cascades, since its high growth rate prevents contamination

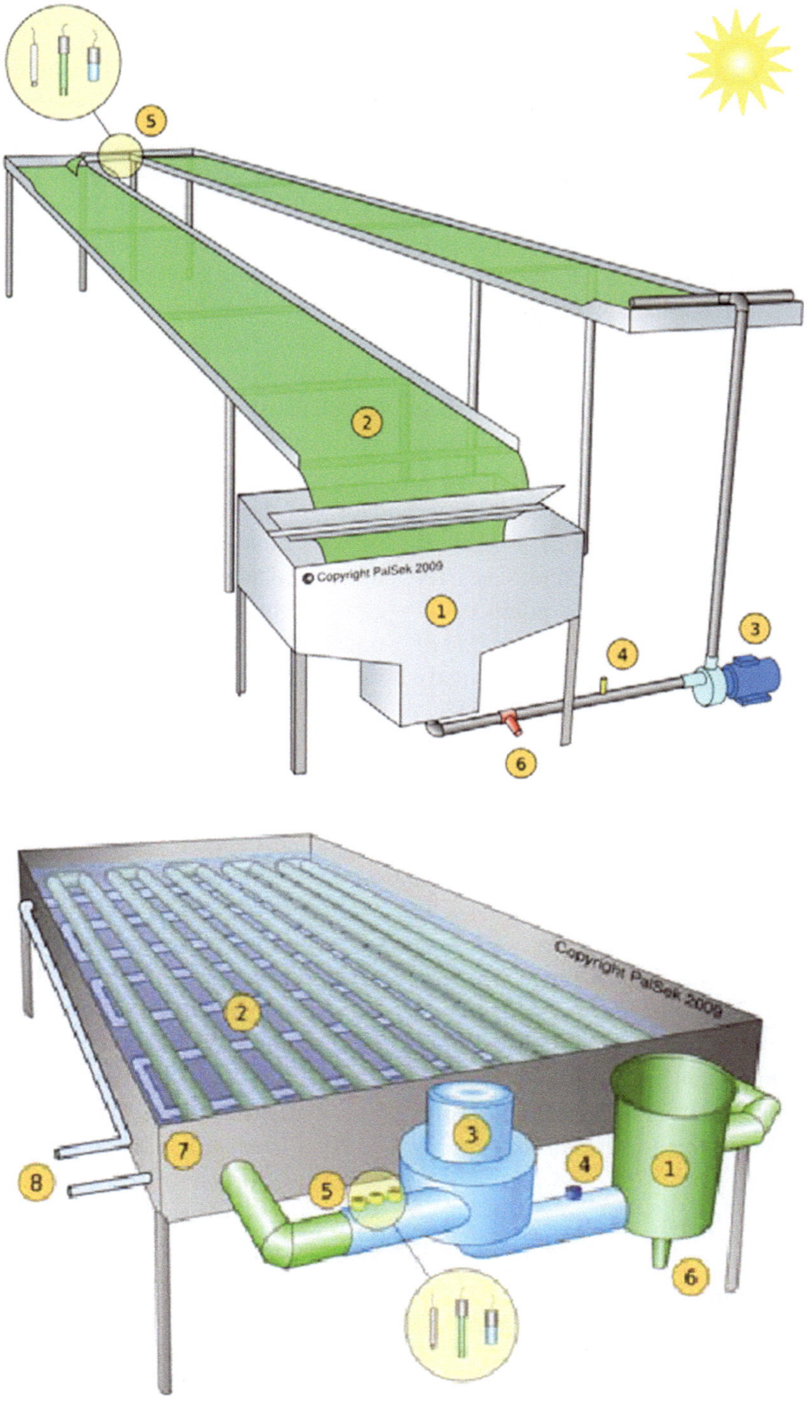

Fig. 1 (*Upper panel*) Schematic diagram of open system for cultivation of microalgae – a cascade where a thin layer (6 mm) of suspension flows along declined surface. The experimental unit (surface of 24 m^2; volume 225 L) consists of two declined cultivation lanes made of glass plates (declination of 1.7%) supported by a scaffolding (Institute of Microbiology, Academy of Sciences, Třeboň , Czech Republic). (*Bottom panel*) Schematic diagram of outdoor closed system – a tubular photobioreactor installed at the Istituto per lo Studio degli Ecosistemi, CNR, in Sesto Fiorentino, Italy. This experimental photobioreactor with the volume of 50 L consists of ten parallel glass tubes (length 2 m, i.d. 4.85 cm) connected by U-bends which are placed in a thermostated stainless steel basin containing water. The cultivation units are made up of several parts: 1 – retention tank (degasser), 2 – declined cultivation surface (loop), 3 – circulation device (e.g. pump), 4 – CO_2 supply, 5 – measurement and control sensors (pH, dissolved oxygen, temperature), 7 – thermostated stainless steel basin containing water, 8 – cooling system. The suspension flows into the retention tank (degasser) from where it is pumped by a pump back to the cultivation area (loop)

by other microalgae. *Chlorella* is the most cultivated eukaryotic alga since it is widely used as a health food and feed supplement, as well as in the pharmaceutical and cosmetics industry. It contains proteins, carotenoids, some immunostimulators, polysaccharides, vitamins and minerals. The bulk of the microalgae biomass market is represented by *Chlorella* and *Arthrospira* with annual production of 3,000 t and 4,000 t, respectively.

Dunaliella salina (Chlorophyta) and similar hypersaline strains have biflagellated, pear-shaped cells (about 10 μm in diameter). *Dunaliella* produces β-carotene in high amounts, up to 12% of dry matter. This microalga is a natural source of carotenoids for some brine shrimp. The high content of β-carotene makes *Dunaliella* attractive to biotechnologists for large-scale production in shallow, open ponds under high solar radiation (Borowitzka 2005). About 25 t of β-carotene enriched health food is produced yearly (Ben-Amotz 2004).

Haematococcus pluvialis (Chlorophyta) is a freshwater, unicellular alga with a rather complex life-cycle. A two-stage process is employed for biomass production. Under stress conditions (nutrient deficiency, salinity, high temperatures in combination with high irradiance), the green vegetative cells (about 10 μm diameter) produce thicker walls and change to large globular cysts (about 50 μm in diameter) with orange-red pigmentation, due to an increased deposition of astaxanthin (up to 5% of dry weight). This pigment is the important natural colorant for salmonoid fish, shrimp, lobster and crayfish and health food market. Annual production of *Haematococcus* biomass is about 100 t.

4 Principles of Microalgae Mass Culturing

Microalgae belong to the fastest-growing photosynthetic organisms since their cell doubling time can be as little as several hours. The necessary cultivation requirements for the growth of microalgal mass cultures are light, a suitable temperature and pH of the growth medium, as well as a sufficient carbon and nutrient supply. As microalgal mass cultures grow in dense suspensions, some kind of mixing is necessary to expose cells to light evenly and to allow for an efficient gas exchange (CO_2 supply/O_2 removal).

Light is the most critical factor for microalgal growth. The amount of photon energy received by each cell is a combination of several factors: irradiance intensity, cell density, length of optical path (thickness of culture layer), spectral quality (light penetration), light absorption, and rate of mixing (Richmond 2004). Two basic factors have to be fulfilled for obtaining maximal photochemical efficiency and productivity: (i) cell density must be optimal, insuring that the average photon irradiance per cell is close to upper end of the linear phase of the growth curve, and (ii) light-dark cycles of the cells caused by culture turbulence must be fast, close to the turnover of the photosynthetic units (Nedbal et al. 1996; Richmond 2004). The light captured by photosynthetic pigments is roughly ten times higher under full sunlight (2000 μmol photon m^{-2} s^{-1}) than that required to saturate growth. In other words, as much as 90% of the photons captured by the Chl antennae can be dissipated as heat and fluorescence. Uncritical acceptance of uncorrected photosynthetic efficiencies of about 10% or even higher (Pirt 1986) inevitably leads to exaggerated estimates of present and future biomass productivity. We can however approach a rather more realistic maximum figure of photosynthetic efficiency (photon energy converted into biomass energy) of about 4.5% for C3 plants or microalgae by using "educated guesswork" and detailed consideration of the partial reactions involved (e.g. Boardman 1980; Benemann and Oswald 1996; Zhu et al. 2008; Walker 2009).

After irradiance, temperature is the most important parameter to control the microalgal culture growth. Some microalgal strains tolerate a broad temperature range between 15°C and 35°C (e.g. *Chlorella*), while *Haematococcus* usually requires much narrower range. However, for the majority of freshwater microalgae the optimum temperature range is within 25–30°C.

Basically, two cultivation regimes are used for the growth of microalgal cultures. In the batch regime, the culture is inoculated and at a certain point of growth it is harvested. In the continuous regime, the culture is harvested continuously according to its growth rate and fresh medium is added to replace nutrients. In practice, semi-continuous or semi-batch regimes are often adopted; i.e. where a part of the culture is harvested at regular intervals, usually every day.

4.1 Culture Maintenance

Successful cultivation requires continuous monitoring of physico-chemical parameters, i.e. irradiance, pH, temperature, dissolved oxygen concentration, and nutrient status. The basic biological method used is a microscopic examination to detect morphological changes and contamination by other microorganisms. CO_2 serves as the main carbon source and is added on demand. Nutrient status can be followed by monitoring the concentration of nitrogen, using it as a measure for adding the proportional amounts of other nutrients. In the mass cultivation of microalgae, monocultures are usually required for biomass exploitation; contaminants often represent a substantial limitation to large-scale production in microalgae cultures. Thus in some cases, e.g. for the cultivation of *Haematococcus*, the use of a closed system becomes mandatory.

Sufficient degassing of the microalgal suspension is necessary in order to prevent the accumulation of oxygen in the culture that can increase photorespiration and promote photoinhibition of photosynthesis resulting in a decline in growth, particularly when the suspension is cultivated in a closed system (Torzillo et al. 1998). On the other hand, excessive mixing can cause hydrodynamic stresses on the cells, and consequently reduce productivity.

Culture growth may be estimated as changes of optical density (O.D.) at 750 nm, or in biomass dry weight in time. Specific growth rate of culture is obtained with the following equation, $\mu = (\ln X_2 - \ln X_1)/(t_2 - t_1)$ expressed in [h^{-1} or day^{-1}].

Biomass productivity is expressed as the areal or volumetric yield per unit time, i.e. in [$g\ m^{-2}\ day^{-1}$] or in [$g\ L^{-1}\ day^{-1}$], respectively. The highest μ is obtained when all requirements for cell growth are optimal and light is saturating (low biomass concentration). Highest productivity, in contrast, is achieved when cells are light-limited (dense cultures), the growth rate of which being about a half of the maximum (Richmond 2004).

Methods of biophysical and biochemical monitoring of activity generally reflect the status of the cells' photosynthetic apparatus and are used to adjust the appropriate cultivation conditions for the production of biomass or certain compounds. For example, photosynthetic oxygen production and Chl fluorescence yield are considered as reliable and sensitive indicators of photosynthetic activity in microalgal cultures.

5 Interpretation of Chl Fluorescence Parameters in Microalgae Mass Cultures

Chl fluorescence measurements make it possible to follow the energy distribution between the photochemical and non-photochemical processes in photosynthesis. Different fluorescence-based parameters are described in the literature for the yields of photochemical energy conversion in PSII, which are complementary with the yields of non-photochemical energy dissipation.

Microalgae mass cultures are well-mixed suspensions representing averaged cell population, dense and homogenous which represent a completely different experimental object as compared to optically-thin phytoplankton populations with chlorophyll concentration lower by several orders of magnitude, or static plant leaves where fluorescence signals are mostly emitted by a surface cell layer. In microalgal cultures, Chl fluorescence reflects the performance of photochemical processes in the photosynthetic apparatus, especially PSII, and subsequently its biomass productivity. The contribution of the PSI emission in the total signal is mostly small and for practical purposes is often neglected. A careful approach is required when measuring fluorescence and evaluating data in cyanobacteria since the fluorescence emission of numerous PSI complexes and phycobilisomes as well as state transition effect contribute significantly to the total signal, which affects the correct determination of certain parameters (Ting and Owens 1992; Büchel and Wilhelm 1993; Schreiber et al. 1995).

At present, basically two Chl fluorescence approaches are used to monitor photosynthetic efficiency in microalgal mass cultures: rapid fluorescence induction or relaxation kinetics (Fig. 2) and the saturation-pulse method (Fig. 3).

At first, and for a long time, starting with the experiments of Kautsky in the 1930s, the most common way of measuring Chl fluorescence was based on the 'conventional' principle using dark-adapted samples. Fluorescence is excited by multi- or single-turnover light (e.g. produced usually by red LED; $\lambda_{max} = 660$ nm

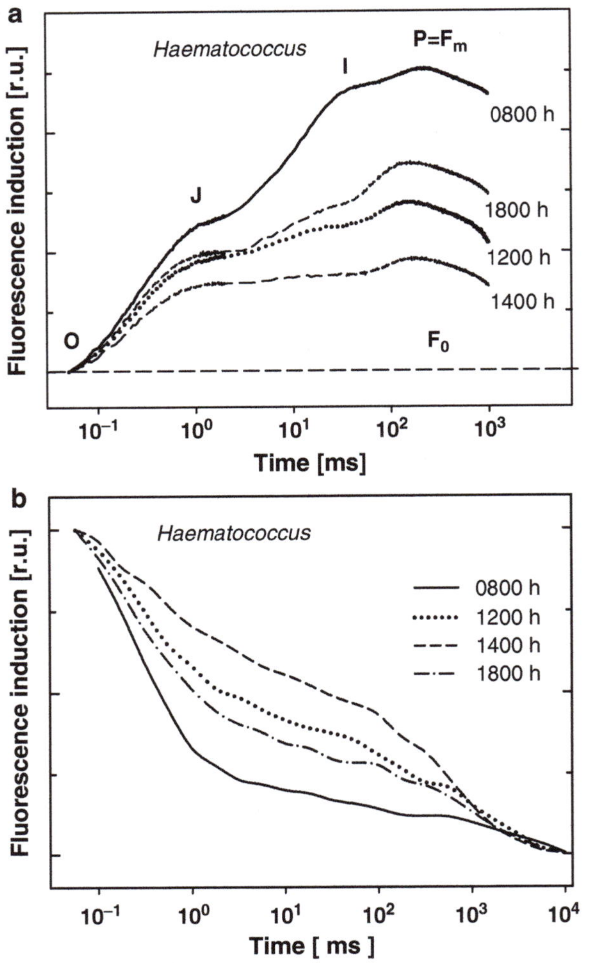

Fig. 2 Rapid Chl fluorescence induction kinetics measured in an outdoor culture of the alga *Haematococcus* at various periods of the diel cycle (0800, 1200, 1400 and 1800 h). Rapid fluorescence induction curves normalised to F_0 level (**panel a**). Intermediate steps of J and I represent various reduction states of the electron carriers of the PSII complex. Q_A reoxidation kinetics after single-turnover light pulse was measured in an outdoor culture of the alga *Haematococcus* at various periods of the diel cycle (0800, 1200, 1400 and 1800 h) (**panel b**)

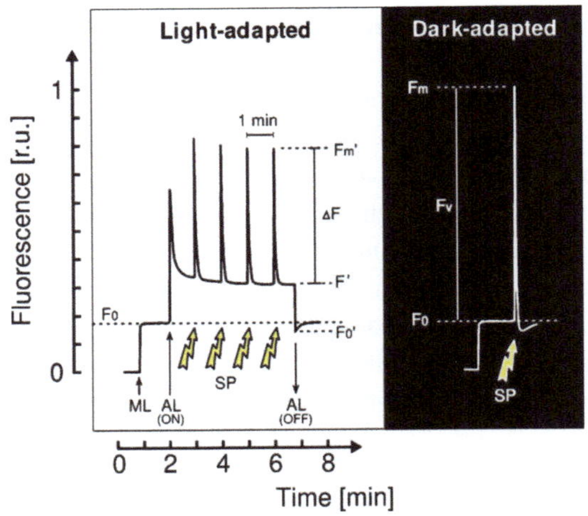

Fig. 3 Fluorescence quenching analysis using saturation-pulse method. The sample (microalgae culture) is illuminated with actinic light (AL) and a series of saturating pulses (SP) in order to reach the steady state F′ and F_m' level. The minimum and maximum fluorescence levels F_0 and F_m are measured in the dark-adapted sample (10–15 min) using modulated measuring light (ML) and SP. In case of cyanobacteria, a PSII electron transport inhibitor, diuron has to be used to achieve F_m. The parameters denoted with a prime (′) are from the sample exposed to light. The parameters without a prime are obtained from the sample in the dark-adapted state. Weak measuring light (ML; < 0.3 μmol photons m^{-2} s^{-1}) has the subsaturating intensity to PSII photochemistry; saturating light pulse (SP; >10 mmol photons m^{-2} s^{-1}; 0.8 s duration)

which is absorbed by chlorophyll) and fluorescence emission is detected at wavelengths beyond 690–710 nm with the help of a suitable photosensor (photomultiplier or photodiode). The rapid Chl fluorescence induction kinetics (the so-called Kautsky curve) of all oxygenic photosynthetic organisms shows a polyphasic rise (Chl fluorescence transient) between the initial (F_0) and the maximum (F_m) fluorescence during the first second of illumination (Neubauer and Schreiber 1987; Schreiber and Neubauer 1987). The phases on the curve were designated as O, J, I and P using a logarithmic time scale (Strasser et al. 1995; Fig. 2a in this chapter). Since the 1990s, the theory has been elaborated to become the so-called 'JIP-test' (Govindjee 1995; Strasser and Strasser 1995; Strasser et al. 1995; Srivastava et al. 1999). The current understanding of the OJIP transient rise is that it reflects the filling-up (i.e. reduction) of the electron acceptor pool of PSII (Ph, Q_A, Q_B and PQ pool) in a four-photon process (Strasser et al. 2004). The inflection J (2 ms) represents the double-reduction of electron carriers Ph, Q_A and Q_B, and because of a limitation in electron acceptance by Q_B, this step usually occurs when cells are exposed to excessive light that increases the degree of reduction of the PQ pool. The step I (30 ms) is connected to a three-electron reduction in the PSII electron carriers to different redox states (e.g. Ph $Q_A^-Q_B^{2-}$, $Ph^-Q_AQ_B^{2-}$ or $Ph^-Q_A^-Q_B^-$) of the reaction centre complex which reduces the PQ pool and probably reflects the heterogeneity of the PSII centres, the Q_A-fast-reducing and slow-reducing ones. The drop in the induction curve of microalgal cells beyond the P step indicates that the PQ pool is being re-oxidized due to the demand of reduction equivalents from the Calvin-Benson cycle.

Thus, the 'JIP' test can be used as a rapid monitor of photosynthetic activity including the effects of inhibitors and mutations on this process.

In microalgae (*Haematococcus*) cultures, the fluorescence induction curves of morning samples (0800 h) showed clear J and I steps in the typical fluorescence Chl induction curve (Fig. 2a). The cultures became light-stressed at mid-day (1200 and 1400 h) corresponding to an over-reduction of the PSII electron acceptors due to excessive irradiance. The presence of the step I was diminished in the 1200 and 1400 h samples, which suggests that the mid-day depression or down-regulation in the Q_B-slowly-reducing centres prevailed. Hence, 'slow' reaction centres may provide a mechanism for quenching excessive energy (Strasser et al. 1995). The situation was partially reversed by late afternoon (1800 h) when PSII over-reduction was released.

The kinetics of Q_A^- re-oxidation, measured as Chl fluorescence decay after single-turnover light pulse, can give detailed information about the rate of electron transport processes on the acceptor side of PSII (Fig. 2b). Analysis of the fluorescence relaxation kinetics is based on the widely-used model of the two-electron gate (Diner 1998). According to this model, the fast decay (in hundreds of microseconds) component reflects Q_A^- re-oxidation via the forward electron transport in reaction centres, which contain bound PQ (in the oxidised or semi-reduced form) at the Q_B site before the pulse. The middle phase (in milliseconds) arises from Q_A^- re-oxidation in centres that had an empty Q_B site at the time of the flash and have to bind a PQ molecule from the PQ pool. Finally, the slow phase (in seconds) reflects Q_A^- re-oxidation via the back reaction with the S-states of the water-oxidising complex. As an example, we show the Q_A re-oxidation curves of the microalga *Haematococcus* during the diel cycle when diluted laboratory cultures were exposed to high irradiance outdoors (Torzillo et al. 2003). The Q_A^- re-oxidation kinetic measurements carried out on dark-adapted samples taken in the morning, mid-day and late afternoon showed significant differences. In the morning sample, the relaxation of fluorescence yield was dominated by the fast (330 μs) phase, the relative amplitude of which was about 74%. The contribution of the medium (13 ms) phase was about 12% and that of the slow phase (about 2.8 s) was 14%. Analysis of decay curve of the sample taken at 1200 h showed mainly an increase of the $t_{1/2}$ of the fast phase to about 417 μs which contribution was about only 35%, while the medium and slow phase abundance prevailed. It indicated a slower electron transfer from Q_A^- to bound Q_B which was possibly also the consequence of the Q_B site modification. At 1800 h, the Q_A re-oxidation kinetics partially relaxed close to the situation in the morning samples as it was dominated by the fast (380 μs) phase, the relative amplitude of which was about 53%.

In the other approach – pulse-amplitude-modulation (PAM) fluorometry – a weak modulated excitation beam is employed, and the fluorescence signal is processed by a selective amplifier which rejects all non-modulated signals (Bradbury and Baker 1984; Schreiber et al. 1986). The principle of *in situ* modulated Chl fluorescence measurement in turbulent microalgal cultures is based on the assumption that signal modulation is two-orders of magnitude faster then the suspension movement in a cultivation unit (Fig. 4). As compared to induction fluorometers, PAM devices have been significant step forward due to the following advantages: low intensities of modulated measuring light do not excite the PSII reaction centres, i.e. the basic F_0-level can be continuously monitored; photochemical and non-photochemical quenching can be clearly separated and quantified using the saturation-pulse method *in situ*; and the signal is not disturbed by ambient light, such that experiments can be carried out continuously, even in full sunlight in the field. A measurement *in-situ* can be made by simply pointing a fluorometer fibre-optics at a photobioreactor or by submerging it into the suspension using an ambient irradiance as actinic light.

Some physiologically-useful expressions have been derived from induction curves and saturation-pulse analysis (Figs. 2a and 3, Tables 1 and 2) which comprehensively describe the fate of the excitation energy in PSII – photochemical and non-photochemical energy conversion (recent reviews by Strasser et al. 2004; Schreiber 2004; Baker 2008; Huot and Babin, Chapter 3, this volume).

The F_v/F_m ratio, designated as the maximum photochemical yield of PSII (variable to maximum fluorescence yield after dark adaptation; usually 10–15 min is sufficient in dense microalgal mass cultures) has been found to be a convenient measure of the performance of photochemical processes in the PSII complex (Kitajima and Butler 1975), and its decline represents a reliable warning signal when microalgal

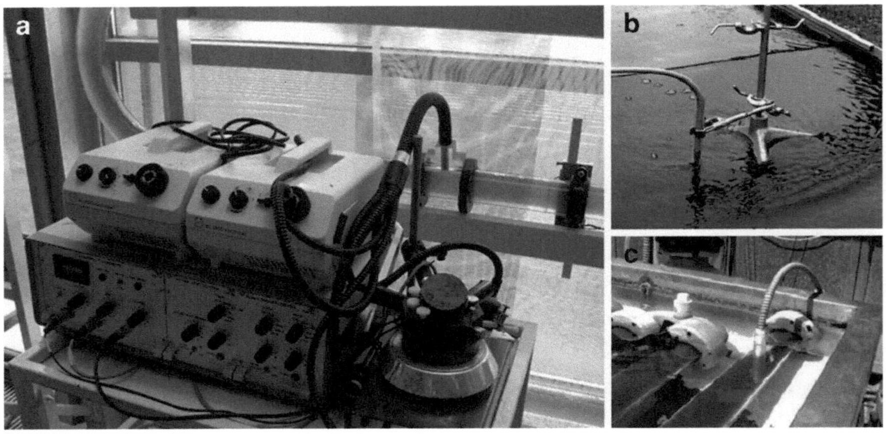

Fig. 4 In-situ measurement of Chl fluorescence quenching using a fibre-optic guide and a pulse-amplitude-modulation fluorometer in a greenhouse tubular photobioreactor (**panel a**). The fibre-optics were protected in a glass test tube and submerged in the microalgal suspension (**panel b**), or placed perpendicularly to the glass wall of the cultivation tube (**panel c**). Distance between the fibre-optics and the culture suspension was about 3 mm and the fibre-optics angle to the sun was about 60°

Table 1 Effect of sub-optimum temperature and high dissolved oxygen concentration on relative electron transport rate rETR and biomass productivity in cultures of the cyanobacterium *Arthrospira* grown in outdoor photobioreactors (Torzillo et al. 1996, 1998)

Temperature	Dissolved oxygen concentration	rETR ($\Delta F/F_m' \times PAR$)		Biomass productivity	
[°C]	[mg L^{-1}]	[mmol $e^- m^{-2} d^{-1}$]	[%]	[g DW $m^{-2} d^{-1}$]	[%]
35	22 ± 2	11100	100	29.0	100
35	60 ± 19	7500	–33	19.4	–33
25	58 ± 16	4300	–60	12.0	–61

Table 2 Selected parameters calculated from the time-course of fluorescence induction curves (Fig. 2a). F_0, F_J, F_I – fluorescence intensity at O-step (0.05 ms), at J-step (~2 ms) and I-step (~20 ms) of fluorescence induction curve

Parameter	Symbol	Formula
Fluorescence yield at J-step	V_J	$V_J = (F_J - F_0)/(F_m - F_0)$
Fluorescence yield at I-step	V_I	$V_I = (F_J - F_0)/(F_m - F_0)$
Maximum photochemical yield of PSII	F_v/F_m	$F_v/F_m = (F_m - F_0)/F_m$

culture growth gets stressed (Vonshak et al. 1994). In outdoor mass cultures, F_v/F_m frequently exhibits a diurnal depression that is roughly symmetric with the irradiance intensity and is mirrored by corresponding changes in the photosynthetic electron transport yield (Neale 1987; Torzillo et al. 1996). In healthy microalgal cultures, F_v/F_m ranges from 0.6 for cyanobacteria, to 0.8 for green algae, and varies significantly during the diurnal cycle, depending on the irradiance regime and treatment which determines the physiological status. The fraction of absorbed light utilized in electron transport is given by the actual PSII quantum yield, designated as Φ_{PSII} or $\Delta F/F_m'$ which might correlate with the reduction in the quantum yield of oxygen evolution or CO_2 uptake (Genty et al. 1989; Torzillo et al. 1998).

The open reaction centre acts as a fluorescence quencher and the fluorescence yield rises proportionally with the level of PSII closure. This phenomenon is called photochemical quenching. On the other hand, the so-called non-photochemical quenching indicates an increased heat dissipation of absorbed energy. In principle, non-photochemical quenching is inversely related to photochemistry, and is considered a safety valve protecting PSII reaction centres from damage by excess irradiance. The non-photochemical quenching coefficient is often calculated as the Stern-Volmer quenching coefficient NPQ (Bilger and Björkman

1990; Gilmore and Yamamoto 1991). Stress-induced damage of the photosynthetic apparatus is often reflected by an increase of NPQ which can compensate for a decrease of Φ_{PSII}. Compared to the qP and qN coefficients, the Φ_{PSII} and NPQ calculation does not need the determination of F_0' which might be problematic in the field.

A relative estimate of the electron transport rate through PSII can be obtained as the product of the actual photochemical yield of PSII and photosynthetic photon flux density, rETR = $\Delta F/F_m' \times$ PPFD expressed in mol photons m^{-2} s^{-1} (Hofstraat et al. 1994; Maxwell and Johnson 2000; Schreiber 2004). For example, possible cyclic electron transport would cause deviations (Prášil et al. 1996). To translate rETR into 'absolute' ETR requires that light absorption specific cross section of PSII, termed a_{PSII}, which is often not available; in such cases relative electron transfer rates rETR are frequently reported (Suggett et al., Chapter 6, this volume). Our experiments with *Spirulina* cultures cultivated under unfavourable conditions of sub-optimum temperature and high dissolved oxygen concentration showed a good correlation between daily sum of rETR and biomass productivity since the trends of these two parameters were similar showing the same percentage of decrease (Table 1).

Recently, saturation-pulse expressions have been updated in order to point out new parameters – Y(NO) = F'/F_m and Y(NPQ) = $F'/F_m' - F'/F_m$ (Klughammer and Schreiber 2008). The validity and usefulness of the last two parameters have to be tested yet in outdoor microalgal mass cultures.

Rapid light-response curves of chlorophyll fluorescence and photosynthetic oxygen production might be simultaneously measured in microalgal cultures at various light intensities, similar to that in phytoplankton populations, in a flow-through regime in a closed chamber (Masojídek et al. 2000, 2001). This can provide information about the acclimation status of the photosynthetic apparatus of microalgal mass cultures and the relationship between the electron transport rate through PSII, respiration and photosynthetic oxygen evolution over a diel cycle, and help explain the involvement of alternate oxygen-consuming electron transfer pathways as a possible explanation for some discrepancies. Although fluorescence parameters can be measured easily, some problems might arise when they are used to predict changes in photosynthetic performance. In particular, questions associated with the accurate estimation of PSII efficiency and the roles of photochemical and non-photochemical quenching as measured by fluorescence and its relationship with the rates of linear electron flux and CO_2 assimilation have to be carefully judged.

6 Chlorophyll Fluorescence Monitoring in Microalgal Mass Cultures

Outdoor dense microalgal cultures may experience large daily variations in light intensity – in the range of one order of magnitude. Outdoor microalgal cultures in cultivation units are usually influenced by the different time-scales of light-dark regimes (Richmond 2004). The first one represents a fast, intermittent light-dark regime which is induced by turbulence of microalgal suspension. As a result, the cells in dense cultures can be shifted between full sunlight, when they are situated in the upper layer, and complete darkness, when they are at the bottom of the culture. The second type of light regime is usually directed by the circulation between the cultivation loop and degasser (dark volume) in tens of seconds to minutes. Thirdly, the slowest light-dark changes are related to diurnal changes in solar radiation.

In most cases, the photosynthetic activity of microalgae becomes saturated within 200 μmol photons m^{-2} s^{-1}, that is 10% of the maximum solar irradiance. Photosynthetic activity at sub-saturating irradiance is rate-limited by light absorption and excitation energy transfer to the PSII reaction centres. Dense microalgal mass cultures are therefore predominantly grown at light limitation and, consequently, their photosynthetic performance would be more dependent on the lineal part of the P/E curve rather than on the light-saturated part (Vonshak and Torzillo 2004). It means that in a dense culture of microalgae the incident light intensity on the surface that penetrates to greater depths is utilized with maximum efficiency because it falls within the limited region of the growth curve.

Conversely, at over-saturating light intensities (in optically thin, low-biomass cultures), photosynthesis is limited by the interplay between electron-transfer processes and the capacity of enzymatic processes in the Calvin-Benson cycle (Sukenik et al. 1987). The rate of consumption of NADPH and ATP are major factors that determine PSII operating efficiency in many situations. Numerous environmental stresses impact on CO_2 assimilation, although the sites of photosynthesis limitation during these stresses can be quite varied. Stress-induced decreases in carbon metabolism, and transport processes can all decrease PSII efficiency.

Sufficient mixing at high biomass concentrations represents the most practical way to reduce the light-saturation effect in outdoor dense cultures and, on the other hand to avoid cells becoming low-light-acclimated, causing an excessive development of antennae pigments which in turn further reduces the light penetration into the culture layer. The mechanisms facilitating photoprotection (down-regulation) adjust the rate of dissipation of the absorbed radiation energy so that the excitation energy density in the PSII antenna is sufficient to drive photosynthesis at the rate needed for assimilatory reactions. The regulation of the PSII output can be performed in several ways – by modulation of its light-harvesting capacity, by changes in the number of active PSII reaction centres (Ramus 1981), or by modulating the activity of downstream energy-consuming processes like Mehler reaction and photorespiration.

In low-biomass cultures (< 1 g dry weight L^{-1}), despite turbulent mixing, photo-stress cannot be avoided. The possibility of over-excitation of the photosynthetic apparatus is strongly increased if other stresses which limit carbon metabolism, are superimposed (suboptimum temperature, nutrient deficiency, or high dissolved-oxygen concentration). Under prolonged supra-optimal irradiance, photosynthetic rates usually decline from the light-saturated value, paralleled by a decline of oxygen production. This phenomenon is commonly referred to as 'photoinhibition' of photosynthesis, illustrated, for example, as a light-induced depression in the maximum quantum yield of PSII photochemistry, measured as F_v/F_m.

Photoinhibition (a mid-day suppression of photobiochemical activity by supra-optimal irradiance) in outdoor cultures of *Arthrospira platensis* has been studied by measuring saturating pulse analysis of fluorescence quenching and the polyphasic rise of Chl fluorescence transients, providing information on the primary photochemistry of PSII (Lu and Vonshak 1999; Torzillo et al. 1996). In the studies, the maximum efficiency of PSII photochemistry (F_v/F_m) and actual PSII photochemical yield, $\Delta F/F_m'$ declined at mid-day depending on biomass density in response to the highest irradiance and recovered as the irradiance decreased in the afternoon. Nevertheless, in some situations, photosynthesis measurements might indicate high light-saturated photosynthesis rates (P_{max}) at midday, along with a lowered PSII photochemical yield, mirrored in a decline of F_v/F_m. The overall electron transport (rETR) can remain virtually unaltered due to increased electron turnover through the remaining functional PSII centres, even after a reduction in the number of active PS II units. In this case, photoinhibition does not directly impact the rate limiting step for photosynthesis at light saturation (Behrenfeld et al. 1998).

Temperature represents another important biological limitation for mass production of microalgae. Even in summer, in moderate climate zones or desert areas, the morning temperature of the culture in open ponds can be as much as 10°C below the optimum value, causing a decrease of the photosynthetic capacity of the microalgal culture for a few morning hours. The main cause of susceptibility to photo-stress at low temperature is usually a decrease of the photosynthesis rate, thus increasing the proportion of closed PSII reaction centres at a given photon fluency rate. Chl fluorescence technique *in situ* has often been applied to study the diurnal synergism of low temperature and high irradiance stress in microalgal cultures grown in outdoor tubular photobioreactors (Vonshak et al. 1994, 1996, 2001; Torzillo et al. 1996). Photoinhibition at sub-optimal temperatures caused a 30%-depression of F_v/F_m of low-density *Arthrospira* cultures grown outdoors at 25°C (i.e. 10°C below optimum value) at midday, while the $\Delta F/F_m'$ratio showed a reduction of up to 50% (Torzillo et al. 1996 and Table 1 in this chapter). The daily productivity of the culture was reduced by 30% with respect to that grown at 35°C. These results showed that measurements of PSII photochemical yield can be used to monitor the physiological status of sub-optimal temperature stressed *Arthrospira* cultures.

Even a relatively short exposure to sub-optimal morning temperatures induced photoinhibitory damage in cultures of the microalga *Monodus* which persisted till the late afternoon (Vonshak et al. 2001). Diel changes in the cultures grown under sub-optimal morning temperature showed a significant decrease in photosynthesis rates measured as oxygen production and Chl fluorescence quenching. Although the two cultures – heated (for several hours in the morning), and non-heated – were maintained most of the day at the same temperature, a diurnal decline of F_v/F_m was faster and of greater extent in the non-heated cultures, reaching a midday value of 0.48, compared to 0.58 in the heated culture. The differences in photosynthetic activity between the two cultures were also reflected in biomass productivity which was much higher in the heated culture (by 60%) compared to the non-heated culture.

Another important environmental factor affecting microalgal cultivation represents the concentration of oxygen dissolved in the suspension, particularly in closed photobioreactors. The saturating pulse fluorescence technique has been applied to study the photoinhibition of photosynthesis in outdoor cultures of *Arthrospira* grown under high oxygen and/or low temperature stress in closed outdoor photobioreactors (Torzillo et al. 1998). Diurnal changes showed that when solar irradiance reached the maximum value (between 1200 and 1300 h), the maximum photochemical quantum yield of PSII in dark-adapted state, F_v/F_m and the actual photochemical quantum yield of PSII in light-adapted state $\Delta F/F_m'$ ratios of the *Arthrospira* cultures grown under high oxygen stress decreased by 35% and 60%, respectively, compared with the morning values. When high oxygen stress was combined with sub-optimum temperature, F_v/F_m and $\Delta F/F_m'$ dropped even more, by 55% and 84%, respectively. Photoinhibition reduced the daily productivity of the culture grown under high oxygen stress by 33%, and that of the culture grown under high oxygen-low temperature stress by about 60%. Changes in the biomass yield of the cultures correlated well with changes in the estimated relative electron transport rate through the PSII complex, rETR (Table 1).

The influence of unfavourable conditions (stressors) such as high temperature and high pH in combination with high irradiance were studied in outdoor cultures of the microalga *Nannochloropsis* in two outdoor production systems – flat panel photobioreactors and raceway ponds (Sukenik et al. 2009). The measurements of the *Nannochloropsis* photosynthetic activity using several chlorophyll fluorescence techniques as well as oxygen production showed that this species was able to withstand high irradiance levels. *Nannochloropsis* coped well with high pH conditions under physiological temperatures. However, a temperature rise above 32°C was detrimental and the repair processes could not keep up with the rate of damage.

As an example we present measurements of Chl fluorescence parameters related to biomass productivity in outdoor *Chlorella* cultures of various biomass densities (Fig. 5) in outdoor thin-layer sloping cascades

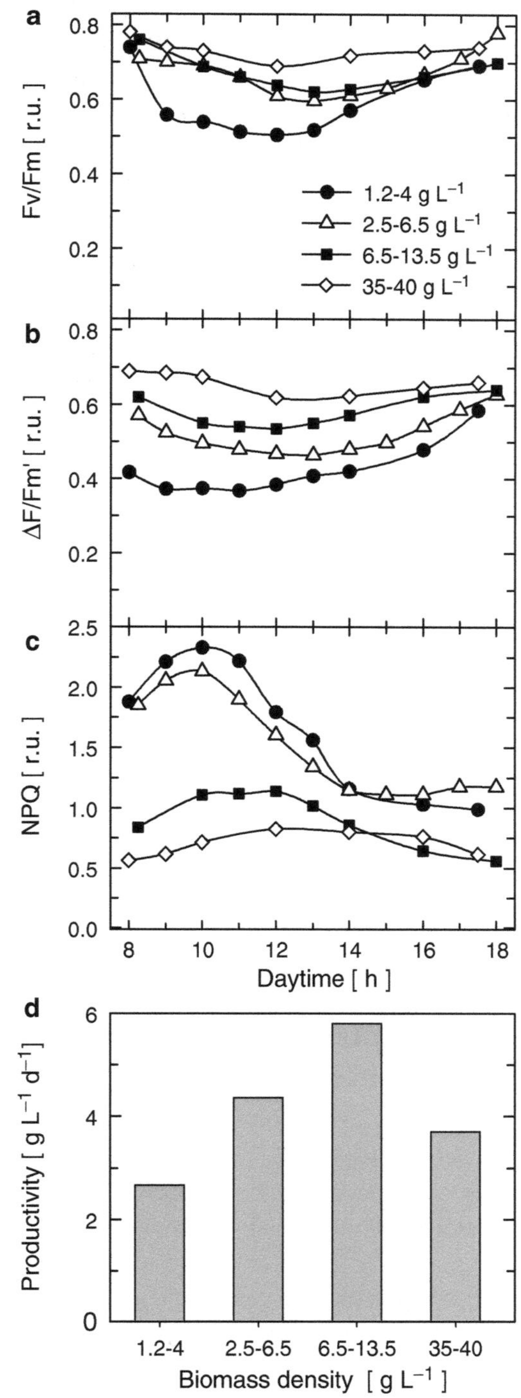

Fig. 5 Diurnal changes in the maximum PSII photochemical yield F_v/F_m (**panel a**), the actual photochemical yield $\Delta F/F_m'$ (**panel b**), non-photochemical quenching NPQ (**panel c**) in correlation with biomass productivity (**panel d**) in *Chlorella* mass cultures grown at different biomass concentrations (1.2–4, 2.5–6.5, 6.5–13.5 and 35–40 g L^{-1}) in outdoor thin-layer cascade units

(Fig. 1). Diel changes of the Chl fluorescence parameters F_v/F_m, $\Delta F/F_m'$ and NPQ, suggested the interplay of cell density vs. irradiance which resulted in correlation between the diel course of variable fluorescence parameters and daily biomass productivity (Fig. 5a–c). In low-biomass cultures, the photochemical yield decreased significantly in the morning which was balanced by the NPQ increase. The results of several experiments demonstrated that a decrease of PSII photochemical yield (F_v/F_m or $\Delta F/F_m'$) of about 20% at mid-day maximum irradiance can be considered physiological and still compatible with high productivity (Fig. 5 in this chapter; Torzillo et al. 1996; Richmond 2000, 2004; Masojídek et al. 2003). If the mid-day values of F_v/F_m or $\Delta F/F_m'$ were by 20% lower or higher than in the morning, the cultures become either photoinhibited or photo-limited which in both cases decreases productivity.

Similarly as biomass productivity, fluorescence measurements might indicate the induction of secondary carotenoid synthesis. The experiments with the *Haematococcus* cultures exposed to supra-high irradiance in a photobioreactor with solar concentrators showed that higher decrease of F_v/F_m and $\Delta F/F_m'$ were counteracted by a corresponding increase of NPQ and these changes indicated faster induction of astaxanthin synthesis as compared to the ambient irradiance intensity (Masojídek et al. 2009).

7 Light Adaptation – Non-photochemical Fluorescence Quenching

Photosynthetic organisms can be exposed to rapid changes of irradiance, often in synergism with other unfavourable environmental conditions. To minimise photoinhibition, photosynthetic organisms have evolved photoprotective mechanisms designated as short- and long-term responses (Krause 1988). These mechanisms serve to balance and optimise the light and dark photosynthetic reactions and to preserve the functioning of the photosynthetic apparatus. The relative extent of the energy dissipation is usually quantified using the so-called non-photochemical quenching parameter NPQ (Table 3). Under most conditions, the major part of NPQ is high-energy-state dependent quenching (referred often as qE; Cosgrove and Borowitzka, this volume) and it is thought to be essential in protecting plant leaves and microalgae from photo-induced damage. Two models were proposed, depending on whether the quenching is assumed to be associated with the PSII reaction centre or with the antenna. In the latter model, the process in the antennae involves the light-induced formation of the carotenoid zeaxanthin (Demmig-Adams 1990). Light dependent conversion of violaxanthin to zeaxanthin *via* the intermediate antheraxanthin, the so-called xanthophyll cycle is supposed to serve as a major, short-term, light-acclimation mechanism of NPQ in higher plants. It is reversible when leaf or microalgal cell is darkened and qE relaxes within minutes. The role of xanthophylls in the thermal dissipation of surplus excitation energy was deduced from the linear relationship between zeaxanthin formation and the magnitude of non-photochemical quenching (Demmig et al. 1987). Unlike in higher plants, the role of the xanthophyll cycle in green algae (Chlorophyta) is ambiguous since its contribution to energy dissipation can significantly vary among species (Casper-Lindley and Björkman 1998; Masojídek et al. 1999; Jin et al. 2003). In our experiments, the xanthophyll cycle was found to be functional in various green microalgae (*Chlorella, Scenedesmus, Haematococcus, Chlorococcum, Spongiochloris*); however its contribution to

Table 3 Selected parameters calculated from saturation pulse method in modulated fluorometers (Fig. 2). F_0, F_v, F_m – minimum, variable and maximum fluorescence in dark-adapted state; F_0', F', F_v', F_m' – minimum, steady-state, variable and maximum fluorescence in light-adapted state; PPFD – photosynthetic photon flux density. The nomenclature used is according to van Kooten and Snel (1990) – see also Cogrove and Borowitzka (Chapter 1, this volume)

Parameter	Symbol	Formula
Maximum photochemical yield of PSII	F_v/F_m	$F_v/F_m = (F_m - F_0)/ F_m$
Actual PSII photochemical yield	Φ_{PSII} or $\Delta F/ F_m'$	$\Phi_{PSII} = (F_m' - F')/ F_m'$
Relative electron transport rate through PSII (rate of photochemistry)	rETR	$rETR = \Phi_{PSII} \times PPFD$
Stern-Volmer coefficient of non-photochemical quenching	NPQ	$NPQ = (F_m - F_m')/F_m'$
Photochemical quenching	qP	$qP = (F_m' - F')/(F_m' - F_0')$
Non-photochemical quenching	qN	$qN = (F_v - F_v')/F_v$

non-photochemical quenching is not as significant as in higher plants, or can vary among species (Masojídek et al. 2004b). This conclusion is supported by two facts: (i) in green algae the content of zeaxanthin normalized per chlorophyll was significantly lower than that reported from higher plants; and (ii) the antheraxanthin + zeaxanthin content displayed different diel kinetics than NPQ. We assume that microalgae rely on other dissipation mechanism(s), which operate along with the xanthophyll cycle-dependent quenching. In microalgae, xanthophylls probably have a preferential role as antioxidants.

In the other model, the qE quenching is located in the PSII reaction centre and is not accompanied by zeaxanthin synthesis. The quenching is associated with a reversible inactivation (quenched state) of a certain fraction of the reaction centres which is probably caused by the transient over-acidification of the thylakoid lumen. Both the fluorescence quenching and PSII inactivation relax in parallel with the activation of the Calvin-Benson cycle (Finazzi et al. 2004).

The second type of non-photochemical quenching, important in microalgae, represents the so-called state transition quenching (qT). This process is induced by changes in the redox state of the plastoquinone pool causing the reversible phosphorylation of antennae proteins which regulate the redistribution of light energy between PSI and PSII (Wollman 2001). The qE and qT fade out in minutes after dark-adaptation.

Processes that relax over a longer-scale (hours) are usually referred to as 'photoinhibition' (qI). Applied to Chl fluorescence analysis, this term generally refers to the protective processes occurring in the light-harvesting antenna as well as destructive processes in the PSII reaction centres.

Cyanobacteria which lack the xanthophyll cycle, have a significant Mehler reaction activity (O_2 uptake by the reducing side of PSI) at light saturation which acts as a sink for electrons when PSII activity exceeds photosynthetic capacity. Recently, a photoprotective mechanism related to quenching in phycobilisomes has also been found in cyanobacteria (Kirilovsky 2007). In this mechanism, the soluble carotenoid-binding protein plays an essential role where the associated carotenoids vary among the cyanobacteria, e.g. zeaxanthin in *Anacystis* and *Lyngbya* or 3′-hydroxyechinenone in *Synechocystis* and *Arthrospira*.

8 Major Achievements in Microalgal Mass Culture Monitoring

Since the mid-1990s, Chl fluorescence has become one of the most feasible and useful techniques in microalgal biotechnology for monitoring the photosynthetic characteristics of a culture and subsequently estimating its biomass productivity. Results have indicated that the Chl fluorescence technique, when used *in situ,* is a useful tool for an immediate assessment of the fitness of outdoor microalgal mass cultures. In this way, we can elucidate the effect of changing environmental factors on the physiology of outdoor microalgal cultures. Chl fluorescence also makes it possible to control microalgal cultivation, using *on-line* monitoring of photobiochemical activities to photo-optimise the cultivation regime (e.g. biomass density, turbulence, CO_2 supply). The analysis of fast Chl fluorescence induction kinetics is used to determine the limiting photochemical processes at the molecular level.

Particularly useful photochemical expressions have been derived from analysis of Chl fluorescence quenching: maximal PS II quantum yield in the dark-adapted sample F_v/F_m, the effective PSII quantum yield of illuminated samples, $\Delta F/F_m'$, and relative electron transport rate rETR, to make the correlation with biomass productivity (Torzillo et al. 1998), or eventually indicate bioactive compound occurrence, e.g. secondary carotenoids (Torzillo et al. 2003; Masojídek et al. 2009). Although fluorescence parameters can be measured easily, some caution is necessary to correlate them with the rates of linear electron flux and CO_2 assimilation.

9 Concluding Remarks

The increased interest in microalgal biotechnology, aimed to the production of biomass, high-value products, or even bio-fuels, has prompted the application of *on-line* measurements for monitoring growth and obtaining rapid evidence of unfavourable conditions affecting the performance of outdoor cultures. For these applications the use of modulated fluorometers, which enable the actual photochemical quantum yield to be measured at a given light intensity during the day, is mandatory.

Finally, some recommendations are necessary. (i) Though Chl fluorescence represents a rapid technique for stress detection in plants and microalgal cultures, it must always be accompanied by other physiological measurements. (ii) Indeed, it is relatively easy to generate fluorescence data, thus care must always be taken to select and calculate sensible parameters. This is particularly true when dealing with microalgal cultures outdoors, where growth limitations, such as light, temperature and other unfavourable factors can occur side by side. As long as this is kept in mind, Chl fluorescence represents a powerful technique which allows rapid monitoring of physiological status, the use of which has been steadily increasing in both the laboratory and field studies of microalgal cultures.

Acknowledgement The authors thank Mr. Pavel Souček for preparation of diagrams and Mr. Steve Ridgill for language corrections.

The Ministry of Education, Youth and Sports and the Czech Academy of Sciences supported this work through the project MSM6007665808 and AVOZ 50200510. Partial funding was also provided by project 522/06/1090 and 521/09/0656 of the Czech Science Foundation, and by project IAA608170601 of the Grant Agency of the Czech Academy of Sciences.

References

Baker NR (2008) Chlorophyll Fluorescence: a probe of photosynthesis in vivo. Annu Rev Plant Biol 59:89–113

Behrenfeld MJ, Prasil O, Kolber ZS, Babin M, Falkowski PG (1998) Compensatory changes in photosystem II electron turnover rates protect photosynthesis from photoinhibition. Photosynth Res 58:259–268

Ben-Amotz A (2004) Industrial production of microalgal cellmass and secondary products – Major industrial species: *Dunaliella*. In: Richmond A (ed) Handbook of microalgal mass cultures. Blackwell Science, Oxford, pp 273–280

Benemann JR, Oswald WJ (1996) Systems and economic analysis of microalgae ponds for conversion of CO_2 to biomass. Final report. US DOE http://www.osti.gov/bridge/servlets/purl/493389-FXQyZ2/webviewable/493389.pdf

Bilger W, Björkman O (1990) Role of xanthophyll in photoprotection elucidated by measurement of light-induced absorbance changes, fluorescence and photosynthesis in leaves of *Hedera canariensis*. Photosynth Res 25:173–185

Boardman NK (1980) Energy from the biological conversion of solar energy. Phil Trans R Soc London A 295:477–489

Borowitzka MA (2005) Carotenoid production using microorganisms. In: Cohen Z, Ratledge C (eds) Single Cell Oils. AOCS Press, Champaign, IL, pp 124–137

Bradbury M, Baker NR (1984) A quantitative determination of photochemical and nonphotochemical quenching during the slow phase of the chlorophyll fluorescence induction curve of bean leaves. Biochim Biophys Acta 765:275–81

Büchel C, Wilhelm C (1993) In vivo analysis of slow chlorophyll fluorescence induction kinetics in algae: progress problems and perspectives. Photochem Photobiol 58:137–148

Casper-Lindley C, Björkman O (1998) Fluorescence quenching in four unicellular algae with different light-harvesting and xanthophyll-cycle pigments. Photosynth Res 56:277–289

Demmig B, Winter K, Krüger A, Czygan FC (1987) Photoinhibition and zeaxanthin formation in intact leaves. Plant Physiol 84:218–224

Demmig-Adams B (1990) Carotenoids and photoprotection in plants. A role for the xanthophyll zeaxanthin. Biochim Biophys Acta 1020:1–24

Diner BA (1998) Photosynthesis: molecular biology of energy capture. Methods Enzymol 297:337–360

Finazzi G, Johnson GN, Dall'Osto L, Joliot P, Wollman F-A, Bassi R (2004) A zeaxanthin-independent nonphotochemical quenching mechanism localized in the photosystem II core complex. Proc Nat Acad Sci USA 101:12375–12380

Genty B, Briantais JM, Baker NR (1989) The relationship between the quantum yield of photosynthetic electron transport and quenching of chlorophyll fluorescence. Biochim Biophys Acta 990:87–92

Gilmore AM, Yamamoto HY (1991) Zeaxanthin formation and energy dependent fluorescence quenching in pea chloroplasts under artificially mediated linear and cyclic electron transport. Plant Physiol 96:635–643

Govindjee (1995) Sixty-three years since Kautsky: chlorophyll *a* fluorescence. Aust J Plant Physiol 22:131–160

Grobbelaar JU (2007) Photosynthetic characteristics of *Spirulina platensis* grown in commercial-scale open outdoor raceway ponds: what do the organisms tell us? J Appl Phycol 19:591–598

Hofstraat JW, Peeters JCH, Snel JFH, Geel C (1994) Simple determination of photosynthetic efficiency and photoinhibition of *Dunaliella tertiolecta* by saturating pulse fluorescence measurements. Mar Ecol Prog Ser 103:187–196

Jin ES, Yokthongwattana K, Polle JEW, Melis A (2003) Role of the reversible xanthophylls cycle in the Photosystem II damage and repair cycle in *Dunaliella salina*. Plant Physiol 132:352–364

Kitajima M, Butler WL (1975) Quenching of chlorophyll fluorescence and primary photochemistry in chloroplasts by dibromothymoquinone. Biochim Biophys Acta 376:105–115

Klughammer C, Schreiber U (2008) Complementary PS II quantum yields calculated from simple fluorescence parameters measured by PAM fluorometry and the saturation pulse method. PAM Application Notes 1: 27–35. http://www.walz.com/e_journal/pdfs/PAN078007.pdf

Krause GH (1988) Photoinhibition of photosynthesis. An evaluation of damaging and protecting mechanisms. Physiol Plant 74: 566–574

Kirilovsky D (2007) Photoprotection in cyanobacteria: the orange carotenoid protein (OCP)-related non-photochemical-quenching mechanism. Photosynth Res 93:7–16

Lu CM, Vonshak A (1999) Photoinhibition in outdoor *Spirulina platensis* cultures assessed by polyphasic chlorophyll fluorescence transients. J Appl Phycol 11:355–359

Masojídek J, Torzillo G, Koblížek M, Kopecký J, Bernardini P, Sacchi A, Komenda J (1999) Photoadaptation of two members of the Chlorophyta (*Scenedesmus* and *Chlorella*) in laboratory and outdoor cultures: changes in chlorophyll fluorescence quenching and the xanthophyll cycle. Planta 209:126–135

Masojídek J, Torzillo G, Kopecký J, Koblížek M, Nidiaci L, Komenda J, Lukavská A, Sacchi A (2000) Changes in chlorophyll fluorescence quenching and pigment composition in the green alga *Chlorococcum* sp. grown under nitrogen deficiency and salinity stress. J Appl Phycol 12:417–426

Masojídek J, Grobbelaar JU, Pechar L, Koblížek M (2001) Photosystem II electron transport rate and oxygen production in natural waterblooms of freshwater cyanobacteria during the diel cycle. J Plankton Res 23:57–66

Masojídek J, Sergejevová M, Rottnerová K, Jirka V, Korečko J, Kopecký J, Zaťková I, Torzillo G, Štys D (2009) A two-stage solar photobioreactor for cultivation of microalgae based on solar concentrators. J Appl Phycol 21:55–63

Masojídek J, Koblížek M, Torzillo G (2004a) Photosynthesis in microalgae. In: Richmond A (ed) Handbook of microalgal mass cultures. Blackwell Science, Oxford, pp 20–39

Masojídek J, Kopecký J, Koblížek M, Torzillo G (2004b) The xanthophyll cycle in green algae (*Chlorophyta*): its role in the photosynthetic apparatus. Plant Biol 6:342–349

Masojídek J, Sergejevová M, Rottnerová K, Jirka V, Korečko J, Kopecký J, Zaťková I, Torzillo G, Štys D (2009) A two-stage solar photobioreactor for cultivation of microalgae based on solar concentrators. J Appl Phycol 21:55–63

Maxwell K, Johnson GN (2000) Chlorophyll fluorescence – a practical guide. J Exp Bot 51:659–668

Neale J (1987) Algal photoinhibition and photosynthesis in the aquatic environment. In: Kyle DJ, Osmond CB, Arntzen CJ (eds) Photoinhibition. Elsevier Science, Amsterdam, pp 39–65

Nedbal L, Tichy V, Xiong F, Grobbelaar JU (1996) Microscopic green algae and cyanobacteria in high-frequency intermittent light. J Appl Phycol 8:325–333

Neubauer C, Schreiber U (1987) The polyphasic rise of chlorophyll fluorescence upon onset of strong continuous illumination. I. Saturation characteristics and partial control by the Photosystem II acceptor side. Z Naturforsch 42c:1246–1254

Pirt SJ (1986) The thermodynamic efficiency (quantum demand) and dynamics of photosynthetic growth. New Phytol 102:3–37

Prášil O, Kolber Z, Berry JA, Falkowski PG (1996) Cyclic electron flow around photosystem II in vivo. Photosynth Res 48:395–410

Pulz O, Scheibenboden K, Gross W (2001) Biotechnology with cyanobacteria and microalgae. In: Reed G (ed) Special processes: biotechnology, vol 10. Wiley-VCH, Weinheim, pp 107–136

Ramus J (1981) The capture and transduction of light energy. In: Lobban CS, Wynne MJ (eds) The biology of seaweeds. Blackwell Scientific, Oxford, pp 458–92

Richmond A (2000) Microalgal biotechnology at the turn of the millenium: a personal view. J Appl Phycol 12:441–451

Richmond A (2004) Biological principles of mass cultivation. In: Richmond A (ed) Handbook of microalgal mass cultures. Blackwell Science, Oxford, pp 125–177

Schreiber U (2004) Pulse-Amplitude-Modulation (PAM) fluorometry and saturation pulse method: an overview. In: Papageorgiou GC, Govindjee (eds) Chlorophyll *a* fluorescence: a signature of photosynthesis. Springer, Dordrecht, pp 279–319

Schreiber U, Neubauer C (1987) The polyphasic rise of chlorophyll fluorescence upon onset of strong continuous illumination: II. Partial control by the photosystem II donor side and possible ways of interpretation. Z Naturforsch 42c:1255–1264

Schreiber U, Schliwa U, Bilger W (1986) Continuous recording of photochemical and nonphotochemical fluorescence quenching with a new type of modulation fluorometer. Photosynth Res 10:51–62

Schreiber U, Endo T, Mi H, Asada K (1995) Quenching analysis of chlorophyll fluorescence by the saturation pulse method: particular aspects relating to the study of eukaryotic algae and cyanobacteria. Plant Cell Physiol 36:873–882

Schreiber U, Bilger W, Hormann H, Neubauer C (1998) Chlorophyll fluorescence as a diagnostic tool: basics and some aspects of practical relevance. In: Raghavendra AS (ed) Photosynthesis: a comprehensive treatise. Cambridge University Press, Cambridge, pp 320–334

Srivastava AM, Strasser RJ, Govindjee (1999) Greening of pea leaves: parallel measurement of 77K emission spectra, OJIP chlorophyll *a* fluorescence transient, period four oscillation of the initial fluorescence level, delayed light emission, and P700. Photosynthetica 37:365–392

Strasser BJ, Strasser RJ (1995) Measuring fast fluorescence transients to address environmental questions: the JIP test. In: Mathis P (ed) Photosynthesis from light to biosphere, vol 5. Kluwer, Dordrecht, pp 977–980

Strasser RJ, Srivastava A, Govindjee (1995) Polyphasic chlorophyll *a* fluorescence transient in plants and cyanobacteria. Photochem Photobiol 61:33–42

Strasser RJ, Tsimili-Michael M, Srivastava A (2004) Analysis of the Chlorophyll *a* fluorescence transient. In: Papageorgiou GC, Govindjee (eds) Chlorophyll *a* fluorescence: a signature of photosynthesis. Springer, Dordrecht, pp 321–362

Sukenik A, Bennet J, Falkowski PG (1987) Light-saturated photosynthesis – limitation by electron transport or carbon fixation? Biochim Biophys Acta 891:205–215

Sukenik A, Beardall J, Kromkamp JC , Kopecký J, Masojídek J, van Bergeijk S, Gabai S, Shaham E, Yamshon A (2009) Photosynthetic performance of outdoor *Nannochloropsis* mass cultures to extreme environmental conditions – assessment by chlorophyll fluorescence techniques. Aquat Microb Ecol 56: 297–308

Ting CS, Owens TG (1992) Limitation of the pulse-modulated technique for measuring the fluorescence characteristics of algae. Plant Physiol 100:367–373

Torzillo G, Pushparaj B, Masojidek J, Vonshak A (2003). Biological constraints in algal biotechnology. Biotechnol Bioprocess Engineering, 8:338–348

Torzillo G, Accolla P, Pinzani E, Masojídek J (1996) *In situ* monitoring of chlorophyll fluorescence to assess the synergistic effect of low temperature and high irradiance stresses in *Spirulina* cultures grown outdoors in photobioreactors. J Appl Phycol 8:283–291

Torzillo G, Bernardini P, Masojídek J (1998) On-line monitoring of chlorophyll fluorescence to assess the extent of photoinhibition of photosynthesis induced by high oxygen concentration and low temperature and its effect on the productivity of outdoor cultures of *Spirulina platensis* (Cyanobacteria). J Phycol 34:504–510

Torzillo G, Goksan T, Faraloni C, Kopecky J, Masojídek J (2003) Interplay between photochemical activities and pigment

composition in an outdoor culture of *Haematococcus pluvialis* during the shift from the green to red stage. J Appl Phycol 15:127–136
Tredici M (2004) Mass production of microalgae: photobioreactors. In: Richmond A (ed) Handbook of microalgal mass cultures. Blackwell Science, Oxford, pp 178–214
Van Kooten O, Snel JFH (1990) The use of chlorophyll fluorescence nomenclature in plant stress physiology. Photosynth Res 25:147–145
Vonshak A, Torzillo G (2004) Environmental stress physiology. In: Richmond A (ed) Handbook of microalgal mass cultures. Blackwell Science, Oxford, pp 57–82
Vonshak A, Torzillo G, Tomaselli L (1994) Use of chlorophyll fluorescence to estimate the effect of photoinhibition in outdoor cultures of *Spirulina platensis*. J Appl Phycol 6:31–34
Vonshak A, Torzillo G, Accolla P, Tomaselli L (1996) Light and oxygen stress in *Spirulina platensis* (cyanobacteria) grown outdoors in tubular reactors. Physiol Plant 97:175–179
Vonshak A, Torzillo G, Masojídek J, Boussiba S (2001) Sub-optimal morning temperature induces photoinhibition in dense outdoor cultures of the alga *Monodus subterraneus* (Eustigmatophyta). Plant Cell Environ 24:1113–1118
Walker DA (2009) Biofuels, facts, fantasy, and feasibility. J Appl Phycol 21:509–517
Wollman F-A (2001) State transitions reveal the dynamics and flexibility of the photosynthetic apparatus. The EMBO Journal 20:3623–3630
Zhu X, Long SP, Ort DR (2008) Converting solar energy into crop production. Curr Opin Biotechnol 19:153–159

Chapter 14
Delayed Fluorescence

Maja Berden-Zrimec, Luka Drinovec, and Alexis Zrimec

1 Introduction

Delayed fluorescence (DF), also termed delayed luminescence or delayed light emission, is a long-lived light emission by plants, algae and cyanobacteria after being illuminated with light and placed in darkness (Strehler and Arnold 1951). It can last from milliseconds to several minutes, which is by itself an odd phenomenon in an otherwise nanosecond world of classical fluorescence. Although a lot of work has been done on a millisecond range, in this chapter we deal only with the seconds range delayed fluorescence. Additionally, we have limited the manuscript to aquatic organisms, which means a lot of studies performed on terrestrial higher plants has been left out (Wang et al. 2005; Yan et al. 2005). This is why this chapter should not be considered as a thorough review of work on delayed fluorescence.

2 Historical Overview

Delayed fluorescence was discovered by coincidence in 1950 when William Arnold and Bernard Strehler performed an experiment with chloroplasts that would presumably lead to what Strehler referred to as "fundamental discovery in plant physiology" (Arnold 1991). Strehler suspected that chloroplasts had to be the ones producing ATP and that the respiration is not its source. But, in a simple experiment based on bioluminescence measurement of ATP production, they measured delayed fluorescence from the chloroplasts instead (Strehler and Arnold 1951). Only years later, when ATP theory has already been proven by Arnon and co-workers, Strehler repeated the experiment by filtering off the delayed fluorescence emission and could measure the bioluminescence caused by the ATP production by chloroplasts (Arnold 1991).

Since Strehler and Arnold described delayed fluorescence in 1951, it has been used for studying photosynthesis, like bicarbonate depletion influence on quinine reactions, possible effects of protein photophosphorylation on the primary photochemistry of PSII (Jursinic 1986) and other studies. Walter Bertsch concluded from his experiments that there had to be two pigment systems linked by an enzyme chain in photosynthetic chains by measuring delayed fluorescence after illumination with two different wavelengths (Bertsch 1962). Paul A. Jursinic has thoroughly reviewed different aspects and theories about delayed fluorescence in the Govindjee and Fork book "Light emission by plants and bacteria" in 1986.

The most important feature of delayed fluorescence is that it is emitted only by a functionally active chlorophyll – in other words, by active photosynthesis (Bertsch 1962). Its emission spectrum resembles the fluorescence emission spectrum of chlorophyll *a* (Arnold and Davidson 1954; Jursinic 1986; Van Wijk et al. 1999). The main source of DF is photosystem II (PSII) in the thylakoid membrane of chloroplasts (Jursinic 1986), whereas photosystem I (PSI) contributes much less to the DF emission. The emission also depends on the number of PSII centers and the rate of back reactions, which are in their turn influenced by

M. Berden-Zrimec (✉), L. Drinovec, and A. Zrimec
Institute of Physical Biology, Toplarniska 19,
SI-1000 Ljubljana, Slovenia
e-mail: maja@ifb.si

D.J. Suggett et al. (eds.), *Chlorophyll a Fluorescence in Aquatic Sciences: Methods and Applications*, Developments in Applied Phycology 4, DOI 10.1007/978-90-481-9268-7_14, © Springer Science+Business Media B.V. 2011

the membrane potential and pH gradient (Avron and Schreiber 1979; Joliot and Joliot 1980).

The difference between the more widely known prompt chlorophyll fluorescence and delayed fluorescence is in the origin of the excited single state of the emitting pigment molecule (Jursinic 1986). Delayed fluorescence originates from the repopulation of excited states of chlorophyll from the stored energy *after* charge separation (Jursinic 1986), whereas prompt fluorescence reflects the radiative de-excitation of excited chlorophyll molecules *before* charge separation. This is why delayed and prompt fluorescence contain information about different fundamental processes of the photosynthetic apparatus (Goltsev et al. 2003).

Two to three percent of the absorbed solar energy is re-emitted from the pigment systems as fluorescence. Yet delayed fluorescence represents only 0.03% of that emission (Jursinic 1986). Although DF reflects an insignificant loss of the total energy stored by photosynthesis, it is a sensitive indicator of the many reactions photosynthesis consists of (Jursinic 1986). This sensitivity makes DF an extremely complex phenomenon, however with awareness and control of the variables, DF becomes an important intrinsic probe (Jursinic 1986).

Delayed fluorescence is affected by many chemical and physical variables, such as ATP (Avron and Schreiber 1979), proton gradient in the thylakoids (Wraight and Crofts 1971), chill stress (Melcarek and Brown 1977), different xenobiotics (Drinovec et al. 2004; Joliot and Joliot 1980), excitation light spectral and intensity characteristics (Wang et al. 2004), cell culture growth stage (Monti et al. 2005), to name only some of them. This variety of influences has in common that they affect the reduction state of the plastoquinone pool (PQ) or its coupling with PSII by modulating the reversed electron flow (Avron and Schreiber 1979; Mellvig and Tillberg 1986).

In field studies, the intensity of delayed fluorescence has been reported to be a measure of photosynthetic activity (Schneckenburger and Schmidt 1996). Krause and Gerhardt (1984) and Wiltshire et al. (1998) have shown that the intensity of DF of phytoplankton is a measure of living algal biomass. Yosef Yacobi and coworkers (Yacobi et al. 1998) have analyzed natural phytoplankton samples using DF excitation spectra and were able to detect taxonomical changes in the algal communities. This technique is now used in monitoring freshwater phytoplankton (Hakanson et al. 2003; Istvanovitcs et al. 2005).

3 Basic Characteristics of Delayed Fluorescence

3.1 *Delayed Fluorescence Decay Kinetics and Intensity*

Delayed fluorescence shows monotonic decay kinetics in the first seconds, sometimes followed by a more or less pronounced transient peak (Bertsch and Azzi 1965) (Fig. 1). DF emission is composed of several components, characterized by different decay rates (Bjorn 1971; Desai et al. 1983). The faster decaying components have been well characterised as they provide information about the fate of energy absorbed by PSII (Desai et al. 1983). The slow components of DF originate in back reactions in the photosynthetic chain, and finally between the S_2 and S_3 states of the oxygen evolving

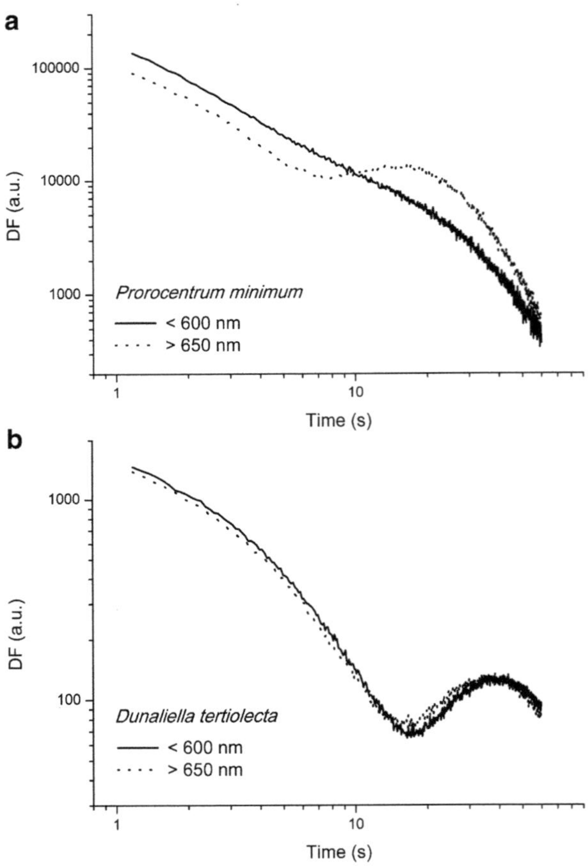

Fig. 1 Delayed fluorescence decay kinetics after illumination with different wavelengths (<600 and >650 nm): (**a**) *Prorocentrum minimum*, (**b**) *Dunaliella tertiolecta*; a.u. – arbitrary units

complex (OEC) and quinones Q_A and Q_B (Joliot et al. 1971). They probably provide information on temporary energy storage during photosynthetic electron transport (Desai et al. 1983), as it was shown that the slow-rate back reactions depend on electron distribution in PQ and PSI (Katsumata et al. 2006). The half-times of the final reactions in isolated chloroplasts are 1.5 s for Q_A+$S_{2/3}$ and 25 s for Q_B+$S_{2/3}$ (Rutherford and Inoue 1984).

DF decay kinetics therefore depends on the rates of back reactions in the electron backflow (Avron and Schreiber 1979; Joliot and Joliot 1980). It can be argued that the organization of the thylakoid membrane is involved in the transient peak phenomenon (Desai et al. 1983). In this chapter, we focus on some characteristics of DF, with special emphasis on the transient peak.

There are several proposed phenomenological models of long-term DF decay kinetics: the "multiexponential models" (Schmidt and Schneckenburger 1992), the "hyperbolic models" (Lavorel and Dennery 1982; Scordino et al. 1993), the "coherence models" (Yan et al. 2005). Recently, Li and co-workers presented a very interesting mathematical–physical analysis where they modeled the electron reflux for photosynthetic electron transport chain (Li et al. 2007). Unfortunately, the presently published models fail to include all the experimental data available in the literature. Especially problematic is the modeling of DF decay kinetics with the transient peaks. Let's see why.

The excitation light spectral composition has been reported to have qualitative influence on the DF decay kinetics (Bjorn 1971; Desai et al. 1983; Hideg et al. 1991; Mellvig and Tillberg 1986). The transient peak is preferentially stimulated by far-red excitation (Desai et al. 1983; Hideg et al. 1991), but in some species it can also be induced by shorter wavelengths (Berden-Zrimec et al. 2008; Zrimec et al. 2005) (Fig. 1a, b). We have observed different appearance of the peak with different algal species (Fig. 2), for example *Dunaliella tertiolecta* Butcher (Chlorophyta) exhibited a peak at illuminations below 600 nm and above 650 nm (Zrimec et al. 2005) (Fig. 1b), whereas *Desmodesmus* (*Scenedesmus*) *subspicatus* Chodat (Chlorophyta) did not exhibit a peak at all (Berden-Zrimec et al. 2007). Bertsch (1962) observed the peak in *Chlorella* (Chlorophyta) at the illumination wavelength of 700 nm, but not at 650 nm. The same is true for *Prorocentrum minimum* (Pavillard) Schiller (Dinophyta), which exhibits a peak only when illuminated with wavelengths above 650 nm (Fig. 1a). DF decay kinetics can

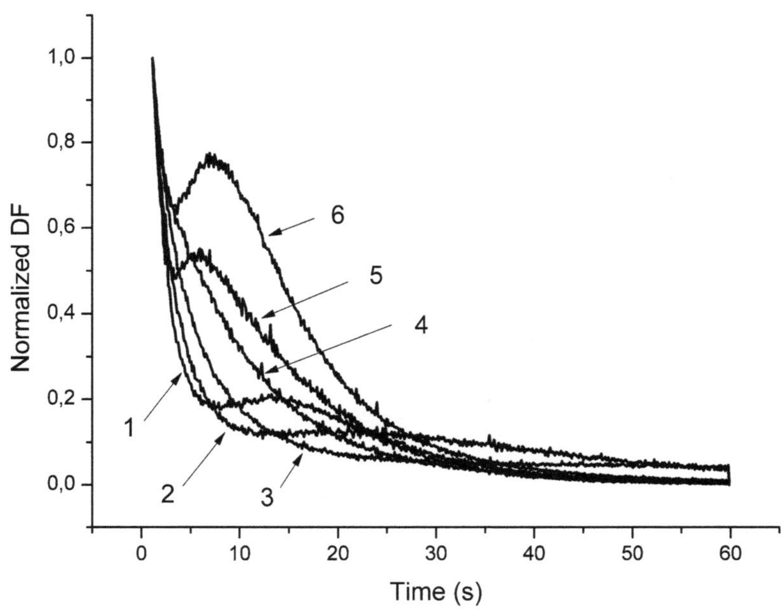

Fig. 2 Normalized delayed fluorescence decay curves of different algal species after a 3 s white-light illumination pulse. (1) *Prorocentrum minimum* (Dinophyta), (2) *Scrippsiella trochoidea* (Dinophyta), (3) *Gyrodinium* sp. (Dinophyta), (4) *Skeletonema costatum* (Bacillariophyceae), (5) *Pyrenomonas* sp. (Chryptophyta), (6) *Isochrysis galbana* (Chrysophyta)

differ even among strains (Monti et al. 2005), all at the same both growth and illumination conditions (Berden-Zrimec et al. 2008), due to the kinetic rate constants of electron back reactions depending on the physiological and organizational state of the entire photosynthetic apparatus. The peak is also culture-state dependent – the peak position and intensity change during culture growth (Monti et al. 2005, Berden-Zrimec et al. 2008), as shown in Fig. 3.

The presence of a peak in DF decay curves after a pulse of light of longer wavelengths indicates PSI involvement in DF generation (Bertsch 1962; Desai et al. 1983; Hideg et al. 1991; Mellvig and Tillberg 1986), because far red light is predominantly absorbed by PSI. If cyclic electron flow produces excess ATP over NADPH, back electron flow from PSI can be generated, resulting in the transient peak a few to tens of seconds after their being illuminated (Mellvig and Tillberg 1986) (Fig. 2).

Delayed fluorescence intensity (DFI) is represented by an integral under the DF decay curve. It is an increasing function of living cell concentration (Figs. 4 and 5) (Berden-Zrimec et al. 2009), the number of PSII centers, the fluorescence yield, and the rate of back reactions, which are influenced by the membrane potential and pH gradient (Avron and Schreiber 1979; Joliot et al. 1971; Joliot and Joliot 1980; Wraight and Crofts 1971).

3.2 Physiology

Delayed fluorescence decay kinetics and intensity depend on several parameters. The variation is especially obvious when observing the transient peak position and intensity, mainly presented in this section. The presented measurements were made with a 3 s long illumination and a sensor with a red-light-sensitive photomultiplier tube (Hamamatsu R1104) with a Hamamatsu C3866 Photon Counting Unit for signal conditioning and amplification (Monti et al. 2005; Zrimec et al. 2005).

3.2.1 Temperature and Illumination Intensity Dependance

The transient peak has an earlier onset and is less extended at higher temperatures (Fig. 6). Average

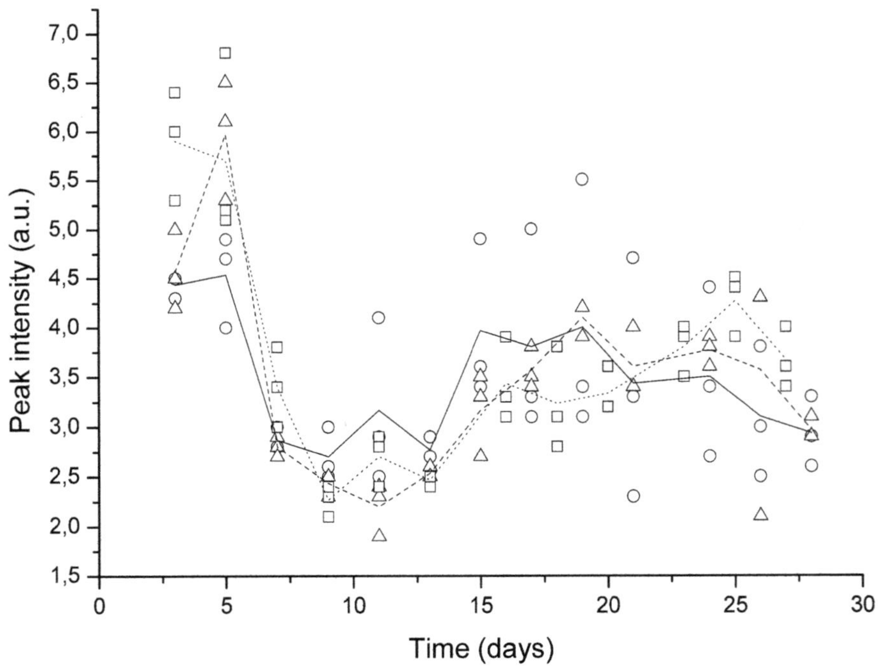

Fig. 3 Changes in transient peak intensity of three *Prorocentrum minimum* (Pavillard) Schiller (Dinophyta) strains during growth at salinity 16 PSU: (···□···) Baltic Sea, (---Δ---) Chesapeake Bay, USA, (—○—) Northern Adriatic Sea (strains described in Monti et al. 2005)

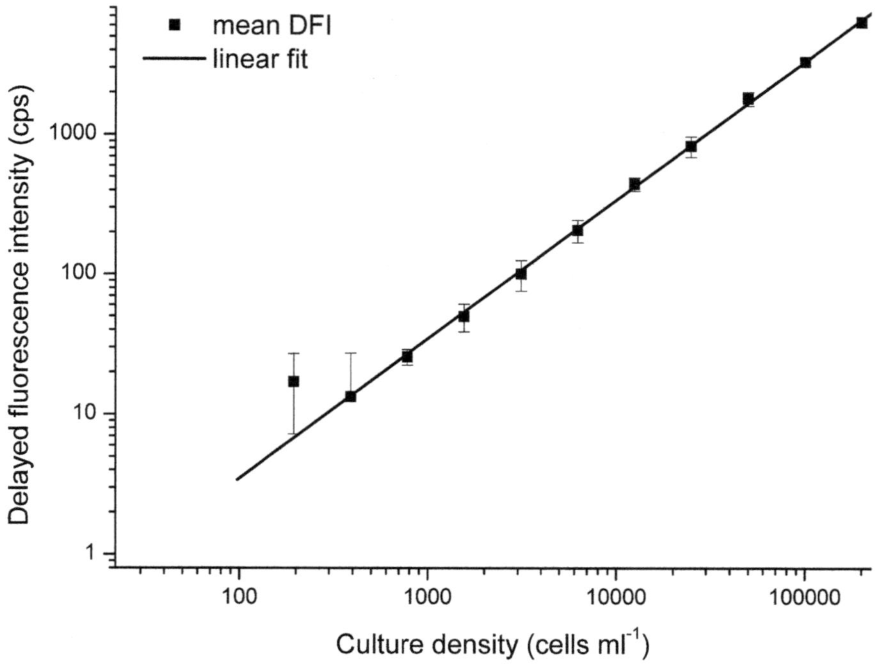

Fig. 4 Delayed fluorescence intensity (DFI) in a dilution series of a *Skeletonema costatum* (Grev.) Cleve (Bacillariophyta) culture in the exponential phase. There is an excellent linear relation between DFI and cell concentration (above several hundred cells with our luminometer setup – see Chapter *Physiology*). cps – counts per second

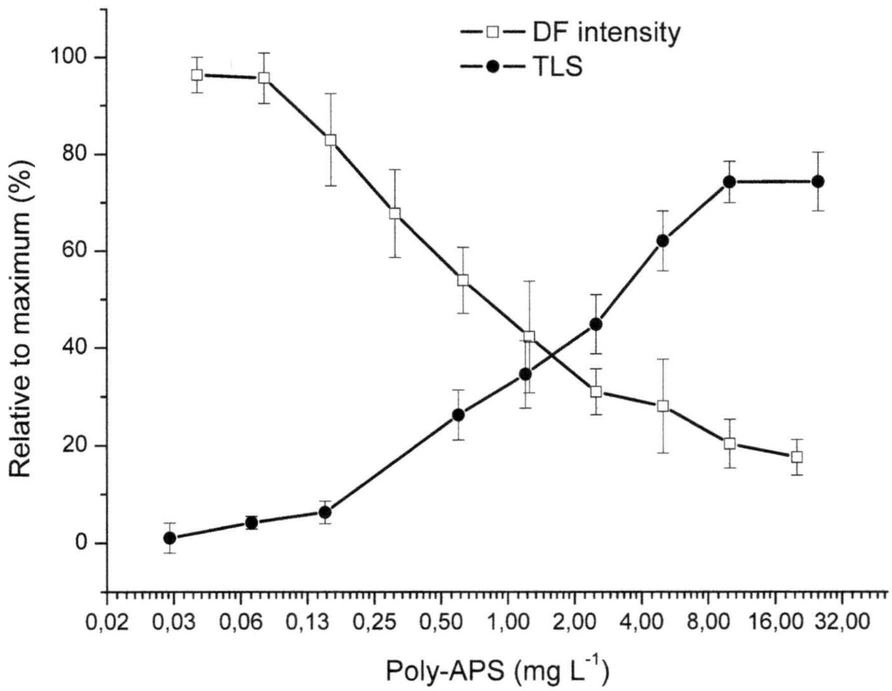

Fig. 5 Dose-response of DF intensity (DFI) and thermal lens spectroscopy (TLS) of *Skeletonema costatum* (Grev.) Cleve (Bacillariophyta) to a series of poly-APS concentrations (Berden-Zrimec et al. 2009). Poly-APS causes cell lysis. TLS measures the concentration of pigments released from the cells due to their lysis and DFI living cell biomass

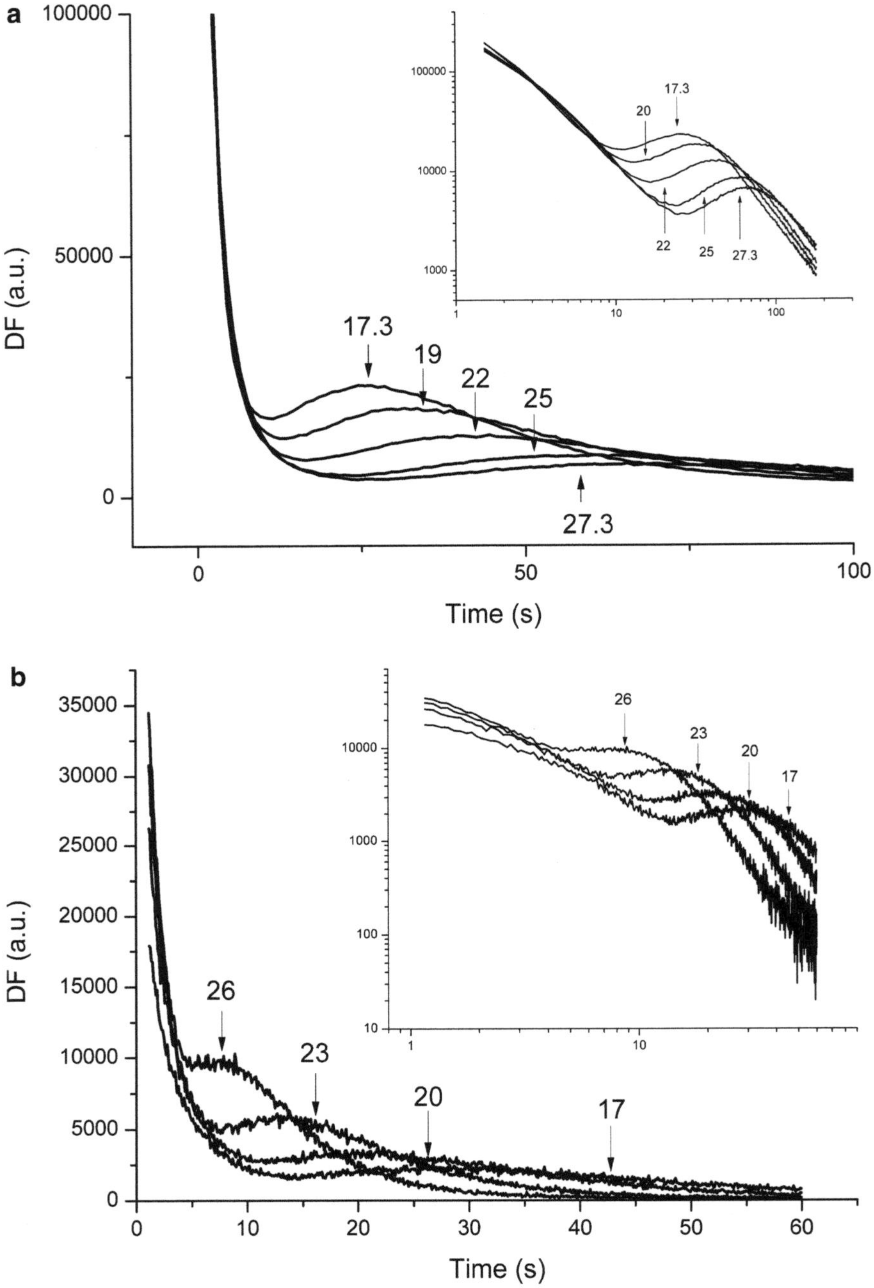

Fig. 6 Delayed fluorescence decay kinetics at different temperatures and white light illumination: (**a**) *Dunaliella tertiolecta* Butcher (Chlorophyta), (**b**) *Prorocentrum minimum* (Pavillard) Schiller (Dinophyta); inset: log-log scale of DF decay kinetics, (**c**) temperature dependence of the peak position: the line represents linear fit

delayed fluorescence intensity increases with temperature with a maximum around 28–30°C (Wang et al. 2004; Yan et al. 2005; Zrimec et al. 2005). However, the temperature dependence of DFI is more complicated because it also depends on the time interval on which it is averaged – due to changes in the kinetics. On the other hand, the peak position is a more reliable measure in the temperature dependence experiment (Zrimec et al. 2005). In the Arrhenius plot the natural logarithm of the peak position is linearly dependent on temperature

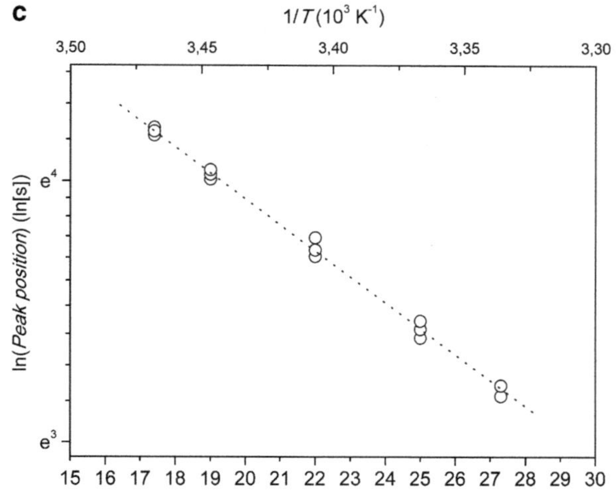

Fig. 6 (continued)

(Zrimec et al. 2005). Q_{10} of 2.6 and the activation energy of 71.5 kJ mol^{-1} calculated from the plot are in the expected range of plastoquinone-PSII reactions (Zrimec et al. 2005).

Delayed fluorescence can already be observed at relatively low illumination intensities (Fig. 7). In the experiment with *Dunaliella tertiolecta*, DF was saturated already by a 3 s excitation pulse of 15 μmol photons m^{-2} s^{-1} PAR (Zrimec et al. 2005), which is even lower than obtained by Wang *et al.* (Wang et al. 2004) for isolated spinach chloroplasts. At the excitation light intensity of 3.75 μmol photons m^{-2} s^{-1} PAR, DF showed slight differences in decay kinetics in the region of the transient peak (Fig. 7), probably due to the changed oxidation state of the plastoquinone pool (Zrimec et al. 2005). DFI at 3.75 μmol photons m^{-2} s^{-1} PAR was only approximately 3% lower than at 15 and 60 μmol photons m^{-2} s^{-1} PAR (Zrimec et al. 2005).

3.2.2 Influence of Toxins

Toxic effects of photosynthesis inhibitors can be measured already after a few minutes (Fig. 8). The effects on the transient peak are easily noticeable as it disappears soon after the addition of photosynthetic inhibitors like 15 μM dicyclohexylcarbodiimide (DCCD) – ATP synthesis inhibitior (Fig. 8a), 2 μM dinitrophenol (DNP), which destroys the proton gradient and inhibits electron flow in thylakoid membranes (Fig. 8b), or 30 mM NaCN, an inhibitior of respiration and photosynthesis (Fig. 8c).

Delayed fluorescence response to toxins is dose-dependent, like in the case of herbicide diuron (DCMU, 3-(3,4-dichlorophenyl)-1,1-dimethylurea), which competes with plastoquinone and plastoquinol for the Q_B binding site, preventing the electron flow between PSII and the plastoquinone pool (Fig. 9), or 3,5-dichlorophenol (3,5-DCP), which is used as an unspecific reference toxicant in toxicity tests (Fig. 10) (Berden-Zrimec et al. 2007).

3.2.3 Nutrients

The nutrient status of algal cells influences delayed fluorescence decay kinetics (Berden-Zrimec et al. 2008; Burger and Schmidt 1988; Mellvig and Tillberg 1986). Mellvig and Tillberg (1986) detected an induction of one or several transient peaks in the DF decay kinetics due to phosphorus starvation in the freshwater green alga *Scenedesmus obtusiusculus*. Bürger and Schmidt (1988) found that nitrogen deprivation caused a DF transition peak cessation in another freshwater green alga – *Scenedesmus obliquus*.

In our experiments with nutrient deprived *Dunaliella tertiolecta*, the best indicator of stress conditions proved to be the peak position (Fig. 11), while the peak intensity did not show any significant differences from replete cells (Berden-Zrimec et al. 2008). In comparison with replete cells, the peaks of nitrogen deprived cells had an earlier onset. The peak of replete cells was increasingly delayed through

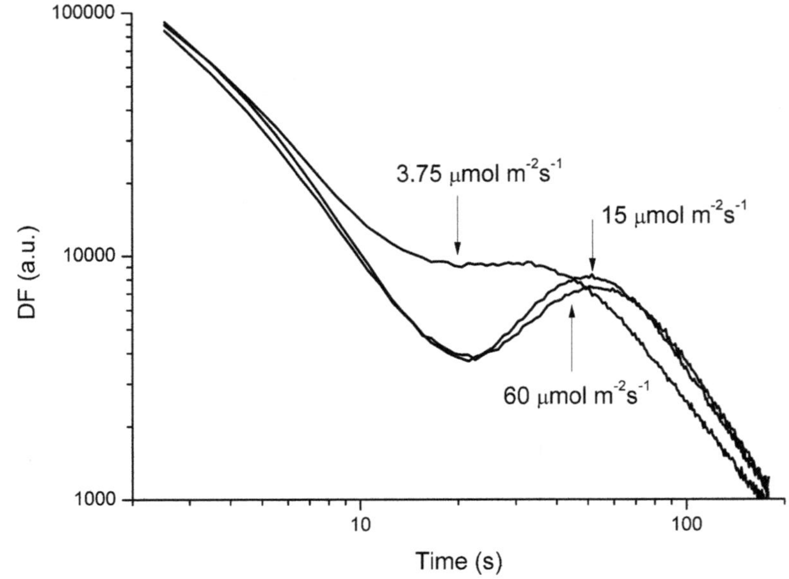

Fig. 7 Delayed fluorescence decay kinetics after 3 s illumination with white light of different intensities

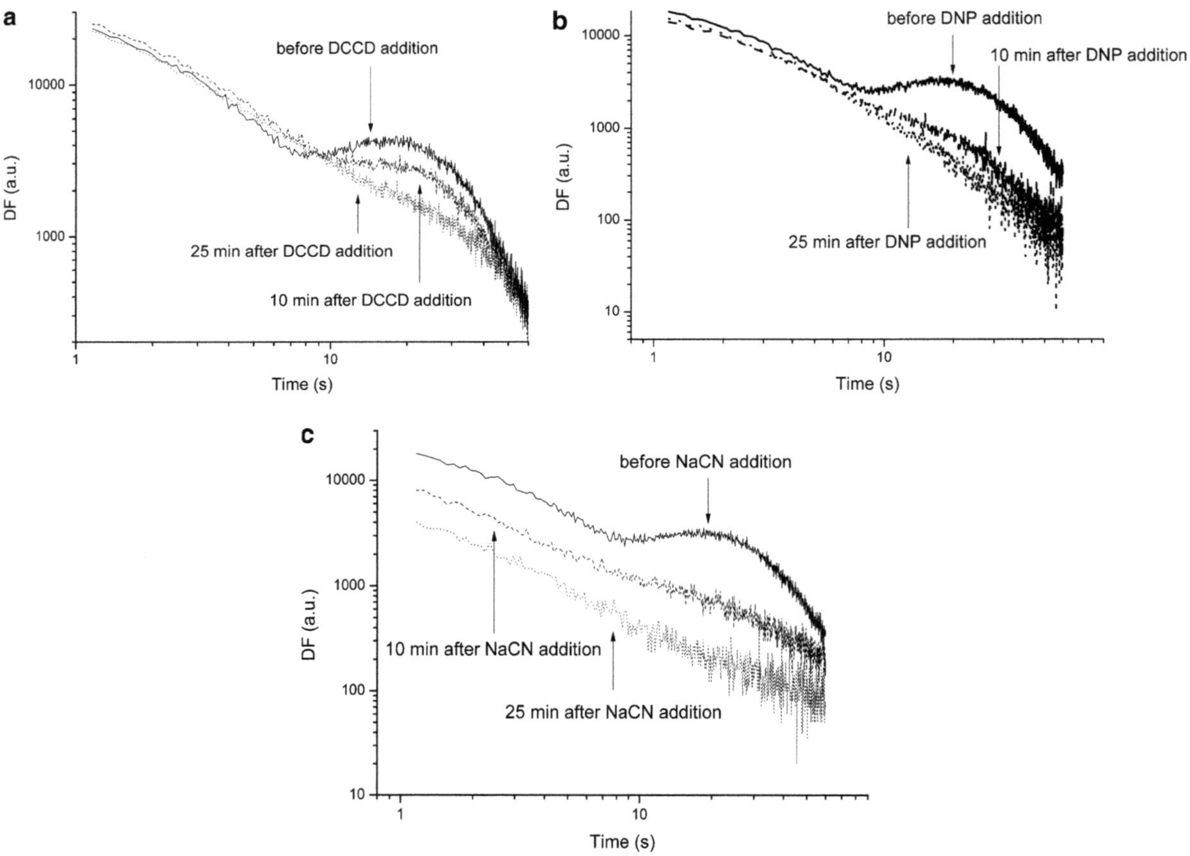

Fig. 8 Influence of photosynthetic inhibitors on *P. minimum* DF decay kinetics: (**a**) dicyclohexylcarbodiimide (DCCD) – inhibitor of ATP synthesis inhibition; (**b**) dinitrophenol (DNP) – destruction of proton gradient and inhibition of electron flow in thylacoid membrane; (**c**) NaCN – inhibition of respiration and photosynthesis

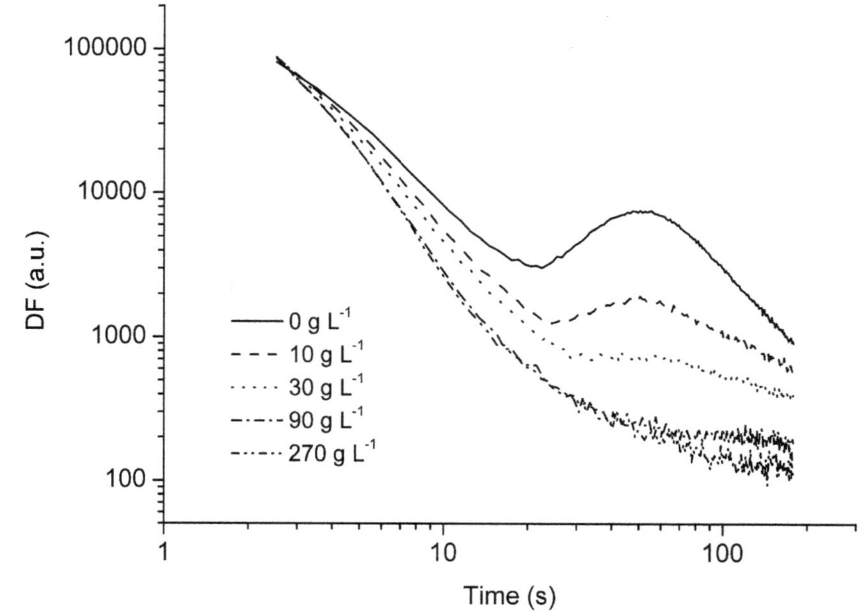

Fig. 9 The response of delayed fluorescence decay kinetics to geometrical series of diuron (DCMU) concentrations

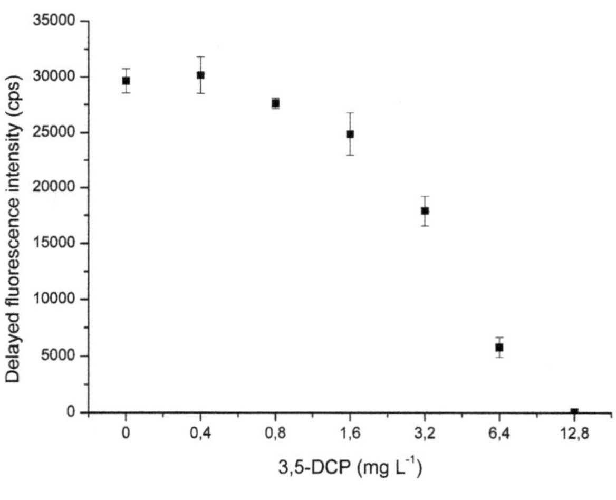

Fig. 10 Dose-response of DF intensity in *Dunaliella tertiolecta* after 24-h exposure to 3,5-dichlorophenol (3,5-DCP)

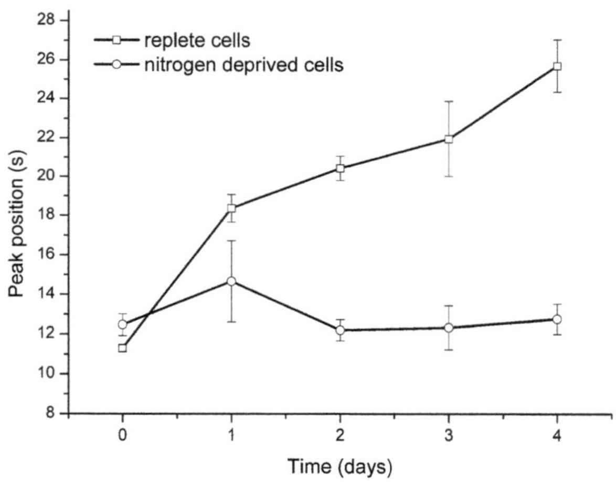

Fig. 11 Peak position dependence on nutrient status of the *Dunaliella tertiolecta* cells

time, whereas the peak position of nitrogen deprived cells was not changing significantly (Fig. 11). Dependence of the DF decay kinetics on the culture age has already been shown with nutrient replete algae and plants (Desai et al. 1983; Monti et al. 2005; Zrimec et al. 2005).

As mentioned earlier, DF decay kinetics depends on the back reactions rates in the electron backflow, which can be changed by nitrogen starvation that causes reduction in the number of active PSII reactive centers and linear electron flow, but does not influence active PSI leading to relatively higher rates of cyclic photophosphorylation (Berges et al. 1996). Nitrogen starvation also influences thylakoid organization, facilitating trapped energy transfer from PSII to PSI (Berges et al. 1996).

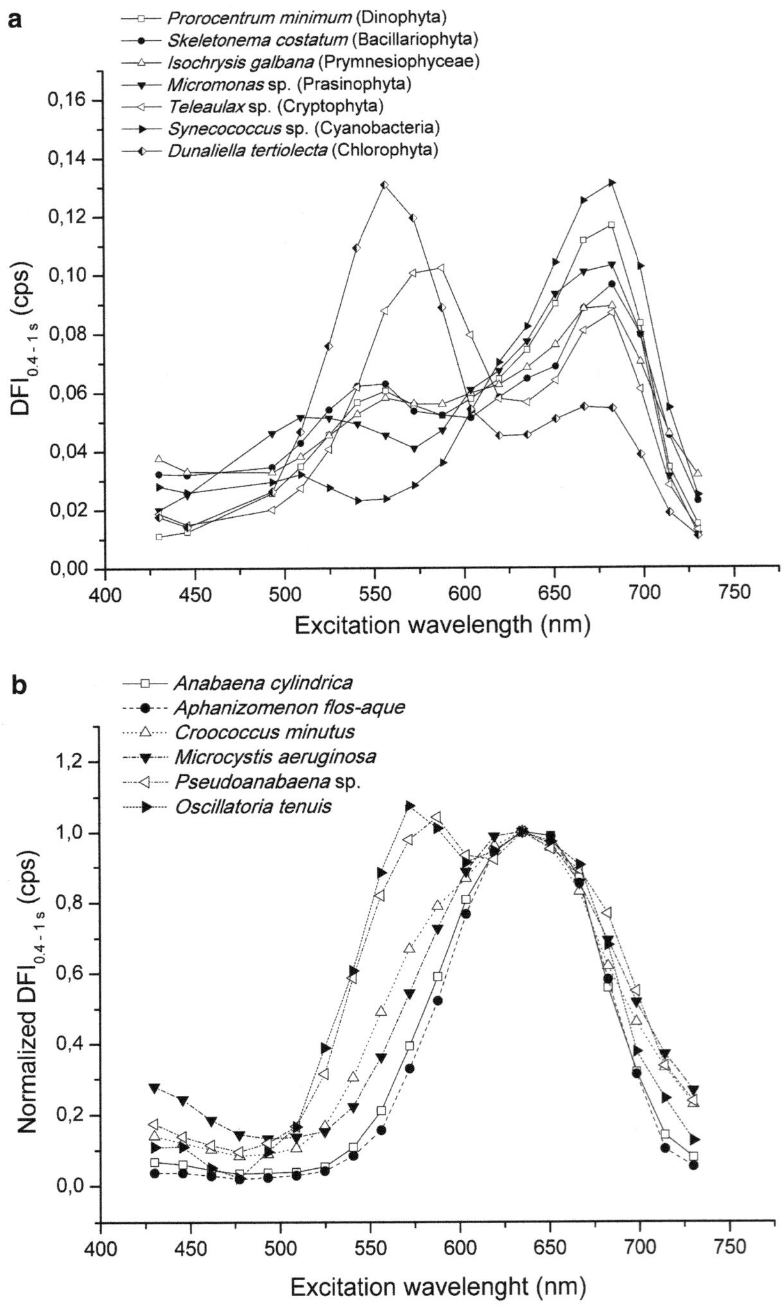

Fig. 12 Delayed fluorescence excitation spectra: (**a**) marine algae, (**b**) freshwater cyanobacteria

3.3 Delayed Fluorescence Excitation Spectroscopy

Delayed fluorescence excitation spectra are obtained by measuring the intensity of delayed fluorescence at different excitation wavelengths. It can be used to determine chlorophyll concentration and phytoplankton composition (Greisberger and Teubner 2007; Istvanovics et al. 2005; Krause and Gerhardt 1984; Yacobi et al. 1998). Algal classes show different excitation spectra when excited by monochromatic light in the wavelength range from 400 to 730 nm due to their different chlorophylls, accessory pigments and phycobiliproteins (Fig. 12). Chlorophyll *a* is present in all algal classes, chlorophyll *b* in Chlorophyta and Euglenophyta; chlorophyll *c* in Bacillariophyta,

Chrysophyta, Dinophyta and Cryptophyta and phycobiliproteins, e.g. phycoerythrin and phycocyanin, in Cryptophyta and Cyanobacteria or allophycocyanin in Cyanobacteria.

3.4 *Photosynthetic Activity Index (PhAI)*

Photosynthetic Activity Index was introduced in order to provide a comparable non-dimensional physiological index of photosynthesis efficiency (Berden-Zrimec et al. 2008). PhAI is calculated by dividing delayed fluorescence intensity (integrated in the interval 1–60 s after the illumination) with F_0:

$$PhAI = \frac{DFI_{1-60}}{F_0}$$

PhAI is very responsive to stress conditions, similarly as maximum variable fluorescence yield F_v/F_m (Berden-Zrimec et al. 2008). F_0 increases with the number of non-active PSII reaction centers, whereas DFI increases with the concentration of active PSII centers (Joliot and Joliot 1980). When the number of active PSII reaction centers is reduced, PhAI values decrease.

In *Dunaliella tertiolecta*, PhAI was more sensitive to the nitrogen deprivation than F_v/F_m (Berden-Zrimec et al. 2008). Whereas nitrogen deprivation was indicated by more than a 50% drop in PhAI, F_v/F_m of nitrogen deprived cells was only approximately 10% lower than in the replete cells (Berden-Zrimec et al. 2008). The decrease resulted primarily from the increase of F_0 (Berden-Zrimec et al. 2008). PhAI is independent of cell concentration and could thus also be used for physiological measurements in field samples.

3.5 *Delayed Fluorescence Visualisation*

The introduction of non-invasive optical techniques into biological research importantly decreased the errors related to manipulation of biological samples and subsequent analysis of results as well as their interpretation. In higher plants, great heterogeneity of photosynthetic activity was measured on the leaf surface. With non-imaging techniques, the signal from the whole leaf surface is averaged and spatial differences overlooked (Nedbal et al. 2000). This is especially important with an early detection of biotic infection. Fluorescence and delayed fluorescence imaging of plant leaves allows detailed information on spatial and kinetic heterogeneity of plant activity. There was, however, not much work done on delayed fluorescence imaging. Bennoun and Beal (1997) successfully screened algal mutant colonies with altered thylakoid electrochemical gradient using delayed fluorescence digital imaging.

The differences of delayed fluorescence imaging from fluorescence imaging can be most easily seen on the leaves of higher plants. While fluorescence of a damaged leaf can be higher (if there is still intact chlorophyll), delayed fluorescence intensity decreases with decreasing photosynthetic activity (Fig. 13).

4 Delayed Fluorescence Applications

4.1 *Description of the Instrument: What is Needed*

Measurement of delayed fluorescence requires sensitive light detectors. Photomultiplier tubes (PMT) in photon counting mode are usually used for DF detection. DF imaging is performed by using image-intensified CCD (ICCD) or electron-multiplication (EMCCD) cameras. Instead of the photo-multiplier tube (PMT), the ICCD concept has also been used as a non-spatial detector for DF measurement. The detector should be tuned with the emission spectra of DF, therefore red light sensitive PMTs with multialkali photocathodes are used. A shutter is necessary to protect the detector during sample illumination.

Intensities from 1 to several hundred μmol photons m^{-2} s^{-1} are used for DF excitation. Higher intensities are not needed since DF yield reaches a plateau at low light levels (Wang et al. 2004). Possible light sources for sample excitation are xenon and halogen lamps, but LEDs and lasers are getting popular due to better spectral composition of the emitted light. During DF measurements the excitation source is turned off and usually a mechanical shutter is used to block the afterglow of the excitation light. This is especially important with LED excitation systems since some of the shorter wavelength LEDs exhibit a weak emission tail which lasts seconds after the switch-off.

There are two main setups for DF measurement: a stationary and a flow through system. In the stationary

Fig. 13 Delayed fluorescence intensity of a potato leaf (*Solanum tuberosum* var. Sante) after 5 h in 10 mM methyl viologen dichloride solution and imaged with an EMCCD camera (L3Vision, e2v, UK). Spreading of the inhibitor through the veins can be seen by a reduction of delayed fluorescence intensity (DFI) – with the highest DFI on the still vital edge of the leaf. Image courtesy of Jaka Razinger

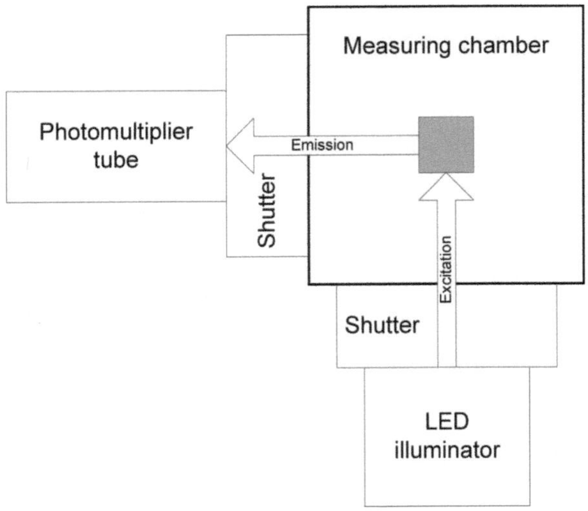

Fig. 14 A stationary setup for DF measurement

setup the sample excitation and measurement is performed in the same sample chamber (Fig. 14). In the flow-through system the sample is illuminated in the excitation chamber, then it is transferred to the measuring chamber (Fig. 15). This system can be used for continuous measurement of DF at a certain time interval after the excitation, and is useful for DF excitation spectra measurements, especially in the waters containing high concentration of phytoplankton (Istvanovics et al. 2005; Yacobi et al. 1998).

4.2 Toxicity Tests

With toxicity tests, simple, fast and reliable protocols are extremely important. Some standard bioassays, like algal growth inhibition tests, require relatively demanding protocols and time-consuming measurements. This is why simpler *in vivo* measurements, which do not demand an extensive handling of algae, are very welcome. The most important features for the quality toxicity tests with photosynthetic organisms are minimal disturbance to the cells, small sample volumes that enable homogenous illumination of all samples, and short test duration minimizing the negative influence of changing physico-chemical properties of the medium on the results. Low density and small volume samples require very sensitive measuring devices, usually used in photosynthesis measurements.

The applicability of measuring delayed fluorescence for toxicity bioassays has been tested by several research groups and few protocols for rapid toxicity bioassays have been proposed (Berden-Zrimec et al. 2007; Katsumata et al. 2006; Katsumata et al. 2008). Toxins influence delayed fluorescence in different ways, but a variety of influences has in common that they affect the reduction state of the plastoquinone pool or its coupling with PSII by inhibiting the reversed electron flow (Wang et al. 2004).

The advantage of the delayed fluorescence is that only living cells are measured, which increases the sensitivity in toxicity tests. Counting of cells under a microscope or spectroscopic chlorophyll measurements includes also the cells that have died because of toxins. That makes toxic effects artificially smaller and sometimes even not significant.

With toxicity tests, living biomass changes are measured with delayed fluorescence intensity (DFI). Standard parameters of toxicity tests (ISO 8692, 2004) can be calculated from DFI. The median effective concentration (EC_{50}) is defined as a toxicant concentration that resulted in a 50% reduction of the growth rate compared to the control (cultures without added toxicants). Similar EC_{50} concentrations as in standard 72-h algal growth inhibition tests (ISO 8692, 2004) can be acquired with delayed fluorescence after 24 h already and in some cases even after 15 or 30 min (Berden-Zrimec et al. 2007; Katsumata et al. 2006).

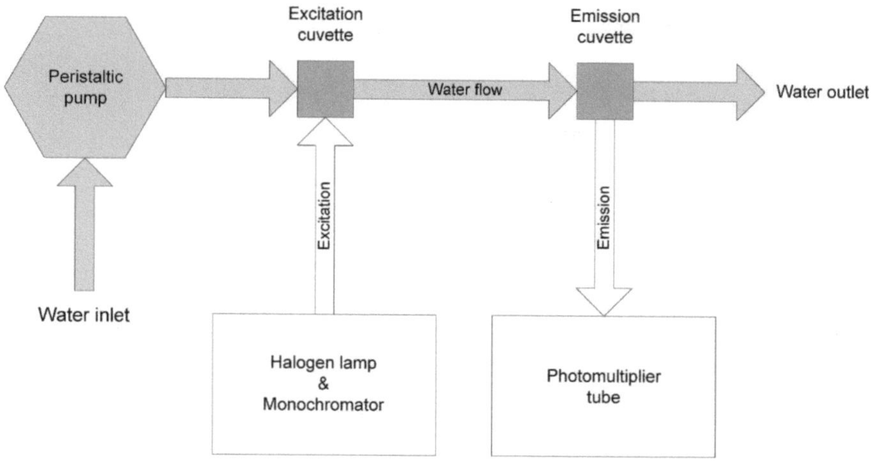

Fig. 15 A schematic presentation of the DF excitation spectrometer: Water is pumped through the excitation cuvette and illuminated by monochromatic light. Delayed fluorescence emitted by phytoplankton is measured in the emission cuvette

Additionally, some toxins (like herbicides) affect delayed fluorescence decay kinetics, which can be used for greater selectivity of bioassays (Berden-Zrimec et al. 2007; Katsumata et al. 2008). Assessing toxicity of complex matrices with DF measurements excludes problems with the fluorescent background (Istvanovics et al. 2005), which can be problematic for example in waste water effluents.

4.3 Primary Production

The gross primary production of aquatic ecosystem is an important parameter, which is commonly measured by the ^{14}C carbon fixation technique. New automatic methods are being developed based on *in vivo* fluorescence: pulse amplitude modulation (PAM) fluorometry (Falkowski and Kiefer 1985) and fast repetition rate fluorometry (FRRF) (Kolber and Falkowski 1993) which measure electron transport in the photosynthetic electron chain. It was shown several times that DF can be used as a proxy for measurement the photosynthetic electron transport rate (Kurzbaum et al. 2007; Wang et al. 2004; Yacobi et al. 1998).

DFI correlates well with algal biomass under controlled conditions, but there are great discrepancies due to environmental light intensity (Fig. 16) (Kurzbaum et al. 2007), nutrient status (Berden-Zrimec et al. 2008), temperature (Wang et al. 2004) and taxonomic composition (Istvanovics et al. 2005).

The correlation between DF and primary production has first been indicated by Yosef Yacobi and coworkers (Yacobi et al. 1998) in the freshwater phytoplankton samples. Water samples were illuminated with light at the saturation level and DFI was compared to the carbon uptake. The method worked in the area of maximal photosynthetic rate where there was no photoinhibition. Kurzbaum et al. (2007) used the same flow-trough apparatus set to lower excitation light intensities to investigate the correlation between quantum efficiency of photosynthesis (QE) and DF intensity during the daily cycle. DFI was greatly reduced in the time of maximal light intensity along with the QE yielding the correlation factor $r^2 = 0.87$ (Fig. 17). In this way the primary production can be calculated as a product of DFI and PAR.

A big advantage of DF for the estimation of primary production is that DF is a measure of charge separation and reflects the quantum efficiency of PSII along with the number of photosynthetic units. With the spectroscopic DF method it is possible to obtain the action spectra for photosynthesis and use it for the calculation of primary production in the actual light environment which rapidly changes with depth.

4.4 Delayed Fluorescence Excitation Spectroscopy

DF excitation spectroscopy is a real-time, non-destructive method for a rapid estimation of photosynthetically

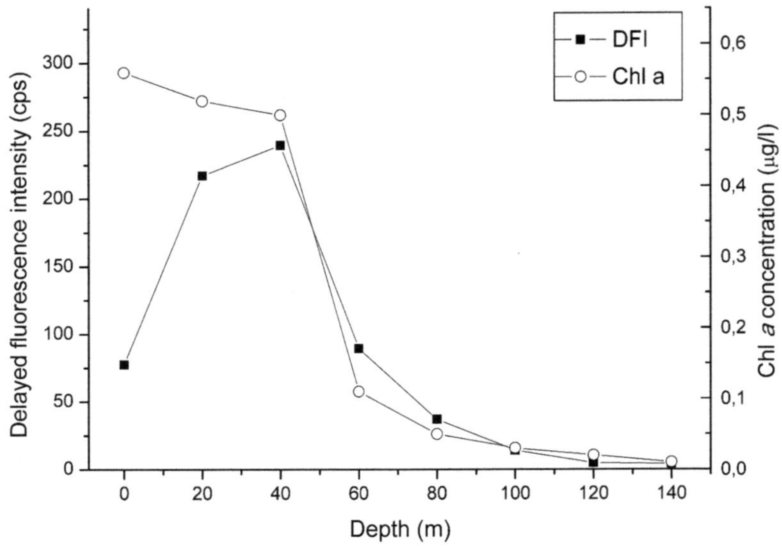

Fig. 16 A comparison between Chlorophyll *a* concentration and delayed fluorescence intensity in the oceanic samples from the Red Sea during the GAP aquatic workshop 2008

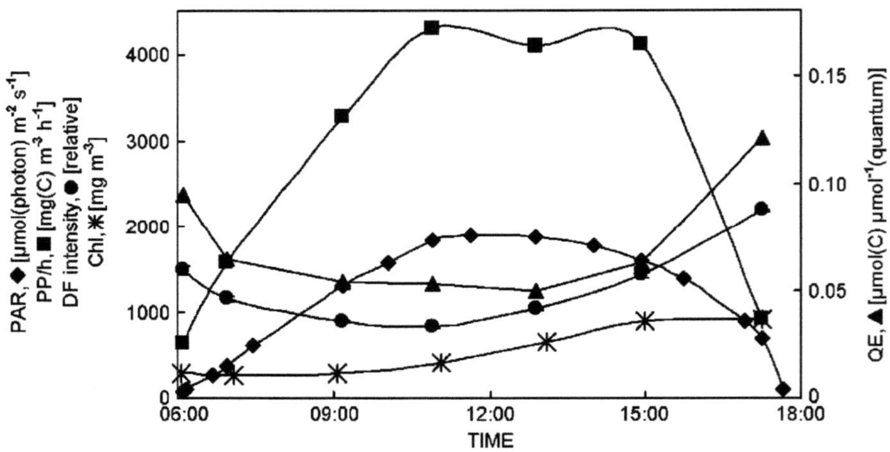

Fig. 17 Diurnal pattern of incident PAR (diamonds), primary production (PP/h) (squares), quantum efficiency (QE) (triangles), DF intensity (circles), and chlorophyll (Chl) content (asterisks) for *Chlorella vulgaris* growing in the experimental tank (From Kurzbaum et al. 2007. With permission)

active pigments. Cellular components absorbing light that do not lead to charge separation do not contribute to DF. Therefore, pigments contained in dead cells and photoprotective pigments like carotenes will not be detected (Gerhardt and Bodemer 2005).

Under a certain saturation level DF intensity is linearly correlated with the excitation light intensity (Fig. 18). Above this level the DFI is reduced due to photoinhibition (Wang et al. 2004). Excitation light intensities in the linear part of the DF/excitation response curve are therefore used for DF spectroscopy.

A monochromatic light source in the range between 400 and 700 nm providing light intensity between 1 and 100 μmol/m^{2}s is used for sample excitation. After sample illumination, the average DFI in the first few seconds is measured. In the stationary setup (Fig. 14), DFI is measured sequentially for each excitation wavelength. In the flow through setup, the sample illumination and measurement are done simultaneously in different compartments allowing for faster scanning of the DF excitation spectra.

For calibration, the DF excitation spectrum of an exponentially growing green algae culture is measured and chlorophyll *a* concentration determined in parallel by a conventional method (Gerhardt and Bodemer 2005). The DF excitation spectrum of a complex sample is the sum of different spectra from the different algal classes present in the sample and can be analysed using a deconvolution program.

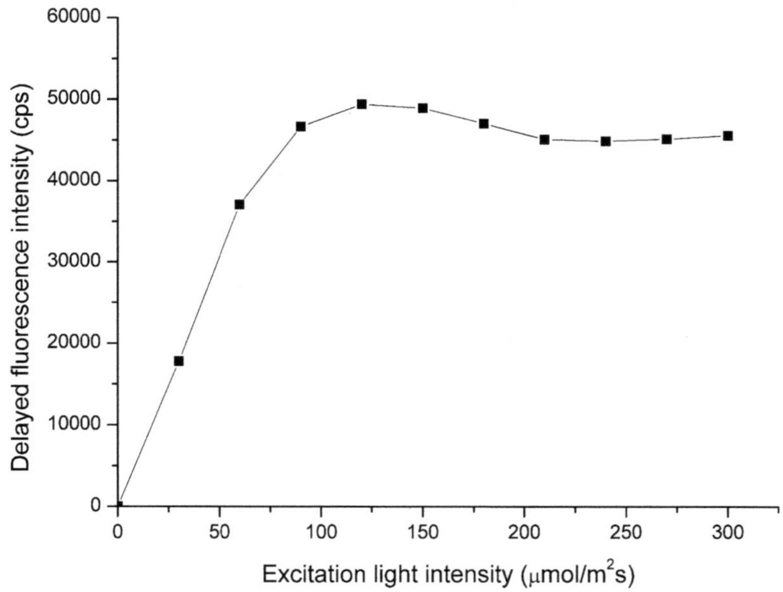

Fig. 18 Delayed fluorescence intensity relation to the excitation light intensity

Currently, four taxonomical classes of phytoplankton could be distinguished from complex spectra of environmental samples: (1) green algae (including Chlorophyta and Euglenophyta), (2) chromophyte algae (including Bacillariophyta, Chrysophyta and Dinophyta), (3) Cryptophyta and, (4) Cyanobacteria (Gerhardt and Bodemer 2005). New approaches to calculate percentage composition of phytoplankton are still under study (Istvanovics et al. 2005).

Acknowledgments We are thankful to Alfred Beran for providing marine algal cultures and Marina Monti for her cooperation in *Prorocentrum minimum* research (both from OGS, Department of Biological Oceanography, Trieste, Italy). Our colleague Jaka Razinger cooperated in PhAI idea and provided delayed fluorescence images. We are also thankful to Lidija Berden for revising the language. This work was financed by the Slovenian Research Agency (grant #P1-0237).

References

Arnold WA (1991) Personal perspective – experiments. Photosynth Res 27:73–82

Arnold W, Davidson JB (1954) The identity of the fluorescent and delayed light emission spectra in *Chlorella*. J Gen Physiol 37:677–684

Avron M, Schreiber U (1979) Properties of ATP induced chlorophyll luminescence in chloroplasts. Biochim Biophys Acta 3:448–454

Bennoun P, Beal D (1997) Screening algal mutant colonies with altered thylakoid electrochemical gradient through fluorescence and delayed luminescence digital imaging. Photosynth Res 51:161–165

Berden-Zrimec M, Drinovec L, Zrimec A, Tisler T (2007) Delayed fluorescence in algal growth inhibition tests. Cent Eur J Biol 2:169–181

Berden-Zrimec M, Drinovec L, Molinari I, Zrimec A, Fonda S, Monti M (2008) Delayed fluorescence as a measure of nutrient limitation in *Dunaliella tertiolecta*. J Photochem Photobiol B 92:13–18

Berden-Zrimec M, Kozar Logar J, Zrimec A, Drinovec L, Franko M, Malej A (2009) New approach in studies of microalgal cell lysis. Cent Eur J Biol 4:313–320

Berges JA, Charlebois DO, Mauzerall DC, Falkowski PG (1996) Differential effects of nitrogen limitation on photosynthetic efficiency of photosystems I and II in microalgae. Plant Physiol 110:689–696

Bertsch WF (1962) Two photoreactions in photosynthesis: evidence from delayed light emission of *Chlorella*. Proc Nat Acad Sci USA 48:2000–2004

Bertsch WF, Azzi JR (1965) A relative maximum in the decay of long-term delayed light emission from the photosynthetic apparatus. Biochim Biophys Acta 94:15–26

Bjorn LO (1971) Far-red induced, long-lived afterglow from photosynthetic cells – Size of afterglow unit and paths of energy accumulation and dissipation. Photochem Photobiol 13:5–20

Burger J, Schmidt W (1988) Long-term delayed luminescence – A possible fast and convenient assay for nutrition deficiencies and environmental pollution damages in plants. Plant Soil 109:79–83

Desai TS, Rane SS, Tatake VG, Sane PV (1983) Identification of far-red-induced relative increase in the decay of delayed light emission from photosynthetic membranes with thermoluminescence peak V appearing at 321 K. Biochim Biophys Acta 724:485–489

Falkowski P, Kiefer DA (1985) Chlorophyll *a* fluorescence in phytoplankton – Relationship to photosynthesis and biomass. J Plankton Res 7:715–731

Gerhardt V, Bodemer U (2005) Delayed fluorescence excitation spectroscopy – an *in vivo* method to determine photosynthetically active pigments and phytoplankton composition. ALGAS 33:4–17

Goltsev V, Zaharieva I, Lambrev P, Yordanov I, Strasser R (2003) Simultaneous analysis of prompt and delayed chlorophyll a fluorescence in leaves during the induction period of dark to light adaptation. J Theor Biol 225:171–183

Greisberger S, Teubner K (2007) Does pigment composition reflect phytoplankton community structure in differing temperature and light conditions in a deep alpine lake? An approach using HPLC and delayed fluorescence techniques. J Phycol 43:1108–1119

Hakanson L, Malmaeus JM, Bodemer U, Gerhardt V (2003) Coefficients of variation for chlorophyll, green algae, diatoms, cryptophytes and blue-greens in rivers as a basis for predictive modelling and aquatic management. Ecol Modell 169:179–196

Hideg E, Kobayashi M, Inaba H (1991) The far red induced slow component of delayed light from chloroplasts is emitted from photosystem II - Evidence from emission spectroscopy. Photosynth Res 29:107–112

ISO 8692 (2004) Water quality – Freshwater algal growth inhibition test with unicellular green algae. International Organization for Standardization, Geneva, Switzerland

Istvanovics V, Honti M, Osztoics A, Shafik HM, Padisak J, Yacobi Y, Eckert W (2005) Continuous monitoring of phytoplankton dynamics in Lake Balaton (Hungary) using on-line delayed fluorescence excitation spectroscopy. Freshw Biol 50:1950–1970

Joliot P, Joliot A (1980) Dependence of delayed luminescence upon adenosine triphosphatase activity in *Chlorella*. Plant Physiol 65:691–696

Joliot P, Joliot A, Bouges B, Barbieri G (1971) Studies of system-II photocenters by comparative measurements of luminescence, fluorescence, and oxygen emission. Photochem Photobiol 14:287–305

Jursinic PA (1986) Delayed fluorescence: current concepts and status. In: Govindjee, Fork DC (eds) Light emission by plant and bacteria. Academic Press, New York, pp 291–328

Katsumata M, Koike T, Nishikawa M, Kazumura K, Tsuchiya H (2006) Rapid ecotoxicological bioassay using delayed fluorescence in the green alga *Pseudokirchneriella subcapitata*. Water Res 40:3393–3400

Katsumata M, Takeuchi A, Kazumura K, Koike T (2008) New feature of delayed luminescence: preillumination-induced concavity and convexity in delayed luminescence decay curve in the green alga *Pseudokirchneriella subcapitata*. J Photochem Photobiol B 90:152–162

Kolber Z, Falkowski PG (1993) Use of active fluorescence to estimate phytoplankton photosynthesis in-situ. Limnol Oceanogr 38:1646–1665

Krause H, Gerhardt V (1984) Application of delayed fluorescence of phytoplankton in limnology and oceanography. J Luminesc 31–32:888–891

Kurzbaum E, Ecker W, Yacobi Y (2007) Delayed fluorescence as a direct indicator of diurnal variation in quantum and radiant energy utilization efficiencies of phytoplankton. Photosynthetica 45:562–567

Lavorel J, Dennery JM (1982) The slow component of photosystem II luminescence – A process with distributed rate-constant. Biochim Biophys Acta 680:281–289

Li Q, Xing D, Ha L, Wang JS (2007) Mechanism study on the origin of delayed fluorescence by an analytic modeling of the electronic reflux for photosynthetic electron transport chain. J Photochem Photobiol B 87:183–190

Melcarek PK, Brown GN (1977) Effects of chill stress on prompt and delayed chlorophyll fluorescence from leaves. Plant Physiol 60:822–825

Mellvig S, Tillberg JE (1986) Transient peaks in the delayed luminescence from *Scenedesmus obtusiusculus* induced by phosphorus starvation and carbon dioxide deficiency. Physiol Plant 68:180–188

Monti M, Zrimec A, Beran A, Berden Zrimec M, Drinovec L, Kosi G, Tamberlich F (2005) Delayed luminescence of *Prorocentrum minimum* under controlled conditions. Harmful Algae 4:643–650

Nedbal L, Soukupova J, Kaftan D, Whitmarsh J, Trtilek M (2000) Kinetic imaging of chlorophyll fluorescence using modulated light. Photosynth Res 66:3–12

Rutherford AW, Inoue Y (1984) Oscillation of delayed luminescence from PSII recombination of S_2Q_B- and S_3Q_B-. FEBS Lett 165:163–170

Schmidt W, Schneckenburger H (1992) Time-resolving luminescence techniques for possible detection of forest decline.1. Long-term delayed luminescence. Radiat Environ Biophys 31:63–72

Schneckenburger H, Schmidt W (1996) Time-resolved chlorophyll fluorescence of spruce needles after different light exposure. J Plant Physiol 148:593–598

Scordino A, Grasso F, Musumeci F, Triglia A (1993) Physical aspects of delayed luminescence in *Acetabularia acetabulum*. Experientia 49:702–705

Strehler BL, Arnold W (1951) Light production by green plants. J Gen Physiol 34:809–820

Van Wijk R, Scordino A, Triglia A, Musumeci F (1999) 'Simultaneous' measurements of delayed luminescence and chloroplast organization in *Acetabularia acetabulum*. J Photochem Photobiol B 49:142–149

Wang C, Xing D, Chen Q (2004) A novel method for measuring photosynthesis using delayed fluorescence of chloroplast. Biosens Bioelectron 20:454–459

Wang CL, Xing D, Zeng LZ, Ding CF, Chen Q (2005) Effect of artificial acid rain and SO_2 on characteristics of delayed light emission. Luminescence 20:51–56

Wiltshire KH, Harsdorf S, Smidt B, Blocker G, Reuter R, Schroeder F (1998) The determination of algal biomass (as chlorophyll) in suspended matter from the Elbe estuary and the German Bight: a comparison of high-performance liquid chromatography, delayed fluorescence and prompt fluorescence methods. J Exp Mar Biol Ecol 222:113–131

Wraight CA, Crofts AR (1971) Delayed fluorescence and the high-energy state of chloroplasts. Eur J Biochem 19:386–397
Yacobi YZ, Gerhardt V, Gonen-Zurgil Y, Sukenik A (1998) Delayed fluorescence excitation spectroscopy: a rapid method for qualitative and quantitative assessment of natural population of phytoplankton. Water Res 32:2577–2582
Yan Y, Popp FA, Sigrist S, Schlesinger D, Yan DA, ZC CS, Chotia A (2005) Further analysis of delayed luminescence of plants. J Photochem Photobiol B 78:235–244
Zrimec A, Drinovec L, Berden-Zrimec M (2005) Influence of chemical and physical factors on long-term delayed fluorescence in *Dunaliella tertiolecta*. Electromagnet Biol Med 24:309–318

Chapter 15
The Study of Phytoplankton Photosynthesis by Photoacoustics

Yulia Pinchasov Grinblat and Zvy Dubinsky

1 Introduction

Life on earth is based on the unique capability of photosynthetic organisms, or primary producers, to harvest solar energy and transmute it into chemical energy stored in oxygen and reduced carbon compounds. These are the substrates for respiration, thereby providing the energy needed for all life processes. Photosynthesis also produced the present O_2 rich atmosphere, which allows oxygenic life forms to thrive.

The biomass and vitality of the phytoplankton population responds rapidly to seasonal changes in the environmental factors such as temperature, light, vertical mixing, and nutrients (Dubinsky 1986). In the course of this succession, the biomass of phytoplankton can fluctuate over several orders of magnitude (Dubinsky and Berman 1979; Dubinsky and Berman 1981).

The seasonal succession of phytoplankton is brought about by the annual rhythms of temperature, light intensity and enrichment or depletion of surface waters by nutrients (Dubinsky 1986). In the course of these time cycles the biomass of phytoplankton may fluctuate over two to three orders of magnitude (Dubinsky and Berman 1979, 1981). Since marine stocks and fisheries in general depend ultimately on the supply of food stemming from the rates of primary photosynthesis in phytoplankton, it is of great importance to follow any changes in the patterns of their distribution and activity.

Y.P. Grinblat (✉) and Z. Dubinsky
The Mina & Everard Goodman Faculty of Life Sciences,
Bar-Ilan University, Ramat-Gan 52900, Israel
e-mail: pinhasy2@zahav.net.il

2 The Photoacoustic Method

As light energy is absorbed by photosynthetic pigments of plant and phytoplankton cells, only a small, variable fraction is stored as a product of photosynthesis. A major fraction that may reach over 60%, is dissipated as heat, while a few percent of the total are lost as fluorescence, mostly emanating from photosystem II (PSII).

This method is based on the conversion of light energy (from a laser pulse, absorbed by the photosynthetic pigments of phytoplankton) into heat energy, leading to rise in temperature and, hence, to an increase in pressure (photothermal effect) (Malkin 1996; Dubinsky et al. 1998; Mauzerall et al. 1998; Pinchasov et al. 2005).

The experimental system is shown schematically in Fig. 1. The basic setup is similar to that represented in detail in Dubinsky et al. (1998); Mauzerall et al. (1998); and Pinchasov et al. (2005).

The laser pulse is incident upon the suspension of algae, whose pigments absorb part of the laser beam. A variable fraction of the absorbed light pulse is stored in the photochemical products of photosynthesis. The remainder of the absorbed light is converted to heat, producing an acoustic wave that is intercepted by a detector. The signal contains a noisy background and later reflections from the walls of the vessel as well as from impedance mismatch within the detector. Hence, the true baseline was determined by subtracting from the signals of a series of 32 pulses in which the sample was exposed to light, an identical series of 32 signals, with the pulsed laser beam blocked. Signal to noise was improved by averaging over 128 pulses, and by taking RMS values over the time of the recorded signal (~10 μs).

D.J. Suggett et al. (eds.), *Chlorophyll* a *Fluorescence in Aquatic Sciences: Methods and Applications*, Developments in Applied Phycology 4, DOI 10.1007/978-90-481-9268-7_15, © Springer Science+Business Media B.V. 2011

Weak (~20 μJ cm^{-2}), 5 ns pulses at 532 nm wavelength, were used as a probe for ongoing photosynthesis. This energy already partially saturated photosynthesis, however it was used to improve signal to noise ratio. The nutrient limitation effects described below were always related to the nutrient replete controls, thereby canceling the effect of partial saturation. The source of the background light was a quartz-halogen illuminator (Cole-Parmer 4971). The intensity of the background light was adjusted to the desired level by neutral density filters and measured with a LiCor light meter equipped with a cosine quantum sensor.

The resulting electric signal PA is stored and subsequently analyzed on a computer. Thus, the light energy storage efficiency Φ is determined according to Eq. 1.

$$\Phi = (PA_{light} - PA_{dark}) / PA_{light} \quad (1)$$

PA_{dark} is the photoacoustic signal generated by the weak laser pulse in the dark and PA_{light} is the signal produced under the same pulse obtained under saturating (~3,000 μmol photons m^{-2} s^{-1}), continuous white light from a quartz-halogen illuminator [for details, see Pinchasov et al. (2005)].

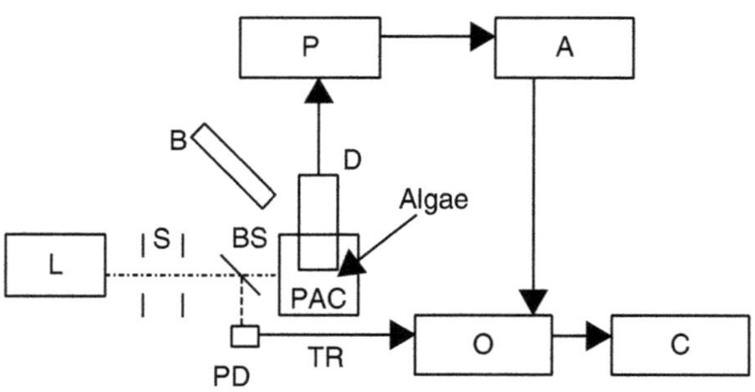

Fig. 1 Schematic of the photoacoustic setup: L – Minilite; Nd:YAG laser, 532 nm; S – beam-shaped slits; BS – beam splitter; PAC – photoacoustic cell with suspension of algae (30 ml); D – stainless-steel photoacoustic detector containing a 10-mm-diameter resonating ceramic disc (BM 500, Sensor, Ontario, Canada); P – low-noise Amptek A-250 preamplifier; A – SRS 560 low-noise amplifier; PD – photodiode; TR – trigger signal; B – background light source, quartz-halogen illuminator (Cole Parmer 4971); O – Tektronix TDS 430A oscilloscope; C – computer

3 Efficiency of Photosynthesis and Photosynthesis Versus Energy Relationship

We were able to determine the light energy storage efficiency under different ambient irradiance levels, resulting in an energy-storage curve. This relationship is similar in shape to the photosynthesis versus energy (or irradiance), (P vs. E, or I) curve (Fig. 2), obtained by the tedious standard measurements of ^{14}C fixation and oxygen evolution (Steeman-Nielsen 1952; Berner et al. 1986; Dubinsky et al. 1987; Ben-Zion and Dubinsky 1988; Grobbelaar et al. 1992) or the indirect results from the measurement of variable fluorescence (Falkowski et al. 1986; Schreiber and Bilger 1993).

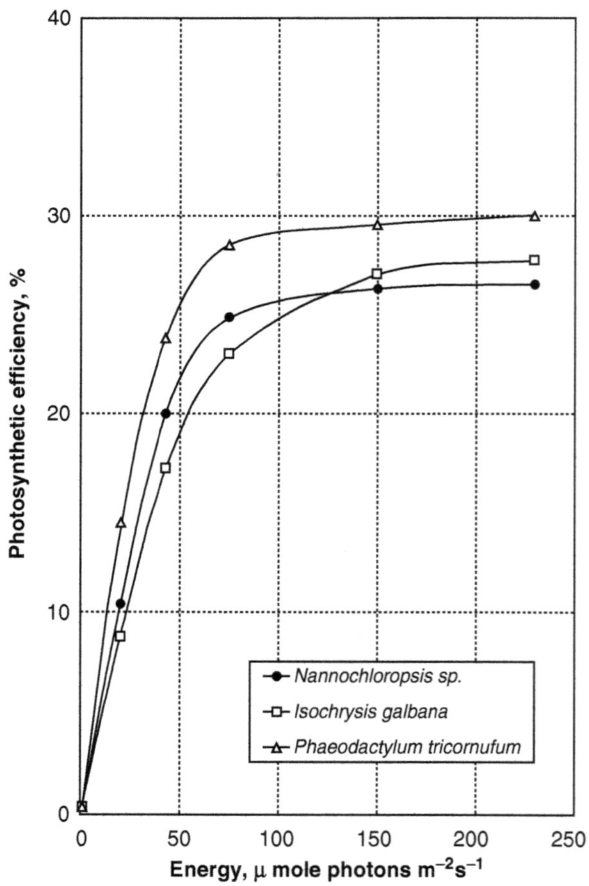

Fig. 2 Measured photosynthetic energy storage versus background light energy curves for the three algae *Nannochloropsis* sp. (○-), *Phaeodactylum tricornutum* (Δ) , and *Isochrysis galbana* (□), nutrient-replete controls

By increasing the continuous background light intensity from zero to saturation of photosynthesis, an increasing fraction of the reaction centers is closed at any time, and a decreasing fraction of the probe laser pulse energy is stored. There is a corresponding increase in the fraction of the pulse energy that is converted to heat that is sensed by the photoacoustic detector. From these detector responses, the photosynthetic energy-storage versus background light-intensity relationship was obtained [for details, see Dubinsky et al. (1998) and Pinchasov et al. (2005)].

4 The Effect of Nutrient Limitation on Photosynthesis

We were able to follow the effects of the key environmental parameter, nutrient status, on the photosynthetic activity of phytoplankton. The nutrients examined were nitrogen, phosphorus, and iron (Pinchasov et al. 2005).

Nitrogen and phosphorus — In order to examine the potential of the photoacoustic method as a tool to assess the health of phytoplankton, we examined the effect of nutrient limitation on their photosynthetic energy-storage efficiency. Three algal species, *Phaeodactylum tricornutum, Isochrysis galbana*, and *Nannochloropsis* sp. were cultured in their respective nutrient-replete media. The algae were grown at 24°C, under white fluorescent lights at ~220 μmol photons m^{-2} s^{-1} PAR. Cell counts were done under the microscope in a Neubauer cytometer. Chlorophyll a was determined spectrophotometrically following overnight extraction in dimethyl formamide of centrifuged cells at room temperature. These cultures were the controls to which nutrient-limited cultures were compared (Fig. 2).

The algae for these cultures were harvested by centrifugation from the nutrient-replete media in which they were grown, and resuspended in media from which N or P was omitted. Cultures were followed over 2 weeks and compared for their photosynthetic energy-storage efficiency.

As seen in Fig. 3, all three algal species showed a sharp decrease in efficiency: by ~50 ± 5% in the P-limited, and ~60 ± 5% in the N-limited cultures, as compared to the nutrient-replete controls (=100%).

Iron — For the iron limitation experiments, the algae were cultured in iron-replete media, under the same conditions as in the nitrogen- and phosphorus-depletion experiments.

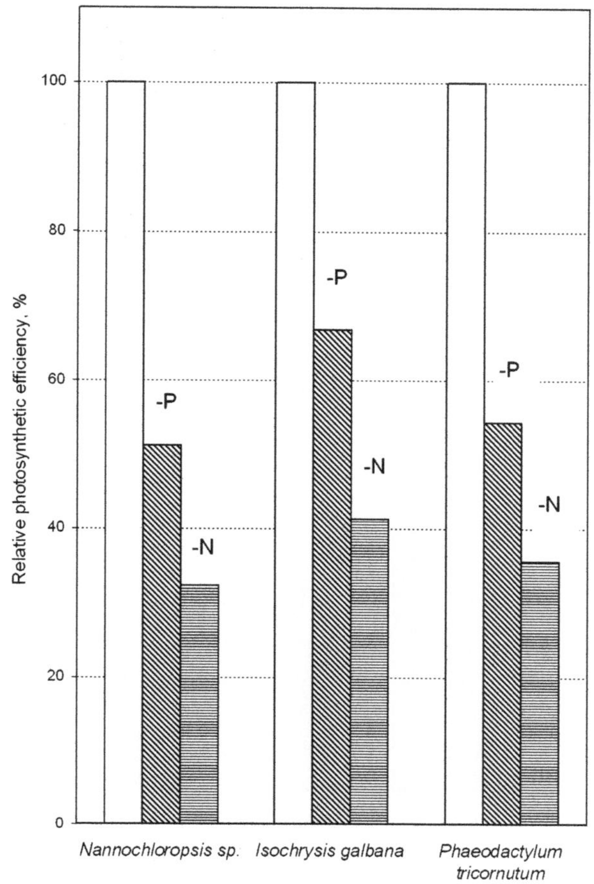

Fig. 3 The effect of nutrient limitation on relative photosynthetic energy storage in the three algae *Nannochloropsis* sp., *Phaeodactylum tricornutum*, and *Isochrysis galbana*. Controls (clear columns) were grown in nutrient-replete media, whereas in the P (oblique line) and N (horizontal line) cultures, phosphorus and nitrogen were omitted from the medium

The algae were grown at 24°C, under white fluorescent lights at ~220 μmol photons m^{-2} s^{-1} PAR. As the iron was progressively depleted, the ability of the three species to store energy decreased (Fig. 4).

5 The Effect of Lead Poisoning on Photosynthesis

In our experiments, the exposure of the cyanobacterium *Synechococcus leopoliensis* to different concentrations of lead resulted in some major changes in chlorophyll

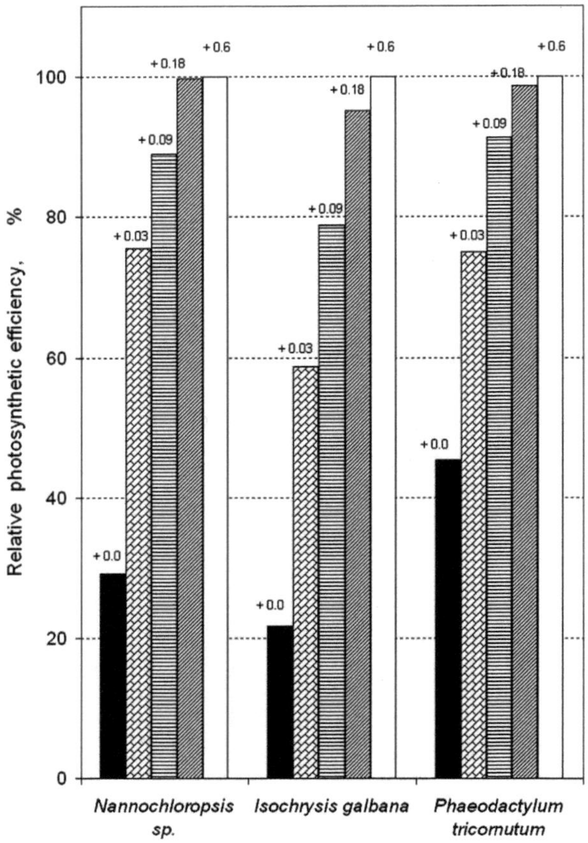

Fig. 4 The effect of different iron concentrations on the relative photosynthetic energy storage in the three algae, *Nannochloropsis* sp., *Phaeodactylum tricornutum*, and *Isochrysis galbana*. Controls were grown in iron-replete media containing 0.6 mg L^{-1}. The iron concentration in the iron-limited cultures was (from left to right), 0, 0.03, 0.09, and 0.18 mg L^{-1}, respectively

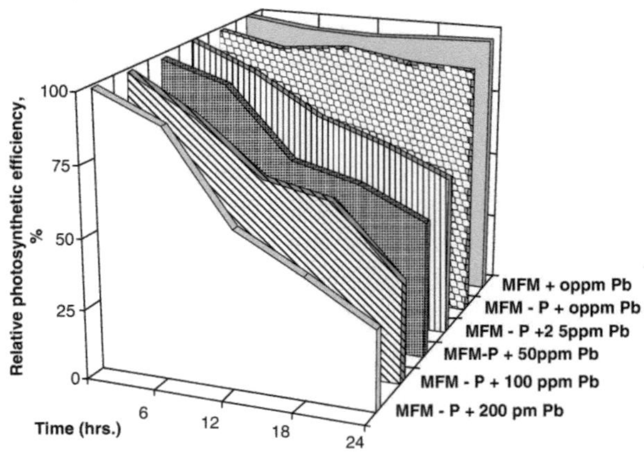

Fig. 5 Relative photosynthetic efficiency following lead application of *Synecococcus leopoliensis*

concentration and photosynthesis (Pinchasov et al. 2006). We had to omit phosphorus from the medium since it binds with lead forming an insoluble precipitate of $Pb_3(PO_4)_2$, leaving no lead in the medium. Cells of *S. leopoliensis* were grown in modified Fitzgerald medium (MFM) at 23 ± 0.1 °C under light intensity of 30 μmol photons m^{-2} s^{-1}.

Figure 5 shows the changes in photosynthetic efficiency following lead application. The reduction of photosynthesis reached ~50% and ~80% with25 ppm and 200 ppm, respectively. It is important to emphasize that these results are similar in trend to the decrease in chlorophyll concentration. Most of the decrease seen after the first 24 h actually took place during the first 40 min, and probably even earlier.

Figure 5 also summarizes the results of the effect of lead on photosynthetic efficiency with time. With increasing lead concentration and duration of exposure, inhibition of photosynthesis increases. Since the photoacoustic method yields photosynthetic energy-storage efficiency results that are independent of chlorophyll concentration, it means that the observed decrease in efficiency is not due to the death of a fraction of the population, but rather due to the impairment of photosynthetic function in all cells, possibly due to the inactivation of increasing fractions of the photosynthetic units.

6 Conclusions

The directness and ease of acquisition of photosynthesis data by the photoacoustic method, and its proven applicability in all aspects of phytoplankton physiology and ecology, make it a promising tool complementary to all the standard methods currently in use. We are conducting work to develop photoacoustics-based *in-situ* profilers and in-line monitoring devices.

References

Ben-Zion M, Dubinsky Z (1988) An on-line system for measuring photosynthetic characteristics via an oxygen-electrode. J Plankton Res 10:555–558

Berner T, Dubinsky Z, Schanz F, Grobbelaar JU, Uehlinger U, Falkowski P (1986) The measurement of primary productivity in a high rate oxidation pond. J Plankton Res 8:659–672

Dubinsky Z (1986) Productivity of algae under natural conditions. In: Richmond A (ed) Handbook for algal mass culture. CRC Press, Boca Raton, FL, pp 107–116

Dubinsky Z, Berman T (1979) Seasonal changes in the spectral composition of downwelling irradiance in Lake Kinneret (Israel). Limnol Oceanogr 24:652–663

Dubinsky Z, Berman T (1981) Light utilization by phytoplankton in Lake Kinneret (Israel). Limnol Oceanogr 26:660–670

Dubinsky Z, Falkowski PG, Post AF, van Hes UM (1987) A system for measuring phytoplankton photosynthesis in a defined light field with an oxygen electrode. J Plankton Res 9:607–612

Dubinsky Z, Feitelson J, Mauzerall DC (1998) Listening to phytoplankton: measuring biomass and photosynthesis by photoacoustics. J Phycol 34:888–892

Falkowski PG, Wyman K, Ley A, Mauzerall D (1986) Relationship of steady state photosynthesis to fluorescence in eucaryotic algae. Biochim Biophys Acta 849:183–192

Grobbelaar JU, Schanz F, Dubinsky Z, Tilzer MM, Burger-Wiersma T, Rijkeboer M, Lemoalle J, Falkowski PG (1992) Photosynthetic characteristics of five high light and low light exposed microalgae as measured with ^{14}C-uptake and oxygen electrode techniques. Mar Microb Food Webs 6:3–19

Malkin S (1996) The photoacoustic method in photosynthesis-monitoring and analysis of phenomena which lead to pressure changes following light excitation. In: Amesz J, Hoff A (eds) Biophysical techniques in photosynthesis. Kluwer, Dordrecht, pp 191–206

Mauzerall DC, Feitelson J, Dubinsky Z (1998) Discriminating between phytoplankton taxa by photoacoustics. Isr J Chem 38:257–260

Pinchasov Y, Kotliarevsky D, Dubinsky Z, Mauzerall DC, Feitelson J (2005) Photoacoustics as a diagnostic tool for probing the physiological status of phytoplankton. Isr J Plant Sci 53:1–10

Pinchasov Y, Berner T, Dubinsky Z (2006) The effect of lead on photosynthesis, as determined by photoacoustics in *Synechococcus leopoliensis* (Cyanobacteria). Water Air Soil Pollut 175:117–125

Schreiber U, Bilger W (1993) Progress in chlorophyll fluorescence research: major developments during the past years in retrospect. Prog Bot 54:151–173

Steeman-Nielsen E (1952) The use of radioactive carbon (C^{14}) for measuring organic production in the sea. J Cons Int Explor Mer 18:117–140

Index

Q

MIX
Papier aus verantwortungsvollen Quellen
Paper from responsible sources
FSC® C105338

If you have any concerns about our products,
you can contact us on
ProductSafety@springernature.com

In case Publisher is established outside the EU,
the EU authorized representative is:
Springer Nature Customer Service Center GmbH
Europaplatz 3, 69115 Heidelberg, Germany

Printed by Libri Plureos GmbH
in Hamburg, Germany